ELEMENTARY PARTICLE PHYSICS
PRESENT AND FUTURE

Proceedings of International Workshop on

ELEMENTARY PARTICLE PHYSICS
PRESENT AND FUTURE

València, Spain *June 5–9, 1995*

editors

J W F Valle

Departamento de Fisica Teórica, IFIC–CSIC
Universidad de València

A Ferrer

Departamento de Fisica Atomica, Molecular & Nuclear
Universidad de València

World Scientific
Singapore • New Jersey • London • Hong Kong

Published by

World Scientific Publishing Co Pte Ltd
5 Toh Tuck Link, Singapore 596224
USA office: 27 Warren Street, Suite 401-402, Hackensack, NJ 07601
UK office: 57 Shelton Street, Covent Garden, London WC2H 9HE

British Library Cataloguing-in-Publication Data
A catalogue record for this book is available from the British Library.

ELEMENTARY PARTICLE PHYSICS: PRESENT AND FUTURE

ISBN-13 9789810225544
ISBN-10 9810225547

Preface

The workshop was held in València, following the earlier meetings organised at UIMP in 1991 and 1993. It addressed the basic issues in elementary particle physics research for the coming decade, with emphasis on the new physics topics pertaining to the electro-weak sector, super-symmetry and neutrino physics, as well as the related experimentation at the new CERN accelerators LEP200 and LHC. Although most participants were workers in the field, some time was devoted to basic reviews. We received the assistance from an International Advisory Committee which consisted of: B. Barish (Caltech), V. Berezinskii (Gran Sasso), F. Dydak (MPI/Munich and CERN), R. L. Mössbauer (Munich), S. Pokorski (Warsaw), G. G. Ross (Oxford), D. N. Schramm (Chicago), J. Steinberger (CERN), D. Treille (CERN), M. Vysotsky (ITEP) and S. Weinberg (Texas).

The material of these lectures/seminars covered the following topics: The Standard Model, Top and Higgs Physics, Neutrinos and Astroparticle Physics, Physics Beyond the Standard Model, including Supersymmetry and Grand Unification, as well as Physics at LEP200, LHC and the Future Underground Experiments.

We thank all our sponsors, first of all, Universidad Internacional Menendez y Pelayo (UIMP), for financial as well as logistic support. We also benefited from the important support provided by the Spanish agencies Comisión Interministerial de Ciencia y Tecnología (CICYT), Dirección General de Investigación Científica y Técnica (DGICYT) and the University of València, who contributed substantially for the success of the event. Last but not least, thanks are due to the great help provided by our scientific secretaries Dr. Fernando de Campos and Miguel Angel García-Jareño without whose patient assistance this meeting would not have been possible.

A. Ferrer and J. W. F. Valle
València, November 14, 1995

CONTENTS

Preface v

Chapter 1

Searches at LEP1 3
F. Richard

Electroweak Results at LEP-I 32
M. Martinez

The LEP200 Programme 52
D. Treille

Chargino Search in DELPHI below the W^+W^- Threshold 71
J. J. Hernández

Search for Heavy Neutral Higgs Bosons in the Four Jet Channel at LEP200 79
C. M. Rivero

Neutrinos and Semileptonic *b*-decays 85
M. Dittmar

Search for Rare *B* Decays 91
M. Wadhwa

Recent Rare *B* Decay Results from CLEO 98
N. Katayama

Study of One-prong Tau Decays with Neutral Kaons at L3 104
T. Paul

Probing Compositeness in $e\gamma$ Collisions 111
S. F. Novaes, O. J. P. Éboli, E. M. Gregores, J. C. Montero and D. Spehler

Trilinear Vector-Boson Couplings: Present and Future 118
Ch. Petridou

Chapter 2

Observation of the Top Quark at the Tevatron 129
J. M. Benlloch

Status of SUSY Searches at DØ 142
D. Norman

Top Quark Physics at the Tevatron 148
C.-P. Yuan

On Gluon Radiation in $t\bar{t}$ Production and Decay 154
L. H. Orr, T. Stelzer and W. J. Stirling

The CMS Detector and Physics at the LHC 160
D. Denegri

Production of Stop and Sbottom at LEP2 and at the LHC 183
A. Bartl, W. Porod, H. Eberl, S. Kraml and W. Majerotto

Searching for B_c Mesons in the ATLAS Experiment at LHC 189
M. Galdón, M. A. Sanchis-Lozano, R. Pérez Ochoa and J. A. Valls

Sphaghetti Calorimeter Spacal 199
J. Hladky

Chapter 3

Topics in String Unification 209
L. E. Ibáñez

Extending the Symmetries of the Standard Model 231
G. G. Ross

Soft Terms from Dilaton/Moduli Sectors 244
C. Muñoz

Low-Energy Supersymmetry: Prospects and Challenges 256
H. E. Haber

Bounds on the Higgs Mass in the Standard Model and the Minimal Supersymmetric Standard Model 270
M. Quiros

Implications of Spontaneous R-Parity Breaking 282
J. C. Romão

Effects of R Parity Breaking in the Higgs Sector 295
J. Rosiek

Natural Flavour Mixing in the MSSM and $\mu \to e, \gamma$ 302
B. De Carlos, J. A. Casas and J. M. Moreno

Fermion Masses in SUSY GUTS 308
B. C. Allanach and S. F. King

Fermion Masses and Mixings from Horizontal Abelian Symmetries 317
C. A. Savoy

Supergravity and the $b \to s, \gamma$ Decay 323
B. De Carlos and J. A. Casas

Dynamical Breaking of Left–Right Symmetry 329
E. Kh. Akhmedov

Mirror World and Present Neutrino Puzzles 335
Z. G. Berezhiani

Light Singlet Fermions and Neutrino Physics 343
E. J. Chun

New Issues in Quark-Lepton Symmetry 354
R. R. Volkas

Low Energy Yukawa Input Parameters for Yukawa Coupling Unification 360
G. Rodrigo

Chapter 4

From the Big Bang to Now, and the Journey Back 371
E. W. Kolb

Solar Neutrinos: Expecting 1996 383
A. Yu. Smirnov

The Atmospheric Neutrino Anomaly and Long Baseline Neutrino Oscillation Experiments 400
B. C. Barish

Problems of Dark Matter 411
A. D. Dolgov

Mixed Dark Matter and Supersymmetry 418
A. Masiero

Double Beta Decay Searches 431
A. Morales

Future ν_τ Oscillation Experiments and Present Data 446
M. C. Gonzalez-Garcia

Correlated Neutrino Oscillations 456
E. Ma

Supersymmetry Effects on Neutrino Oscillations 462
E. Roulet

Neutrino Propagation in Medium with Random Magnetic Field 467
V. B. Semikoz

Status of the CHORUS Experiment 473
The CHORUS Collaboration (presented by C. Mommaert)

The San Onofre Neutrino Oscillation Experiment 479
M. Mascarenhas, F. Boehm, M. Chen, B. Cook, H. Henrikson, R. Hertenberger, K. Kou, V. Novikov, A. Piepke, P. Vogel, G. Gratta, J. Hanson and D. Michael

Cosmology of the Next-to-Minimal Supersymmetric Standard Model 490
S. A. Abel, S. Sarkar and P. L. White

Electroweak Baryogenesis with Neutral Heavy Leptons 496
N. Rius

Renormalization Group Calculation of the Galaxy-Galaxy Correlation Function 504
J. Pérez-Mercader, T. Goldman, D. Hochberg and R. Laflamme

Chapter 1

SEARCHES at LEP1

F. Richard

Laboratoire de l'Accélérateur Linéaire, IN2P3-CNRS et Université de Paris-Sud, F-91405 Orsay Cedex, France

LEP data are used to set lower mass limits on new particles, primarily in the SUSY sector. It is pointed out that, in the MSSM scenario, the Higgs boson mass limits depend critically on theoretical assumptions. Using the MSSM-GUT scheme and combining the various searches performed at LEP, it is shown that one has $m_h > 61$ GeV and $m_A > 80$ GeV. It is also shown that one can set a limit on the lightest neutralino mass $m_\chi > 12$ GeV.

1 Introduction

After analyzing about 10 million Z decays, the present conclusions on LEP searches are:

- No significant anomaly has been observed (or confirmed...) in any of the very many final states selected for searches
- Most searches have reached the kinematical limit or "saturate" (e.g. in the case of the SM Higgs search)

This talk will therefore focus on topics still under discussion:

- How conclusive is the Higgs-SUSY search ?
- Which mass limits are legitimate and under which assumptions ?
- Can we exclude the existence of a very light SUSY particle ?

2 Methods used at LEP1

2.1 Indirect Limits

A particle X with weak isospin I_3^X and/or electric charge Q^X different from zero will be produced through the process $Z \rightarrow X\bar{X}$. It will therefore contribute either to the total width of the Z resonance Γ_T or to its invisible width Γ_{inv}. From the LEP measurements[1] one may obtain tight upper bounds on these contributions.

Taking into account the present uncertainties on the top mass, the strong coupling constant and the Higgs mass, one has, at 95% C.L., $\Gamma(Z \rightarrow X\bar{X}) < 20$ MeV for a visible mode and $\Gamma(Z \rightarrow X\bar{X}) < 5$ MeV for an invisible mode. The first result was obtained by comparing Γ_T^{Min}=2.481 GeV, the minimal value for the theoretical expectation of the total width (assuming m_t=160 GeV, m_H=1 TeV and α_s=0.110), to the LEP measurement: Γ_T=2.496±0.003±0.001 GeV. For the limit on Γ_{inv}, one uses the result on the number of neutrinos N_ν=2.99±0.016.

In the SUSY scenario, the maximum allowed value allowed for m_H is $\sim$ 140 GeV, giving $\Delta\Gamma_T^{Min}$=+8 MeV. One may argue that virtual corrections due to light SUSY particles will modify the estimate of the minimal theoretical width. For instance, the experimental excess of about 2% reported on $\Gamma(Z \rightarrow b\bar{b})$ could be due to such effects and would also increase Γ_T^{Min} by $\sim$ 8 MeV. A very nearby chargino would induce a negative variation[2], roughly cancelling the previous benefits. I therefore conclude that the upper limits used for the SM apply, to a good approximation, to the SUSY case.

The following table gives the corresponding mass limits derived for various particles. Remarkably enough, these limits are already covering a large part of the range allowed kinematically.

2.2 Direct Searches

Direct searches have reached the kinematic limit in all the channels presented in the table with the exception of invisible sneutrinos. They are however more model-dependent since a definite decay pattern is assumed to allow a tight selection on the final states.

As an example, the slepton channel is searched in the decay mode $\tilde{\ell} \rightarrow \ell\chi$ where χ the neutralino χ is assumed stable and undetected. The expected final state has two leptons plus large missing transverse momentum.

If R-parity is not conserved, one can have $\chi \rightarrow \ell\bar{\ell}'\nu$ such that the final state is completely different and requires new dedicated searches. This scenario has been recently explored by ALEPH in[3] showing that the kinematical limit can also be reached.

Particle type	Lower bound (GeV)
Top quark	45
b' quark	45
$L^{\pm}$	36
L^0_{Dirac}	45
$L^0_{Dirac} \Gamma_{inv}$	45
$L^0_{Majorana}$	40
$L^0_{Majorana} \Gamma_{inv}$	43
u-type squark (L + R)	41
d-type squark (L + R)	42
squarks (L + R) (5 flavours)	45
sleptons (L + R)	29
sneutrino (1 flavour)	41(36)
LSP sneutrino Γ_{inv} (1 flavour)	43(41)
chargino	45
$H^{\pm}$	17
u*	45
d*	45
$L^{\pm *}$	36
ν^*	45

2.3 Particular Cases

One can have:

- Z→ XY if there are non-diagonal couplings of the Z particle.

  DELPHI has recently[4] searched for Z→ νN, where ν is an ordinary neutrino and N is a heavy unstable neutrino. Upper limits on BR(Z→ νN) have been set at the few 10^{-6} level.

- Mixed states

  SUSY particles can be a mixture of the weak eigenstates in such a way that the effective coupling to Z becomes strongly suppressed. As an example the top squark $\tilde{t}$ can be a mixture of $\tilde{t}_L$ and $\tilde{t}_R$. Figure 1 shows the results of two searches[5] which cover this scenario. Even when the mixing angle gives a complete suppression of Z→ $\tilde{t}\bar{\tilde{t}}$, there remains a finite cross-section due to the coupling through the photon.

Similarly the lightest neutralino χ is a mixture of the two gaugino states ,$\tilde{\gamma}$ and $\tilde{Z}$, and of the two higgsino states $\tilde{h}$ and $\tilde{H}$. Only the transition Z→ $\tilde{H}\tilde{h}$ is allowed. It turns out that when tan$\beta \to 1$ and $m_\chi \to 0$, $\Gamma(Z\to \chi\chi) \to 0$. It becomes therefore impossible to set any lower limit on the neutralino mass from the invisible width measurement. Figure 2 shows, in the case of L3[6], the absence of mass limits for the 2 lightest neutralinos at low tanβ.

The gluino searches[7] (figure 3) have also missed region I and region II and cannot exclude the existence of a very light gluino. The mass relations given by GUT (Grand Unified Theory), $m_\chi \sim 0.14 m_{\tilde{g}}$, can therefore not be used to exclude a light neutralino. It seems unlikely that LEP will be able to fill these holes in spite of various suggestions[8] to attack the problem of the light gluino.

3 SM Higgs

3.1 Introduction

Let me recall the basic arguments:

- Symmetry breaking in the SM occurs using an auxiliary iso-doublet complex field Φ, which has four real components. Z and $W^\pm$ become massive and "eat" three of the Φ components to acquire longitudinal polarization. The remaining scalar field corresponds to the Higgs boson.

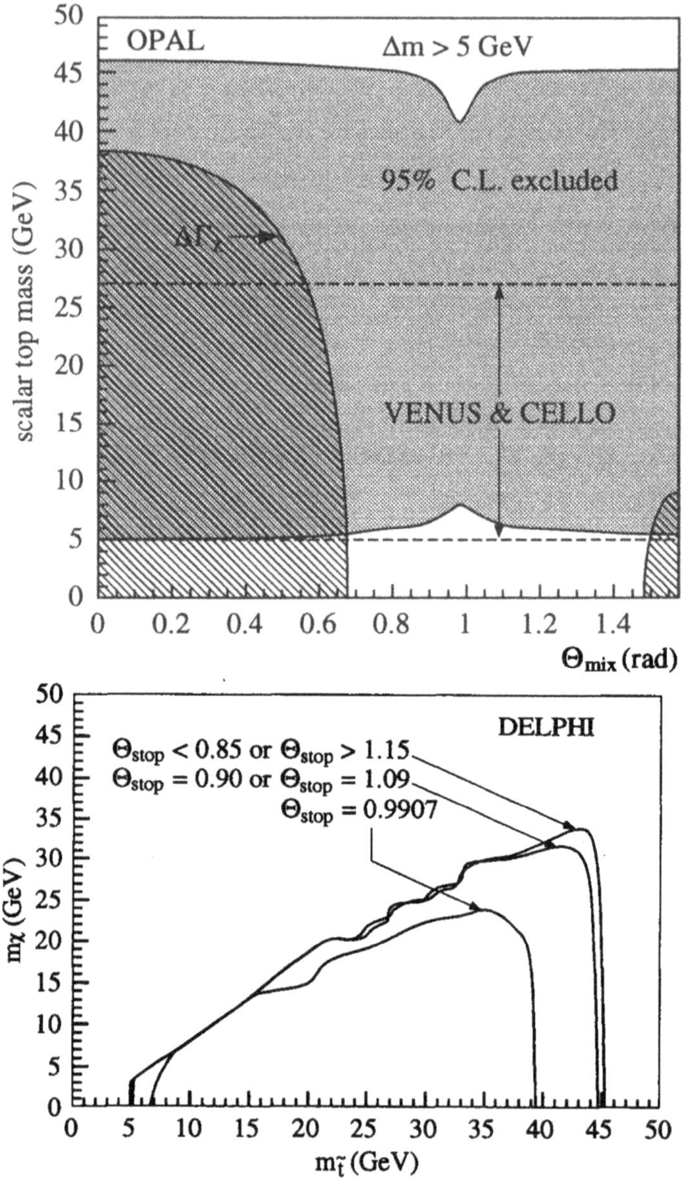

Figure 1: Top squark exclusion plots

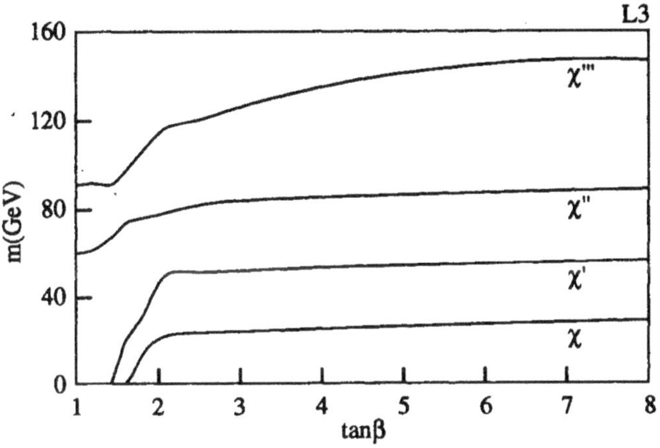

Figure 2: Neutralinos from L3

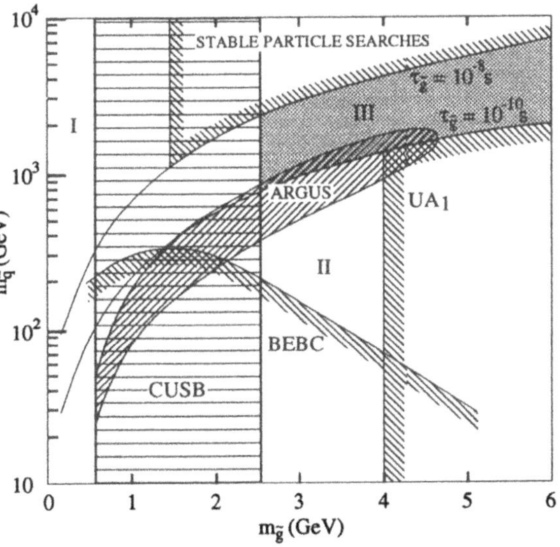

Figure 3: Gluinos

- One has $m_H^2 \sim \lambda v^2$ where λ is the coefficient of the quartic term in the Higgs potential and v is the vacuum expectation of Φ related to the W mass. λ is a running constant $\lambda(\Lambda)$ which should remain positive until the SM breaks down. If this occurs at the GUT scale, one should have[9,10] $m_H > 110$-130 GeV with $m_t \sim 170$ GeV. One can therefore conclude that the SM Higgs boson search is irrelevant at LEP unless new physics appears at a scale $\Lambda \leq$ few TeV.

3.2 SM Higgs search

- The standard process Z→HZ* gives 5 events for m_H=70 GeV with 7 million hadronic events analyzed at LEP as shown in figure 4. One assumes that only the modes $Z^* \rightarrow \nu\bar{\nu}, \ell^+\ell^-$ are usable (primarily $\mu^+\mu^-$ and e^+e^-). At high mass, H decays into $b\bar{b}$ in $\sim$ 85 % of the cases which can be exploited as an additional signature.

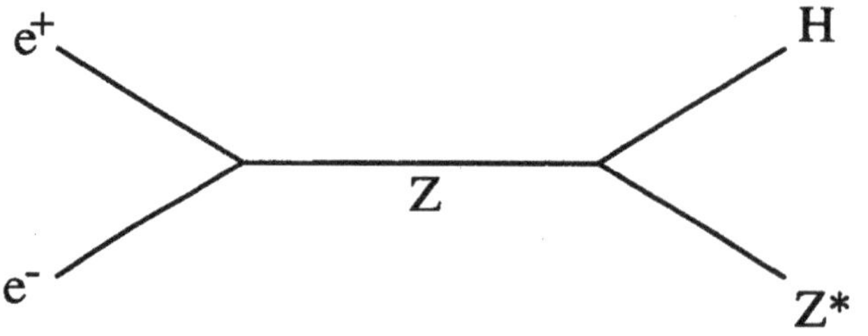

- The radiative process Z→Hγ is almost hopeless for large m_H (figure 4). As discussed in the next section, this process can be greatly enhanced if new physics appears at low Λ.

The LEP results are based on the HZ* search and depend primarily on detector properties. For the $H\nu\bar{\nu}$ channel, the relevant features are the hermeticity of the detector (e.g. there are dead

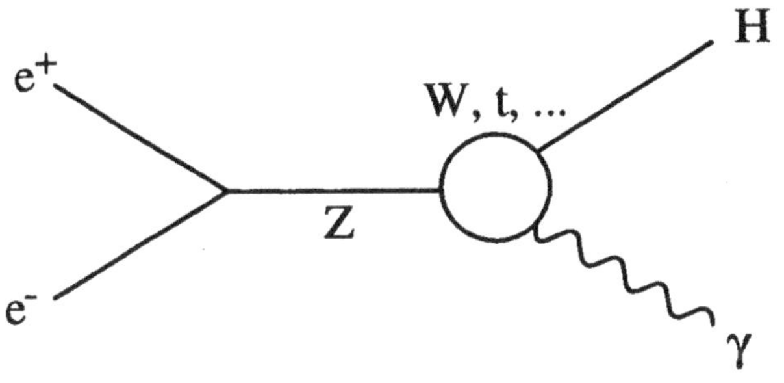

regions not covered by the calorimeters) and the energy resolution of the hadronic final states. The following table[11] summarizes the main parameters relevant for this analysis.

Experiment	DATA	ϵ (%) (60GeV)	Signal (60GeV)	Background
ALEPH	1.94M	42.0	2.32	0.26+0.15-0.10
DELPHI	1.62M	34.9	1.58	1.12±0.28
L3	1.60M	31.6	1.42	0.5±0.5
OPAL	1.90M	32.4	1.75	0.22+0.08-0.05
SUM	7.06M		7.07	2±0.6

Seven Higgs events are expected for a mass of 60 GeV, while no candidate remains in the data. The predicted background of 2±0.6 is not sufficiently precise to cause any worry.

The $H\ell^+\ell^-$ channel is unavoidably contaminated by the four fermion final state. It is therefore suitable to use b-tagging which should reduce this background by about one order of magnitude. In addition it is possible to reconstruct very precisely the Higgs mass (σ_M ~0.5-1 GeV) using the leptons.

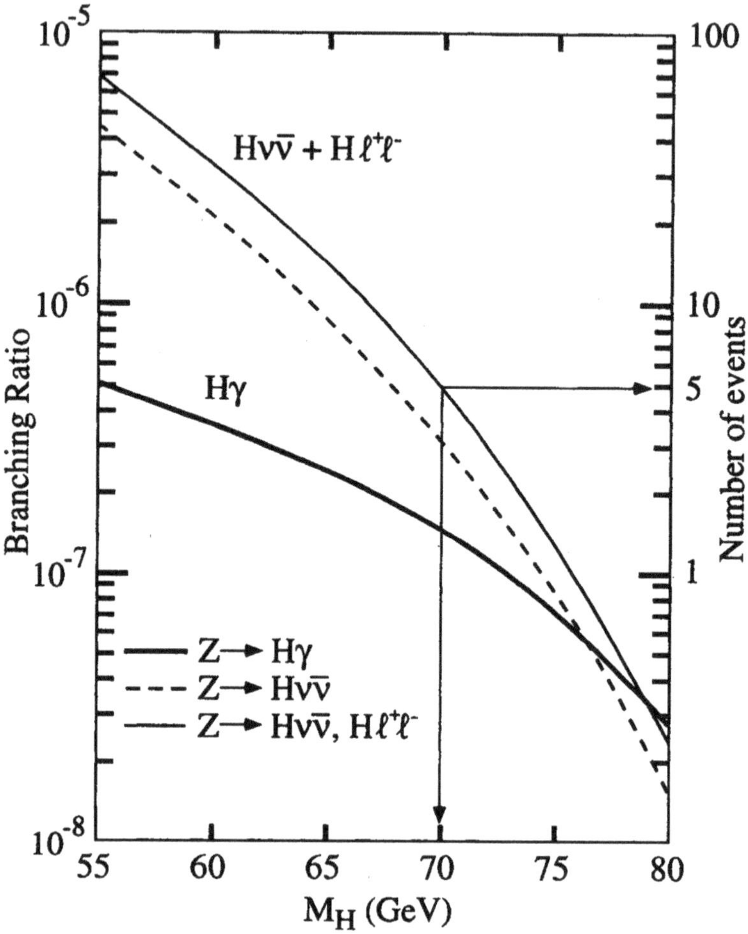

Figure 4: Higgs rate

The various candidates are displayed in figure 5 which show no significant mass accumulation. Two of the candidates labelled (b) have been tagged as $b\bar{b}$ using the microvertex information, one, labelled (μ), shows a muon with large transverse momentum in one of the hadronic jets. Note that the OPAL mass limit is significantly affected by a $H\ell^+\ell^-$ candidate at 61 GeV.

There is not yet(?) a standard procedure agreed by the four experiments to combine these limits. The procedure used here is described in[11]. Figure 5 summarizes the results obtained by the 4 experiments. The combined limit is 64.5 GeV taking into account a reduction in efficiency.

The 95 % C.L. line limit is modulated by the presence of $H\ell^+\ell^-$ candidates. All selected candidates show an indication of b-tagging except the L3 candidate at 67.6 GeV (events without b-tagging are removed in the ALEPH analysis).

4 Z→Hγ search

This process has a large radiative background which can be somewhat reduced by demanding an energetic photon well isolated from the hadronic jets. In the DELPHI[12] analysis, based on 3 million hadronic events, the isolation angle varies between 20° and 120° depending on the mass of the hadronic system. The mass resolution, essential for this analysis, is almost constant $\sim$ 2 GeV, using kinematical constraints in conjunction with the photon energy measurement.

The mass spectrum given in figure 7 is obtained using b-tagging. In figure 8, the resulting limits on BR(Z→Hγ) are shown. For masses below the beauty threshold, the channel $\tau^+\tau^-\gamma$ has been used. These limits can exclude the presence of strong anomalous couplings[13] and are not far from the SM predictions. In the SUSY case, BR(Z→Hγ) can be modified by a factor 2 or 3 [14] and therefore this search is relevant for masses below 30 GeV.

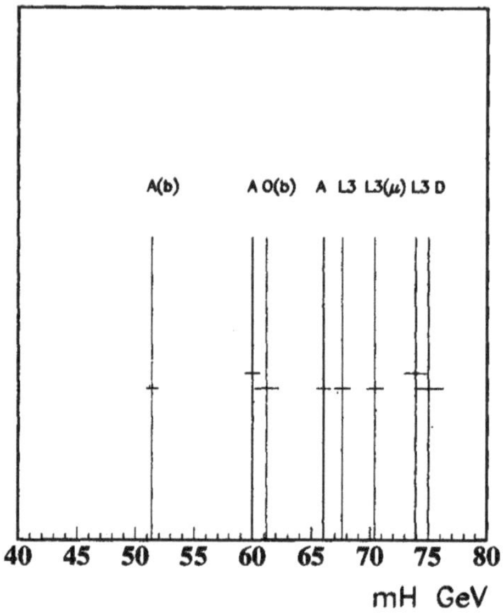

Figure 5: Leptonic candidates

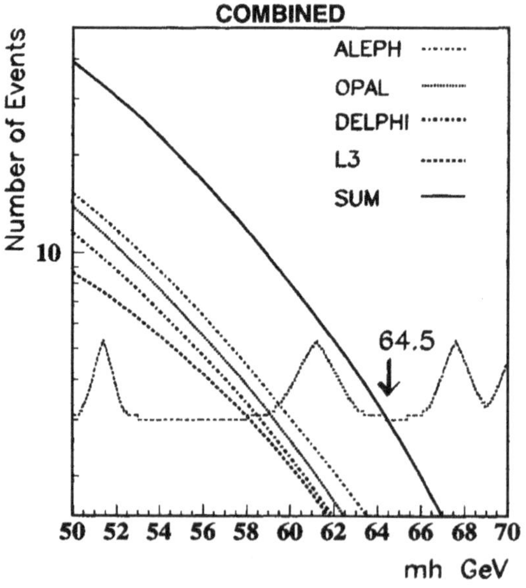

Figure 6: SM Higgs limits

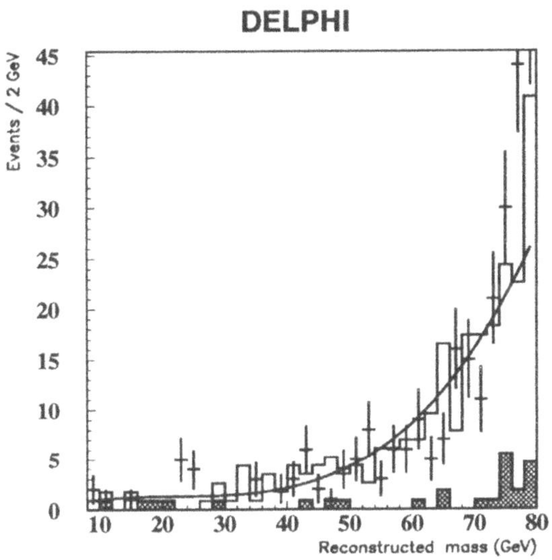

Figure 7: Mass distribution of radiative candidates (data are crosses, histograms give the expected background with fake photons in hatched style)

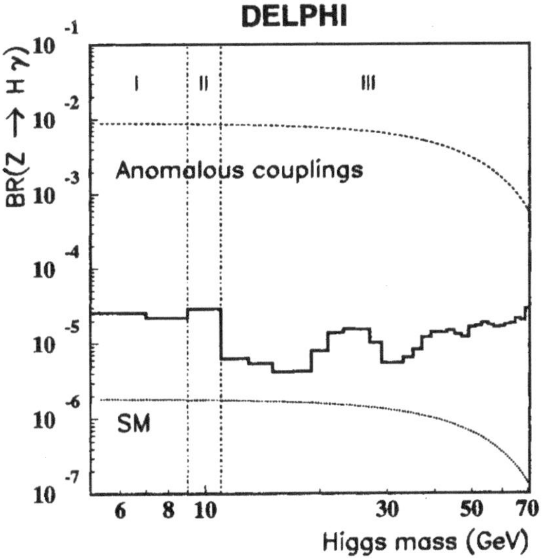

Figure 8: Limits on radiative branching ratio

It may happen that $H \to \gamma\gamma$ becomes dominant[13] in the presence of an anomalous coupling. In this case one can use the limit on $BR(Z \to 3\gamma)$ given in[15] which is $7x10^{-6}$.

5 SUSY Higgs

5.1 *Introduction*

- Hierarchy problem

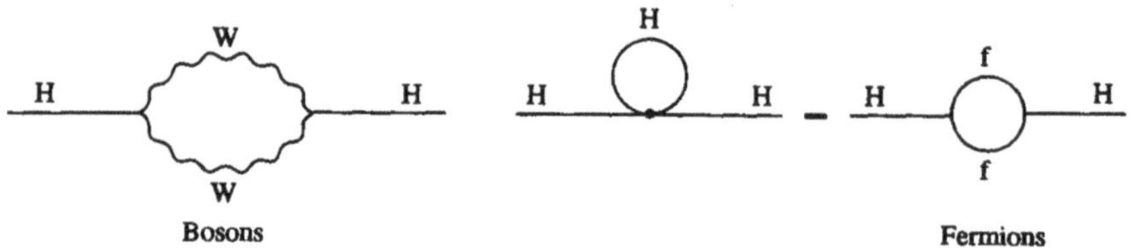

In the SM model, quantum corrections to the Higgs boson mass are ultra-violet divergent. A natural cut-off is provided by Λ, the scale at which the SM breaks down. One has $\delta m_H^2 \sim g^2 C \Lambda^2$. Although there is some cancellation between the boson and the fermion loops, C is different from zero. If no new physics occurs below a few TeV, there is therefore no natural way to have a Higgs boson mass at the EW scale.

- SUSY

  This problem is naturally cured in the SUSY scheme where each fermion(boson) has a mirror boson(fermion) partner. One has $\delta m_H^2 \sim g^2(m_F^2 - m_B^2)$. Provided that supersymmetric particles are not heavier than a few TeV, the hierarchy problem gets solved.

- Two doublet MSSM

In the SUSY scheme, the Higgs boson should have a fermionic partner $\tilde{H}$. This partner cannot be "left alone" to avoid triangle anomalies (theory not renormalizable). This is cured by adding a new fermionic partner $\tilde{H}'$ with opposite hypercharge: consequently one gets extra mirror scalar Higgs bosons. MSSM is the minimal scheme in which there are two Higgs doublets with opposite hypercharges.

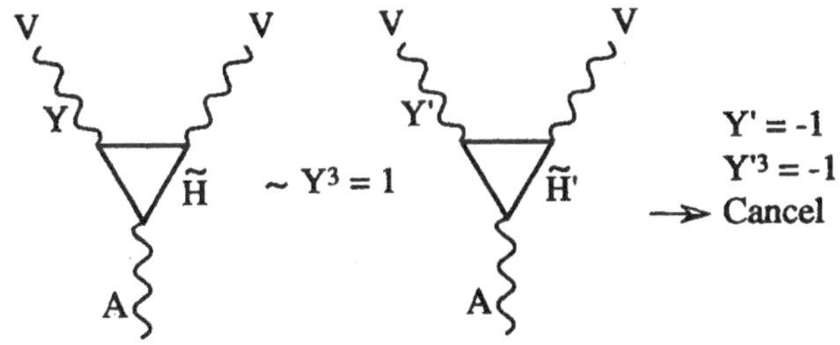

In this scheme[16], one has 5 degrees of freedom leftover corresponding to h and H the CP-even , A the CP-odd, and $H^{\pm}$ the charged Higgs bosons. A mixing angle α defines the two CP-even Higgs physical states. The two doublets acquire vacuum expectations v_1 and v_2 with $tg\beta = \frac{v_2}{v_1}$. Within SUSY, $\lambda \sim (g^2 + g'^2)$ and the mass of h, the lightest Higgs boson, is simply given by:

$$m_h^2 = m_Z^2 cos^2 2\beta$$

assuming m_A large. This means that LEP should be able to observe h. One has also:

$$m_{H\pm}^2 = m_A^2 + m_W^2$$

which means that $H^{\pm}$ is not relevant for LEP.

5.2 Radiative Corrections

Previous mass formulae were ignoring radiative corrections[17]. Assuming m_A large, one has:

$$m_h^2 = m_Z^2 cos^2 2\beta + \frac{3g^2}{8\pi^2 m_W^2}\Delta$$

where:

$$\Delta = m_t^4[ln\frac{m_{\tilde{t}_1} m_{\tilde{t}_2}}{m_t^2} + A_t(A_t + \mu cotg\beta)F(m_{\tilde{t}_1}, m_{\tilde{t}_2}) + ...]$$

These parameters appear in the top squark mass matrix:

$$\begin{pmatrix} \tilde{t_L}^2 + m_t^2 & (A_t + \mu cotg\beta)m_t \\ (A_t + \mu cotg\beta)m_t & \tilde{t_R}^2 + m_t^2 \end{pmatrix}$$

Very large effects are possible due to the m_t^4 coefficient: the correction term may reach values of order m_Z^2. These corrections are maximum for large tgβ.

Two questions therefore arise:

- How reliable are these radiative corrections and in particular what is the effect of H.O. corrections[18] ? This point is actively discussed in the LEP2 workshop and accurate (± 5 GeV) mass formulae are now available[19].

- How can we deal with so many arbitrary parameters of SUSY ? At the tree level, two parameters are sufficient like, for instance, tgβ and m_A. With radiative correction one has to fix the squark masses, the mixing parameter A_t and μ.

5.3 Combined LEP searches and their interpretations

In MSSM, two final states have to be considered:

- Z→HZ* with $\sigma = \sigma_{SM} sin^2(\alpha - \beta)$ dominant at low tgβ

- Z→hA with $\sigma = \sigma_{\nu\bar{\nu}}\lambda^{3/2}cos^2(\alpha - \beta)$ dominant at large tgβ with $m_h \sim m_A$ (λ is a phase-space term).

Heavy Higgs bosons h and A, relevant at this stage of the LEP searches, decay $\sim$ 90% into $b\bar{b}$ and the rest goes into $\tau^+\tau^-$. DELPHI has searched for 4b states[20] and ALEPH[21] gives the best limits in $\tau^+\tau^-$+hadrons. Combining the four LEP experiments for the hZ* channel and using the DELPHI result for the hA search, one obtains the limits given in figure 9. These results take into account radiative corrections with the simplistic but usual assumptions: $m\tilde{t}_1 = m\tilde{t}_2$=1 TeV ,$\mu = A_t$=0 and with a top mass of 175 GeV.

The kinematical limit, imposed by the hA channel, is reached at LEP1:

$$m_h > 45GeV$$

$$m_A > 45GeV$$

What happens when μ or/and A_t differ from zero ? In[22], it has been pointed out that the LEP1 limit can be significantly modified. By tuning these parameters one can create holes in a mass region where h is as light as 20 GeV as shown in figure 10 obtained with the L3 data. These holes correspond to a large reduction of the cross-section for the process hZ* in a mass region where the complementary process hA is kinematically closed at LEP1. The suppression factor is about 25 with respect to the SM cross-section. As shown in figure 11, this is at the edge of the best available limits but is probably not inaccessible if one combines the 4 LEP experiments.

Technically speaking, the result from[22] was obtained in a diagrammatic approach. The same effect can be reproduced in the potential approach. As an example, if one takes: μ=-500 GeV, A_t= -1000 GeV and tgβ=3.62, one gets $\sin^2(\alpha - \beta)=10^{-5}$ with m_A=98 GeV, m_h=29 GeV. In this example one has $m\tilde{t}_1$=39GeV and $m\tilde{t}_2$=546 GeV.

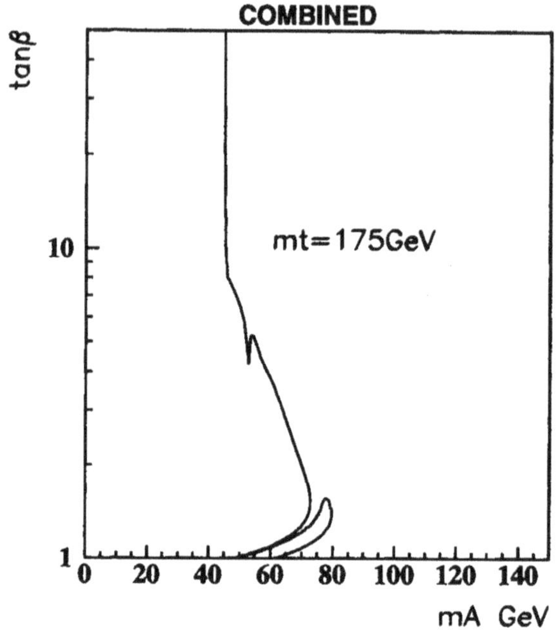

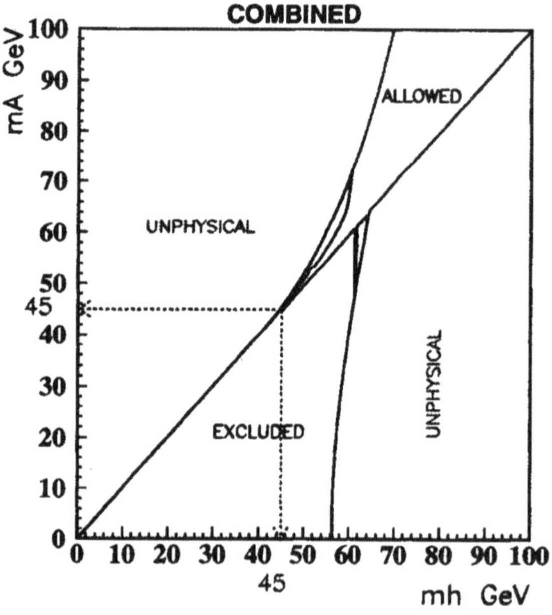

Figure 9: MSSM limits with $\mu = A_t = 0$

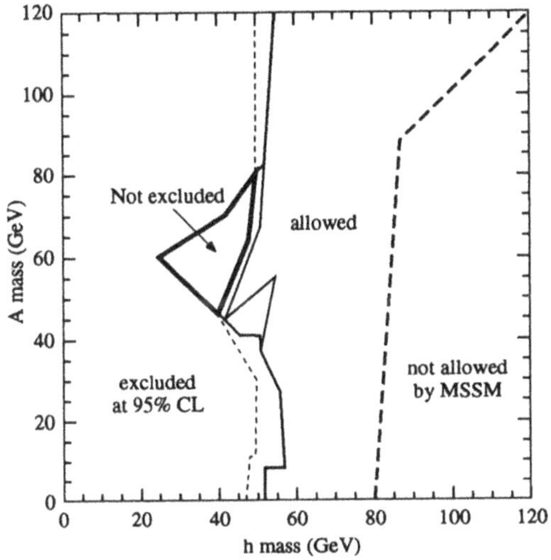

Figure 10: MSSM results from L3 with free SUSY parameters

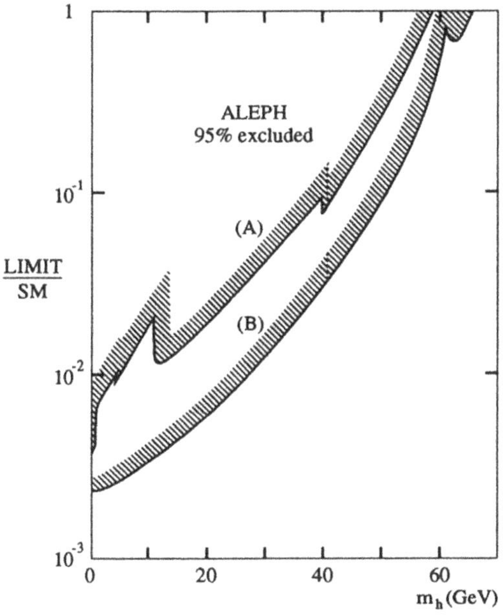

Figure 11: Limits from ALEPH normalized to the SM expectations. The curve (B) assumes an invisible decay of the Higgs boson

From this discussion, I conclude that there is no clear limit on m_h when the SUSY parameters vary arbitrarily ! In the next section, I will show how the GUT approach can clear this mess.

5.4 *Results within MSSM-GUT*

In reference[22], the SUSY parameters are varied independently in contrast to the GUT approach where they are interrelated. For instance one has[23]:

$$m_A^2 sin^2\beta = \mu^2 + m_{\tilde{\nu}}^2 + g^2\mu^2 f(m_t, \tilde{t_1}, \tilde{t_2})$$

where the last term is due to radiative corrections. The set of quantities used by[22] to create the "hole" in the Higgs search therefore becomes irrelevant (m_A should be as large as μ, i.e. 500 GeV, in the example given at the end of section 5.3). One can also argue that μ is related to the chargino/neutralino masses and therefore cannot be chosen arbitrarily. In figure 12, the recent results from[6] show that when tgβ differs from 1, μ=0 is excluded. In addition, as pointed out in section 2.1, there is a significant limit on the sneutrino mass. This clearly shows that, in a GUT scheme, the Higgs sector can be constrained by other searches.

To treat this problem in a systematic way, I have used the genuine GUT approach which assumes that there is a common gaugino and scalar mass at the GUT scale (M,m_0). Five quantities are then sufficient to compute the SUSY spectrum (assuming that tgβ is not too large, one can neglect the effect of the bottom quark): μ, m_0, M, tgβ, A_t. There is an additional constraint, since m_Z is also given in terms of these quantities:

$$m_Z^2 = F(\mu, m_0, M, tg\beta, A_t)$$

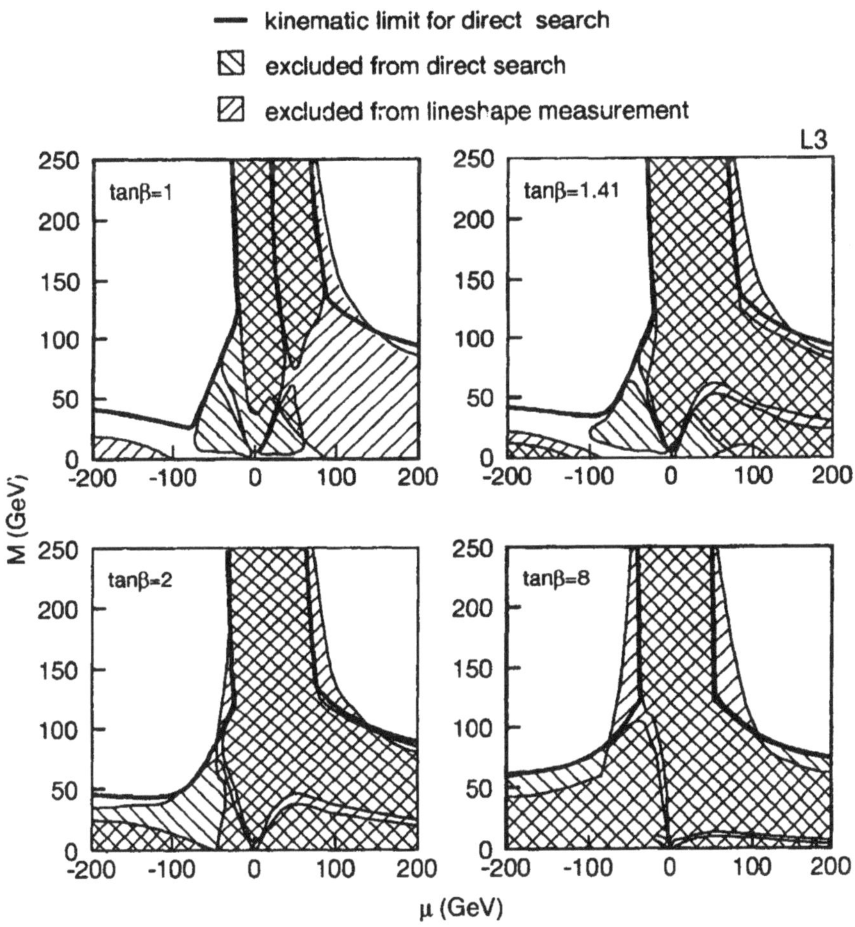

Figure 12: Exclusion contours from L3 using the chargino/neutralino searches

The method used is very simple: the five parameters are varied simultaneously taking into account the previous constraint and including the restrictions due to the LEP searches on charginos, sleptons, sneutrinos and top squarks. The minimal acceptable value for m_h and m_A are then obtained. This work is done using the latest code given by[19] which includes H.O. corrections for the Higgs masses. The results are shown in figure 13. The dashed lines are the limits obtained a priori, i.e. without any use of the Higgs searches. One can see that there is already a strict bound on m_A while m_h has no bound at low tgβ. Direct searches go beyond these indirect bounds for tgβ <4. The net result is improved with respect to the "naive" one:

$$m_h > 61GeV$$

$$m_A > 80GeV$$

What would happen for very large tgβ ? One has $m_A \sim m_h$ and therefore the bound on m_A coming from $m_A^2 \sim \mu^2 + m_{\tilde{\nu}}^2$ still operates and induces the same limit on m_h. This intuitive conclusion has been checked, including the effect of the bottom sector in the radiative correction term in the equation relating m_A to the SUSY parameters.

The same treatment can be applied to the lightest neutralino. Figure 14 shows that while in the absence of the Higgs search there is no lower mass limit, the combined searches give:

$$m_\chi > 12GeV$$

Using the same method, which one can also deduce that $m_{\tilde{g}} > 90GeV$. These results are clearly model dependent but this is also true for those obtained with $\mu = A_t = 0$!

Precision measurements[24] are consistent with the MSSM-GUT scheme provided tgβ is not too large. When tg$\beta \sim m_t/m_b$, one can still interpret the data but at the expense of "fine tuning" the SUSY parameters or by relaxing the unification of the masses in the scalar sector[25].

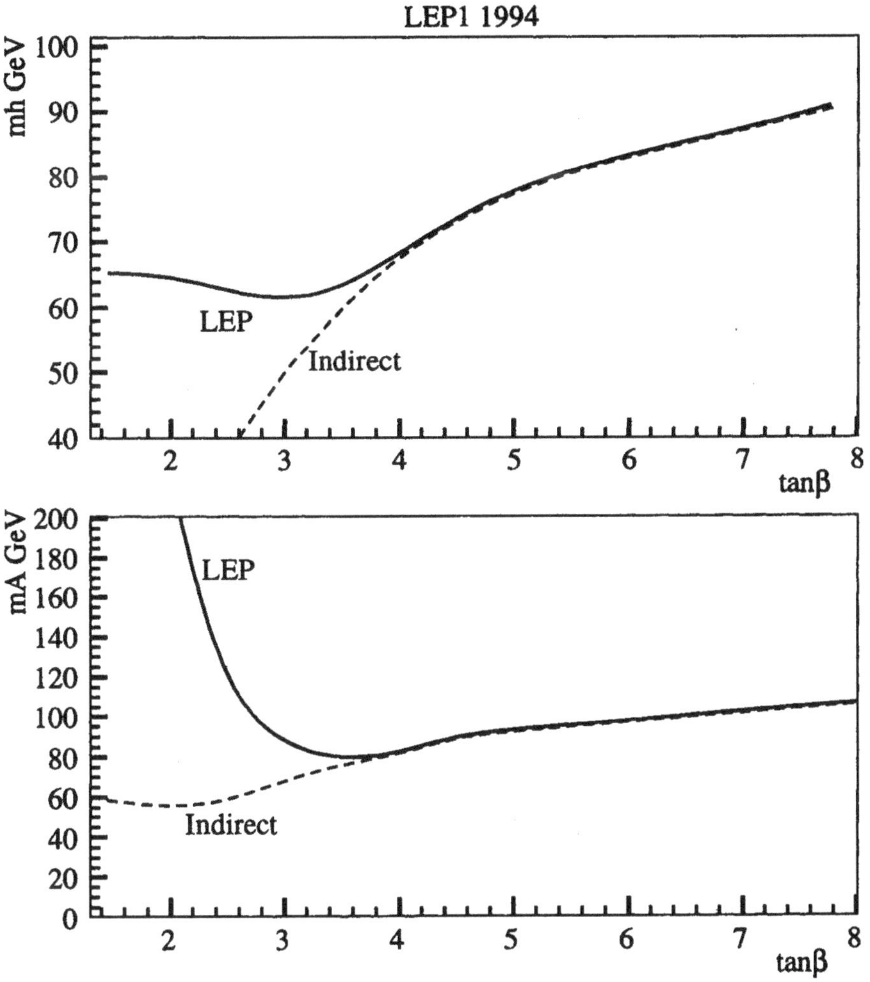

Figure 13: MSSM-GUT limits on m_h and m_A from LEP. The dashed curves are obtained without the direct Higgs searches

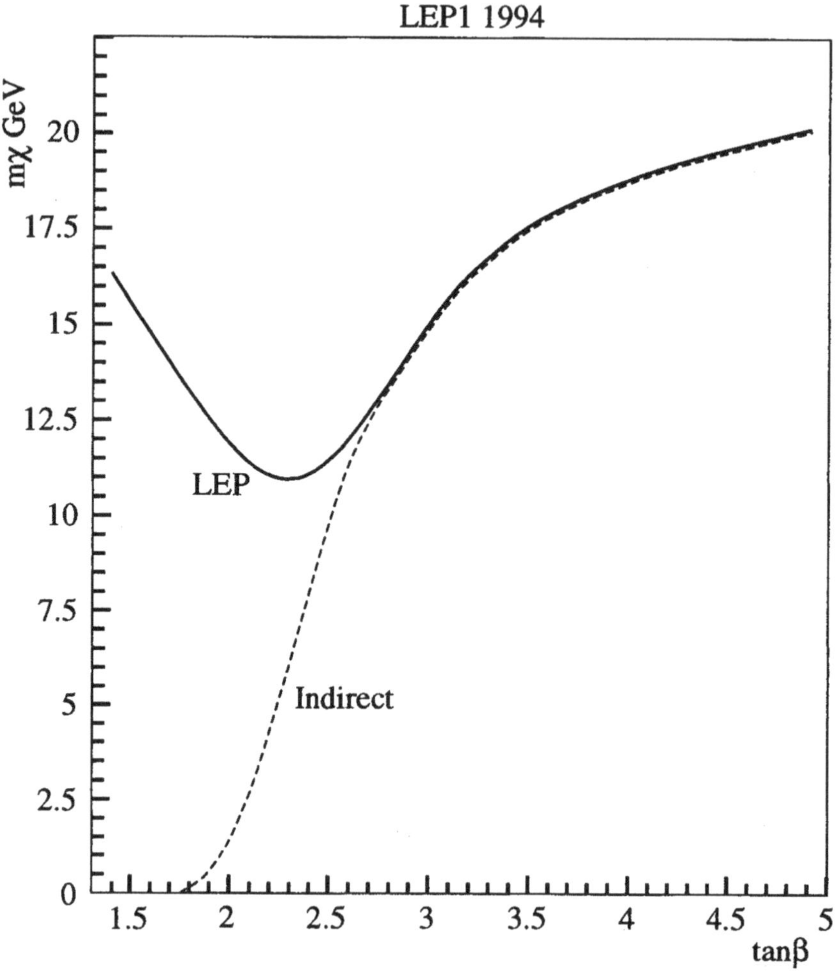

Figure 14: MSSM-GUT limits on the lightest neutralino from LEP. The dashed curves are obtained without the Higgs searches

5.5 NMSSM

In the SUSY superpotential, the μ parameter appears explicitly in the "mixing" term $\mu H_1 H_2$. As for the SM case, one needs a mechanism to stabilize μ against quantum corrections which are of the order of the GUT scale. This is usually done by introducing an isoscalar Higgs field N and by replacing $\mu H_1 H_2$ by $N H_1 H_2$. By generating a vacuum expectation for N, one obtains the appropriate contribution to the superpotential.

As a result one has three CP-even Higgs bosons, one of them being sterile, i.e. not coupled to fermions nor to vector bosons. As shown in[26], the lightest Higgs h is likely to be dominantly isoscalar and therefore, as can be seen in figure 15, become invisible at LEP if it has a mass below 60 GeV.

A similar effect could occur in the neutralino sector which is under investigation.

There is apparently not much to be done in such a scenario except look for heavier Higgs bosons which could be accessible at LEP1 or LEP2.

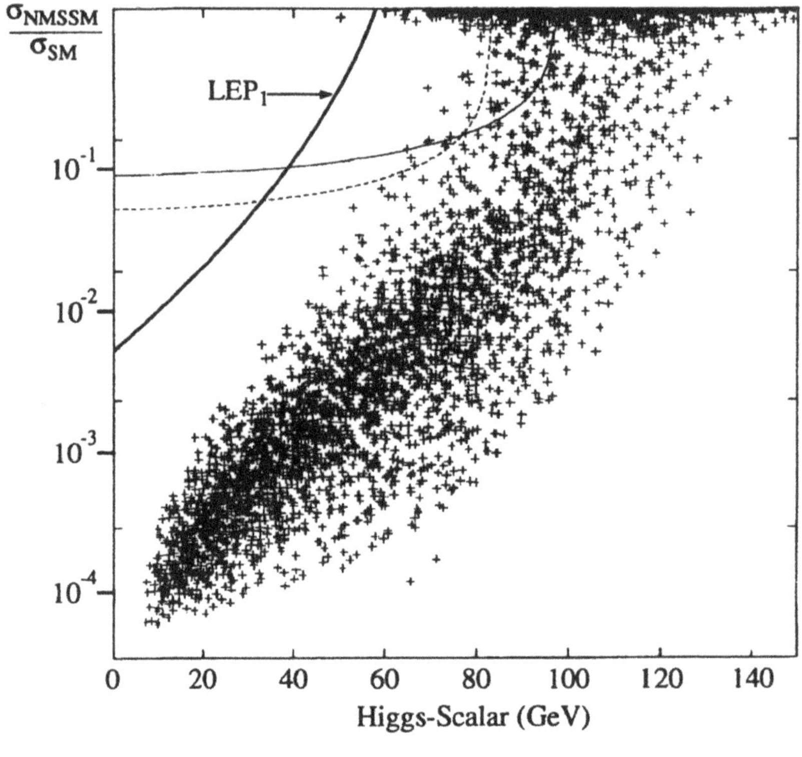

Figure 15: The ratio of the Higgs NMSSM cross-sections normalized to the SM is expected in the region with crosses. The LEP1 line corresponds to the what is excluded by ALEPH. The other lines correspond to what can be covered at LEP2.

6 SUMMARY

Charginos, sfermions are excluded up to $\sim$ the kinematical limit of LEP1 even in the case of R-parity violation.

- SM limit (at 95 % C.L.):

$$m_H > 60 GeV\, ALEPH$$

$$m_H > 64.5 GeV\, combined$$

- MSSM limit (at 95 % C.L.) , assuming $\mu = A_t = 0$:

$$m_h > 45 GeV$$

$$m_A > 45 GeV$$

  There is no lower mass limit for neutralinos nor for gluinos.

- MSSM-GUT limit (at 95 % C.L.):

$$m_h > 61 GeV$$

$$m_A > 80 GeV$$

$$m_\chi > 12 GeV$$

$$m_{\tilde{g}} > 90 GeV$$

1. LEP EW working group, presented at Brussels 1995.
2. G. Altarelli, R. Barbieri and F. Caravaglios. *Phys. Lett.* **B314** (1993) 357.
3. ALEPH Collaboration (D. Buskulic et al.). CERN-PPE-95-015, Feb 1995.
4. DELPHI Collaboration, Search for Monojets and Isosinglet Neutral Heavy Leptons (DELPHI 95-75 PHYS 510). Presented at Bruxelles 1995. Published in *Phys. Lett.* **B349** (1995) 238-252.
5. OPAL Collaboration (R. Akers et al.). CERN-PPE-94-103, Jul 1994. Published in *Phys. Lett.* **B337** (1994) 207-218. DELPHI Collaboration, DELPHI 92-172, Phys 255.
6. L3 Collaboration (M. Acciarri et al.). CERN-PPE-95-014, Feb 1995. Published in *Phys. Lett.* **B350** (1995) 109-119.
7. UA1 Collaboration, *Phys. Lett.* **B198** (1987) 261.
8. See e.g. B. Kileng, P. Osland. BERGEN-1994-10, Jul 1994. Published in *Z . Phys.* **C66** (1995) 503-512.
9. J.R. Espinosa, M. Quiros. CERN-TH-95-18, Mar 1995.
10. G. Altarelli, G. Isidori. CERN-TH-7351-94, Jul 1994. Published in *Phys. Lett.* **B337** (1994) 141-144.
11. F. Richard. LAL-94-50, Sep 1994. Talk given at 27th International Conference on High Energy Physics (ICHEP), Glasgow, Scotland, 20-27 Jul 1994.
12. DELPHI Collaboration, Update to the search for the Z^0 decay into a Higgs boson and a Photon (DELPHI 95-73 PHYS 508). Presented at Bruxelles 1995.
13. K. Hagiwara et al. *Phys. Lett.* **B318** (1993) 155.
14. G. Gamberini et al. *Nucl. Phys.* **B292** (1987) 237.
15. DELPHI Collaboration, An updated analysis of the $e^+e^- \to \gamma\gamma(\gamma)$ reaction at LEP energies (DELPHI 95-71 PHYS 506) Presented at Bruxelles 1995.

16. For a review on the Higgs sector in SUSY, see J. Gunion, H. Haber, G. Kane and S. Dawson, The Higgs Hunter's Guide, Addison-Wesley, Reading 1990.
17. Y. Okada, M. Yamaguchi and T. Yanagida, Prog. Theor. *Phys. Lett.* 85 (1991) 1;
J. Ellis, G. Ridolfi and F. Zwirner, *Phys. Lett.* **B257** (1991) 83 ;
H.E. Haber and R. Hempfling, *Phys. Rev. Lett.* **66** (1991) 1815.
18. R. Hempfling and A.H. Hoang, *Phys. Lett.* **B331** (1994) 99. J. Kodaira et al., HUPD-9316 (Nov 1993).
19. M. Carena, J.R. Espinosa, M. Quiros, C.E.M. Wagner. CERN-TH-95-45, Mar 1995.
20. DELPHI Collaboration, DELPHI 93-59 PHYS 287.
21. S. Simion, Thèse LAL 95-14 (april 1995).
22. J. Rosiek, A. Sopczak. CERN-PPE-94-134, Sep 1994. 24pp. Contributed to Int. Conf. on High Energy Physics, Glasgow, Scotland, Jul 20-27, 1994. Published in *Phys. Lett.* **B341** (1995) 419-430.
23. Formulae used here are from W. de Boer, R. Ehret, D.I. Kazakov. IEKP-KA-94-05, May 1994.
24. P. H. Chankowski, S. Pokorski. IFT-UW-95-5, Mar 1995. 10pp. Talk given at 4th International Conference on Physics Beyond the Standard Model, Lake Tahoe, CA, 13-18 Dec 1994.
25. F.M. Borzumati, M. Olechowski and S. Pokorski. CERN-TH-7515-94, Dec 1994. Published in *Phys. Lett.* **B349** (1995) 311-318.
26. U. Ellwanger, M. Rausch de Traubenberg, C. A. Savoy. LPTHE-ORSAY-95-04, Jan 1995.

ELECTROWEAK RESULTS AT LEP-I

Manel MARTINEZ

Institut de Fisica d'Altes Energies, Edifici Cn, U. A. B.
E-08193 Bellaterra (Barcelona)
Spain

The most relevant LEP electroweak measurements obtained during 1989–1994 are briefly reviewed and their consistency with the Minimal Standard Model is discussed. It is shown that only the Z partial width into b and c quark measurements show significative discrepancies with respect to their MSM expectation. Assuming the MSM, the precision LEP data alone predicts $m_t = \left(170 \pm 10^{+17}_{-19}\right)$ GeV, in good agreement with the direct observation at Tevatron. The inclusion of the CDF m_t evidence in the analysis is shown to open the window for the inference of the Higgs mass from precision data.

1 Introduction

Since its start-up in 1989, LEP has proven to be the most powerful machine to probe the theory of the electroweak interactions, the Minimal Standard Model (MSM). A full bunch of high precision measurements performed by the four LEP experiments allow now the extraction of many important parameters to an unprecedented level of accuracy.

With the upcoming of the LEP-II program, the LEP-I program is arriving to its end and 1995 will likely be the last year of Z Physics at LEP. Having this in mind, our goal will be to make a short review of the present situation of the MSM tests performed at LEP-I and a critical inspection of the open questions.

In this report, the most relevant LEP Electroweak measurements performed so far are briefly presented, paying special attention to their limitations and still possible improvements and their impact in the Standard Model is discussed. The results presented here come, for most observables, from all data taken between 1989 and 1994 by the four LEP experiments: about twelve million hadronic Z decays and about twelve-hundred charged-lepton Z decays. Most of the results discussed are still preliminary although they have been already presented in public [1].

The observables are classified into three groups: Z lineshape and lepton F-B asymmetry measurements, tau polarization measurements and finally electroweak results with quarks, where the present results show a clear discrepancy with the Standard Model expectations. They are discussed in turn in the following.

2 The LEP measurements.

2.1 Z lineshape and lepton F-B asymmetry measurements.

The cross section for the production of a fermion pair $f\bar{f}$ can be written as

$$\sigma_{f\bar{f}}(s) = \int_{4m_f^2}^{s} ds' H(s,s')\hat{\sigma}_{f\bar{f}}(s') \quad (1)$$

where $H(s,s')$ is the so-called radiator function which takes care of initial state radiation corrections and the reduced cross section $\hat{\sigma}$ is written as

$$\hat{\sigma}_{f\bar{f}}(s) = \sigma^0_{f\bar{f}} \cdot \frac{s\Gamma_Z^2}{\left(s - M_Z^2\right)^2 + \left(s\Gamma_Z/M_Z\right)^2} + (\gamma - Z) + |\gamma|^2 \quad (2)$$

where $\sigma^0_{f\bar{f}}$ is called the peak cross section, $\Gamma_Z = \Gamma_h + \Gamma_e + \Gamma_\mu + \Gamma_\tau + \Gamma_{inv}$ stands for the total Z width and M_Z is the Z mass. This parameterization assumes the validity of QED for the photon exchange part and also takes from the Minimal Standard Model the interference between the photon- and Z-mediated amplitudes.[a] This interference is very small around the Z pole. In the case of Bhabha scattering, $f = e$, one can either subtract for the data the t-channel contributions, also taken from the theory or add them to the previous expression. The cross section at the peak can be written in turn in terms of the Z mass and width and the Z partial widths to the initial state Γ_e and the final state Γ_f:

$$\sigma^0_{f\bar{f}} = \frac{12\pi}{M_Z^2} \cdot \frac{\Gamma_e\Gamma_f}{\Gamma_Z^2} \quad (3)$$

where the partial widths can be written in terms of effective vector and axial couplings of the fermions to the Z:

$$\Gamma_l = \frac{G_F M_Z^3}{6\pi\sqrt{2}}({g_{V_l}}^2 + {g_{A_l}}^2)\left(1 + \frac{3\alpha}{4\pi}\right) \quad (4)$$

Assuming lepton universality, only four parameters are needed to describe the s-dependence of the hadronic and leptonic cross sections: M_Z, Γ_Z, σ_h^0 and the ratio of hadronic to leptonic partial widths ($R_l = \Gamma_h/\Gamma_l = \sigma_h^0/\sigma_l^0$), where the lepton is taken to be massless. If lepton universality is not assumed, R_l is substituted by three analogous quantities, R_e, R_μ, R_τ. This choice of parameters, given the fact that they are directly related to geometrical characteristics of the lineshape, has two advantages: on the one hand, their correlation are small, and on the other hand they simplify the task of disentangling common uncertainties:

[a] For some experiments this is only the case for hadronic final states.

- M_Z and G_Z are the only measurements in the energy scale and they are strongly affected by common LEP energy uncertainties which, thanks to the use of the resonant depolarization technique for the determination of the LEP energy, have been reduced to 1.5 MeV and 1.7 MeV respectively [2]. The LEP results can be seen in fig. 1.

- σ_h^0 is the only lineshape observable in which the overall normalization, and hence the luminosity measurement uncertainties, enter. Presently this measurement is already dominated by the common systematical error due to the theoretical uncertainty in the low-angle Bhabha cross section. The uncertainty presently claimed is of 0.16% [3]. Since R_l is a ratio of cross sections, the luminosity measurement cancels and there is, so far, no relevant source of common systematics affecting sizably it. The LEP results can be seen in fig. 2.

The forward-backward asymmetry is defined as

$$A_{FB} = \frac{\sigma_F - \sigma_B}{\sigma_F + \sigma_B}, \tag{5}$$

where F and B indicates the forward or backward hemisphere. Normally it is obtained by fitting the measured angular distribution to the formula

$$\frac{d\sigma^l}{d\cos\theta}(s) = \frac{3}{8}\sigma^l(s) \cdot \left(1 + \cos^2\theta + \frac{8}{3}A^l_{FB}(s)\cos\theta\right). \tag{6}$$

In the case of e^+e^- final state, the t-channel contribution is either subtracted from the observed asymmetry or added to the previous expression.

Once the different $A^l_{FB}(s_i)$ are obtained, they are fitted together with the lineshape data to get the the lineshape parameters mentioned above and the peak asymmetry, $A^{0,l}_{FB}$:

$$\begin{aligned} A^{0,l}_{FB} &= \frac{3}{4}A_e A_l \\ A_f &= \frac{2g_{V_f}/g_{A_f}}{1 + \left(g_{V_f}/g_{A_f}\right)^2}. \end{aligned} \tag{7}$$

from which the effective weak mixing angle is measured. The results of the four collaborations are shown in fig. 3. The main error is still statistical. Experimental systematics can only come from simultaneous charge and forward-backward asymmetries in the detector, which are bound to be very small. The knowledge of the beam energy contributes a non-negligible 0.0005 to $\Delta A^{0,l}_{FB}$.

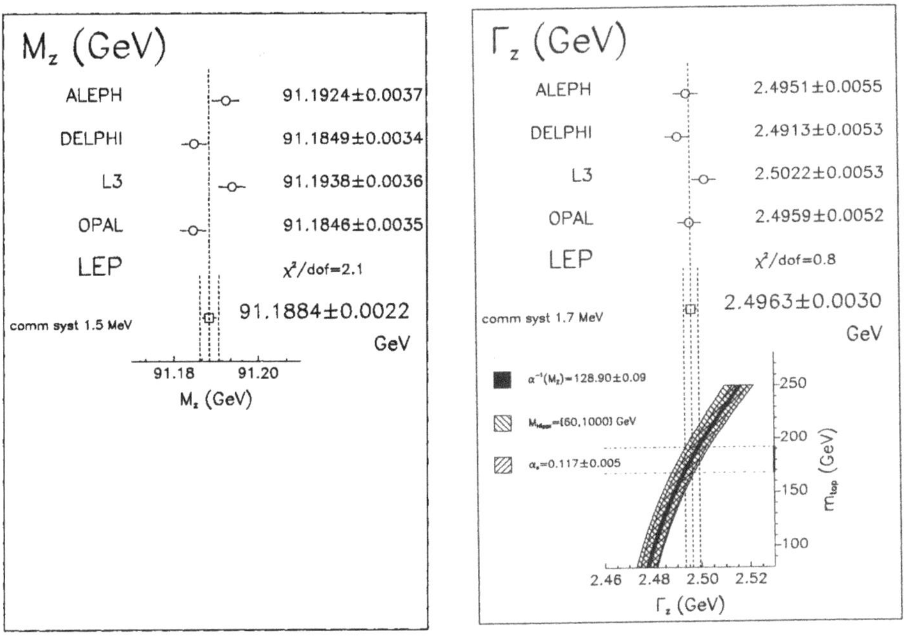

Figure 1: The Z mass and width measured by the four experiments together with the mean. The bands show the MSM prediction as a function of the top quark mass. The numerical values given include all uncertainties while the error bars include only non-common uncertainties to allow a visual inspection of the actual agreement between them. Unless explicitly stated, the same convention applies to all the figures of this kind in this report.

The use of the sketched procedure to study the asymmetries does not allow, for instance, to check the consistency of their energy dependence because, being it proportional to the axial coupling, its actual value is wiped out by the much more precise determination coming from the leptonic peak cross sections which, in the procedure presented, are fitted simultaneously with the asymmetries to provide consistently constraints on the values of M_Z and Γ_Z. To avoid this problem, one needs a more general description of these observables, as the one provided by the s-matrix ansatz [4]. In that approach, there is a pa-

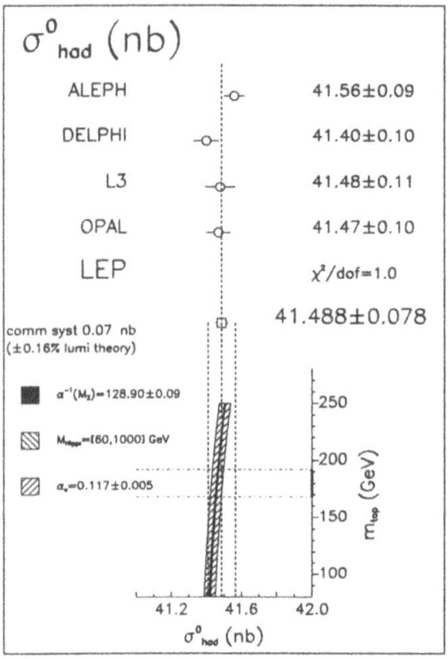

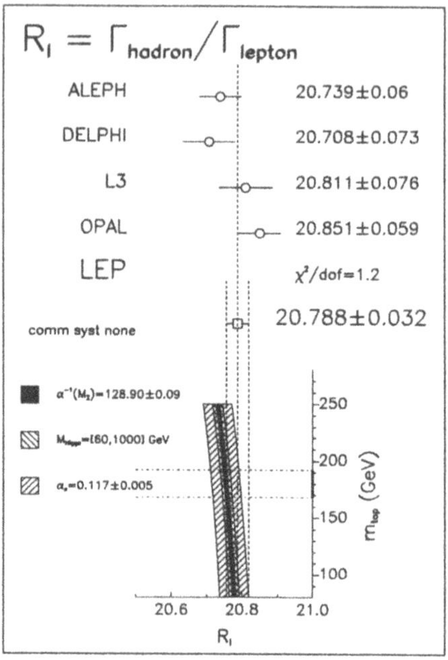

Figure 2: The hadronic peak cross section and its ratio to the leptonic one measured by the four experiments together with the mean and the MSM prediction as a function of the top quark mass.

rameter called j_{FB} which specifically quantifies the energy dependence of the forward-backward cross section. Unexpectedly, the use of this new approach, has brought to the light some, so far hidden, puzzles as the one shown in fig. 3 which are still under study [5].

2.2 Tau Polarization

The measurement of the final state longitudinal polarization asymmetry for tau leptons:

$$< P_\tau >= \frac{\sigma_R - \sigma_L}{\sigma_R + \sigma_L} = -A_\tau, \tag{8}$$

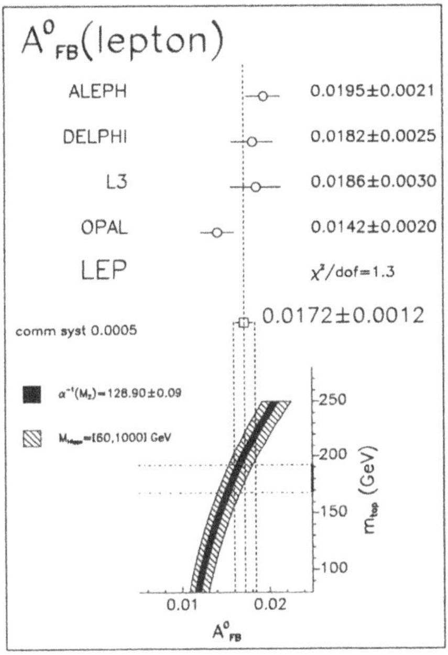

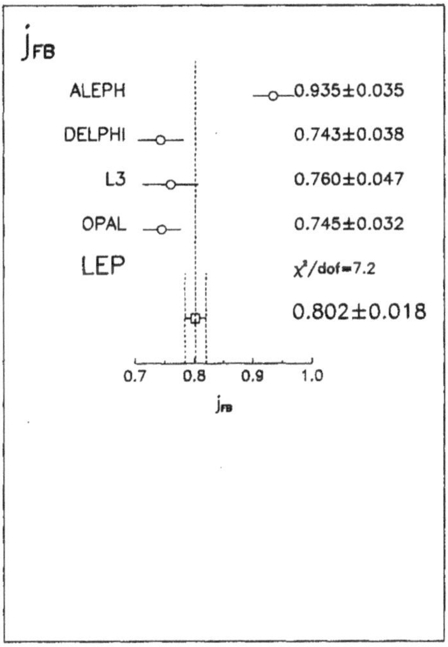

Figure 3: The leptonic forward-backward asymmetry measured by the LEP experiments together with the MSM prediction as a function of the top quark mass, and the s-matrix parameter j_{FB} which measures the energy dependence of the lepton forward-backward asymmetry as measured by the 4 LEP experiments. The MSM expectation for a top mass of 175 GeV and a Higgs mass of 300 GeV is $j_{FB} = 0.79996$.

where $\sigma_{R(L)}$ is the integrated cross section for right (left) handed tau's, provides a means to measure the tau couplings to the Z directly. Since the tau decay inside the detectors, their helicity information can be obtained by using their parity-violating weak decays as a spin analyzer. The main decays are used: to electron, muon, pion, rho, and a_1. Maximal sensitivity is obtained for the semileptonic modes (pion and rho) since then, just one neutrino escapes detection. In this case, the decay angle of the hadron in the tau rest frame (or equivalently its scaled energy) is measured and the integrated tau polarization is extracted from

$$\frac{1}{N}\frac{dN}{dx} = 1 + \alpha < P_\tau > \cos\theta^* \tag{9}$$

where α is a sensitivity coefficient linked to the spin of the hadron. For pions (s=0) $\alpha = 1$ and for rho and a_1 (s=1) $\alpha < 1$ but can be improved by studying the hadron helicity through the analysis of its decay products.

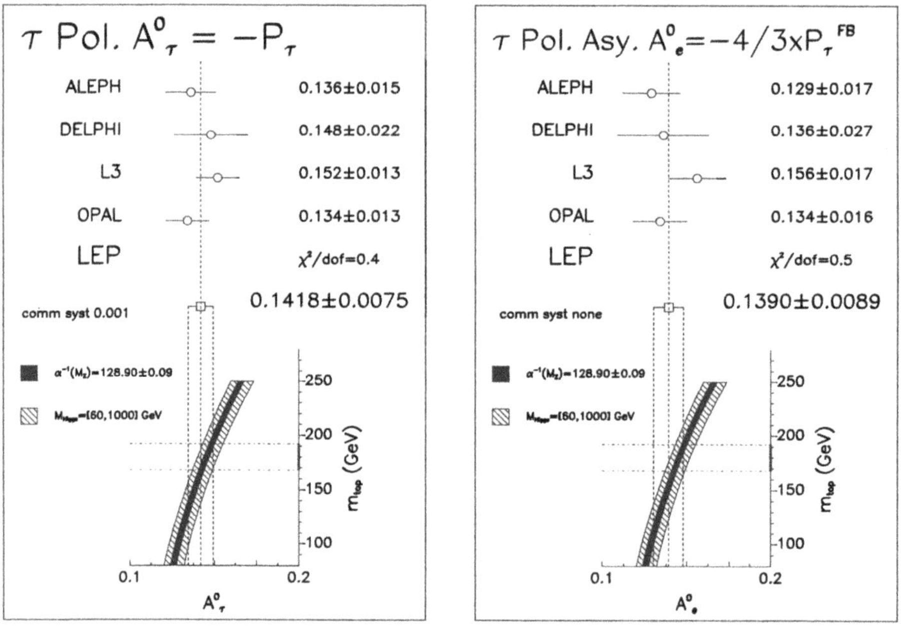

Figure 4: The tau polarization and its forward-backward asymmetry measured by the four experiments together with the mean and the MSM prediction as a function of the top quark mass.

The results are given in fig. 4. The systematics are in this case comparable to the statistical errors and their reduction would require a lot of effort in the understanding of the calorimeters.

By measuring the tau polarization as a function of the the tau production angle, θ one can write

$$P_\tau(\cos\theta) = -\frac{A_\tau + A_e \cdot \frac{2\cos\theta}{1+\cos^2\theta}}{1 + A_e A_\tau \cdot \frac{2\cos\theta}{1+\cos^2\theta}} \, . \tag{10}$$

From this expression its is apparent that, while the integrated polarization measures A_τ, its forward-backward asymmetry measures the electron coupling, A_e.

The results are shown in fig. 4. The error is still mainly statistical, since most systematic effects cancel out when computing the forward-backward asymmetry.

This measurement can be directly compared with the left-right initial state polarization asymmetry from SLD [7] since, in a rather model independent way, both determine the same quantity: $A^0_{LR} = A_e$. The SLD measurement $A_e = 0.1551 \pm 0.0040$ turns out to be at about 1.6 σ from the LEP average which, given the statistical origin of the uncertainties claimed, should be assigned to an statistical fluctuation.

2.3 Electroweak results with quarks.

Given the lack of techniques to tag efficiently the light quark flavors, most of the LEP experiments compute inclusively the forward-backward asymmetry in all hadronic events by estimating the average quark charge via a momentum-weighted mean of charges of the hadrons belonging to each quark's jet. This way, the inclusive forward-backward asymmetry for the actual Z decay mixture of quark flavors, or Quark Charge Asymmetry $< Q_{FB} >$ can be extracted leading to a LEP average of $\sin^2\theta^{lept}_{eff} = 0.2315 \pm 0.0013$. This uncertainty is dominated by the Monte Carlo modelling for light quarks.

The measurement of the Z decay width into b hadrons is especially important because, within the Minimal Standard Model, it receives a vertex correction involving the top quark which is absent from any other final state. By taking the ratio $R_b = \Gamma_b/\Gamma_h$, most of the vacuum polarization corrections depending on the top quark and the Higgs mass cancel out, and one is left with the following approximate expression:

$$R_b \simeq R_d \cdot \left[1 - \frac{20}{13}\frac{\alpha}{\pi}\left(\frac{m_t^2}{M_Z^2} + \frac{13}{6}\log\frac{m_t^2}{M_Z^2}\right)\right] \tag{11}$$

Therefore, R_b has a singular rôle since its accurate measurement should provide an indirect determination of m_t independent on M_H, which is something that none of the rest of precision measurements can do. Anyway, the effect of the top quark vertex corrections is only of order 2% for a top mass of 150 GeV. Therefore only a precise measurement, to better than 1%, is useful to get information on the top mass.

The relatively large b mass ($\sim$ 4.7 GeV) and lifetime ($\sim 1.5 \times 10^{-12}$ s), makes possible the use of its decay kinematics to have the largest identification efficiency and purities among all the quarks. Several methods have been used to tag b events at LEP:

- Lepton tag: It uses high p, high p_t leptons from b decays. High purity can be achieved but one has to pay for the small semileptonic b branching ratio.

- Event shape tag: High mass, high momentum b mesons or baryons give raise to particular event shapes which have been used to tag b events with high efficiency although rather modest purity. Recognition has been optimized using Neural Network techniques.

- Lifetime tag: The long lifetime of the b quark can be used, using silicon microvertex detectors, to tag b events by looking for tracks not coming from the Z production vertex. This is currently the best performing method with both high purity and efficiency.

The main systematic errors come from the evaluation of the efficiency and the background of the selection. The best option is to try to use data to estimate both. In the case of the efficiency, the techniques mentioned above can be used to tag only one hemisphere and look at the other one to measure the tag efficiency. Similar techniques could also be used for the backgrounds. The results from the four collaborations are shown in fig. 5.

The c quarks can be tagged in two different ways:

- using b tag techniques (lepton tag, event shape) extending them to the lower p and p_t regions and fitting then simultaneously the b and c information,

- through the reconstruction on charmed meson decays. The cleanest one is $D^{*+} \to D^0\pi^+$ and then $D^0 \to K^-\pi^+$ and its charge conjugate, because the low $D^* - D^0$ mass difference produces the signature of a soft π acompaning a D^0 with opposite charge to the K.

In either case, the efficiencies and purities are much lower than for b quarks and the dependence on external input for the product and decay branching ratios very important. The results from the four collaborations are shown in

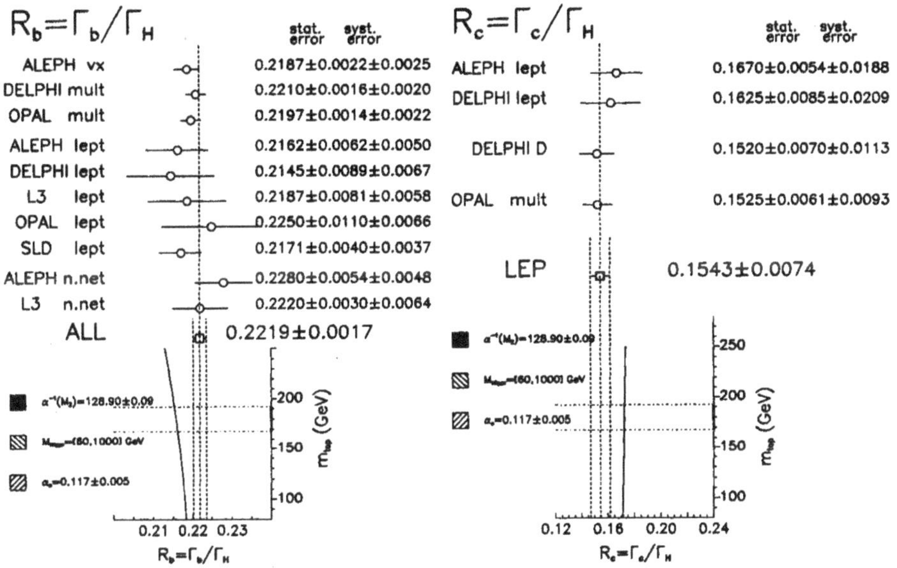

Figure 5: The ratios of the Z partial widths to b and to c hadrons to the one to all hadrons measured by the four experiments together with the mean and the MSM prediction as a function of the top quark mass. The error bars reflect the full uncertainty.

fig. 5, where one can see that the agreement between the different measurements of both quantities is excellent. Usually the selection cuts are chosen to optimize the ratio between statistical and systematical uncertainties and, in this sense, these measurements are not yet systematically limited. At any rate, the weight of correlated systematical uncertainties is very important in the final errors and therefore, the proper averaging of the measurements from the different experiments and using different techniques is non-trivial. A detailed discussion, which also applies to the heavy flavour asymmetries to be presented later, can be found in [6]. One relevant conclusion of such study is that, given the contamination of b events into the c sample and viceversa the correlation between the final averages of R_b and R_c is as big as ~ -0.35.

For both measurements, the agreement with the MSM prediction is bad. In the case of R_b, the current central value of the mean corresponds to a negative value for m_t^2 and it is more than 3.5σ higher than the MSM for $m_t = 180$ GeV, as directly measured at Tevatron. In the case of R_c the measurement is about 2.5σ lower than the MSM prediction, which is extremely precise. Given the anticorrelation of both measurements, due to the intercontamination of both samples, their discrepancies could be explained if, for any reason, the border defining which heavy flavor events should be classified as b and which as c would be placed inconsistently in data and in theory. Nevertheless, there is no evidence showing that this is the case.

The heavy quark forward-backward asymmetries provide presently the most precise determination of the effective weak mixing angle at LEP. What is measured is

$$A_{FB}^{0,q} = \frac{3}{4} A_e A_q. \tag{12}$$

where A_q is $\sim 0.66, 0.93$ for c- and b-type quarks, respectively and depends only mildly on $\sin^2 \theta_{eff}^{lept}$. Therefore, the asymmetry is quite large and mainly sensitive to the $\sin^2 \theta_{eff}^{lept}$ dependence of A_e. The main difficulties measuring the forward-backward asymmetry for quark final states are the flavour identification and the charge assignment.

For b quarks, the charge assignment is done in two different ways: either the charge of a high p, high p_t lepton from semileptonic decays which identify the b events is also used to extract the charge of the parent quark, or the lifetime information in one hemisphere is used to tag the event while the weighted mean charge (jet charge) in the other hemisphere measures the quark charge. The two methods lead to samples almost completely statistically independent. The systematical errors differ as well: in the first method, the knowledge of the lepton purities and of the semi-leptonic branching ratios is crucial; in the second, the charm background in the b sample is the main worry.

For c quarks, the charge is extracted either from the lepton charge, like for b's, in global fits using the low p_t leptons from semileptonic decays or, from the charge of the $D^{*\pm}$ meson. Results using these methods are shown in fig. 6 for b and c quarks. The results given in fig. 6 correspond to the "row" asymmetries measured at the average LEP energy and therefore, they have to be corrected for QED, QCD and energy effects to obtain $A_{FB}^{0,q}$ as appearing in eq. 12. After the corrections, the results are

$$\begin{aligned} A_{FB}^{0,b} &= 0.0999 \pm 0.0031 \\ A_{FB}^{0,c} &= 0.0725 \pm 0.0058 \end{aligned} \tag{13}$$

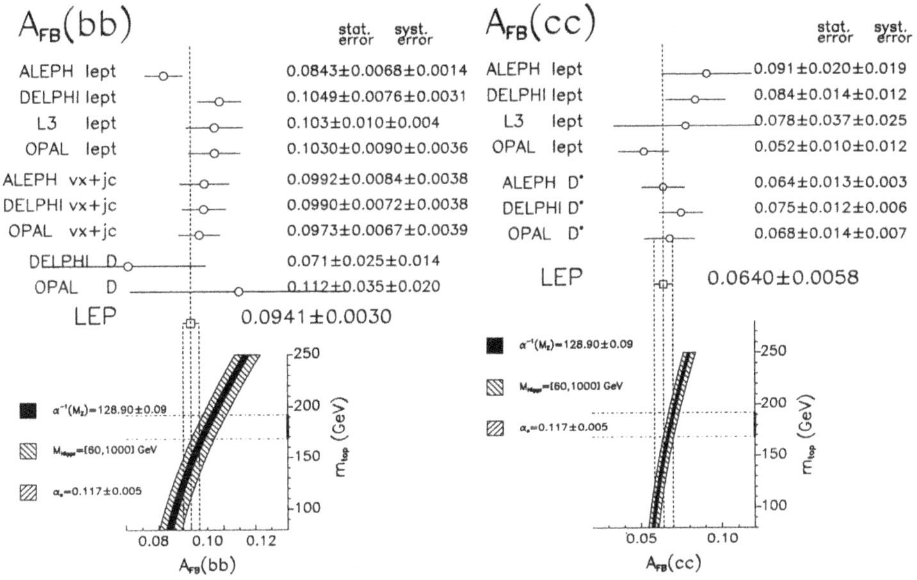

Figure 6: The b and c quark forward-backward asymmetry measured by the four experiments. The error bars reflect the full uncertainty

3 Interpretation of the measurements

3.1 Direct results

Tests of lepton universality.
Lepton universality in the Z couplings can be tested by analyzing simultaneously the information coming from the lepton partial widths and the leptonic forward-backward and tau polarization asymmetries. Then, following eqs.4 and 7 the vector and axial couplings of the Z to each lepton specie can be disentangled. The results (see fig. 7) show perfect agreement with the hypothesis of lepton universality for both, the vector and axial couplings.

$$g_{A_\mu}/g_{A_e}=1.0000\pm 0.0018\ ,\ g_{A_\tau}/g_{A_e}=1.0007\pm 0.0020\ ,$$

$$g_{V_\mu}/g_{V_e}=1.01\pm 0.14\qquad ,\ g_{V_\tau}/g_{V_e}=1.008\pm 0.071\ .$$

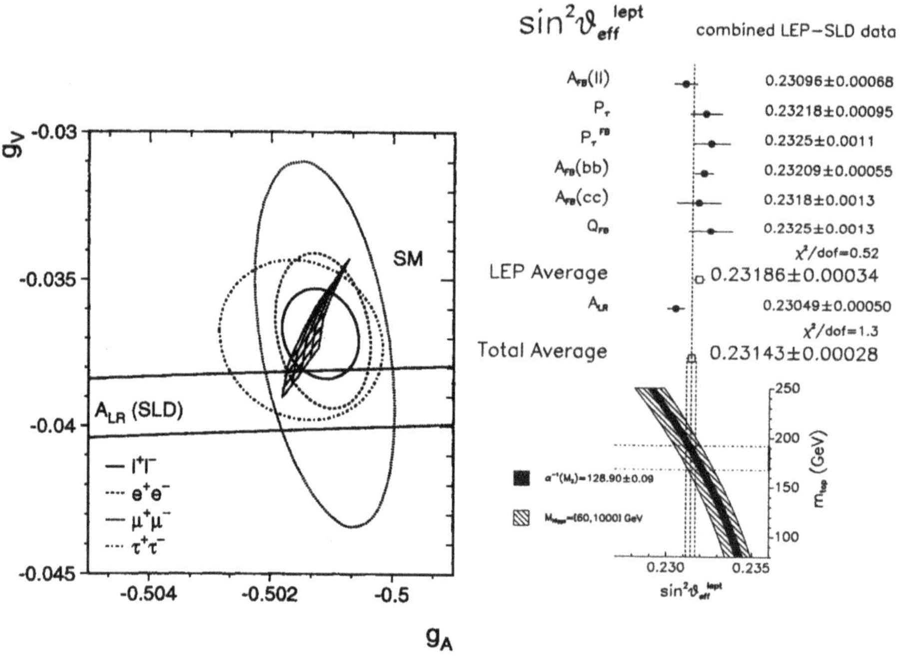

Figure 7: The 68% probability contours in the $g_{V_l}-g_{A_l}$ plane compared to the MSM predictions and the LEP averages for $\sin^2\theta_{eff}^{lept}$ obtained from the different asymmetry observables discussed in the text, together with SLD measurement and the total mean.

The effective sinus.

Figure 7 shows the compilation of all the values of $\sin^2\theta_{eff}^{lept}$ obtained from the asymmetry measurements at LEP presented in the previous sections. The SLD determination coming from the measurement of the left-right polarization asymmetry [7] is also shown. Its discrepancy with the average of the LEP measurements alone is of about 2.3 σ. At any rate, the overall agreement of all the measurements is still acceptable and the mean value provides a very precise determination of the effective weak mixing angle, which is very sensitive to the top quark mass.

Quantities derived from ratios.

The ratios of partial widths, due to the cancellation of universal $\Delta\rho$ corrections between numerator and denominator, allow the direct determination of important parameters of the theory without relying, in principle, too much on the validity of the MSM. Extensions of the MSM which would manifest themselves mainly via effects in vacuum polarization would produce very similar predictions for ratios.

As can be seen from fig. 2, R_l depends little on m_t and M_H whereas it has a large dependence on the strong coupling constant $R_l \sim R_l^0(1 + \alpha_s(M_Z)/\pi)$. Therefore it allows a direct determination of $\alpha_s(M_Z)$ with minimum theoretical uncertainties. From the LEP average of R_l and using the formulae suggested in [8], which relate R_l with the QCD prediction, known to $\mathcal{O}(\alpha_s^3)$, one gets

$$\begin{aligned} \alpha_s(M_Z^2) &= 0.126 \pm 0.006 \pm 0.002_{ew} \pm 0.002_{QCD} \pm 0.002_{M_H} \\ &= 0.126 \pm 0.007 \end{aligned} \tag{14}$$

where the second and third error reflect uncertainties on the electroweak and QCD parts of the theoretical prediction respectively, and the last one comes from the lack of knowledge on the Higgs mass.

The invisible width, Γ_{inv} can be derived from the direct lineshape measurements by subtracting from the total width the one due to visible channels. If one assumes that all the invisible width is due to neutrino final states, then one can derive the number of light neutrino species N_ν by writing

$$\frac{\Gamma_{inv}}{\Gamma_l} = N_\nu \cdot \frac{\Gamma_\nu}{\Gamma_l} \tag{15}$$

and taking the ratio Γ_ν over Γ_l from the MSM: $\Gamma_\nu/\Gamma_l = 1.991 \pm 0.001$. It should be noted the small error in the Minimal Standard Model prediction for this ratio, which does not depend on α_s and in which the top and Higgs mass dependences largely cancel. Using the LEP average for $\Gamma_{inv}/\Gamma_l = 5.956 \pm 0.031$ one obtains

$$N_\nu = 2.991 \pm 0.016$$

Since the result favours three species without any doubt, this measurement is actually a test of the MSM, a test of the assumptions made: that all invisible decays are to neutrinos, and of the value Γ_ν/Γ_l.

3.2 Standard Model Fits

The results presented in the previous sections can be interpreted in the context of the Minimal Standard Model allowing a check of its validity in describing all the observations, and enabling the inference of some of the MSM parameters.

The comparison is made through a fit of the measurements shown in table 1 to their MSM predictions in terms of the top quark mass m_t and the value of $\alpha_s(M_Z)$. The Higgs boson mass is fixed to 300 GeV and, afterwards, varied in the range between 60 and 1000 GeV.

The most up-to-date MSM calculations have been used [9] and their estimated theoretical uncertainties [10] have been also propagated in the fitting procedure, being the uncertainty in $1/\alpha(M_Z^2) = 128.79 \pm 0.090$ [11], due to the contribution of light quarks in the photon vacuum-polarization, the one that dominates by far.

The results of the fit can be gleaned from table 2. As can be seen in that table, in spite of the high precision of LEP data, non-LEP data (specially the SLD measurement on A_{LR}) does improve sizably the precision on m_t. Nevertheless, since the quality of the fits is bad ($\chi^2/n.d.f. \simeq 2$), the actual meaning of the fitted parameters should be questioned. By looking at the pulls of the different measurements with respect to their predictions for the MSM best fit in table 1, it can be seen that three measurements are the responsible of such a high χ^2: on the one hand, the LEP R_b and R_c measurements and on the other hand, the SLD measurement of A_{LR}. For the confrontation with the MSM predictions, the case of A_{LR} can be considered as "normal" in the sense that, in a rather model-independent way, its value can be confronted with several other direct measurements and then, as shown before, there is no obvious signal of a problem with the MSM. This is not the case for R_b and R_c since, in their case, any confrontation with other direct measurements requires an assumption on the value of $\alpha_s(M_Z^2)$. In the absence of any convincing explanation of the present discrepancy for these two quantities, any interpretation of the data in the MSM has to be taken with a grain of salt.

To circumvent this problem, for instance, the following exercise can be made: attributing the deviation of R_b (or also the one of R_c) to new physics acting only in $\Gamma_{b\bar{b}} = \Gamma_{b\bar{b}}(SM) + X_b$ (or also in $\Gamma_{c\bar{c}} = \Gamma_{c\bar{c}}(SM) + X_c$), but forgetting about the possible origin of this X, the data can still be fitted within this adhoc "extended" MSM. Since the b and c quarks are also present in the hadronic width and in the total width, the predictions are affected accordingly. The results of such fits [1] show that $\alpha_s(M_Z^2)$ changes drastically whereas the results for m_t do not change sizably. Therefore, the indirect determination of the top mass in the present situation can still be considered robust whereas the one of $\alpha_s(M_Z^2)$ cannot. Moreover the m_t value resulting from these fits is in excellent agreement with the direct search evidence reported by CDF and D0 of $m_t = 180 \pm 12$ GeV proving that the bulk of non-trivial quantum effects in precision electroweak observables is indeed due to the top quark as predicted by the MSM.

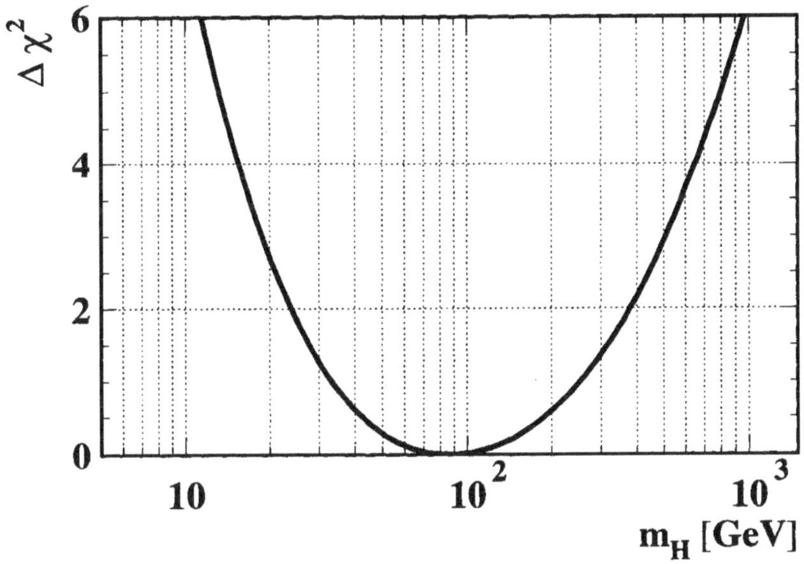

Figure 8: $\Delta\chi^2$ vs $M_{\rm H}$ using all data and the Tevatron measurement of m_t.

Assuming that the problem of the R_b and R_c measurements is of experimental origin, if the Tevatron determination of m_t is used as an additional constraint, then $\log(M_H)$ can be extracted reliably from the data as can be seen in figure 8. At any rate, since the measurements are sensitive to $\log(M_H)$ and not directly to M_H, fluctuations in the position of the minimum (even if small in the $\log(M_H)$ scale in comparison to the width of the χ^2 parabola) correspond to very important changes in M_H. Therefore, in spite of the reliability of the present determination of $\log(M_H)$, its actual translation in terms of a measurement of M_H leads to conclusions which strongly depend on the chosen input data and on their fluctuations. Therefore, being cautious, the only conclusion that should be stressed from fig. 8 is that the data seem to prefer a light Higgs and that the M_H value preferred by the data within the MSM is consistent with the validity of perturbation theory.

4 Summary

Using the preliminary analysis of the precision Electroweak data accumulated by the LEP experiments until the end of 1994 the accurate determination of most of the relevant Z boson observables has been presented.

The confrontation of the measurements with the Minimal Standard Model predictions shows good agreement for all parameters but for the ratios R_b and R_c which show significative discrepancies with respect to their Standard Model expectations. Forgetting about the possible origin of this discrepancy, the interpretation of the measurements in the MSM framework, still allows a determination of the top quark mass with a $\sim$ 20 GeV accuracy which agrees with the CDF and D0 measurements. On the contrary, the determination of the strong coupling constant $\alpha_s(M_Z^2)$ for these data has been shown to be strongly dependent on the way these two measurements are treated. Therefore, the indirect determination of the top mass in the present situation can be considered robust whereas the one of $\alpha_s(M_Z^2)$ cannot.

Analyzing the precision Electroweak measurements together with the Tevatron m_t measurements within the MSM and assuming the applicability of perturbation theory in the Higgs sector, stable errors can be estimated for $\log(M_H)$ and we can conclude that the data prefers a M_H value which is consistent with the validity of perturbation theory.

References

1. LEP Electroweak Working Group, Internal Note LEPEWWG/95-02, 1 August 1995. Prepared from Contributions of the LEP experiments to the 1995 Summer Conferences.
2. The Working Group on LEP Energy, private communication.
3. S.Jadach et al. CERN-TH/95-38, February 1995.
4. A.Leike,T.Riemann and J.Rose, Phys. Lett. B273 (1991) 513. T.Riemann, Phys. Lett. B293 (1992) 451.
5. The S-Matrix Subgroup of the LEP Electroweak Working Group, Internal Note LEPEWWG/LS/95-01, 1 August 1995. Prepared from Contributions of the LEP experiments to the 1995 Summer Conferences.
6. The LEP Electroweak Heavy Flavors Working Group, LEPHF/95-02, July 1995; see also ref. [1]

7. K. Abe et al., SLD Coll., Phys. Rev. Lett. 73 (1994) 25 and Contributed Papers to EPS-HEP-95 Brussels: eps0654, eps0222, eps0248, eps0249 and eps0250.
8. T. Hebbeker, M. Martinez, G. Passarino, G. Quast, Phys. Lett. B331 (1994) 165.
9. BHM: G. Burgers, W. Hollik and M. Martinez; M. Consoli, W. Hollik and F. Jegerlehner: Proceedings of the Workshop on Z physics at LEP I, CERN Report 89-08 Vol.I,7 and G. Burgers, F. Jegerlehner, B. Kniehl and J. Kühn: the same proceedings, CERN Report 89-08 Vol.I, 55.
10. D. Bardin et al.,"Precision Calculation Working Group" (PCWG), CERN 95-03 Yellow report, 31 March 1995.
11. S. Eidelmann and F. Jegerlehner, PSI-PR-95-1, January 1995 // H. Burkhardt and B. Pietrzyk, LAPP-EXP-95.05, June 1995.
12. Average of the results presented in the 95' Summer conferences following the method in: M. Demarteau et al., *Combining W mass measurements*, CDF/PHYS/2552 and D0NOTE 2115.
13. C. G. Arroyo et al., CCFR Coll., Phys. Rev. Lett.72 (1994) 3452 and Contributed Paper to EPS-HEP-95 Brussels, eps0041.
H. Abramowicz et al., CDHS Coll., Phys. Rev. Lett. 57 (1986) 298; A. Blondel et al., Z. Phys. C45 (1990) 361.
J. V. Allaby et al., CHARM Coll., Phys. Lett. B177 (1986) 446; Z. Phys. C36 (1987) 611.

Table 1: Summary of measurements included in the combined analysis of Standard Model parameters. Section a) summarizes LEP averages, section b) electroweak precision measurements from SLD [7] and section c) gives the results from hadron colliders [12] and νN-scattering [13]. The Standard Model fit result in column 3 and the pulls in column 4 are derived from the fit including all data (Table 2, column 3) for a fixed value of $M_H = 300$ GeV.

	measurement	Correlation matrix	Standard Model fit	pull
a) LEP				
lineshape and lepton asym.:				
M_Z [GeV]	91.1884 ± 0.0022	1.00	91.1882	0.1
Γ_Z [GeV]	2.4963 ± 0.0032	-0.08 1.00	2.4973	-0.3
σ_h^0 [nb]	41.488 ± 0.078	0.02 -0.12 1.00	41.450	0.5
R_ℓ	20.788 ± 0.032	0.00 -0.01 0.15 1.00	20.773	0.5
$A_{FB}^{0,\ell}$	0.0172 ± 0.0012	0.08 0.00 0.01 0.00 1.00	0.0159	1.1
τ polarization:				
$\mathcal{A}_\tau$	0.1418 ± 0.0075		0.1455	-0.5
$\mathcal{A}_e$	0.1390 ± 0.0089		0.1455	-0.7
b and c quark:				
$R_b = \Gamma_{b\bar{b}}/\Gamma_{had}$	0.2219 ± 0.0017	1.00	0.2156	3.7
$R_c = \Gamma_{c\bar{c}}/\Gamma_{had}$	0.1543 ± 0.0074	-0.34 1.00	0.1724	-2.5
$A_{FB}^{0,b}$	0.0999 ± 0.0031	0.00 0.09 1.00	0.1020	-0.7
$A_{FB}^{0,c}$	0.0725 ± 0.0058	0.06 -0.07 0.10 1.00	0.0728	0.0
$q\bar{q}$ charge asym:				
$\sin^2\theta_{eff}^{lept}$	0.2325 ± 0.0013		0.23172	0.6
b) SLD				
$\sin^2\theta_{eff}^{lept}$ ($\mathcal{A}_e$)	0.23049 ± 0.00050		0.23172	-2.5
$R_b = \Gamma_{b\bar{b}}/\Gamma_{had}$	0.2171 ± 0.0054		0.2156	0.3
A_b	0.841 ± 0.053		0.935	-1.8
A_c	0.606 ± 0.090		0.667	-0.7
c) $p\bar{p}$ and νN				
M_W [GeV] (CDF,D0,UA2)	80.26 ± 0.16		80.35	-0.5
$1 - M_W^2/M_Z^2$ (νN)	0.2257 ± 0.0047		0.2237	0.4

Table 2: Results of fits to LEP and other data for m_t and $\alpha_s(M_Z^2)$. No external constraint on $\alpha_s(M_Z^2)$ has been imposed. The central values and the first errors quoted refer to M_H = 300 GeV. The second errors correspond to the variation of the central value when varying M_H from 60 GeV to 1 TeV.

	LEP	LEP + SLD + Collider and ν data
m_t (GeV)	$170 \pm 10 \, ^{+17}_{-19}$	$178 \pm 8 \, ^{+17}_{-20}$
$\alpha_s(M_Z^2)$	$0.125 \pm 0.004 \, \pm 0.002$	$0.123 \pm 0.004 \, \pm 0.002$
$\chi^2/(d.o.f.)$	18/9	28/14
$\sin^2\theta_{eff}^{\rm lept}$	$0.23206 \pm 0.00028 \, ^{+0.00008}_{-0.00017}$	$0.23172 \pm 0.00024 \, ^{+0.00007}_{-0.00014}$
$1 - M_W^2/M_Z^2$	$0.2247 \pm 0.0010 \, ^{+0.0004}_{-0.0002}$	$0.2237 \pm 0.0009 \, ^{+0.0004}_{-0.0002}$
M_W (GeV)	$80.295 \pm 0.057 \, ^{+0.011}_{-0.019}$	$80.346 \pm 0.046 \, ^{+0.012}_{-0.021}$

THE LEP200 PROGRAMME

Daniel TREILLE
CERN, PPE Division, CH-1211 Geneva 23, Switzerland
E-mail: treille@cernvm.cern.ch

ABSTRACT

In the light of an ongoing workshop, the LEP200 programme is reviewed.

The potential of LEP200 has been studied and described since long[1]. I will only quote here the new ideas and facts, and in particular the main preliminary conclusions of a presently ongoing workshop[2].

After considering in detail alternative programmes for LEP, like the exploitation of longitudinal polarization[3] or the potential of increased luminosity at the Z^0 [4], it became clear that the energy rise was the most promising, specific and urgent option to be pursued. One has still however to define more precisely the actual goal, in terms of energy and integrated luminosity, and this is the main task of the workshop.

LEP200 offers a vast programme of physics, both for searches (Higgs, SUSY,...) and for measurements (M_W, triple boson couplings, ...). The former is particularly exciting because one enters a crucial domain of accessible masses, if supersymmetry is the truth. For the later, it is the first machine giving direct access to the self coupling of electroweak bosons.

LHC, and a linear collider, will open new worlds. The energy rise of LEP, or the Tevatron upgrade, do not pretend to offer such a potential. However their promises, in a subset of possible and even likely scenarios, are outstanding and their full exploitation in the next decade is obviously mandatory.

LEP200 is also a very challenging programme from the machine point of view, and this will be the first point I will discuss. From the experimental side, its strength relies on the existence of four high quality experiments, well understood after several years of LEP I: minor upgrade programmes, already done or under way, have optimized their potential for the LEP200 era.

1. The Machine

The overall problem is well known. The energy loss per turn due to synchrotron radiation goes like the fourth power of the energy and will reach 2.5 GeV for a beam energy of 100 GeV. Besides the increased background problems, solved by appropriate masking of the experiments[5], one has to compensate for that loss by providing the needed accelerating voltage around the ring. This is done by RF cavities: the warm copper cavities of LEP I are totally insufficient and one had to develop and realize a large set of supraconducting ones, made of copper with an internal layer of sputtered niobium[6].

This programme turned out to be very challenging. Besides the cavities themselves, the main couplers, feeding in the RF power, and the higher order mode couplers, filtering out selectively the bad harmonics, were critical items. After solving a series of problems, it is now foreseen that an accelerating field of 6 MV/m with a quality factor of at least 3.2×10^9 should be achieved by the set of cavities. Further improvements may be possible ultimately.

One can then deduce the number of cavities needed to reach a given beam energy. Besides the financial ones, there exist practical limitations to the amount of cavities one can install around LEP. It is out of question to undertake new massive civil engineering and one must then cope with the room available in the existing galleries of the four even straight sections: the corresponding limit is around 352 cavities, and can be raised to around 384 cavities by removing the separators needed for bunch train operation. Furthermore, at each point, there is presently a cryogenic limit amounting to about 12 kW: the exact number of cavities which can be accommodated locally will depend of the cryogenic load they represent. Beyond that number (72 to 80 cavities?) one will have to increase the cryogenic power, a step anyway needed for LHC.

Table 1 [2] gives the details of the possible scenarios. Scenario IV is presently under serious discussion and seems to be favourably considered by the scientific committees. One sees from the table that up to $\sqrt{s} \sim 205$ (i.e. except for the ultimate Z scenario) one can work with 4 x 2 bunch trains, providing a peak luminosity of nearly 10^{32} $cm^{-2}s^{-1}$.

Should one multiply this value by 10^7 sec (a "year") one would expect close to 1000 pb^{-1}/"year"/experiment. This is however an irrealistic number and, guided by the present LEP achievements and the conditions expected at LEP200, one has adopted a value of ~170 pb^{-1}/"year"/experiment which leads to ~500 pb^{-1} in 3 years.

Since for many physics channels it is likely that the four experiments will combine their results, one can also consider the "quantum" of 1 fb^{-1}, which represents the total luminosity delivered during around 1.5 year.

Table 1. Foreseen and possible phases of LEP200

	Cavities										Missing cavities	Total MV (MV)	I_{max} (mA)	ξ_{bb}	Beam power (MW)	E_{max} (GeV)	L_{max} (10^{31})*	HOM (W)
Point No.	2			4	6		8	Totals										
Type / Phase	Cu	Nb	NbCu	NbCu	Cu	NbCu	NbCu	Cu	Nb	NbCu								
II	60	32	0	64	60	32	64	120	32	160	0	2243	6	0.0335	24.3	92.0	8.0	704
	Original LEP2 'Design'										8	2161	6	0.0340	23.3	91.2	8.1	704
											16	2080	6	0.0350	22.7	90.4	8.2	704
III	28	32	16	64	28	48	64	56	32	192	0	2390	6.6	0.0355	28.6	93.5	9.5	855
	Upgrade for increased luminosity										8	2308	6.9	0.0380	28.6	92.5	10.5	933
											16	2227	7.2	0.0400	28.6	91.5	11.6	1017
IIIa	26	32	24	56	26	56	56	52	32	192	0							
	Modifications for bunch trains										8	Almost identical to previous phase						
											16							
IIIb	26	32	32	56	26	64	56	52	32	208	0	2542	6.7	0.0345	30.5	94.8	9.5	890
	Use of 16 active spares (yet to be approved)										8	2460	6.9	0.0360	30.5	94.0	10.1	936
											16	2379	7.1	0.0370	30.5	93.4	10.6	977
IV	26	32	32	72	26	64	72	52	32	240	0	2869	6.8	0.0310	34.5	97.5	8.8	894
	Maximum energy upgrade with 4 · 12 kW cryogenics										8	2787	6.9	0.0320	34.5	97.0	9.3	932
											16	2706	7.05	0.0340	34.5	96.5	10.0	971
X1	26	32	32	88	26	64	88	52	32	272	0	3196	6.7	0.0280	38.5	100.4	8.1	874
	18 kW cryo at pts 4 and 8										16	3032	7. 1	0.0310	38.5	98.7	9.4	985
X2	0	32	48	88	0	80	88	0	32	304	0	3377	6.9	0.0280	42.0	102.0	8.5	922
	18 kW at pts 2 and 6										16	3213	7.3	0.0310	42.0	100.5	9.8	1039
Y	0	32	56	88	0	88	88	0	32	320	0	3540	6.92	0.0270	44.0	103.0	8.3	936
	Symmetrical LEP with all even points identical										16	3377	7.34	0.0300	44.0	101.7	9.6	1053
Z	0	32	64	96	0	96	96	0	32	352	0	3866	4.0	0.0300	27.5	105.2	5.5	625
	Ultimate number of cavities with ZL's removed (4 bunches)										16	3703	4.0	0.0303	26.4	104.2	5.5	626

* in cm^{-2} s^{-1}

2. The Scenarios

It is not necessary to recall once more the successes of the Standard Model, nor its theoretical shortcomings[7].

The two main roads beyond the SM are some types of composite scenarios, like Technicolor, in which the existence of new types of constituents and interactions is postulated, or to resort to a higher level of symmetry, as in the case of supersymmetry.

In the former case, which seems to meet some difficulties when confronted to the accurate measurements of LEP I [8], one does not foresee the existence of elementary Higgs bosons: one expects instead the appearance at high energy of a new type of strong interaction between the longitudinal part of the intermediate vector bosons. Depending whether this interaction leads to resonance or not, it will be more or less difficult to observe it at the future colliders. In any case it is unlikely that LEP 200 has much to say about this scenario.

The situation is a priori very different if SUSY is the truth. Its rich phenomenology and in particular its very constrained Higgs sector could, as we shall see, be already partly revealed at LEP200, provided energy and luminosity are sufficient.

It may be relevant to spend some time defining which type of supersymmetric theories one is actually considering. The minimal model (MSSM) postulates the minimal set of partners and a Higgs sector made of two doublets, which, after electroweak symmetry breaking (EWSB), amounts to 5 bosons: the scalars h^0 and H^0, the pseudoscalar A^0 and the charged ones, $H^{\pm}$: its phenomenology will be reviewed below.

SUSY must be broken and this is classically achieved by soft SUSY breaking (SSB) terms, which avoid the re-introduction of quadratic divergence. In the absence of further restrictive assumptions, the number of such parameters is high. If however Grand Unification and gravity inspired universality are assumed, one is left with the familiar set of 5 free parameters[9]

$$M, m, A, B, \mu$$

M and m are the common gaugino and scalar masses, A and B the trilinear and bilinear couplings of SSB and μ the Higgsino mixing parameter.

The requirement of a correct EWSB allows to trade away B and $|\mu|$, while $tg\beta \equiv v_2/v_1$ appears, and one then deals with another usual set:

$$M, m, A, \text{sign}(\mu), tg\beta$$

Five independent parameters are still a lot and one can reduce further their number and their possible range in different ways.

One way is to impose phenomenological constraints, experimental or cosmological: this leads to various constrained models (CMSSM).

Another way is to get further inspiration on the nature of the soft SUSY breaking terms from Supergravity (SUGRA) and superstrings. This leads to models which, like in the so-called dilaton version of SSB, have finally only two independent parameters[10]. It is fair to say that the corresponding assumptions are far from being proven and one can only consider such models as an interesting and convenient set, in particular to compare the potentials of various machines in exploring the allowed parameter space.

A third possibility, which is linked with the assumed grand unification of b and τ Yukawa couplings, is the so-called "m_{top} fixed point" scenario, in which a relation between the top mass and tgβ is established. With the value found for the mass, one version of this scenario is favouring a small value for tgβ, between 1 and ~3 [11]. This likely realization of the MSSM will be considered below.

The virtues of SUSY are well known[12]. Quite spectacular is the fact that in SUSY the EWSB is "built in", once the top is heavy enough. Figure 1 shows the results of a CMSSM[13], which is only one among several, but nevertheless indicates clearly what are the strong points on which one should focus at LEP200: a general and unavoidable fact is the lightness of h^0, the lightest scalar, and this will be quantified in the next chapter. Another striking fact is that the gauginos, charginos and neutralinos, may be light and within reach although this is not guaranteed. Another possibility is that, through a large mixing in the stop sector, the lightest stop mass eigenstate is quite low as well.

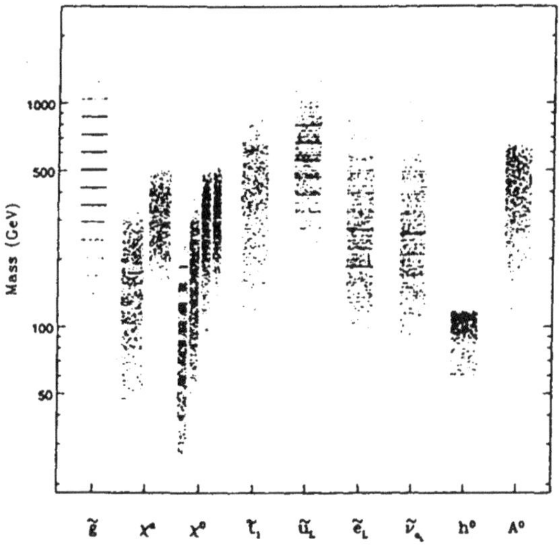

Fig. 1. Scatter plot of mass vs. particle type for all acceptable solutions of the constrained model of ref. 13

3. The Legacy of LEP I

The impact of accurate measurements, presently dominated by LEP I, has been discussed at length[8]. Table 2 summarizes the situation concerning $\sin^2\Theta_W^{eff}$, m_{top} and m_W. It gives the results of a SM fit to the LEP data, with m_H = 300 GeV. One sees that the direct m_{top} measurements (CDF only, CDF+DO, same with the cross section information included) agree well with the LEP value. This is still true if a light Higgs mass is adopted, as it should be the case for SUSY.

Table 2. Some results of accurate measurements

	$\sin^2\Theta_W^{eff}$	M_{top}	M_W
LEP SM fit	0.2320 ±0.0003 +0.0001 -0.0002	$M_H = 300$ GeV $176 \pm 10\ ^{+17}_{-19}$ MSSM 160 ± 10	$80.32 \pm 0.06\ ^{+0.01}_{-0.01}$
Direct measurement		CDF 176 ± 13 CDF+DO 180 ± 12 idem, with cross section information 170 ± 9	UA2+CDF+DO 80.26 ± 0.16

$\alpha_s\ (M_Z)$	
$\Gamma(Z^0 \to \text{had})$	0.127 ± 0.006
ev. shapes (SLD)	0.120 ± 0.008
Scal. viol. (ALEPH)	0.127 ± 0.011
World average	0.117 ± 0.006
Value needed for GU	0.129 ± 0.010

Similarly the indirect W mass determination from LEP agrees with the still imprecise direct measurements. In para.8 we discuss the mass accuracy one can expect from LEP200.

As often stated, the accurate ew measurements agree well with the SM, but agree as well also with the MSSM. We will not consider here the possible meaning of the discrepancy concerning R_b, which is presently a 3 σ effect.

From the accurate measurements of the three basic couplings at LEP I and the estimate of their evolution with energy, it has been shown that the MSSM leads to an exact convergence at $\sim 10^{16}$ GeV, while this does not happen in the SM[14]. In spite of the very simplified character of this demonstration, the conclusion is still valid after more elaborate considerations are introduced. Table 2 gives the value of α_s (M_Z) which is needed, according to ref. 15 to ensure the Grand Unification of the couplings: it agrees with the LEP α_s data, and also with the world average. There is however a potential problem not shown here, the slight discrepancy between low and high energy measurements of α_s (M_Z).

4. Higgs Phenomenology

If, in spite of its shortcomings, one tries to stick to the SM it is natural to study the bounds this model announces for the Higgs mass. Most relevant for LEP200 is the lower limit that one can deduce from the requirement of stability, or metastability, of the vacuum[16]. This limit depends on the scale at which the perturbative character of the SM is expected to break down. If the model is supposed to be valid up to the GUT or Planck scale, one finds: $m_h \geq 135$ GeV. If, on the contrary, the model breaks down at low energy (1 to 100 TeV) the limit is lower and the boson could be in the region accessible to LEP200:

however, in such a case, the upper limit is ~600 GeV or so, and there is no particular reason to expect the Higgs boson in the 100 GeV region, nor any strong argument to try to gain 10-20 GeV of accessibility there.

So, at LEP200, the SM Higgs boson search does not present any particular interest. Actually, strictly speaking, the SM boson is not expected to be in the LEP200 domain.

The situation is totally opposite in the case of SUSY where the lightest boson h^0 is bound to exist in this region. The tree level upper bound on its mass is raised by loop corrections which depend on the 4th power of m_t and, logarithmically, on the masses in the stop sector and therefore on the mixing which determines them

$$m_h \leq m_Z \, |\cos 2\beta| + \frac{1}{\sin^2\beta} F\left(m_t^4, \ln\frac{m_{\tilde{t}}}{m_t}, \ldots\right)$$

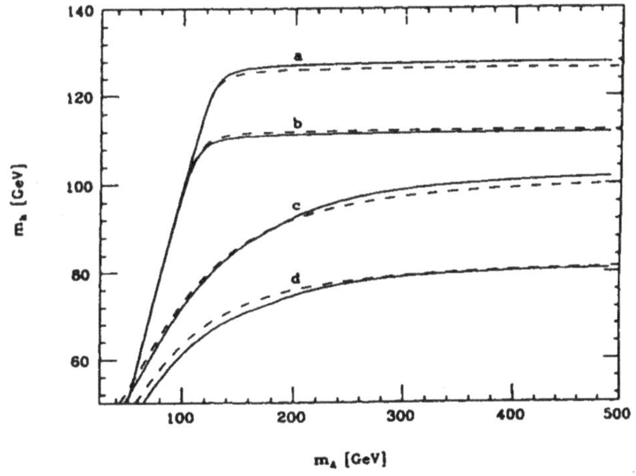

Fig. 2. The lightest CP-even Higgs mass as a function of the CP-odd Higgs mass for a physical top-quark mass M_t = 175 GeV and M_{SUSY} = 1 TeV, as obtained from the one-loop improved RG evolution (solid lines) and the analytical formulae (dashed lines) of ref. 17. The four sets of curves correspond to: a) $\tan\beta$ = 15 with large squark mixing; b) $\tan\beta$ = 15 with zero mixing; c) the minimal value of $\tan\beta$ allowed by perturbativity constraints for the given value of M_t (IR fixed point), $\tan\beta$ = 1.6, with large squark mixing; d) $\tan\beta$ = 1.6 with zero mixing.

The dependence in $tg\beta$ is made explicit as well. In the last few years, the value of m_t has been sharpened and the computation of m_h in the MSSM has been refined. One can summarize the results by fig. 2 [17] which shows four extreme cases: large and small $tg\beta$ (i.e. IR fixed point), large and small mixing. It is thus possible to tell which scenarios will be covered by a given mass reach. For instance reaching $m_h \cong 110$ GeV would allow to cover the small $tg\beta$ cases, whatever be the mixing, and the large $tg\beta$ case with small mixing, while the large $tg\beta$ situation with large mixing stays partly uncovered.

We recall that, as soon as m_A is beyond ~100 GeV, the h^0 boson is essentially SM like, both for its bremsstrahlung production mode and in its decay (Fig. 3). So the description of the SM boson search which will follow is actually done having the h^0 in mind.

A last point, illustrated by Fig. 4, is the quasi stationarity of m_h in the upper right part of the $tg\beta - m_A$ plane: one understands then that a small change in $\sqrt{s}$ can lead to a large one in the coverage of the plane.

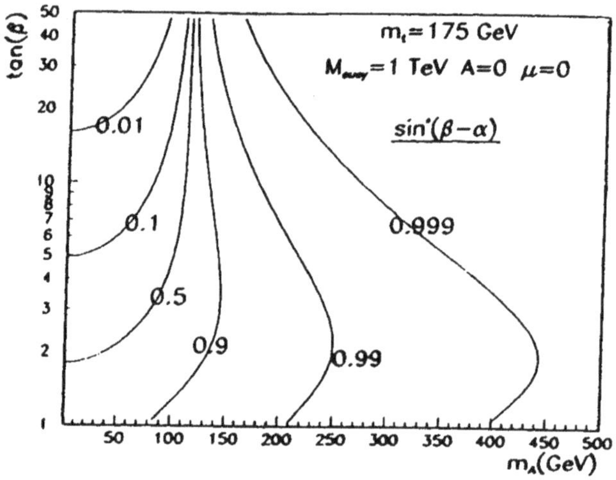

Fig. 3. The ratio of hZ to $H_{SM}Z$ rates in the $tg\beta$ - m_A plane

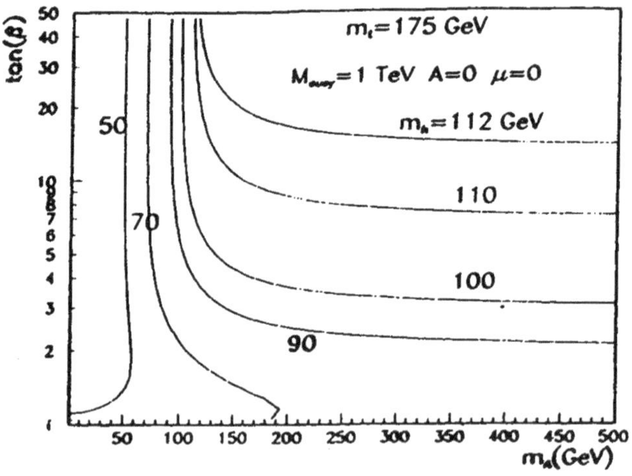

Fig. 4. h isomass curves in the $tg\beta$ - m_A plane

5. Search for the SM Higgs Boson

Figure 5 gives the cross section of the Higgs production process versus energy and shows that above m_h ~55 GeV it is more profitable to search it at LEP200 than on the Z^0 since the cross section is bigger and the background smaller.

The sharp threshold occurs at $\sim m_Z + m_h$ and one needs ~10 GeV CM more to get close to the maximum of the cross section: hence the old rule of thumb giving the reach:

$$m_h \sim \sqrt{s} - 100 \text{ GeV}$$

For the relevant LEP200 mass region the Higgs boson decays mostly into $b\bar{b}$ and this explains the overall importance of b tagging in its search. The $b\bar{b}$ decay is a tree level process, at the difference of the $\gamma\gamma$ mode, and its branching ratio is unambiguously calculable. The Z^0 is observed in the usual ways: $q\bar{q}$, $\nu\bar{\nu}$ and l^+l^-.

Although it is still modest, the fusion channel $e^+e^- \rightarrow \nu_e\bar{\nu}_eH$ does exist and interferes with the main one ee $\rightarrow$ HZ ($Z \rightarrow \nu_e\bar{\nu}_e$) [18]. This channel is not bound to the kinematical limit expressed above: the possibility that it allows to gain a few GeV in mass reach is under study.

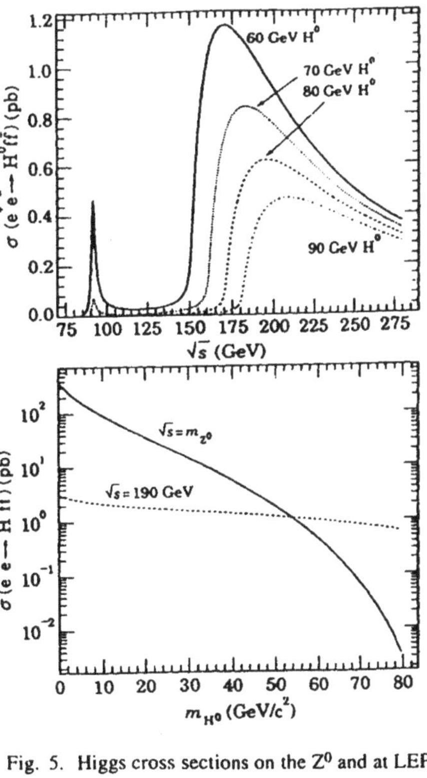

Fig. 5. Higgs cross sections on the Z^0 and at LEP200

The background channels, shown in Fig. 6, are mostly $q\bar{q}\gamma$ (half being radiative return to the Z region), WW and ZZ. Combinations of kinematics and b tagging allow to reduce them to manageable levels and all HZ modes turn out to be exploitable. The most difficult case is when $m_H \approx m_Z$, since the ZZ final state is then kinematically indistinguishable and one rests fully on b tag.

The b tag efficiency and purity needed have already been achieved in the present LEP experiments. Figure 7 shows their effect on the LEP200 channels: one can keep ~50% efficiency to the HZ channel and reject WW by ~70 since there are no b's in W decay.

Even in the most difficult case, the signal/background figures are ≥ 1. Figure 8 [ref. 2] gives for the three energies the exclusion and discovery curves, which have the behaviour expected from our previous considerations. To interpret them in terms of running time, one must make

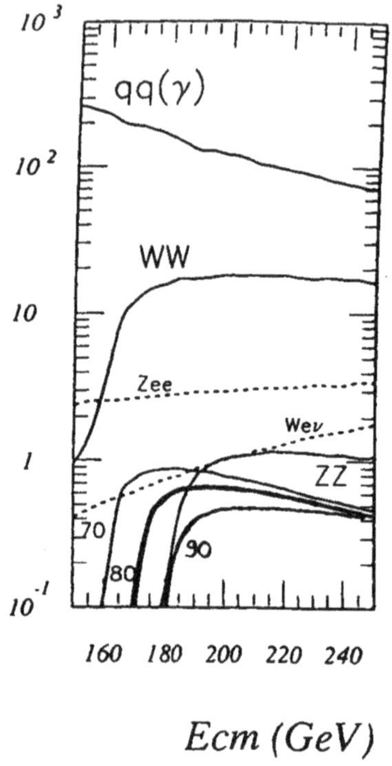

Fig. 6. The HZ and background cross sections in pb (courtesy from E. Gross).

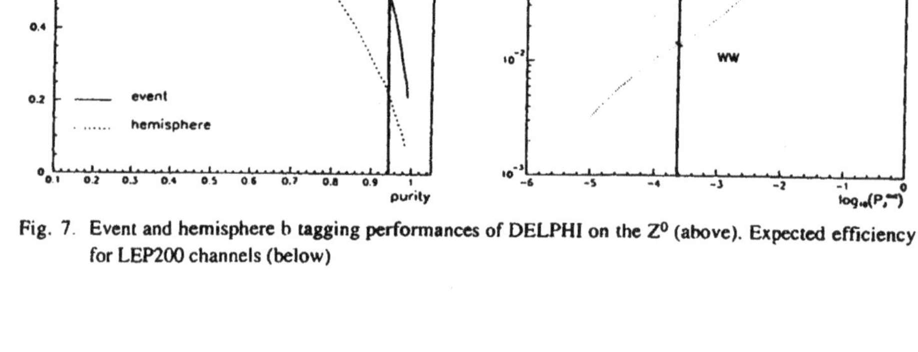

Fig. 7. Event and hemisphere b tagging performances of DELPHI on the Z⁰ (above). Expected efficiency for LEP200 channels (below)

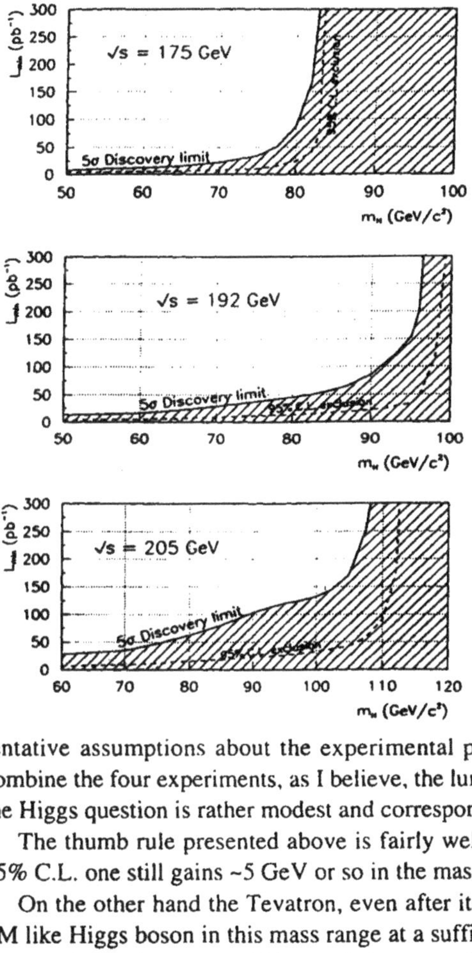

Fig.8. Minimum luminosity needed per experiment, in pb^{-1}, for the combined 5σ discovery (full line) or the combined 95% C.L. exclusion (dashed line) of the Higgs boson as a function of its mass, at the three centre-of-mass energies.

tentative assumptions about the experimental procedure at LEP200: if it is possible to combine the four experiments, as I believe, the luminosity needed per experiment to answer the Higgs question is rather modest and corresponds to at most one year of running.

The thumb rule presented above is fairly well verified for discovery: for exclusion at 95% C.L. one still gains ~5 GeV or so in the mass reach.

On the other hand the Tevatron, even after its upgrade, will not be able to produce an SM like Higgs boson in this mass range at a sufficient rate. At least an order of magnitude increase in luminosity would be needed to do so [19].

6. The MSSM Higgses

As we said, the previous chapter devoted to the SM Higgs describes as well the search for h^0 through the Bremsstrahlung mechanism. However one can also exploit at LEP200 the associated production (AP):

$$e^+e^- \rightarrow hA$$

The complementarity between the two mechanisms is well known, although the P wave AP has the handicap of a $(\text{velocity})^3$ factor. Fig. 9 shows that, at large $tg\beta$ and modest m_A, the AP dominates, while it is the contrary at small $tg\beta$.

The AP process, leading to 4 b's, is particularly prone to b tagging: this is welcome to allow the elimination of the much larger WW background, which is kinetically identical to the signal when $m_A \sim m_h \sim m_W$.

The exclusion/discovery domains in the tgβ - m_A plot are shown in Fig. 10 [2]. One distinguishes clearly the regions where the two search channels dominate. The coverage obtained through the Bremsstrahlung channel depends critically, as we saw previously, on the available $\sqrt{s}$, and also, for a given m_{top}, on the level of mixing in the stop sector.

At $\sqrt{s}$ = 205 GeV, small mixing and 1 fb^{-1}, the plane is nearly fully covered. This can be seen also in an alternative representation using the (m_h, tgβ) plane (Fig. 11).

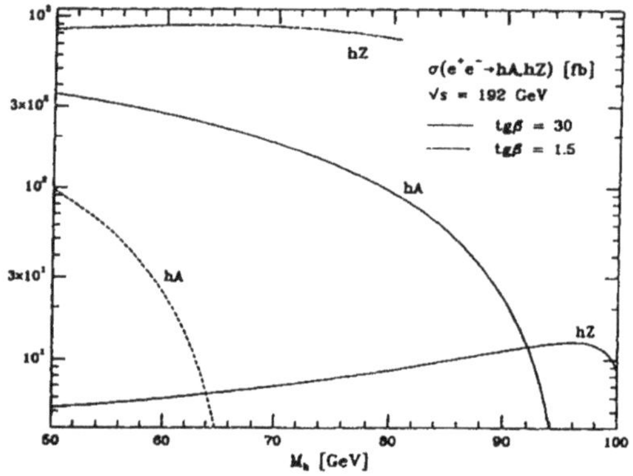

Fig. 9. Cross sections for the production of the lightest CP-even Higgs boson h and the CP-odd Higgs boson A for two representative values of the parameter tgβ = 1.5 and 30, M_t = 175 GeV, M_S = 1 TeV and zero mixing.

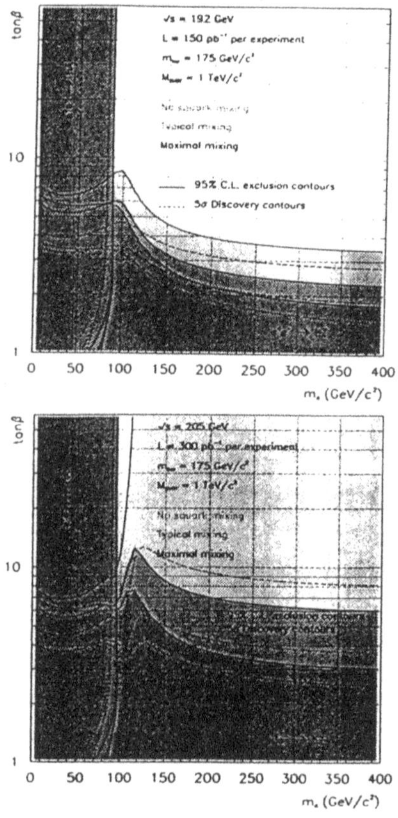

Fig. 10. Exclusion and discovery limits in the (m_A, tgβ) plane for $\sqrt{s}$ = 192 and 205 GeV and various stop mixing conditions

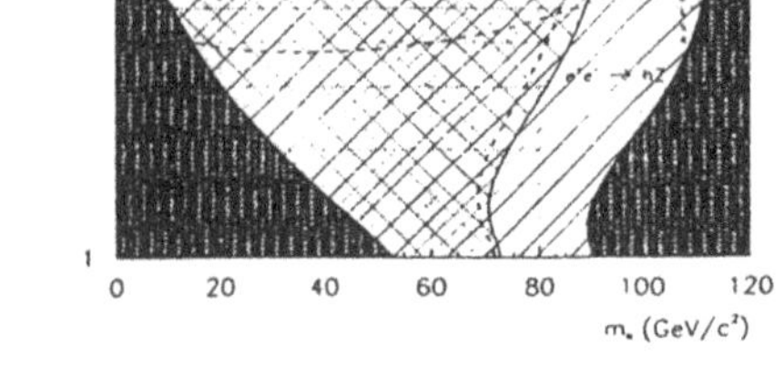

Fig. 11. Exclusion and discovery limits in the [m_h, tanβ] plane, for M_t = 175 GeV and A = $-\mu$ = M_S = 1 TeV for each of the c.m. energies. The dark shaded areas are excluded theoretically.

7. SUSY Particles

In a decade or so the LHC will open a vast domain of search and should solve relatively rapidly the problem of the existence of strongly interacting partners, squarks and gluinos.

But the situation will be more difficult for weakly interacting ones and more generally for light objects. Meanwhile it is important to exploit the potential of LEP200, more limited but promising in a large subset of all possible scenarios. One should also emphasize that the Tevatron, in its upgraded version and even in the present one is, for SUSY matters, a potentially rewarding machine. We will underline the complementarity/competition between both machines.

7.1. Gauginos

For charginos the situation was very simple at LEP I where the only production diagram is Z/γ exchange (Fig. 12): the scan allowed to set on its mass a limit of 47 GeV, provided that the lightest neutralino is below 41 GeV.

At LEP200 the $\tilde{\nu}_e$ t-exchange amplitude can interfere destructively with the previous one, and some set of parameters, with small $m_{\tilde{\nu}_e}$, may in principle lead to small cross sections.

A systematic scan of the MSSM parameters was performed and the result is shown in Fig. 13 [ref. 2]. Only a small fraction of pathological cases correspond to a cross section below ~1 pb at LEP200. One should add the condition that the chargino mass is at least 5-10 Gev above the LSP one, so that visibility is guaranteed. The conclusion is that most of the possible cases lead to a clearly observable situation, where chargino discovery, through a set of kinematical cuts rejecting WW background, is achievable with the luminosity foreseen.

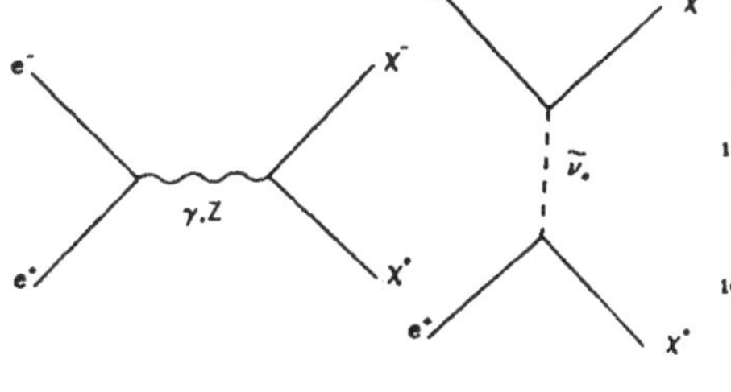

Fig. 12. Chargino pair production

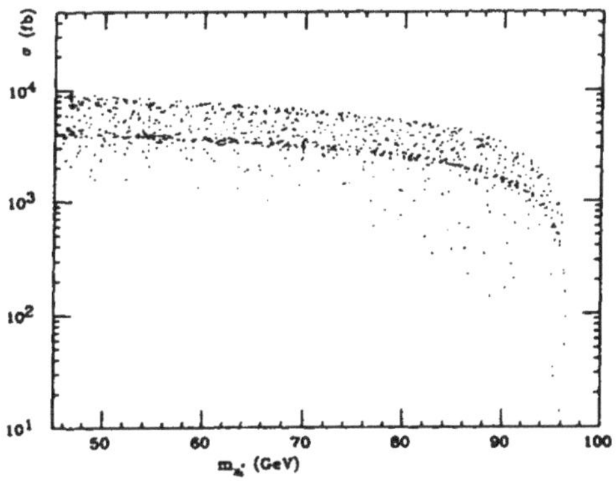

Fig. 13. Chargino production cross sections at LEP2, $\sqrt{s}$ = 192 GeV, as a function of $m_{\chi_1^\pm}$. The parameters M, μ, $\tan\beta$ and $m_{\tilde{\nu}_e}$ were chosen within the ranges specified in ref. 2.

Could neutralinos add some relevant information?

At LEP I the $Z \rightarrow \chi_i\chi_j$ coupling in the MSSM is large if χ_i and χ_j are higgsino-like, vanishing if χ_i or χ_j are pure gaugino. Since long the Z lineshape measurement allowed to exclude a large region, shown in the M, μ plane, in the former case (Fig. 14). But even direct searches for $Z \rightarrow \chi^0\chi'^0$ (since $\chi^0\chi^0$ is not accessible) are not sufficient at small tgβ: for tgβ < 2 one does not get a limit on the χ^0 mass (Fig. 15) [20]. One must then invoke the gluino mass limit of hadron colliders as a substitute.

At LEP200 the neutralino search may allow to cover small regions of the M, μ parameter space which are not kinematically accessible through charginos, provided other parameters are such that the rates are large enough [21].

What about the potential impact of the Tevatron for gaugino searches? The cleanest observable there is the three-lepton final state from $\chi^0\chi^\pm$ associated production. The present mass reach quoted is slightly above the LEP one and it will rise rapidly with increased statistics. But, as shown in Fig. 16 [ref. 2] which represents a scan of parameters, the situation is very model dependent: with the fb^{-1} foreseen at the upgraded Tevatron one can reach mass values as high as 150 GeV, with a reasonable probability of discovery, but in case of a negative result it will be impossible to set a lower limit.

In the particular frame of SUGRA and string inspired models one can similarly compare the potential of LEP200 and the Tevatron. Figure 17 [22] shows the reach of both machines in the plane of the two free parameters left (chosen to be tgβ, $m_{\tilde{g}}$). It illustrates the gain in

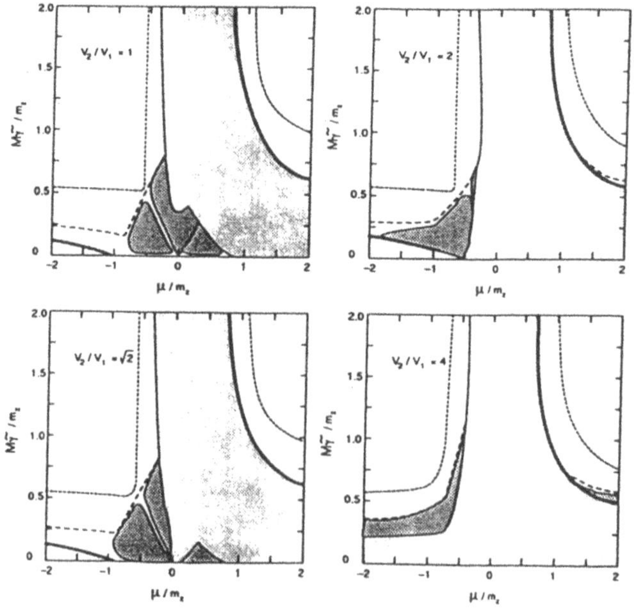

Fig. 14. For various values of tgβ, 95% c.l. excluded domains in the ($M_{\tilde{\gamma}}$, μ) plane of the MSSM (from ALEPH, ref. 20).

coverage obtained by increasing the LEP200 top energy, as well as the obvious competition between the two machines.

Any increase in LEP energy above the Z^0 will improve the mass reach for the charginos and, until the W pair threshold is crossed, a very modest exposure is sufficient to conclude. It is therefore foreseen to raise the energy as soon as possible and perform by the end of 95 such an exposure of ~10 pb-1 at $\sqrt{s}$ = 140-150 GeV, which, barring a tiny fraction of pathological situations, will give access to mass values close to the kinematics limit.

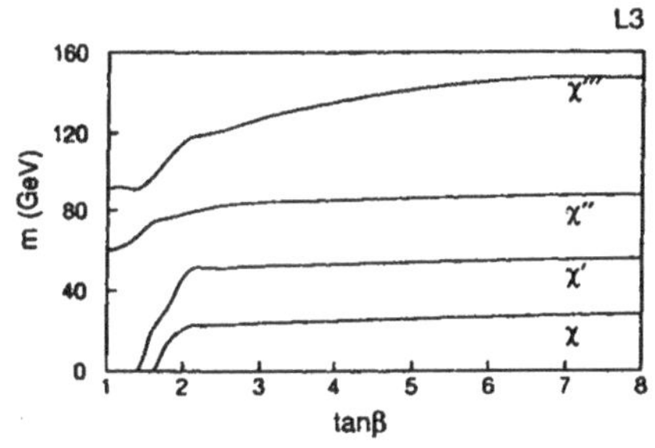

Fig. 15. Neutralino lower mass limits (95% C.L.). The lines correspond to the lower mass limit for the different neutralinos. The regions below the lines are excluded (from L3, ref. 20).

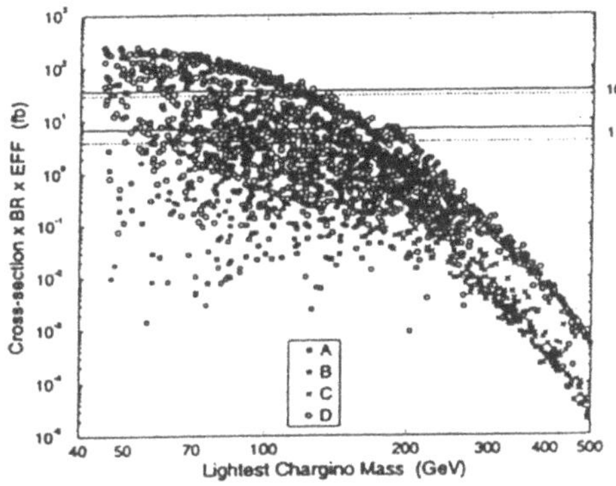

Fig.16. Trilepton observable cross sections and chargino discovery reach in Minimal Supergravity models at theTevatron. Each point corresponds to a different choice of parameters. See ref. 2 for details.

7.2. Sleptons

This is a domain where e^+e^- machines have unique possibilities. On the other hand the modest $\sqrt{s}$ of LEP200 and the absence of any good reason in general to have very light sleptons make it less attractive. This can nevertheless happen in models with a favourable set of soft SUSY breaking parameters.

Experimentally the situation will be quite clear: the reach foreseen is $\sqrt{s}/2$ - (5 to 10) GeV depending on the integrated luminosity and provided that the LSP into which the slepton is assumed to decay ($\tilde{l} \rightarrow \chi l$) is not too close in mass.

7.3. A light Stop

The possibility of a large mixing in the stop sector, leading to a low value for the mass of the lightest eigenstate, has been much emphasized [23].

At LEP it is likely that the decay mode to be searched for is $c\chi^0$, from a loop process, rather that the tree process $b\chi^{\pm}$ probably kinematically inaccessible. One must therefore look for low visible energy charmed events and distinguish them, in particular, from two-gamma hadronic events.

A similar search was already performed on the Z^0 at LEP I [24] by DELPHI and OPAL. The situation there was slightly complicated by the possibility that, for an unfavourable value of the mixing angle, a decoupling occurred from the Z, leaving only the photon mediated process. At higher energies no such dramatic possibility can occur and the production cross section is comfortable as shown in Fig. 18.

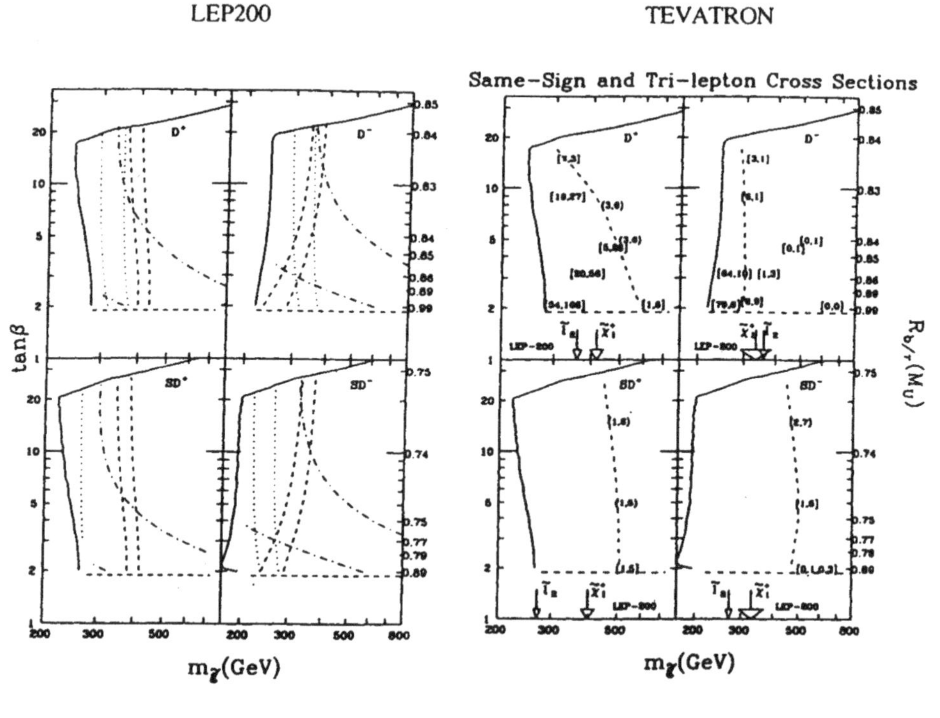

Fig. 17. For various versions of the dilaton scenario, the allowed domain and the reach of:
a) LEP200 through h^0 (dash dot), $\tilde{l}$ (dot) and charginos (dash) at $\sqrt{s} = 176$ and $\sqrt{s} = 200$ GeV
b) the Tevatron with 1 fb^{-1} through charginos (left of dashed line)
From ref. 22.

However there too the Tevatron has a strong potential and may well give the answer before LEP goes up in energy.

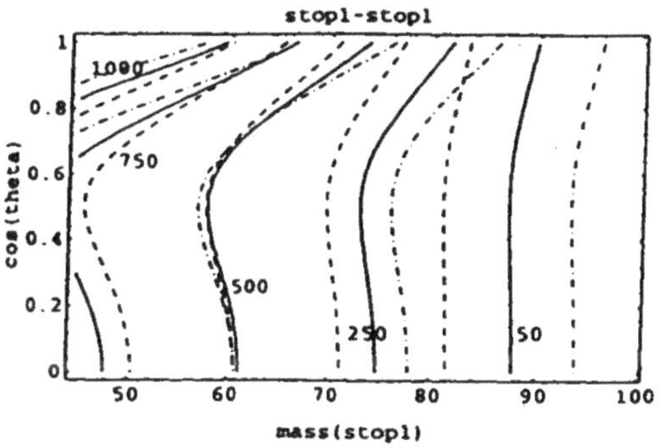

Fig. 18. Stop production cross section at LEP200 in fb as a function of $m_{\tilde{t}}$ and the mixing angle (see ref. 2)

8. Electroweak Measurements at LEP200

Let us now turn to topics whose existence is guaranteed and which will stem from the analysis of W pair production

8.1. M_W

Previous studies have concluded that the most promising method at LEP200 was the reconstruction method, exploiting the set of kinematic constraints available.

A statistical accuracy of 55 MeV per experiment (500 pb^{-1}) is foreseen. The systematic amounts to ~24 MeV, due to the beam energy determination (~17 MeV), the mastering of radiative corrections in the initial state (~10 MeV) and instrumental problems, partly solved by the technique of superimposed and boosted two-Z events[25].

However the most powerful channel is the 4-jet one (statistical error ~75 MeV) and in the last year some basic doubts arose about the clarity of its message. Both QCD colour rearrangement effects and Bose Einstein correlation phenomena were considered, as possibilities of a "cross talk" between the two W's, which would slightly invalidate the methodology of the reconstruction method[26]. Non perturbative QCD effects are the most dangerous and do not seem to offer any convincing way to monitor them. BE interference effects are quantum mechanical ones which cannot be properly taken into account by the probabilistic language of the available generators. A risky and doubtful approximate procedure is then needed to try to reproduce their effect on the mass determination.

All that may cast some doubts about an optimal exploitation of 4 jets and needs more work. In the most pessimistic scenario, one would then be left with the mixed 2J 2l decay mode, offering a statistical accuracy of ~77 MeV/expt (500 pb^{-1}) and whose systematics needs also a closer look.

Meanwhile a totally different method of M_W measurement has emerged[27]: near threshold a specific $\sqrt{s}$ value (161 GeV) corresponds to an optimum between statistics and sensitivity to the W mass. The computations of the W pair cross section and of the four fermion backgrounds on one hand, the experimental procedures on the other hand, are safe enough to guarantee a high quality measurement of M_W with

~56 MeV of statistics (100 pb^{-1}/expt, all four added)

~36 MeV of systematics

Unfortunately any time spent at that low energy is lost for most of the search programme.

The impact of the M_W measurement can be a major one in case of a discrepancy with the SM. On the other hand, should the SM still be valid for a while, the information on M_W will not add to the one of $\sin^2\Theta_w^{eff}$ and m_{top} (Fig. 19)[28].

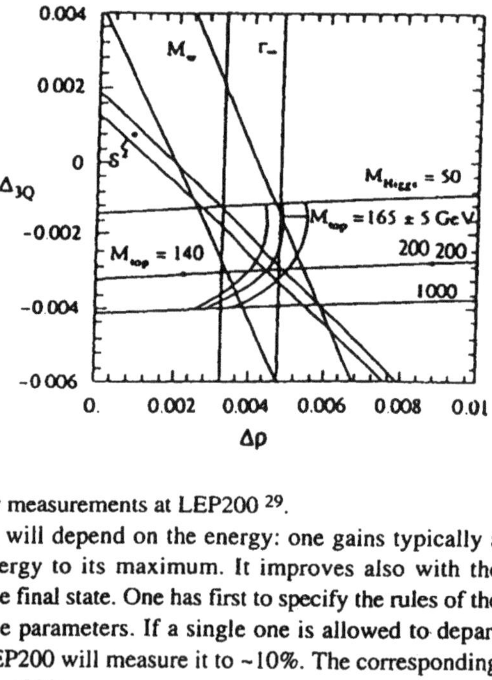

Fig. 19. The impact of various accurate measurements in the plane Δ_{3Q} and Δ_ρ. The oblique band labelled S^2 corresponds to $\Delta\sin^2\theta_W = 10^{-4}$. See ref. 28 for details.

8.2. Triple boson couplings

This is the other classical topic of ew measurements at LEP200 [29].

The accuracy of their measurement will depend on the energy: one gains typically a factor 2 by increasing the LEP200 energy to its maximum. It improves also with the quantity of information available from the final state. One has first to specify the rules of the game, i.e. if one keeps 1, 2 or more free parameters. If a single one is allowed to depart from the SM values, one can say that LEP200 will measure it to ~10%. The corresponding present figure from Hadron Colliders is ~100%.

The measurement of anomalous couplings has raised many questions in the past[30]:

a. is any anomaly, at the level of sensitivity of LEP200, already precluded by accurate measurements at lower energy, for instance at LEP I? The answer is, in principle, no : if one corrects the SM Lagrangian by a sum of allowed correction terms ("grafts"), one can find some which intervene in the lower energy observables only at the loop level (i.e. not at tree level) and whose existence is only weakly constrained by the present accurate measurements. These are called "blind directions". On the other hand, other ones do intervene at tree level and are already severely constrained.

b. one can then ask whether there exists likely new physics scenarios which would give a privileged role to "blind directions": the answer is again no, as far as one can see. It would then be "unnatural" (although not strictly forbidden) to assume such a dichotomy.

c. the third natural question is to ask what is, for likely scenarios, the magnitude expected for anomalous couplings. This is a controversial question[31], but the values foreseen are generally tiny: 10^{-2} - 10^{-3}, well beyond LEP200 reach, and may be the reach of other machines (NLC, LHC). A natural goal would then be to acquire enough sensitivity to attain for sure the expected ew radiative correction level (few ‰): this may require a TeV Linear Collider.

In summary one should, at LEP200, perform the best possible measurement of triple boson couplings. But the odds to discover an anomaly there are low.

9. Conclusion

To conclude, LEP200, if it is adequately exploited, is a very promising machine. In case of SUSY any increase of energy is of importance and multiplies the chance of discovery, especially in the Higgs sector. In some sectors (gauginos, light stop, ...) a serious competition comes from the Tevatron.

Any discovery would signal the opening of a new world for the future machines. It is also true that an absence of discovery, because of the incomplete coverage of LEP200, does not close the hope that such a new world exists.

10.Acknowledgements

I would like to thank the actors of the LEP200 Worshop, in particular G. Altarelli, F. Richard and P. Janot who provided several of the plots shown here. I warmly thank L. Braize for her efficiency in editing this report. I am grateful to the organizers of the Valencia meeting for their kindness and the exciting scientific atmosphere.

11.References

1. *ECFA Workshop on LEP200,* CERN Preprint 87-08, ECFA, 87-108
   LEPC Workshop summarized in D. Treille, Preprint CERN-PPE/93-54 Rev.
2. Interim Report on *the Physics Motivations for an energy upgrade of LEP2,* CERN-TH/95-151, CERN-PPE/95-78
3. *Polarization at LEP*, Preprint CERN 86-06: G. Alexander et al., editors
4. *High luminosity at LEP* Preprint CERN 91-02: E. Blucher et al., editors
5. DELPHI Collaboration, *Proposal for the upgrade of DELPHI in the forward region,* CERN/LEPC/92-13/P2/Add2, DN 92-142 GEN135
6. Ph. Bernard et al, *ECFA Workshop on LEP200,* CERN Preprint 87-08, vol. 1, p.29.

7. See for instance P. Langacker in *Precision Tests of the SM*, P. Langacker editor, World Scientific Advanced Series on Directions in HEP, Vol. 14
8. G. Altarelli, CERN TH/95-3. Data are from the preprint LEP-EWWG/95-10
9. For a comprehensive review see W. de Boer, CERN-PPE/93-31 (March 94).
10. J.L. Lopez et al., CERN TH7401/94.
11. P. Langacker and N. Polonsky, Preprint UPR-0594T (Feb. 94).
M. Carena and C.E.M. Wagner, CERN TH-7320/94 (June 94).
12. F. Zwirner, *The SUSY World*, CERN TH-6951/93 (July 93).
13. G.L. Kane et al., UM-TH/93-24.
14. U. Amaldi et al., *Phys. Lett.* **B260** (1991) 447, and references therein.
15. P. Langacker and N. Polonsky, Preprint UPR-0642T.
16. G. Altarelli and G. Isodori, CERN TH-7351/94 (July 94).
J.R. Espinosa and M. Quiros, CERN TH/95-18; DESY 95-039 (March 95).
17. M. Carena et al., CERN TH/95-45/DESY 95-038; IEM-FT-103/95 (March 95).
18. M. Dubinin et al., KEK 93-138 (1993). E. Boos et al., DESY 93-183 (Dec. 1983)
19. A. Stange et al., *Phys. Rev. D* , vol **50** nr. 7 (1994) 4491.
20. ALEPH Collaboration: Phys. Rep. **216** (1992) 253
L3 Collaboration: CERN-PPE/95-14 (1995).
21. S. Ambrosiano and B. Mele, ROME 1-1094/95, HEP-PH-9503362 (March 95).
22. H. Baer et al., UCD 94-19, FSU-HEP-940501, May 94.
23. M. Drees and K. Hikasa, KEK-TH-262, CERN TH 5854/90 (1990).
24. R. Keranen et al., *Search for Light Stop*, DELPHI 92-172 PHYS 255 (1992).
The OPAL Collaboration, CERN PPE/94-103 (July 94).
25. S. Katsanevas et al., *Report from the Working Group on LEP200 Physics*, DELPHI 92-166 PHYS250 (1992) and references therein.
26 T. Sjöstrand and V.A. Khoze, *Z. Phys.* **C62** (1994) 281, *Phys. Rev. Lett.* **72** (1994) 28, *Phys. Lett.* **B328** (1994) 466 (for color rearrangement).
L. Lönnblad and T. Sjöstrand, CERN TH/95-17 (for Bose Einstein).
27. W.J. Stirling, DTP 95-24 (Feb. 95)
28. A. Blondel et al., Preprint LAPP-TH.343/91
29. M. Bilenky et al., *Nucl. Phys.* **B409** (1993) 22.
30. A. de Rujula et al., *Nucl. Phys.* **B384** (1992), K. Hagiwara et al., MAD/PH/690 KEK-TH-314, UT-597 (Feb. 92), R. Szalapski and D. Zeppenfeld, MAD/PH/719 (Sep. 92).
31. M.B. Einhorn and J. Wudka, UM-TH/92-25 (1992).

Chargino Search in DELPHI below the W^+W^- threshold

J.J. Hernández

IFIC, CSIC - Universitat de València, Burjassot, Valencia, Spain

We show that with an integrated luminosity of 5–20 pb^{-1} at an energy of 150 GeV the lightest chargino, if present at this energy, can be detected in DELPHI. A chargino signal, essentially background free, can be observed in DELPHI in the two jet plus lepton and four jet topologies for most of the accessible parameter space. We also point out some of the specific regions where special problems in the detection of this signal might be encountered.

1 Introduction

An early run of LEP at energies well above the Z^0 peak, but below the W^+W^- production threshold $(130-150$ GeV) will probably take place in autumn this year (1995). If this is the case, it will allow to extend substantially the search for new particles and in particular those of the supersymmetric spectrum. Charginos are among the most interesting particles to look for at this energy for several reasons. First, they are expected to be among the lightest supersymmetric partners; secondly, in the context of the MSSM the chargino pair production cross-section is in general rather sizeable in e^+e^- collisions ($\sim 1\ pb$) and finally, W^+W^- production, the most dangerous background for chargino search, is highly supressed at this energy.

2 Simulation of the signal and background

Chargino events were generated with the program SUSYGEN [1]. This Monte Carlo generator is based on the formulae of Bartl *et al.* [3], which give the cross-section for the reaction $e^+e^- \rightarrow \tilde{\chi}^+\tilde{\chi}^-$ together with the chargino decay branching ratios in terms of the parameters of the MSSM.

For the sake of simplicity, we make the hypothesis that the masses of squarks, gluinos, sleptons and sneutrinos are much higher than those of the first chargino and neutralino and therefore do not have a critical influence neither on the cross-section nor on the decay rates of the charginos.

We have generated chargino events for several specific points in the SUSY parameter space and processed them through the DELPHI full chain of simulation and reconstruction programs [2]. We have then performed a detailed analysis of the detection conditions and the influence of the selection criteria on the efficiency and purity of our signal in these points. We have finally generalized our conclusions to the full parameter space, pointing out the difficult regions.

The points representative of the different regions in the parameter space are given in Table 1. In all these points we have taken $tan\beta = 1.5$ according to present theoretical and phenomenological prejudices [4].

N	M (GeV)	μ (GeV)	$m_{\tilde{\chi}_1^\pm}$ (GeV)	$m_{\tilde{\chi}_1^0}$ (GeV)	σ (pb)	Expected Events	Generated Events
$m_{\tilde{\nu}} = 500$ GeV							
1	120	200	72	39	3.6	72	4000
2	190	105	59	35	6.5	130	3000
3	800	80	72	66	2.1	42	3000
4	700	70	61	54	4.8	96	3000
5	300	100	73	56	1.4	28	3000
$m_{\tilde{\nu}} = 80$ GeV							
6	25	-100	60	15	2.7	54	3000

Table 1: Selected points in the (M, μ) plane and the corresponding lightest chargino and lightest neutralino masses. For each point, the chargino production cross-section, the expected number of events for an integrated luminosity of 20 pb^{-1} and the number of generated Monte Carlo events are given.

Concerning the decays and assuming that squarks, gluinos, sleptons and sneutrinos are heavier than the first chargino, the only possible decay modes of the lightest chargino are $\tilde{\chi}_1^\pm \rightarrow \tilde{\chi}_1^0 l^\pm \nu$ and $\tilde{\chi}_1^\pm \rightarrow \tilde{\chi}_1^0 q \bar{q}'$, which take place through the interchange of a virtual W in the s-channel. Therefore, the chargino decay branching ratios are those of the W boson. Since neutralinos escape detection, we will observe events with either two leptons (*ll*), two jets plus one lepton (*jjl*) or four jets (*jjjj*) at an approximate rate of 9%, 42% or 49%, respectively. In this work, we study the *jjl* and *jjjj* topologies. The main backgrounds for chargino production for the *jjl* and *jjjj* topologies are listed in Table 2. Non-$\gamma\gamma$ background events were generated by means of the Monte Carlo program PYTHIA [5] and $\gamma\gamma$ events with the TWOGAM Monte Carlo program [6]. All events were processed through the DELPHI full chain of simulation and reconstruction programs [2]. For the $\gamma\gamma$ background,

Background Event Type	σ (pb)	Expected events	Generated events
$e^+e^- \to q\bar{q}(\gamma)$	190	3800	14750
$e^+e^- \to W^+W^-(\gamma)$	1.01	20	2000
$e^+e^- \to Z^0Z^0(\gamma)$	0.49	10	2000
$e^+e^- \to We\nu(\gamma)$	0.80	16	2000
$e^+e^- \to Z^0ee(\gamma)$	2.00	40	2000
"filtered $\gamma\gamma$" (for jjl)			
$e^+e^- \to e^+e^-X$ (VDM)	11.3	226	3750
$e^+e^- \to e^+e^-X$ (QPM)	10.9	218	11250
"filtered $\gamma\gamma$" (for $jjjj$)			
$e^+e^- \to e^+e^-X$ (VDM)	25.6	512	4000
$e^+e^- \to e^+e^-X$ (QPM)	11.0	220	3950

Table 2: Simulated backgrounds. 1) Background type, 2) Cross section in pb ($\sqrt{s}$=150 GeV), 3) Number of expected events for a luminosity of 20 pb^{-1} and 4) number of Monte Carlo events generated. For the $\gamma\gamma$ backgrounds, see text for explanations.

the extremely high cross-sections make it impossible to generate a statistical significant sample of Monte Carlo events. Actually, it is unpractical as well, since most of the $\gamma\gamma$ events are easily cut away by the appropiate selection criteria. What is crucial in this case is to have a good control of the tails, i.e. of those events which are more likely to pass the selection criteria. Therefore, we first generated small samples of events (a few thousands), which were used to optimize our selection criteria and afterwards we generated samples of "filtered" events, i.e. events with features more similar to the signal but with manageable cross-sections. For the jjl topology the "filtered" events were required to have a total multiplicity higher than 5 and one electron at a polar angle larger than 6°. For the $jjjj$ topology, "filtered" events have a total multiplicity higher than 5 and a total missing p_T larger than 2 GeV.

3 Selection Criteria and efficiencies

In Tables 3 and 4 we show the criteria used to select the chargino events in the jjl and $jjjj$ topology, respectively.

	multiplicity	≥ 10
	Presence of an isolated lepton: $\theta^{iso} > 20^0$ y $P^{iso} > 3$ GeV	
	P_t	> 5 GeV
$20^0 <$	θ^{miss}	$< 160^0$
	$E^{20^0}_{cone}$	$< 20\%$
	P^{iso}	< 30 GeV
	M_{had}	< 45 GeV
	M_{vis}	< 70 GeV
	Acollinearity	$> 10^0$

Table 3: Selection criteria for the topology *jjl*.

	multiplicity	≥ 10
	Absence of isolated lepton: $\theta^{iso} < 20^0$ ó $P^{iso} < 3$ GeV	
	P_t	> 10 GeV
$20^0 <$	θ^{miss}	$< 160^0$
	$E^{20^0}_{cone}$	$< 20\%$
	E_{vis}	< 45 GeV
	M_{vis}	< 70 GeV
	$\lvert E_{miss} - P_{miss}\rvert$	> 40 GeV
	E_γ	< 5 GeV
$40^0 <$	Acoplanarity	$< 170^0$
	Acollinearity	$> 10^0$

Table 4: Selection criteria for the topology *jjjj*.

For both topologies we require a minimum multiplicity of ten. In the jjl topology, we required the presence of an isolated lepton (an electron or muon according to DELPHI's identification algorithms). The second bunch of cuts in Table 3 are mainly devised to reject $\gamma\gamma$ events. We cut on p_T, we demand that the missing momentum does not point to the forward-backward region and at most 20% of the total energy should be in that region. Finally, we cut on hadronic and visible mass, on the momentum of the isolated lepton and on acollinearity.

For the $jjjj$ topology the cuts are similar, except that they are somewhat more stringent for the non-$\gamma\gamma$ backgrounds, where apart from cuts in the visible mass and energy we also require that the difference between the missing energy and momentum be higher than 40 GeV.

A summary of the results for the six different scenarios is given in Table 5. In this table we show the efficiency, purity and significance (in number of standard deviations) of the signal for each scenario assuming a cross-section of 1 pb and an integrated luminosity of 20 pb^{-1}.

			jjl			$jjjj$		
N	$m_{\tilde{\chi}_1^\pm}$ (GeV)	$m_{\tilde{\chi}_1^0}$ (GeV)	ϵ (%)	δ (%)	S (σ)	ϵ (%)	δ (%)	S (σ)
1	72	39	46.1	94.4	4.4	71.5	78.7	3.8
2	59	35	40.6	93.7	4.1	53.1	73.4	3.1
3	72	66	2.3	45.8	1.5	0.5	2.6	1.0
4	61	54	4.7	62.3	1.7	1.8	9.8	1.1
5	73	56	37.0	93.1	4.0	56.9	75.2	3.3
6	60	15	35.3	92.2	3.6	29.2	62.3	2.4

Table 5: Efficiency, ϵ, Purity, δ and Significance, S (in number of standard deviations) of the signal selected by our criteria for both jjl and $jjjj$ topologies for the corresponding SUSY point. All the results are normalized to a chargino production cross-section of 1 pb and are given for an integrated luminosity of 20 pb^{-1}.

We can see in Table 5 that, except for the points where there is a small chargino-neutralino mass difference, an integrated luminosity smaller than 20 pb^{-1} will be enough to detect (as an effect always larger than 3σ) any chargino signal having $\sigma \cdot B$ larger than 1 pb. In the case of a lightest chargino and neutralino almost degenerate in mass, the visible and hadronic mass distributions tend to zero and the total missing momentum and missing p_T also get lower, so that the cuts tend to reject the signal.

But apart from the degenerate case (points 3 and 4), the efficiency remains non-negligible and stable even when the neutralino is very light. Therefore, our results do not depend strongly on the chargino and neutralino masses, i.e. are independent of the SUSY space parameters.

4 Discovery limits

In Figure 1 we show the minimal luminosity required to discover charginos at the 3σ level as a function of the center of mass energy for the jjl topology. The curves were computed assuming the same rejection efficiency for each background as the one estimated at 150 GeV and rescaling the number of events by the corresponding background cross-section at each energy. The chargino production cross-section was fixed at 1 pb, so that the extrapolation to other values of $\sigma \cdot B$ is straightforward. The scenarios with the first chargino and neutralino very close in mass were excluded.

As can be seen, in the range 120-150 GeV around 10 pb^{-1} are enough to discover the lightest chargino, provided it is kinematically accesible.

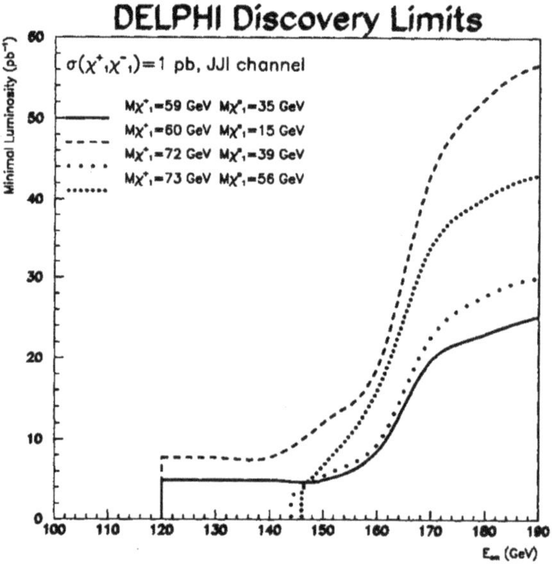

Figure 1: Minimal luminosity needed to discover charginos (at 3σ level) as a function of the center of mass energy. The different lines correspond to different scenarios, as indicated. The results are for the jjl topology.

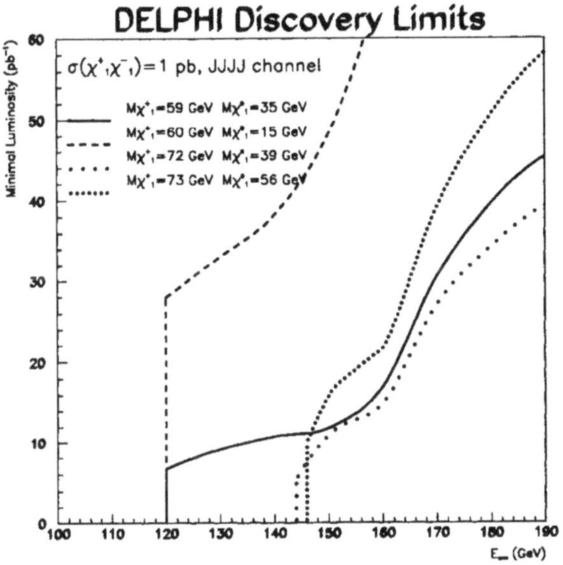

Figure 2: Minimal luminosity needed to discover charginos (at 3σ level) as a function of the center of mass energy. The different lines correspond to different scenarios, as indicated. The results are for the *jjjj* topology.

In Figure 2 the same curves can be seen for the $jjjj$ topology. Again, less than 20 pb^{-1} are enough to see a 3σ effect, although a very light neutralino scenario will require higher luminosity in this case.

5 Conclusions

Chargino events in the lepton plus hadrons and purely hadronic final states have been studied at LEP energies around 150 GeV. We conclude that below the W^+W^- threshold the search for the lightest chargino is almost background free in DELPHI.

An integrated luminosity of 10 pb^{-1} should be enough to detect charginos up to the kinematical limit, provided that they are produced with a $\sigma \cdot B$ larger than $\sim$ 1 pb. Our selection criteria give results almost independent of the chargino and neutralino masses down to chargino-neutralino mass differences of $\sim 10\ GeV$.

Acknowledgments

I would like to thank my colleagues Pierpaolo Rebecchi, Sergio Navas and François Richard for the nice and fruitful collaboration we had during this work. I would also like to congratulate the organizers for the kind atmosphere we had during the meeting and the interesting lectures we had the opportunity to attend.

This work benefited from the network ERBCHRXCT930345 of the Human Capital and Mobility Program of the Commission of the European Communities.

References

1. SUSYGEN Manual, S. Katsanevas
2. DELSIM, DELPHI Event generation and detector simulation,
   User's guide, DELPHI 89-67 PROG 142
   Reference Manual, DELPHI 89-68 PROG 142
   DELANA, DELPHI Data analysis program,
   User's guide, DELPHI 89-44 PROG 137
3. A. Bartl,H. Fraas and W. Majerotto, *Z. Phys.* C **30**, 441 (1986)
   A. Bartl,H. Fraas and W. Majerotto, *Z. Phys.* C **41**, 475 (1988)
4. See for instance:
   M. Carena, S. Pokorski and C.E.M. Wagner, *Nucl. Phys.* B **406**, 59 (1993)
   P. Langacker and N. Polonsky, PRD **49**, 1454 (1994)
   Wim de Boer, R. Ehret and D.I. Kazakov, IEKP-KA/94-05
5. T. Sjöstrand, PHYTHIA 5.6 and JETSET7.3, Physics and Manual, CERNLIB W5035/W5044, CERN-TH-6488/92
6. S. Nova, A. Olshevski and T. Todorov, "Monte Carlo event generator for two photon physics ", DELPHI 90-35 PROG 152

SEARCH FOR HEAVY NEUTRAL HIGGS BOSONS IN THE FOUR JET CHANNEL AT LEP200

CELSO MARTINEZ RIVERO

Dpto. Física Moderna, Univ. de Cantabria, Avda. Los Castros, s/n 39005-Santander, Spain

The new phase of the LEP machine will achieve an energy around 200 GeV in the next future which will allow the extension of the search of the neutral Higgs bosons predicted by the MSSM. The experimental signature is the presence of b flavoured jets, having the four LEP experiments vertex detectors specially suited for the 4 jet channel. The results of a detailed simulation are presented giving the expected limits for exclusion and discovery in the $(mA, \tan\beta)$ plane.

1 Minimal Supersymmetric Standard Model Higgs

The Minimal Supersymmetric Standard Model (MSSM) extension of the SM has a rich Higgs sector. It contains one charged ($H^{\pm}$) and three neutral (h^0, H^0, and A^0) physical states. The couplings of these Higgs bosons are suppressed with respect to the SM ones by the following factors :

$$\sigma_{h^0 Z^0} = \sin^2(\alpha - \beta)\sigma_{H^0_{SM} Z^0}$$
$$\sigma_{h^0 A^0} = \frac{1}{2}\lambda^3 \cos^2(\alpha - \beta)\sigma_{\nu\bar{\nu}}$$

where $(\alpha - \beta)$ is a mixing parameter and λ^3 is the usual phase space factor.

The mass of the lightest neutral Higgs particle is predicted, at tree level, to be smaller than the Z mass. This bound is, however, shifted upwards by radiative corrections which grow as the fourth power of the top quark mass and the logarithm of the ratio of the stop mass to the top mass. These corrections have been calculated up to next–to–leading order which can effectively be incorporated in a one–loop renormalization group analysis by choosing the top quark mass as the scale for the evaluation of the running Higgs mass. In addition, the left-right stop mixing which has a large impact on the upper bound of the lightest CP-even Higgs mass, must be included.

We will present results using a detailed simulation of the DELPHI detector for a top mass of 175 GeV, a SUSY scale of 1 TeV, and a SUSY Higgs mass parameter μ and soft SUSY breaking parameter $A_t = A_b = A$ for two different scenarios :$A = 0$, $\mu \ll M_S$ and $A = \sqrt{6}M_S$, $\mu \ll M_S$

In figure 1 one can see the upper limit in the mass of the h boson as function of the mass of the top quark for two different cases of mixing.

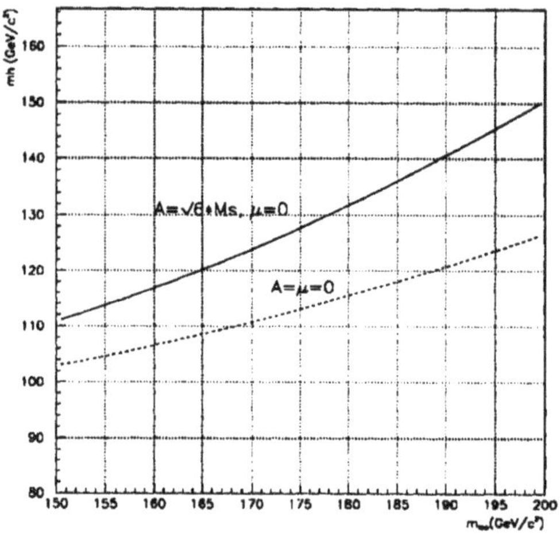

Figure 1: Upper limit on mh as function of the mass of the top for two different mixings.

For small values of $\tan\beta$ the h^0 Higgs boson is produced primarily through Higgs-strahlung while the associated A^0h^0 production becomes increasingly important with rising $\tan\beta$. The dominant decay modes of the h^0, A^0 Higgs bossons are decays into b and τ pairs. In figure 2 are presented the number of expected events for two different CMS energies and a luminosity of $300pb^{-1}$ in the $(mA, \tan\beta)$ plane.

For the $e^+e^- \rightarrow h^0Z$ process the analyses developed for the SM Higgs boson can be used without modifications; the efficiencies are similar, provided the Higgs boson decay channels into supersymmetric particles, such as charginos and neutralinos, are not open.

2 The hA channel

The $e^+e^- \rightarrow h^0A^0$ associated production leads mainly to two final states, $b\bar{b}b\bar{b}$ in 83% of the cases and $\tau^+\tau^-q\bar{q}$ in 16% of the cases, if supersymmetric decays are absent. Thus, the four-jet analysis of the standard model should be slightly modified to exploit the four-b-jet signature. This is done by tightening

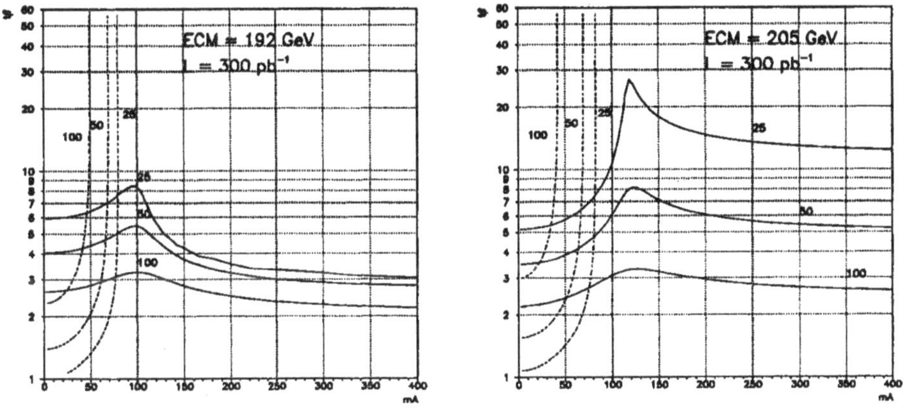

Figure 2: Number of expected events for hA and hZ channels.

the b-tagging criterion, while the Z mass constraint and the requirement of incompatibility with a WW final state are removed.

Standard quality tracks are selected and we require that the event should be of hadronic type, with high multiplicity ($N_{\text{charged}} >= 25$, $E_{\text{total}} >$ 100 GeV/c). Then, tracks with at least two hits in the VD are used to reconstruct the main vertex, which is required to have at least four associated tracks.

Jets are reconstructed using the JADE algorithm, forcing four jets and selecting events with y_{min} above 0.015. In the background, genuine 4 jet events (like WW and ZZ) survive the cut in the same proportion, while a big reduction factor is achieved for $Z^0 \rightarrow f\bar{f}$ as expected.

A detailed procedure for each (mA, $\tan\beta$) point has been applied. Two different cases are treated separately. In the first case $\tan\beta$ is high enough, so that mh is close to mA. A kinematical fit is then applied , testing for 2 dijets with equal mass hypothesis. The fit returns the best combination found and the corresponding mass. The χ^2 is required to be less than 30 and the fitted mass is to be within a ± 6 GeV window around the true value. In the second case a direct rescaling is employed and the sum of the adequate dijet masses is required to be close to $m_h + m_A$.

Here too, two methods for b-tagging both based on the vertex detector have been employed, giving similar results. In the case of probability b-tagging hA events are selected by requiring $-log(Prob) > 10$ for at least three jets with three or more tracks in them. This typically leaves about 20% of the signal, and reaches a high rejection factor against the WW and $Z^0 \to qq$ events. A similar result is obtained if the direct offset method is employed. With a cut on the number of offsets of 13, compatible results are obtained.

Figure 3 shows the final mass distribution for the case of $m_A = 80GeV$, $\tan\beta = 10$. To extract the signal, the number of events in a window of $2\sigma_{M_H}$ centered at the expected A mass are counted. This gives the final numbers of selected events for the signal and background summarised in table 1.

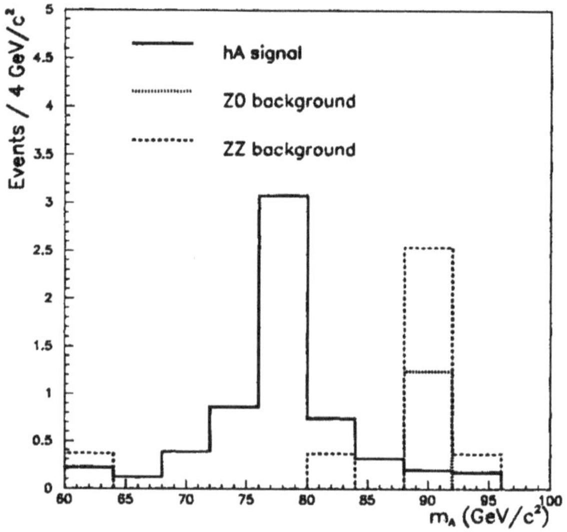

Figure 3: Mass distribution for large $\tan\beta$ ($m_h = m_A$)

3 MSSM Higgs discovery/exclusion plots

Exclusion and discovery limits were calculated taking into account the hZ and hA channels.

The results are summarised in a series of figures which display the areas in the $[m_A, \tan\beta]$ plane that can be covered for a given energy of LEP2, $\sqrt{s} = 192$

or 205 GeV, at the integrated luminosities 300 and 1000 pb^{-1}, respectively, and for a top quark of mass 175 GeV. The different mixing scenarios studied are described above. We note that :

- In the case of zero mixing, ($A = \mu = 0$) the theoretical upper limit on the mass of lightest scalar (h) is quite low, and becomes kinematically accessible to the highest energy considered here (205 GeV). Provided the integrated luminosity is sufficient (see figure 5), and provided the top quark mass does not lie close to its experimental upper bound, the whole $m_A \tan\beta$ plane can be covered with a weak area around $m_A = 90$ GeV and high $\tan\beta$.
- In the case of non-zero mixing (the other case) a rise in energy would allow the exclusion of the low $\tan\beta$ part of the plane. This would also be of great interest in the light of predictions from many theoretical studies embedding the MSSM in GUT assumptions.

Table 1: Cross sections, signal efficiencies and expected signal and background event rates in 300pb^{-1} of integrated luminosity, for the $hA \to b\bar{b}b\bar{b}$ channel.

$Ecms = 192\ GeV$

$(m_A, tan\beta)$	(70,10)	(80,3)	(80,10)
m_h	68.8	67.1	78.2
σ(pb)	0.158	0.069	0.080
$\sigma(hA \to b\bar{b}b\bar{b})$(pb)	0.134	0.059	0.068
events at 300 pb^-1	40	18	20
eff. (%)	18	23	24
# signal	7.2 ± 1.7	4.1 ± 0.3	4.8 ± 0.3
# background	0.4 ± 0.2	0.4 ± 0.2	0.4 ± 0.2

$Ecms = 205\ GeV$

$(m_A, tan\beta)$	(80,5)	(85,5)	(90,3)	(90,10)
m_h	73.9	77.7	72.7	87.3
σ(pb)	0.087	0.063	0.040	0.042
$\sigma(hA \to b\bar{b}b\bar{b})$(pb)	0.074	0.054	0.034	0.036
events at 300 pb^-1	22	16	10	11
eff. (%)	25	25	20	23
# signal	6 ± 1	4 ± 1	2.0 ± 0.1	2.5 ± 0.2
# background	0	0.4 ± 0.2	1.3 ± 0.5	1.3 ± 0.5

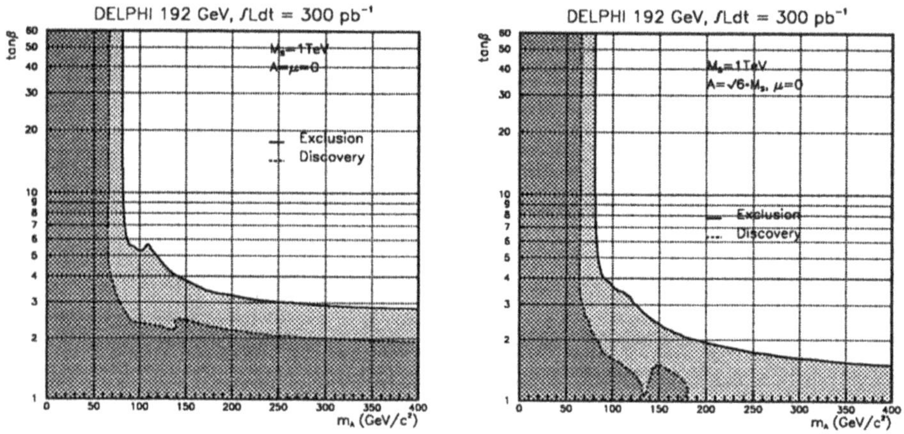

Figure 4: Exclusion/discovery plane in the $m_A \tan\beta$ plane.

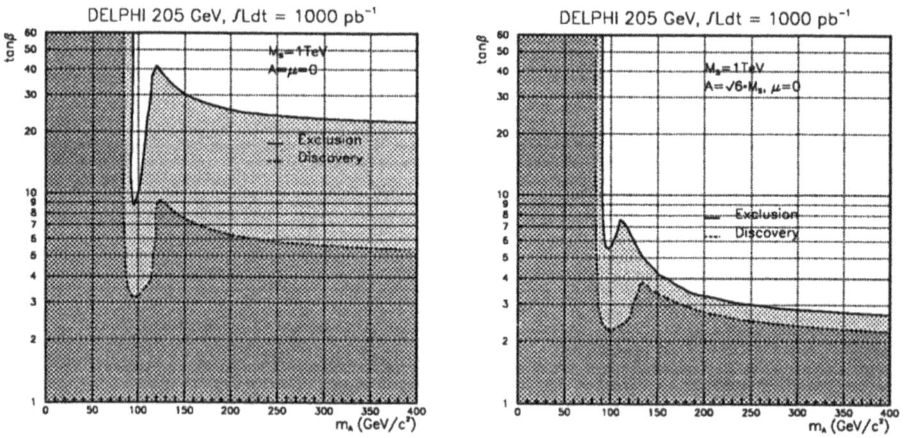

Figure 5: Exclusion/discovery plane in the $m_A \tan\beta$ plane.

Neutrinos and semileptonic b–decays

M. Dittmar

Institute for Particle Physics (IPP), ETH Zürich, CH-8093 Zürich, Switzerland

The missing energy spectrum of jets, associated to the neutrino from semileptonic b–hadron decays has been measured with the L3 experiment at LEP. The neutrino energy spectrum is found to be well described by the free b–quark decay spectator model with a $W^\pm$ according to the (V–A)×(V–A) decay structure. The alternative of a (V+A)×(V–A) decay structure is excluded with a significance of more than 6 standard deviations. The possibility that hadronic corrections to the b–hadron decay destroy any $W^\pm$ polarization is disfavored by more than 3 standard deviations. Furthermore, the BR(b→ νX) has been measured from the missing energy spectrum associated to b–flavored jet and is found to be (23.08±0.77±1.24)%, corresponding to a BR($b \rightarrow e(\mu)\nu X$) of (10.26±0.35±0.55)%.

1 Semileptonic b–decays and neutrinos

The dynamics of semileptonic decays of b–flavored hadrons (b→ $\ell\nu X$) are studied with identified electrons, muons and charmed hadrons. However, the errors for inclusive and exclusive BR for b–hadron decays are still too large and sizable exotic contributions from right handed charged currents are still possible. Right handed couplings of the b–quark could be seen from a measurement of the parity violation strength in semileptonic b–decays. Recently, Z. Was and myself have proposed a new method to extract the parity violation strength in b–decays using the missing energy spectrum of jets from hadronic Z decays[1]. The method exploits the relative energy sharing between the charged lepton and the neutrino to extract the parity violation strength. For a (V–A)×(V–A) b–decay structure, expected within the framework of the Standard Model and the free b–quark decay model[2], the charged lepton obtains on average more energy than the neutrino. In contrast, one expects for an exotic (V+A)×(V–A) decay structure that the neutrino gets on average more energy than the charged lepton. The neutrino energy spectrum can be extracted indirectly from the missing energy of jets with identified semileptonic electron or muon candidates. A simulation of such an experiment using realistic jet energy resolutions, as obtained for example by the L3 experiment at LEP, indicated a high sensitivity to the polarization of the weak charged current. Such a measurement of the parity violation in semileptonic b–decays has now been performed by the L3 collaboration[3]. This measurement and a new analysis of the semileptonic BR(b$\rightarrow \nu X$) will be described in the following. For both measurements, only well measured hadronic "two-jet" events are used.

2 The W polarization in semileptonic b–decays

Semileptonic B–hadron decays are selected using electrons (muons) candidates with an energy above 3 GeV (4 GeV) and a transverse momentum of more than 1.4 GeV with respect to the closest hadronic jet. With these criteria, about 5000 inclusive electron and 10000 muon events are selected using about 350k well measured two-jet hadronic Z decays. The obtained purity of correctly identified semileptonic B–hadron decays is found to be 75±4% for the electron sample and 64±3% for the muon sample.

The observed p and p_t charged lepton spectra in the data are reproduced by the simulation, if the weak semileptonic B–hadrons are generated with an average energy of 72% of the beam energy using the Peterson fragmentation function[4] and a simulation of B–hadron decays with a virtual W polarization according to the (V–A)×(V–A) structure. Alternatively, the charged lepton spectra can also be described with a (V+A)×(V–A) B–hadron decay structure and a roughly 5% harder fragmentation function.

The jet energy scale in the data and in the Monte Carlo has been determined from a Gaussian fit in the energy range between 40–65 GeV, a region which is essentially independent of neutrinos. The data are calibrated such that this average jet energy agrees with the one in the Monte Carlo. The stability of this calibration is found to be better than 50 MeV using different data taking periods. A possible difference of the flavor dependence between data and the simulation for the jet energy response has been studied with b–quark enriched or depleted event samples and was found to be smaller than 150 MeV. The neutrino energy spectra obtained from the missing energy of the jet associated to the charged lepton, are shown in figure 1a for the electron tagged events and in figure 1b for muon tagged events.

The data are well described by a Monte Carlo with a (V–A)×(V–A) decay structure while the predicted neutrino spectrum with a (V+A)×(V–A) structure is much too hard. Systematic uncertainties of the measured neutrino spectrum arise mainly from three sources. These are the uncertainties due to the jet energy calibration, the purity of the inclusive charged lepton sample and efficiency uncertainties of the charged lepton selection. The resulting total systematic error of the average neutrino energy, adding the different contributions in quadrature, are found to be 235 (215) MeV for the neutrino energy associated to the $b \rightarrow e(\mu)\nu X$ decays. Comparing the data with a (V+A)×(V–A) decay simulation, one finds that the average neutrino energy in the data is

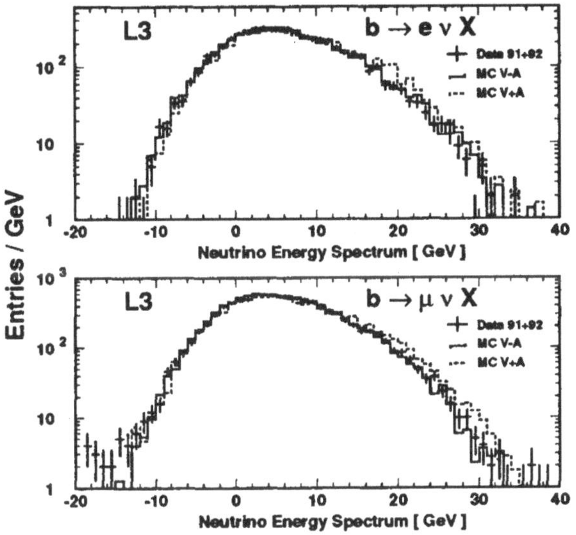

Figure 1: The missing energy spectrum of b→X$\ell\nu$ candidates in the data and in the Monte Carlo with a (V+A)×(V−A) and a (V−A)×(V−A) b–hadron decay structure, (a) is for ν_e candidates and (b) for ν_μ candidates. The systematic uncertainties due the ν energy resolution and the background are not included.

too soft by 1020±120(stat.) MeV (for the electron sample) and 590±85 MeV (for the muon sample) with the systematic errors given above. Taking only the number of observed events with neutrino energies above 16 GeV, the data are found to be in agreement with the (V−A)×(V−A) b–decay model and disagree with a significance of 5.7 and 3.7 standard deviations from the (V+A)×(V−A) model for the b→ X$e\nu_e$ and b→ X$\mu\nu_\mu$ candidates respectively. The possibility of a b–decay model with a kaon–like V×(V−A) decay structure can be excluded by more than 3.5 standard deviations from the number of high energy neutrino events for the b→ X$e\nu_e$ candidates. The neutrino energy spectrum from the b→ X$\mu\nu_\mu$ candidates is found to be equally consistent with the (V−A)×(V−A) and the V×(V−A) case.
The obtained accuracy of this inclusive $W^\pm$ polarization measurement is more significant than measurements of the parity violation strength in exclusive $B_{u,d} \to D^* \ell\nu$ decays[5].

3 Measurement of the BR(b→ νX)

The semileptonic BR(b→ νX) is measured from the rate of b–flavored jets with large missing energy. To enhance the signal over background, a b–flavor enriched event sample (using a lifetime tag), with a b–flavor purity of (59.4±1.5)%, is used. With the requirement that the measured polar angle of the jet fulfills the condition $|\cos\theta_{\rm Jet}| < 0.65$, about 41k two jet events are analyzed. The Monte Carlo neutrino spectrum for each semileptonic decay is found with the assumption of a (V–A)×(V–A) b–hadron decay structure. The neutrino spectra estimated for the different semileptonic b–hadron decays are combined, assuming a ratio for the charged leptons (e:μ:τ) of 1:1:0.25±0.05, in agreement with theoretical estimates[6]. To be as much as possible independent from the Monte Carlo, the background in the missing energy spectrum is determined from the data. This background is from jets with purely hadronic b–decays and from the remaining background from light quark flavored jets. The background determination uses the constraint that the jet energy spectra of the two–jet event sample (b–flavor purity of 22.8%) and the lifetime tagged b–flavor enriched events have to be described simultaneously with the same BR(b→ νX). This method of estimating the background has been developed and tested with the Monte Carlo. Using different input semileptonic b–decay branching ratios and b–flavor purities, the background from mismeasured jets is found to be accurately described with this method. It is found that with a known purity of the selected event samples, the MC input branching ratio BR(b→ νX) can be determined with a relative systematic accuracy of better than ±2%. Using a maximum likelihood fit to describe the jet energy spectra in both event samples, the BR($b \to \nu X$) is measured to be (23.08±0.77(stat.)±1.24(syst.))%. The jet energy spectrum for the lifetime tagged events and the fit results are shown in figure 2. The systematic error is obtained by adding the different uncertainties in quadrature. In addition to the systematic error introduced by the method to determine the background, ±0.61%, the error due to the b–flavor purity uncertainty, ±0.90%, the uncertainties of the assumed neutrino energy spectra (this includes uncertainties due to the b–fragmentation function), ±0.56%, and the uncertainty of the contribution from the BR(b→ $\tau\nu X$), ±0.20%, are found to be important. The uncertainty of the neutrino energy scale error (±200 MeV) and the resulting neutrino energy spectra has been estimated from the measurement of the neutrino energy spectrum described in section 2 and the modeling of b→ $\tau\nu X$ and semileptonic charm decays. To compare the semileptonic BR result obtained with neutrinos with the ones from electrons and muons the BR(b→ νX) has to be divided by 2.25. The result for BR(b→ $e(\mu)\nu X$) obtained from the neutrino measurement is (10.26±0.34(stat.)±0.55(syst.))%. This result has comparable errors to the direct measurements with electrons and muons with an average BR(b→ $e(\mu)\nu X$) of (10.85±0.12±0.47)% [7]. The agreement with the direct BR results obtained with electrons and muons indicates a consistent treatment of the energy spectra of the charged leptons and the corresponding neutrinos. It provides further evidence for parity violation due to the V–A structure in b–decays. In the case of a V+A b–decay structure, the average neutrino energy would be about 1.5 GeV larger. As a result of the wrong efficiency estimate, an apparent BR of 27±1% would have been seen in the case of a V+A b–decay structure. A value which is excluded from the neutrino BR measurement.

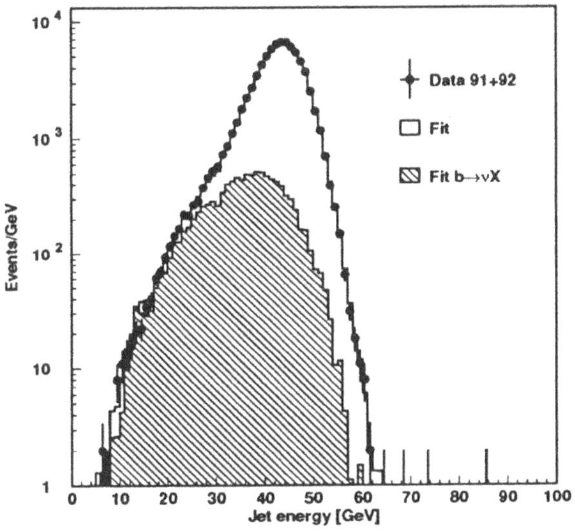

Figure 2: The jet energy spectrum in a b–lifetime enriched sample and the result of the fit for the BR(b→ νX) measurement.

4 Summary

The inclusive neutrino energy spectrum in semileptonic b–hadron decays has been measured from the missing energy of jets which contain energetic high p_t electron or muon candidates. The neutrino energy is measured with a reso-

lution of about 4.2 GeV and an absolute energy scale accuracy of better than ±150 MeV. The observed number of neutrino candidates with energies above 16 GeV excludes b–hadron decay models with a (V+A)×(V–A) structure by more than 6 standard deviations and excludes the possibility that hadronic corrections to semileptonic b–hadron decays destroy any polarization of the virtual $W^{\pm}$ by more than 3 standard deviations.
In contrast to these exotic possibilities, the measured inclusive neutrino energy spectra are found to be in good agreement with the virtual $W^{\pm}$ polarization expected from a (V–A)×(V–A) b–hadron decay structure.
The semileptonic branching ratio BR($b \to \nu X$) has been measured using a two–jet hadronic event sample. The efficiency and backgrounds have been cross checked with b–enriched and b–depleted event subsamples.
The BR(b$\to \nu X$) is measured to be: **BR(b$\to \nu X$) = (23.08±0.77±1.26)%.** and corresponds to **BR(b$\to e(\mu)\nu X$)=(10.26±0.34±0.55%)**. This value is in agreement with the direct L3 BR measurements with inclusive electron and muon candidates with an average BR(b$\to e(\mu)\nu X$) of (10.85±0.12±0.47)%. This good agreement between these measurements provides further evidence for parity violation in semileptonic b–decays.

Acknowledgments

It is a pleasure to thank the organizers and especially J. W. F. Valle for the interesting program of the workshop and the nice atmosphere.

References

1. M. Dittmar and Z. Was, *Phys. Lett.* B **332**, 168 (1994).
2. G. Altarelli *et al.*, *Nucl. Phys.* B **208**, 365 (1982).
3. L3 Collab., M. Acciarri *et al.*, *Phys. Lett.* B **351**, 375 (1995).
4. C. Peterson *et al.*, *Phys. Rev.* D **27**, 105 (1983).
5. CLEO Collab., S. Sanghera *et al.*, *Phys. Rev.* D **47**, 791 (1993) and ARGUS Collab., H. Albrecht *et al.*, *Z. Phys.* C **57**, 533 (1993).
6. P. Heiliger and L. M. Sehgal, *Phys. Lett.* B **229**, 409 (1989), and A. Falk *et al.*, *Phys. Lett.* B **326**, 145 (1994),
7. L3 Collab., Contribution to the EPS Conference Brussels 1995, EPS 112.

Search For Rare B Decays

MANEESH WADHWA

Institute of Physics, University of Basel,
CH-4056 Basel, Switzerland

A search for rare decays of B hadrons has been performed in Z hadronic decays using the L3 detector at LEP. The exclusive channels $B^0_{d,s} \to \gamma\gamma$, $B^0_{d,s} \to \eta\eta$, $B^0_{d,s} \to \pi^0\pi^0$ and $B^0_{d,s} \to \pi^0\eta$ have been studied. No candidate events are observed in the signal region. Consequently, upper limits at 90% CL on the branching ratios have been set. The $B^0_{d,s} \to \gamma\gamma$, $B^0_d \to \eta\eta$ and all the $B^0_s \to \pi^0\pi^0/\pi^0\eta/\eta\eta$ are the first experimental limits set on these exclusive decays.

1 Introduction

Rare B hadron decays are important discriminators for the higher order structures of the Standard Model (SM) as well as physics beyond. Of particular interest are the decays $B^0_{d,s} \to \gamma\gamma$ for which there are no experimental efforts as yet. The decays correspond to a second order weak transition, including gluonic penguins, followed by annihilation, often referred as "effective" flavor changing weak neutral current process as shown by some simple examples in Figure 1.

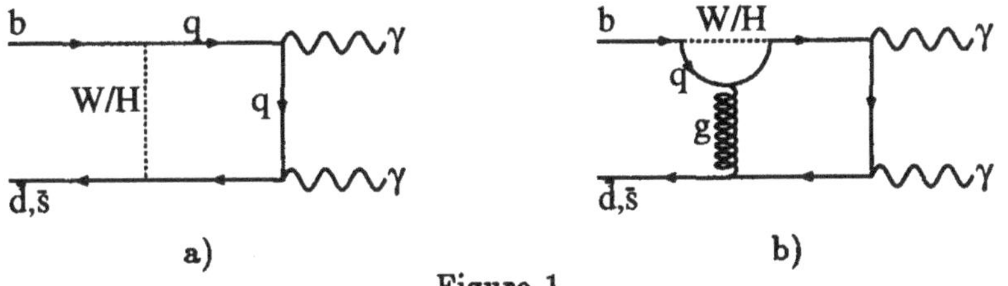

Figure 1

The expected branching ratio for this decay mode is $\approx 10^{-7}$ in the SM, however there are large theoretical uncertainties[1,2,3,4]. In contrast to the theoretical situation, experimentally $B^0_{d,s} \to \gamma\gamma$ is relatively clean channel to study.

There has also been a search performed for neutral charmless hadronic decays of B^0_d and B^0_s mesons in the neutral exclusive final states:

$$B^0_d \to \eta\eta,\ B^0_d \to \pi^0\eta,\ B^0_d \to \pi^0\pi^0,\ B^0_s \to \eta\eta,\ B^0_s \to \pi^0\eta,\ B^0_s \to \pi^0\pi^0.$$

where η's and π^0's decays into two photons. Within the SM, the neutral charmless $B^0_{d,s}$ decays can occur through a variety of processes such as Cabibbo-suppressed $b \to u$ transition[5,6] with a further color suppression with respect to the charged modes[7] or one loop diagrams with a heavy quark and a virtual $W^{\pm}$ boson [8,9,7]. Contributions can also arise from electroweak penguins[10].

It is believed that for some non-standard scenarios, such as minimal supersymmetric and two-Higgs-doublet models, these decay rates can be enhanced significantly. Thus any observation of the decay at an unusual rate could be an indication of new physics.

The present work is an effort to look for these decay modes exploiting the high resolution BGO electromagnetic calorimeter of the L3[11] detector which is particularly well suited for this purpose. As will be demonstrated later in this paper, the resolution of the detector is appropriate to reconstruct exclusively these decay modes. The analysis is performed on data recorded in 1991 through 1994 at $\sqrt{s} \approx M_Z$, corresponding to a sample of 3.0 million $e^+e^- \to$ hadron events. The mixed sample of B hadrons available in Z decays provides an opportunity to study B^0_s meson decays which are not accessible at the center-of-mass energy of the $\Upsilon(4S)$.

2 Event Selection

Hadronic events selected with a standard set of cuts[13,12] are used for the analysis described in the following sections. Candidate photons in an event are selected with $|cos\theta| \leq 0.75$, where θ is the polar angle, and are recognized as isolated energy clusters in the electromagnetic calorimeter with a narrow shower shape consistent with that of electromagnetic particles. To ensure that the electromagnetic cluster corresponds to a photon, there should be no charged track within 10 mrad of the electromagnetic cluster in the plane transverse to the beam direction.

3 Search for $B^0_{d,s} \to \gamma\gamma$

Due the hard fragmentation and high mass of the b quark, the energy carried by the two photons from a B^0 decay is large. The sum of the two photon energies is required to be more than 25 GeV, where combinatorial background is peaked at small energies in contrast to the $B^0 \to \gamma\gamma$ Monte Carlo events. In addition it is required that the opening angle between the two photons be less than 40°. In this case combinatorial background tends to be at large angles. These cuts reduce the combinatorial background either from the low energy photons or from the opposite hemisphere. The invariant mass of the $\gamma\gamma$ system is reconstructed with photons having a minimum energy of 2 GeV. In order to identify the $b\bar{b}$ events from other hadronic events, the multidimensional analysis based on a neural network approach described elsewhere[13] is used. Requiring a large value of network response, an enriched sample of $b\bar{b}$ events can be obtained ($\approx$ 35% efficiency with $\approx$ 60% b - purity). It has been checked that the neural network response for $B^0_{d,s} \to \gamma\gamma$ has the same shape as for other $b\bar{b}$ events. The efficiency calculation is based on a full simulation of the L3 detector[14]. We use JETSET[15] simulated b-flavor events, requiring the decay chain $B^0_{d,s} \to \gamma\gamma$, for one of the $B^0_{d,s}$ mesons. For the background studies a sample of approximately 2.1 million hadronic Z events were generated and simulated using the JETSET Monte Carlo program.

The mass resolutions obtained from a Gaussian fit on the $\gamma\gamma$ invariant mass of the $B^0_{d,s}$ are $\sim$ 72 MeV(see Figure 2(a,b)). The efficiency for this process is calculated to be $0.12 \pm 0.005 \pm 0.007$ with the $b\bar{b}$ selection requirement, where the first error is statistical and the second is systematic. The systematic uncertainties are evaluated by varying the selection criteria and the fragmentation function. Figure 2(c,d) shows the invariant mass distribution of the $\gamma\gamma$ system for all events and for the $b\bar{b}$ enriched sample in data and five-flavor $Z \to q\bar{q}$ Monte Carlo events. As one can see from Figure 2c the shape of the background is well reproduced by the five-flavor $q\bar{q}$ Monte Carlo events. With the additional $b\bar{b}$ selection requirement, no candidates are observed within $\pm 2\sigma$ mass window of $B^0_{d,s}$ (see Figure 2d).

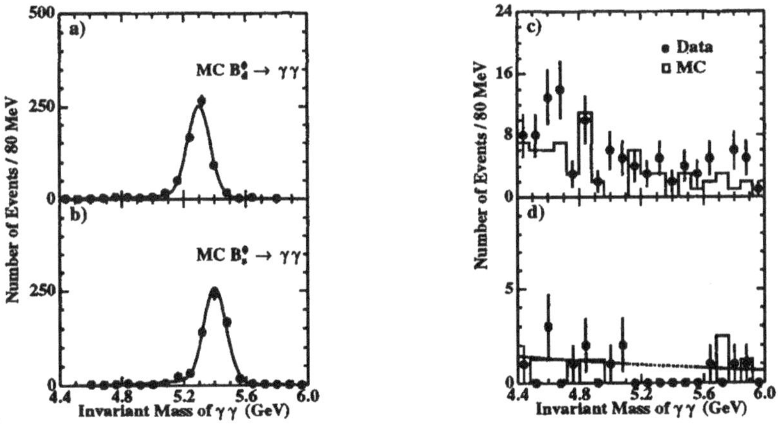

Figure 2

Upper limits on the exclusive branching ratio for the B^0_d and B^0_s mesons decaying into two photons is obtained using a binned likelihood fit to the $\gamma\gamma$ invariant mass distribution (For more details see[16]). The background is estimated from the shape of the invariant mass distribution in the data shown in Figure 2d. An exponential function is fitted to the invariant mass distribution in the range 4.4 GeV to 6.0 GeV, excluding the $B^0_{d,s}$ mass window. For the signal there are two components, corresponding to B^0_d and B^0_s. The overlapping region of B^0_d and B^0_s is taken into account using the Monte Carlo $\gamma\gamma$ mass distributions. Following numerical values used to set upper limit at 90% confidence level: $R_{b\bar{b}}$ is taken from the L3 measurement[13], and $f_{b\to d,s}$ is the fraction of $b \to d = 39.5 \pm 4\%$ and $b \to s = 12.0 \pm 3\%$ used, in agreement with the available measurements[18]. The likelihood is mapped as a function of the branching ratios $Br(B^0_d \to \gamma\gamma)$ and $Br(B^0_s \to \gamma\gamma)$, taking into account Gaussian errors on the efficiency, $R_{b\bar{b}}$ and $f_{b\to d,s}$. By varying $Br(B^0_d \to \gamma\gamma)$ and $Br(B^0_s \to \gamma\gamma)$ simultaneously we obtain the two dimensional contours for different confidence levels. The one dimensional 90% confidence level upper limits on the B^0_d (B^0_s) branching ratios are obtained by integrating the likelihood over all the values of the B^0_s (B^0_d) branching ratios. The 90% confidence level limits obtained are [a]

$$Br(B^0_d \to \gamma\gamma) < 3.9 \times 10^{-5} \quad \text{at } 90\% \text{ CL},$$
$$Br(B^0_s \to \gamma\gamma) < 14.8 \times 10^{-5} \quad \text{at } 90\% \text{ CL}.$$

4 Search for $B^0_{d,s} \to \eta\eta/\pi^0\eta/\pi^0\pi^0$

The searches for the neutral charmless B decays have been performed in four different final state configurations:

- $B^0_{d,s} \to \eta\eta$: four detected photons in the final state,

  one η giving two detected photons and the other detected as a cluster,
- $B^0_{d,s} \to \pi^0\eta$: the η detected as two photons and the π^0 as a cluster,
- $B^0_{d,s} \to \pi^0\pi^0$: both π^0's detected as clusters in the final state.

In contrast to the previous section, global kinematics of the $B^0_{d,s}$ meson candidate are used to search for these modes rather than $b\bar{b}$ selection based on neural network. Cuts are made on the opening angle of the two light mesons and the total energy of the selected photons in the final state. In addition, angle between the direction of one decay product in the $B^0_{d,s}$ candidate rest frame and the $B^0_{d,s}$ candidate flight direction, $\cos\theta^*$, is defined. Since signal events have isotropic distribution of $cos\theta^*$ whereas for background it is peaked. In the decay mode where an η is detected as a photon pair, a cut on the invariant

[a] The contributions from $B^0 \to \pi^0\pi^0$ is negligible.

mass $M_{\gamma\gamma}$ of these photons can be applied. A flat invariant mass spectrum is expected for random combinations. A constrained fit to the $B^0_{d(s)}$ mass has been performed for the $B^0_{d,s} \to \pi^0\pi^0$ search. The χ^2 of this fit shows high values for background and low ones for signal. A cut at the value of 1.6 has been chosen.

These cuts were optimized for the B^0_d and B^0_s exclusive modes. First by examining the selected variables on the loose set of cuts for the Monte Carlo and then comparing with the data. Satisfactory agreement was found. The final sets of cuts chosen for all the final state configuration are discussed in detail in reference[19]. The JETSET Monte Carlo was used to generate all the eight decay channels on one side of b quark, other side of the event was left free to hadronize and decay. The distribution of the invariant mass for events surviving the cuts in the signal Monte Carlo was fit with a Gaussian of width σ. Events in a $\pm 2\sigma$ window around the fit mass of the $B^0_{d,s}$ were used to calculate the efficiency of $B^0_{d,s}$ in different final state configurations. The invariant mass spectra for the data and the Gaussians fit to the B^0_d and B^0_s Monte Carlo samples after the application of the final cuts are shown in Figure 3.

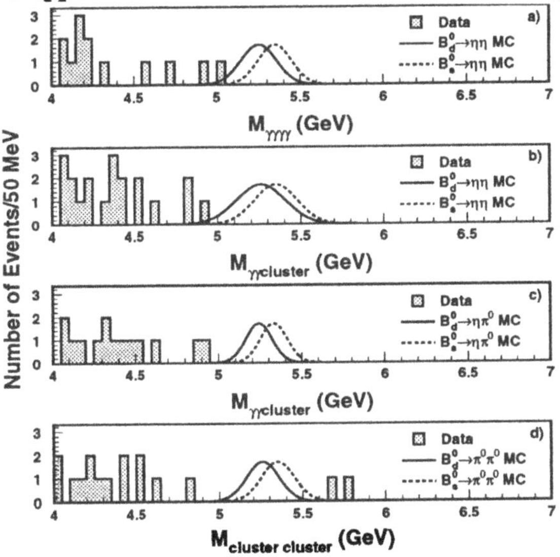

Figure 3

The efficiencies are given in Table 1 together with their statistical and systematic errors. Since no candidate event has been found in data for any of the eight final configurations, upper limits at 90% confidence level have been set using the same numerical values as discussed in the previous section. The errors on the numerical values and on the efficiencies were taken into account by folding their Gaussian distribution with the Poisson distribution describing the number of expected events. In the analysis of the $B^0_{d,s} \rightarrow \pi^0\pi^0$ the branching ratio of the decay $B^0_{d,s} \rightarrow \gamma\gamma$ is assumed to be negligible.

In Table 1 the σ's of the Gaussian fits to the signal Monte Carlo, the efficiencies and the upper limits set with the procedure described above are reported, for all the considered decay modes.

Process	Resolution	Efficiency	Upper limit 90% C.L.
$B^0_d \rightarrow \eta\eta$ (I)	107 ± 10 MeV	$2.5 \pm 0.2^{+0.2}_{-0.2}\%$	
$B^0_d \rightarrow \eta\eta$ (II)	146 ± 11 MeV	$4.6 \pm 0.3^{+0.02}_{-0.2}\%$	
$B^0_d \rightarrow \eta\eta$			$< 4.1 \times 10^{-4}$
$B^0_d \rightarrow \pi^0\eta$	79 ± 5 MeV	$4.5 \pm 0.3^{+0.05}_{-0.03}\%$	$< 2.5 \times 10^{-4}$
$B^0_d \rightarrow \pi^0\pi^0$	97 ± 4 MeV	$7.6 \pm 0.4^{+0.2}_{-0.5}\%$	$< 6.0 \times 10^{-5}$
$B^0_s \rightarrow \eta\eta$ (I)	101 ± 10 MeV	$2.4 \pm 0.2^{+0.2}_{-0.2}\%$	
$B^0_s \rightarrow \eta\eta$ (II)	129 ± 8 MeV	$4.8 \pm 0.3^{+0.2}_{-0.3}\%$	
$B^0_s \rightarrow \eta\eta$			$< 1.5 \times 10^{-3}$
$B^0_s \rightarrow \pi^0\eta$	81 ± 1 MeV	$4.3 \pm 0.3^{+0.02}_{-0.1}\%$	$< 1.0 \times 10^{-3}$
$B^0_s \rightarrow \pi^0\pi^0$	99 ± 4 MeV	$8.3 \pm 0.4^{+0.4}_{-0.7}\%$	$< 2.1 \times 10^{-4}$

Table 1: Resolutions (σ of a Gaussian fit to the signal Monte Carlo invariant mass distribution), efficiencies and experimental limits for B^0_d and B^0_s branching ratios. The (I) and (II) modes refer respectively to the search for a four photon final state or one with a photon pair for one η and a single cluster for the other one. The first error on the efficiencies is statistical, the second systematic.

Conclusions

A search for the exclusive rare decays with photons in the final state in Z decays has been performed. No candidates are observed and upper limits on the branching ratios at 90% confidence level were set as:

$$Br(B^0_d \rightarrow \gamma\gamma) < 3.9 \times 10^{-5},\ Br(B^0_s \rightarrow \gamma\gamma) < 14.8 \times 10^{-5},$$

$$Br(B^0_d \rightarrow \eta\eta) < 4.1 \times 10^{-4},\ Br(B^0_s \rightarrow \eta\eta) < 1.5 \times 10^{-3},$$

$$Br(B^0_d \rightarrow \pi^0\eta) < 2.5 \times 10^{-4},\ Br(B^0_s \rightarrow \pi^0\eta) < 1.0 \times 10^{-3},$$

$$Br(B^0_d \rightarrow \pi^0\pi^0) < 6.0 \times 10^{-5},\ Br(B^0_s \rightarrow \pi^0\pi^0) < 2.1 \times 10^{-4}.$$

It is noted that there are very few theoretical calculations within the Standard Model. On general considerations large enhancements in these branching fractions are expected from new physics. The $B^0_{d,s} \rightarrow \gamma\gamma$, $B^0_d \rightarrow \eta\eta$ and all the $B^0_s \rightarrow \pi^0\pi^0/\pi^0\eta/\eta\eta$ are the first experimental limits set on these exclusive decays.

Acknowledgments

I would like to thank T. Aziz, S.Mele and L. Taylor for their collaboration. I am grateful to Prof. L. Tauscher for his valuable advise. This work was supported by the Swiss National Science Foundation.

References

1. G. Lin *et al*, Phys. Rev. **D 42** (1990) 2314.
2. S. Herrlich and J. Kalinowski, Nucl. Phys. **B 381** (1992) 501.
3. J. O. Eeg and I. Picek, Phys. Lett. **B 336** (1994) 549.
4. J. Trampetic, FIZIKA **B 2** (1993) 121.
5. N. Cabibbo, Phys. Rev. Lett. **10** (1963) 531.
6. M. Kobayashi, T. Maskawa, Prog. Theo. Phys. **49** (1973) 652.
7. L. L. Chau *et al*, Phys. Rev. **D43** (1991) 2176
8. M. Gronau *et al*, Preprint TECHNION-PH-95-10.
9. M. Gronau *et al*, Preprint TECHNION-PH-94-8.
10. M. Gronau *et al*, Preprint TECHNION-PH-95-11.
11. L3 Collaboration, B. Adeva *et al.*, Nucl. Instr. Meth. **A 289** (1990) 35; L3 Collaboration, O. Adriani *et al.*, Phys. Rep. **236** (1993) 1.
12. L3 Collaboration, B. Adeva *et al.*, Phys. Lett. **B 252** (1990) 35.
13. L3 Collaboration, O. Adriani *et al.*, Phys. Lett. **B 307** (1993) 237.
14. The L3 detector simulation is based on GEANT Version 3.15. See R. Brun et al., "GEANT 3", CERN DD/EE/84-1 (Revised), September 1987. The GHEISHA program (H. Fesefeldt, RWTH Aachen Report PITHA 85/02 (1985)) is used to simulate hadronic interactions.
15. T. Sjöstrand and M. Bengtsson, Comput. Phys. Commun. **43** (1987) 367;
    "PYTHIA 5.6 and JETSET 7.3: Physics and manual", preprint CERN-TH.6488/92.
16. L3 Collaboration, M. Acciarri *et al.* Search for $B^0 \to \gamma\gamma$ using the L3 detector: In preparation.
17. Particle Data Group, Phys. Rev. **D 50** (1994).
18. OPAL Collaboration, R. Akers *et al.*, Z. Phys. **C 66** (1995) 555;
    OPAL Collaboration, R. Akers *et al.*, Z. Phys. **C 67** (1995) 57;
    ALEPH Collaboration, D. Buskulic *et al.*, Preprint CERN-PPE/95-092;
    ALEPH Collaboration, D. Buskulic *et al.*, Preprint CERN-PPE/95-094.
19. L3 Collaboration, M. Acciarri *et al.* Search for Neutral Charmless B Decays at LEP : In preparation.

RECENT RARE B DECAY RESULTS FROM CLEO

Nobu Katayama

Wilson Lab, Cornell University
Ithaca, NY 14853, USA

This paper reports on a preliminary measurement of the branching fractions for $B \to \pi\ell\nu$ and $B \to \rho\ell\nu$. B decays to the final states $\pi^{\mp}\ell^{\pm}\nu$, $\pi^0\ell^{\pm}\nu$, $\rho^{\mp}\ell^{\pm}\nu$, $\rho^0\ell^{\pm}\nu$ and $\omega\ell^{\pm}\nu$ are identified via full reconstruction, using a measurement of the missing energy and momentum in the event to infer the neutrino momentum. We find branching ratios $\mathcal{B}(B^0 \to \pi^-\ell^+\nu) = (1.34 \pm 0.35 \pm 0.28) \times 10^{-4}$, and $\mathcal{B}(B^0 \to \rho^-\ell^+\nu) = (2.28 \pm 0.36 \pm 0.59^{+0.00}_{-0.46}) \times 10^{-4}$. The ratio of widths $\Gamma(B^0 \to \rho^-\ell^+\nu)/\Gamma(B^0 \to \pi^-\ell^+\nu) = 1.70^{+0.81}_{-0.50} \pm 0.58^{+0.00}_{-0.34}$. These results use the ISGW model for efficiencies. If instead one uses the WSB model, one finds $\mathcal{B}(B^0 \to \pi^-\ell^+\nu) = (1.63 \pm 0.46 \pm 0.34) \times 10^{-4}$, $\mathcal{B}(B^0 \to \rho^-\ell^+\nu) = (3.88 \pm 0.54 \pm 1.01^{+0.00}_{-0.78}) \times 10^{-4}$, and $\Gamma(B^0 \to \rho^-\ell^+\nu)/\Gamma(B^0 \to \pi^-\ell^+\nu) = 2.39^{+1.18}_{-0.69} \pm 0.81^{+0.00}_{-0.48}$.

1 Introduction

ARGUS[1] and CLEO[2] have made inclusive measurements of $b \to u\ell\nu$ transition and proved that $|V_{ub}| > 0$. In this paper we present a preliminary CLEO analysis of $b \to u\ell\nu$ decays to the exclusive final states $\pi\ell\nu$, $\rho\ell\nu$ and $\omega\ell\nu$.[3] The analysis presented here studies the two pseudoscalar modes $\pi^-\ell^+\nu$ and $\pi^0\ell^+\nu$, the three vector modes $\rho^-\ell^+\nu$, $\rho^0\ell^+\nu$ and $\omega\ell^+\nu$.

2 Neutrino Measurement and Exclusive $b \to u\ell\nu$

Experimentally, semileptonic decays are troublesome because of the undetected neutrino. This analysis takes advantage of the excellent hermeticity and resolution of the CLEO detector located at the Cornell Electron Storage Ring (CESR). Three concentric tracking devices provide a momentum resolution of $\sigma_p/p = 0.005 \oplus 0.0015p$ (p in GeV/c), while covering 95% of the 4π solid angle. The CsI calorimeter located inside of the CLEO solenoid provides an energy resolution well approximated by $\sigma_E/E = 0.019 + 0.0035/E^{0.75} - 0.001E$ (E in GeV), while covering 98% of 4π. The detector is described in detail elsewhere.[4] This analysis is based on a data sample with a luminosity of 2.46 fb^{-1} (2.64×10^6 $B\bar{B}$ decays).

The $B\bar{B}$ system is produced at rest in CLEO and the beam energy is known very precisely, so we can measure the neutrino four momentum by measuring the missing energy and momentum of an event. We define

$$E_{miss} \equiv 2E_{beam} - \sum_i E_i \quad (1)$$

$$\vec{p}_{miss} \equiv -\sum_i \vec{p}_i, \quad (2)$$

where the index i runs over all charged tracks and all showers in the calorimeter that pass cuts designed to reject false tracks and spurious showers from hadronic interactions.

In events with no extra missing particles, $\vec{p}_{miss}$ can be reliably associated with the momentum $\vec{p}_\nu$ of the neutrino. The $b \to u\ell\nu$ decay can then be fully reconstructed: the energy difference $\Delta E \equiv E_{beam} - (E_h + E_\ell + |\vec{p}_\nu|)$, where h is the candidate hadron, should be zero, and the beam-energy constrained mass $M_B \equiv \sqrt{E_{beam}^2 - |\vec{p}_h + \vec{p}_\ell + \vec{p}_\nu|^2}$ should reconstruct at the B mass. Signal events without K_L^0's or extra neutrinos show resolutions of approximately 260 MeV on E_{miss} and 110 MeV on $|\vec{p}_{miss}|$.

Continuum background is suppressed using standard event shape variables. We employ the event sphericity (characterized by the second Fox-Wolfram moment[6]) and the net momentum transverse to the momentum of the candidate lepton outside of forward and backward cones.

Both electrons and muons are used in this analysis. We combine information from specific ionization, energy/momentum measurements from the calorimeter and tracking systems, and position matching from these two systems. Muon candidates must register hits in muon counters at least 5 interaction lengths deep, limiting the muon momentum range to approximately 1.5 GeV/c. The probability that a hadron is misidentified as a lepton (a "fake lepton") is of the order 0.1% (1%) for electron (muon) identification. The $b \to u\ell\nu$ processes are enhanced over $b \to c$ by requiring the leptons to have momenta larger than 1.5 GeV/c (2.0 GeV/c) in the $\pi\ell\nu$ (vector) modes.

The vector meson candidates are approached by examining the 2π and 3π invariant mass distributions. For the ω, we impose the additional requirement that the $\omega \to 3\pi$ decay lie within the central region of the Dalitz plot.

The dominant background in both the π and the vector modes comes from $B\bar{B}$ events in which there is at least one $b \to c\ell\nu$ decay and either an undetected K_L, a second neutrino, or both. Monte Carlo studies indicate that feed-across among the five modes contributes the next largest background. In the vector modes, $b \to u\ell\nu$ decays to higher mass final states result in a small background component. The backgrounds from fake leptons and from continuum processes are measured with the data and are small in every mode.

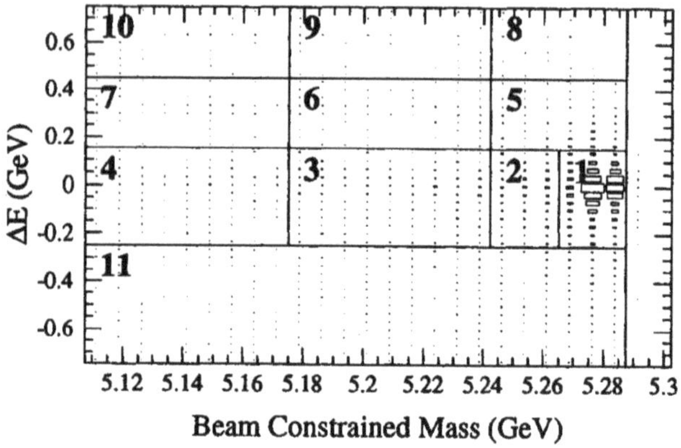

Figure 1: Bins in $M_B - \Delta E$ Space

3 Extracting the yields

The general approach taken to determine the signal yields is to fit to the observed distribution of events in the $M_B - \Delta E$ plane in each of the five modes. The feed-across background from the vector modes into the $\pi\ell\nu$ modes is non-negligible, but *a priori* we do not know the relative rate between these modes. By fitting the distributions in $M_B - \Delta E$ in the five reconstructed $b \to u\ell\nu$ modes simultaneously, the data in the vector modes constrains the $\rho\ell\nu$ background in the $\pi\ell\nu$ modes.

For each of the five decay modes, for each event passing the cuts, we compute M_B and ΔE. We obtain yields in 11 bins in $M_B - \Delta E$ space, as shown in Fig. 1. The signal peaks in Bin 1. The efficiency for signal to fall in Bin 1 is 2.9% for $\pi^-\ell^+\nu$, 1.9% for $\pi^0\ell^+\nu$, 0.8% for $\rho^-\ell^+\nu$, 1.4% for $\rho^0\ell^+\nu$, and 0.6% for $\omega\ell^+\nu$.

After subtracting the continuum and fake lepton backgrounds, we fit the $M_B - \Delta E$ distributions in our five reconstructed $b \to u\ell\nu$ modes simultaneously, 55 bins in all. In addition to the signal shapes and the feed-across shapes between the five modes, the fit includes $b \to c$ and other $b \to u\ell\nu$ background components. The relations $\frac{1}{2}\Gamma(B^0 \to \pi^-\ell^+\nu) = \Gamma(B^+ \to \pi^0\ell^+\nu)$ and $\frac{1}{2}\Gamma(B^0 \to \rho^-\ell^+\nu) = \Gamma(B^+ \to \rho^0\ell^+\nu) \approx \Gamma(B^+ \to \omega\ell^+\nu)$ constrain the neutral meson rates relative to the charged meson rates. We therefore obtain two yields, $N_{\pi^\pm\ell^\mp\nu}$ and $N_{\rho^\pm\ell^\mp\nu}$, from the fit.

The signal, $b \to c\ell\nu$ background and feed-across background distributions in M_B versus ΔE are obtained from a Monte Carlo simulation. For this preliminary analysis we have used signal event generators based on the ISGW and

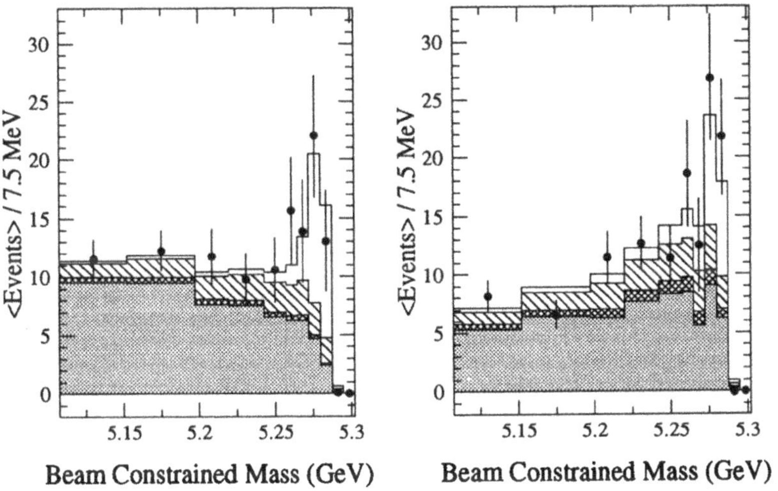

Figure 2: Beam constrained mass distributions for the combined $\pi^- \ell^+ \nu$ and $\pi^0 \ell^+ \nu$ modes (left) and the combined vector modes (right), for the nalysis with simple $\pi\pi/3\pi$ mass cuts. The points are continuum- and fake-subtracted data. The histograms show the contribution from $b \to c$ (shaded), feeddown from higher mass $b \to u\ell\nu$ (cross-hatched), signal mode crossfeed (hatched) and signal (hollow).

WSB models. The $b \to c\ell\nu$ background level is taken as a single fitting parameter, the same for all of the five modes, while the feed-across rates between the five modes are tied to the signal yields $N_{\pi^\pm \ell^\mp \nu}$ and $N_{\rho^\pm \ell^\mp \nu}$.

Thus we fit 55 data points with 3 parameters: the $B^0 \to \pi^- \ell^+ \nu$ rate; $B^0 \to \rho^- \ell^+ \nu$ rate; and the $b \to c\ell\nu$ background scale factor. The fit finds an efficiency-corrected π^+ yield of 728 $\pm$ 195 events, a ρ^+ yield of 1183 $\pm$ 269 events, and a $b \to c$ scale factor of 1.06 $\pm$ 0.04. It has a χ^2 of 64.7 for 52 degrees of freedom. In Fig. 2 we compare the M_B distributions for the data and for the fit in the ΔE range -250 MeV $< \Delta E <$ 150 MeV (the ΔE signal band) in both the combined $\pi\ell\nu$ and the combined vector modes. A signal is apparent, both in $\pi\ell\nu$ and $\rho/\omega\ell\nu$. With the procedure just described, the $\rho/\omega\ell\nu$ yield includes both true $\rho/\omega\ell\nu$ and any nonresonant $\pi\pi/3\pi$ yield in the mass region, since we have *not* included low-mass nonresonant $\pi\pi/3\pi$ in our $b \to u\ell\nu$ background term.

We then include the $\pi\pi$ mass distribution within the range $\pm$475 MeV of the nominal ρ mass (corresponding to 5 mass bins). The 3π mass distribution is included within the range $^{+150}_{-90}$ MeV of the nominal ω mass (4 bins). Higher masses can then be viewed as an unbiased indicator of the fit quality.

Table 1: Summary of systematic uncertainties in the $\pi\ell\nu$ and $\rho\ell\nu$ modes, and in the ρ/π ratio.

	$\pi\ell\nu$	$\rho\ell\nu$	ρ/π ratio
$D/D^*/D^{**}$	$\pm5.5\%$	$\pm1.7\%$	$\pm5.2\%$
$b \to u\ell\nu$ feeddown	$\pm2.4\%$	$\pm8.2\%$	$\pm10.6\%$
Detector simulation and second B	$\pm18.0\%$	$\pm18.0\%$	$\pm24.2\%$
ρ/ω polarization	$\pm5.0\%$	$\pm14.0\%$	$\pm19.0\%$
$b \to u\ell\nu$ crosstalk	$\pm3.0\%$	$\pm7.0\%$	$\pm5.0\%$
nonresonant $\pi\pi$	-	$+0\%, -20\%$	$+0\%, -20\%$
lepton finding	$\pm2.0\%$	$\pm2.0\%$	$\pm2.0\%$
lepton fakes, cont. BG	$\pm6.0\%$	$\pm6.0\%$	$\pm6.0\%$
luminosity	$\pm2.0\%$	$\pm2.0\%$	$\pm0.0\%$
Total, excluding nonresonant $\pi\pi$	$\pm20.9\%$	$\pm26.1\%$	$\pm33.9\%$

4 Systematic Errors

The changes in fitted yield due to reasonable variations in the mix of $D/D^*/D^{**}$ assumed for $b \to c\ell\nu$ in the $B\bar{B}$ Monte Carlo ($\pm$ 30% variations) are small. The changes due to $\pm50\%$ variations in the amount of high mass $b \to u\ell\nu$ assumed in the Monte Carlo are also small, ±20 events for $B \to \pi\ell\nu$ and ±90 events for $B \to \rho\ell\nu$. Systematic errors from variations in the assumed decay of the second B in the event and from the detector simulation have also been investigated, by carrying out variations in K_L^0 fraction, charm semileptonic decay branching fraction, charged particle and photon finding efficiencies, false charged particle and false photon rejection effieiencies, charged particle momentum resolution, and photon energy resolution. The systematic uncertainties are summarized in Table 1.

5 Results and Conclusions

With the fit that includes $\pi\pi/3\pi$ mass spectra we have a $B^0 \to \pi^-\ell^+\nu$ yield of 708 $\pm$ 183 events, and a $B^0 \to \rho^-\ell^+\nu$ yield of 1202 $\pm$ 189 events. Dividing by 2 (for μ and e), and by 2.64×10^6 (the number of neutral B's), we obtain branching ratios:

$$\mathcal{B}(B^0 \to \pi^-\ell^+\nu) = (1.34 \pm 0.35 \pm 0.28) \times 10^{-4},$$

$$\mathcal{B}(B^0 \to \rho^- \ell^+ \nu) = (2.28 \pm 0.36 \pm 0.59^{+0.00}_{-0.46}) \times 10^{-4},$$

The first errors is statistical, the second is systematic, and the third, where given, is the uncertainty from a nonresonant $\pi\pi$ contribution. The ρ/π ratio is:

$$\frac{\Gamma(B^0 \to \rho^- \ell^+ \nu)}{\Gamma(B^0 \to \pi^- \ell^+ \nu)} = 1.70^{+0.81}_{-0.50} \pm 0.58^{+0.00}_{-0.34}.$$

These results were obtained using the ISGW model for $B \to \pi\ell\nu$ and $B \to \rho/\omega\ell\nu$ decays. If instead we use WSB, we find:

$$\begin{aligned}
\mathcal{B}(B^0 \to \pi^- \ell^+ \nu) &= (1.63 \pm 0.46 \pm 0.34) \times 10^{-4}, \\
\mathcal{B}(B^0 \to \rho^- \ell^+ \nu) &= (3.88 \pm 0.54 \pm 1.01^{+0.00}_{-0.78}) \times 10^{-4}, \\
\frac{\Gamma(B^0 \to \rho^- \ell^+ \nu)}{\Gamma(B^0 \to \pi^- \ell^+ \nu)} &= 2.39^{+1.18}_{-0.69} \pm 0.81^{+0.00}_{-0.48}.
\end{aligned}$$

The ratio of rates found using the ISGW model is inconsistent with the large ratio predicted by that model. The ratio found using the WSB model is consistent with that model. A more careful evaluation of model dependence is needed before any precise statement about V_{ub} can be obtained.

These first measurements of exclusive $b \to u\ell\nu$ branching fractions signal the beginning of a new era for the determination of V_{ub}. Improvements in these measurements, in model calculations and lattice calculation will provide useful information for extracting new values for V_{ub}. Techniques developed in this analysis are also being applied to inclusive $b \to u\ell\nu$ analyses in the hopes of obtaining a more model-independent inclusive determination of V_{ub} as well.

References

1. H. Albrecht *et al.* (ARGUS collaboration), Phys. Lett. **B234**, 409 (1990); *Ibid*, Phys. Lett. **B255**, 297 (1991).
2. R. Fulton *et al.* (CLEO collaboration), Phys. Rev. Lett. **64**, 16 (1990); J. Bartelt *et al.* (CLEO collaboration), Phys. Rev. Lett. **71**, 4111 (1993).
3. Previously CLEO have published upper limits on thes decay modes. A.Bean *et al.* (CLEO collaboration), Phys. Rev. Lett. **70**, 2681 (1993).
4. Y. Kubota *et al.* (CLEO collaboration), Nucl. Instrum. Methods Phys. Res. Sect. A **320**, 66 (1992).
5. N.Isgur, D.Scora, B.Grinstein and M.Wise, Phys. Rev. **D39**, 799 (1989).
6. G. Fox and S. Wolfram, Phys. Rev. Lett. **41**19781581.

Study of One-prong Tau Decays with Neutral Kaons at L3

T. Paul

INFN-Sezione di Perugia, I-06100 Perugia, Italy

Using a sample of 43 500 $Z \to \tau^+\tau^-(\gamma)$ events collected during 1991-93 by the L3 detector at LEP, we have analyzed one-prong τ decays involving neutral kaons. The technique exploits a fine-grained hadron calorimeter. The following branching fractions are determined: $\mathcal{B}(\tau^- \to \pi^-\bar{K}^0\nu_\tau) = 0.0095 \pm 0.0015 \pm 0.0006$, $\mathcal{B}(\tau^- \to \pi^-\pi^0\bar{K}^0\nu_\tau) = 0.0041 \pm 0.0021 \pm 0.0003$, and $\mathcal{B}(\tau^- \to \pi^-\bar{K}^0K^0\nu_\tau) = 0.0031 \pm 0.0012 \pm 0.0004$.

1 Introduction

Measurements of the branching fractions of τ decays with neutral kaons are important both for understanding the dynamics of neutral kaon production and as input to the historically problematic cross check of inclusive against exclusive branching fractions. Although the more glaring inconsistencies, such as the one-prong deficit, have largely disappeared, improved precision now necessitates careful treatment of contributions from modes with neutral kaons [1]. Furthermore, the interplay among factors like Cabibbo suppression for modes with an odd number of kaons, large phase-space suppression for modes with two kaons, and uncertain resonant structures make firm theoretical predictions elusive [2].

Here we report on a measurement of the branching fractions for the modes $\mathcal{B}(\tau^- \to \pi^-\bar{K}^0\nu_\tau)$, $\mathcal{B}(\tau^- \to \pi^-\pi^0\bar{K}^0\nu_\tau)$, and $\mathcal{B}(\tau^- \to \pi^-\bar{K}^0K^0\nu_\tau)$, using a sample of 43 000 τ pairs collected at L3 during 1991-93. Neutral kaon identification is based on the existance of a low energy charged track in the central tracker accompanied by deposition of a large amount of energy in the calorimeters. The L3 detector is suited to this measurement because of its fine-grained hadron calorimeter and relatively small tracking volume, which allows comparable efficiencies for detecting the K^0_L and K^0_S components in the calorimeters.

2 The L3 Detector

The L3 detector consists of a central tracking chamber, high resolution electromagnetic calorimeter, cylindrical shell of scintillators, hadron calorimeter,

[a] Charge conjugate decays are implied throughout.

and muon chamber system. The entire detector is immersed in a uniform 0.5 T magnetic field. A detailed description of L3 is given elsewhere [3].

The hadron calorimeter (HCAL) covers the polar angle range $5^\circ < \theta < 175^\circ$ and consists of uranium absorber plates interleaved with proportional wire chambers oriented both along and perpendicular to the beam direction. The energy resolution of the HCAL for charged pions is $55\%/\sqrt{E(\mathrm{GeV})} + 8\%$, and the angular resolution is better than 40 mrad in both the polar and azimuthal projections for hadronic showers above 6 GeV, which is roughly the minimum energy of neutral kaons produced in τ decays at LEP.

The barrel region of the electromagnetic calorimeter (ECAL) covers the angular range $42^\circ < \theta < 138^\circ$ and is composed of 7680 crystals of bismuth germanate in the shape of truncated pyramids pointing to the interaction region.

The central tracking detector consists of time expansion chamber (TEC) with high spatial resolution perpendicular to the beam direction, and a Z-chamber which supplements the TEC $r-\phi$ measurement with a $z-$coordinate measurement. The transverse momentum resolution of the TEC is $\sigma(1/P_\mathrm{T}) = 0.018(\mathrm{GeV/c})^{-1}$ in the barrel region. The position resolution of a track extrapolated to the ECAL is 0.2 mm in the plane transverse to the beam direction and 1.5 mm along the beam direction.

3 Selection

3.1 Preselection of $\mathrm{Z} \rightarrow \tau^+\tau^-(\gamma)$ Events

A sample of dilepton events is first selected by requiring low multiplicity and back-to-back topology. Each event is then divided into hemispheres by a plane perpendicular to the event thrust axis, and particle identification algorithms are applied separately in each hemisphere. Finally, background from Bhabha and dimuon events, cosmics, and two-photon interactions are suppressed by placing requirements on the side opposite the selected τ decay. The selection procedure is described in detail elsewhere [4].

The efficiency for selecting $\mathrm{Z} \rightarrow \tau^+\tau^-(\gamma)$ events is 57.8%, with geometrical acceptance the main source of inefficiency. The background contamination from all non-τ sources is estimated from data to be 3.7%, and the total number of $\mathrm{Z} \rightarrow \tau^+\tau^-(\gamma)$ events in the sample is estimated to be $43\,497 \pm 60$.

3.2 Selection of τ Decays with Neutral Kaons

Hemispheres with exactly one charged track originating from the e^+e^- impact point are considered for $\tau^- \to K^0X^-\nu_\tau$ selection.[b] The momentum of the track measured by the TEC and the associated energy in the ECAL are used to predict the energy deposition in the HCAL. Any energy exceeding this prediction may be due to a fluctuation, or it may be due to the presence of neutral kaons. In order to reject fluctuations, the quantity $\Delta E_{HC}/\sqrt{P_\pi}$ is required to exceed 5 GeV^{-1}, where ΔE_{HC} is the excess energy observed in the HCAL and P_π is the momentum of the charged track. This cut rejects a substantial fraction of τ decays with no K^0, as shown in Figure 3a. The dominant background at this point comes from $\tau^- \to \rho^-\nu_\tau$ and $\tau^- \to a_1^-\nu_\tau$ decays in which photons from the π^0 leak into the HCAL. This background is suppressed by requiring the energy in the first 18 X_0 of the HCAL to be less than 80% of the total energy deposition (Figure 3b). Further purity is attained by cutting on the angular separation between the π^- or K^- and the center of gravity of the K^0 shower (Figure 3c).

The final selected sample consists of 274 $\tau^- \to K^0X^-\nu_\tau$ candidates with an estimated 27.7 background events. The modes $\tau^- \to \pi^-\bar{K}^0\nu_\tau$, $\tau^- \to K^-\bar{K}^0\nu_\tau$, $\tau^- \to \pi^-\pi^0\bar{K}^0\nu_\tau$, $\tau^- \to K^-\pi^0\bar{K}^0\nu_\tau$, and $\tau^- \to \pi^-\bar{K}^0K^0\nu_\tau$ contribute to the signal with selection efficiencies estimated to be 13.0%, 5.3%, 8.8%, 3.7%, and 19.8% respectively. The sample is composed of 68% K_L^0 and 32% K_S^0.

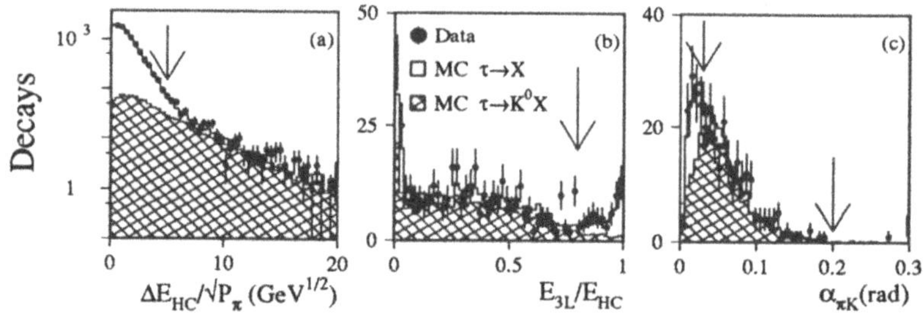

Figure 1: (a) Distribution of excess energy in the HCAL. The hatched area shows the contribution from modes with K^0's. (b) Fraction of energy deposited in the first 3 layers (18 X_0) of the HCAL. (c) Distribution of opening angle between the π^- and K^0 shower.

[b] X^- means one charged hadron and any number of neutral hadrons.

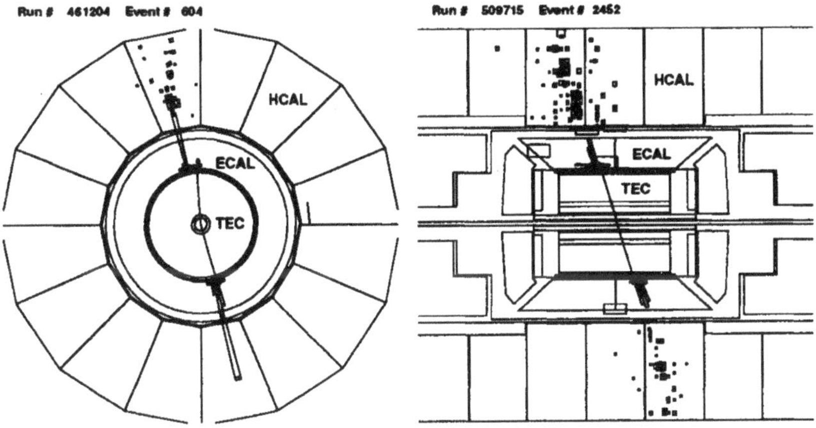

(a) A $\tau^- \to \pi^-\pi^0\bar{K}^0\nu_\tau$ candidate. In the upper hemisphere a 1.3 GeV charged track is separated in space from a 12.0 GeV π^0 detected in the ECAL and a 12.6 GeV neutral hadron shower in the HCAL.

(b) A $\tau^- \to \pi^-\bar{K}^0K^0\nu_\tau$ candidate. In the upper hemisphere, a 2.39 GeV charged track is separated in space from a 9.4 GeV cluster in the ECAL and a 33.6 GeV neutral hadron shower in the HCAL, for wich two distinct local maxima are evident.

Figure 2: Event pictures of selected $\tau^- \to K^0X^-\nu_\tau$ decays.

4 Determination of Branching Fractions

In order to separate the various modes the selected sample is divided into three subsamples, each enriched by one of the decays $\tau^- \to \pi^-\bar{K}^0\nu_\tau$, $\tau^- \to \pi^-\pi^0\bar{K}^0\nu_\tau$, or $\tau^- \to \pi^-\bar{K}^0K^0\nu_\tau$. The $\tau^- \to \pi^-\pi^0\bar{K}^0\nu_\tau$ enriched subsample is produced by requiring an energy cluster in the ECAL of greater than 6 GeV with a transverse shape consistent with a π^0 (Figure 2a). Next, the $\tau^- \to \pi^-\bar{K}^0K^0\nu_\tau$ subsample is produced by requiring the energy deposition in the ECAL and in the front and back parts of the HCAL to each exceed 10% of the total kaon energy. This cut exploits the fact that events with two energetic kaons will generally produce a more uniform energy distribution in the calorimeters than events with a single kaon. An example of such an event is shown in Figure 2b. Remaining events are assigned to the $\tau^- \rightarrow \pi^-\bar{K}^0\nu_\tau$ subsample.

To determine the cross-feed among the various signal and background contributions to the three subsamples, the energy and/or invariant mass spectra of each subsample is fitted with a linear combination of the Monte Carlo distributions for the signal channels and the background. For the $\tau^- \rightarrow \pi^-\pi^0\bar{K}^0\nu_\tau$ sample, a 10-bin histogram of the π^0 energy is used. For the $\tau^- \rightarrow \pi^-\bar{K}^0\nu_\tau$ and $\tau^- \rightarrow \pi^-\bar{K}^0K^0\nu_\tau$ samples, a 5×3 histogram of total decay energy versus invariant mass is fitted (Figure 3). Only the coefficients for the $\tau^- \rightarrow \pi^-\bar{K}^0\nu_\tau$, $\tau^- \rightarrow \pi^-\pi^0\bar{K}^0\nu_\tau$, and $\tau^- \rightarrow \pi^-\bar{K}^0K^0\nu_\tau$ modes are allowed to vary, while the coefficients for $\tau^- \rightarrow K^-\bar{K}^0\nu_\tau$ and $\tau^- \rightarrow K^-\pi^0\bar{K}^0\nu_\tau$ are fixed to recently published values [5].

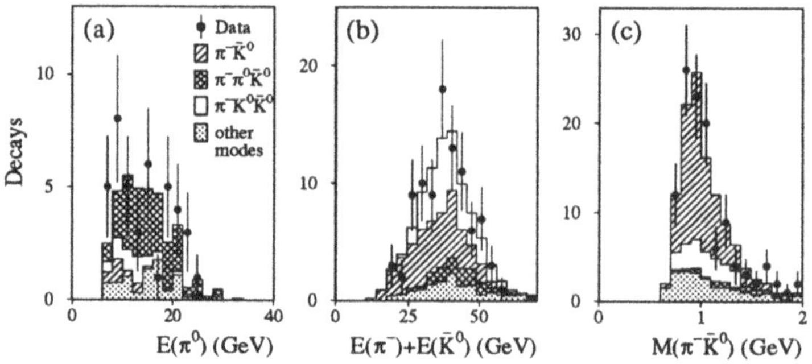

Figure 3: (a) The π^0 energy spectrum for the sample enriched with $\tau^- \rightarrow \pi^-\pi^0\bar{K}^0\nu_\tau$. (b) Energy spectrum for the sample enriched with $\tau^- \rightarrow \pi^-\bar{K}^0K^0\nu_\tau$. (c) Invariant mass spectrum for the sample enriched with $\tau^- \rightarrow \pi^-\bar{K}^0\nu_\tau$.

From the fit, the numbers of decays in the signal channels are found to be: $N(\tau^- \rightarrow \pi^-\bar{K}^0\nu_\tau) = 112.0 \pm 17.7$, $N(\tau^- \rightarrow \pi^-\pi^0\bar{K}^0\nu_\tau) = 33.1 \pm 9.7$, and $N(\tau^- \rightarrow \pi^-\bar{K}^0K^0\nu_\tau) = 55.8 \pm 21.4$.

5 Systematic Errors

The dominant systematic errors arise from uncertainties in the energy scale, resolution, background estimation and theoretical uncertainties. Uncertainty on the energy scale and resolution is determined by studying processes with known energy distributions, and from test beam data. Systematic error from uncertainty in the rates of $\tau^- \to K^-\bar{K}^0\nu_\tau$ and $\tau^- \to K^-\pi^0\bar{K}^0\nu_\tau$ is determined using the uncertainties in these rates given in reference [5]. Uncertainties in background from π^0 leakage into the HCAL are estimated from a sample enriched with π^0's with abnormally high leakage into the first three layers of the HCAL. Errors associated with assumptions made in the Monte Carlo about structure of the weak hadronic current and resonant structures are estimated by varying these assumptions in the Monte Carlo. There is also a systematic error in the $\tau^- \to \pi^-\bar{K}^0K^0\nu_\tau$ channel from uncertainty in the CP phase of the $K^0\bar{K}^0$ system. The error is assigned according to the difference in fit results assuming the $K^0\bar{K}^0$ system consists of only CP-even or only CP-odd components; the central value is computed assuming $K^0_LK^0_L : K^0_SK^0_S : K^0_LK^0_S = 0.25 : 0.25 : 0.5$. A breakdown of contributions to the systematic error is given in Table 1.

Source		$\Delta\mathcal{N}_{\pi^-\bar{K}^0}$	$\Delta\mathcal{N}_{\pi^-\pi^0\bar{K}^0}$	$\Delta\mathcal{N}_{\pi^-K^0\bar{K}^0}$
	HCAL	3.5	0.9	4.3
Calibration	ECAL	1.0	1.6	2.5
	TEC	1.0	0.8	0.6
	π^0	2.0	0.8	0.6
Background	$\pi^\pm/K^\pm$	1.5	0.5	0.6
	$\Delta\mathcal{N}_{K^-K^0}$	3.4	0.2	2.4
	$\Delta\mathcal{N}_{K^-\pi^0K^0}$	0.2	0.8	0.9
Theory	matrix element	3.9	1.2	4.0
	$CP(K^0\bar{K}^0)$	-	-	2.7
Total	$\Delta\mathcal{N}^{syst}$	6.9	2.6	7.4

Table 1: Systematic uncertainties in the number of fitted decays.

6 Summary

Using the numbers of observed decays for each channel as determined by the fit, the detection efficiencies, and the total number of observed τ decays, the following branching fractions are determined:

$$\begin{aligned}
\mathcal{B}(\tau^- \to \pi^-\bar{K}^0\nu_\tau) &= 0.0095 \pm 0.0015 \pm 0.0006 \\
\mathcal{B}(\tau^- \to \pi^-\pi^0\bar{K}^0\nu_\tau) &= 0.0041 \pm 0.0012 \pm 0.0003 \\
\mathcal{B}(\tau^- \to \pi^-\bar{K}^0K^0\nu_\tau) &= 0.0031 \pm 0.0012 \pm 0.0004
\end{aligned}$$

where the first error is statistical and the second is a combination of all systematics.

The invariant mass spectrum of the $\pi^- K^0$ system in the $\tau^- \to \pi^- \bar{K}^0 \nu_\tau$ sample is consistent with the model that this final state comes from the decay $\tau^- \to K^{*-} \nu_\tau \to \pi^- \bar{K}^0 \nu_\tau$. Assuming this is the only contribution to the final state, the following branching fraction is determined:

$$\mathcal{B}(\tau^- \to K^{*-} \nu_\tau) = 0.0142 \pm 0.0022 \pm 0.009$$

These results are consistent with other recent measurements [6 7]. The measurement of the branching fraction in the $\tau^- \to \pi^- \bar{K}^0 K^0 \nu_\tau$ mode is uniquely insensitive to assumptions about the CP phase of the $K^0 \bar{K}^0$ system, and is consistent with measurements of the related decay [7] $\tau^- \to \pi^- K^- K^+ \nu_\tau$.

Acknowledgments

It is a pleasure to thank Andrei Kunin, James Gerald, and Doris Kim, and of course the organizers who arranged for a very comfortable stay in Valencia.

References

1. see for eg. Particle Data Group, L. Montanet *et. al.*, Review of Particle Properties, *Phys. Ref.* **D50**
2. R. Decker *et al.*, *Phys. Rev.* **D50** (1994) 6863.;
   E. Braaten *et al.*, *Int. J. Phys.* **A5** (1990) 2737.
3. L3 Collaboation, B. Adeva *et al.*, *Nucl. Instr. and Meth.* **A289** (1990) 35;
   O. Adriani *et al.*, *Nucl. Instr. and Meth.* **A302** (1991) 53;
   K. Deiters *et al.*, *Nucl. instr. and Meth.* **A323** (1992) 162;
   J.Bakken, *Nucl. Instr. and Meth.* **A275** (1989) 81.
4. L3 Collaboration, O.Adriana *et al.*, *Phys. Lett.* **B294** (1992) 466;
   L3 Collaboration, O.Acciarri *et al.*, *Phys. Lett.* **B341** (1994) 245.
5. ALEPH Collab., D. Buskulic *et al.*, *Phys. Lett.* **B332** (1994) 219.
6. ALEPH Collab., D. Buskulic *et al.*, *Phys. Lett.* **B332** (1994) 219.;
   OPAL Collab., R. Akers *et al.*, *Phys. Lett.* **B339** (1994) 278.
7. DELCO Collab., G.B. Mills *et al.*, *Phys. Rev. Lett.* **54** (1985) 624;
   TPC/2γ Collab., D.A. Bauer *et al.*, *Phys. Rev.* **D50** (1994) 13.

PROBING COMPOSITENESS IN $e\gamma$ COLLISIONS

S. F. NOVAES[a]

Instituto de Física Corpuscular - IFIC/C.S.I.C.
Dept. de Física Teórica, Universidade de València
46100 Burjassot, València, Spain

O. J. P. ÉBOLI

Instituto de Física, Universidade de São Paulo,
C.P. 66318, 05389-970 São Paulo, Brazil

E. M. GREGORES, J. C. MONTERO, D. SPEHLER

Instituto de Física Teórica, Universidade Estadual Paulista
Rua Pamplona 145, 01405-900 São Paulo, Brazil

We investigate the potentiality of the next generation of linear electron–positron colliders to probe the spin and couplings of excited fermions predicted by composite models. We stress that these colliders, operating in the $e\gamma$ mode with polarized beams, is able to further increase the reach of these machines.

The standard model of electroweak interactions (SM) is not able to give a satisfactory explanation to family repetition and the complex pattern of the fermion masses. We expect that our understanding of these problems can have a considerable improvement if we consider an underlying fermionic substructure where the usual leptons and quarks share some common constituents (preons) [1]. In this sense, the SM would be just the low energy limit of a more fundamental theory, characterized by a large mass scale Λ.

Unfortunately, we do not have yet a satisfactory model that could reproduce the whole family spectrum. This lack of a predictive model requires that we make a phenomenological analysis of the effects of fermion compositeness. One of the most undeniable predictions of the composite models is the existence of excited states of the usual fermions, with different spins. On a phenomenological ground, we can employ a model independent treatment, in order to describe the couplings of these excited states through the use of the effective Lagrangian approach. We assume a magnetic moment type coupling

[a]Permanent Address: Instituto de Física Teórica, Universidade Estadual Paulista.

among the excited spin-$\frac{1}{2}$ fermion ($\Psi^*_{1/2}$), the ground state fermion (ψ), and the photon [2,3],

$$\mathcal{L}^{1/2}_{\text{eff}} = \frac{e}{2\Lambda}\bar{\Psi}^*_{1/2}\sigma^{\mu\nu}(A + B\gamma_5)\psi \; F_{\mu\nu} + \text{h.c.} \; , \tag{1}$$

where Λ is the compositeness scale and $F_{\mu\nu}$ is the electromagnetic field strength tensor.

Composite models require, in a natural way, the existence of higher-spin states. We also investigate the possibility of a spin-$\frac{3}{2}$ excited state ($\Psi^{*\mu}_{3/2}$) which couples to usual fermions and photon through the $U(1)$em gauge-invariant effective Lagrangian [4,5]

$$\mathcal{L}^{3/2}_{\text{eff}} = \frac{e}{\Lambda}\bar{\Psi}^{*\mu}_{3/2}\gamma^{\nu}(C + D\gamma_5)\psi \; F_{\mu\nu} + \text{h.c.} \; , \tag{2}$$

where $\sigma_{\mu\nu} = \frac{i}{2}[\gamma_\mu, \gamma_\nu]$. The invariance under CP requires that A, B, C, and D are real. Furthermore, we will assume a pure right-handed (RH) or left-handed (LH) structure for these couplings in order to comply with the strong bounds coming from the measurement of the anomalous magnetic moment of leptons [2,6]. The effective Lagrangians (1) and (2) can be embodied in a $SU(2)_L \times U(1)_Y$ gauge invariant model [7], however, since we intend to investigate the influence of these particle in $e\gamma$ collisions, we ignore the possible coupling of the excited states with the weak gauge bosons and assume that the branching ratio of the excited states into the usual lepton and a photon is of the order 1.

There have been several phenomenological studies that analysed the possible signatures of these excited fermions in pp [5,8], e^+e^- [5,7,9,10,11], and ep [7,10] collisions at high energies. On the experimental side, several searches for these particles have been carried out, including the ones at LEP [12] and at HERA [13]. However, all the direct searches for compositeness have failed up to this moment, and we expect that the next generation of accelerators, working at higher center-of-mass energies, will be able to unravel the existence of these composite states.

A particularly interesting machine for probing the substructure of the electron is the Next Linear e^+e^- Collider (NLC) that should operate with a center-of-mass energy of at least 500 GeV and an integrated luminosity around 10 fb^{-1} [14]. At the NLC, it will be possible to convert electrons into photons by focusing a laser on the incoming electron beam. By Compton scattering, high energy photons are produced along the electron direction carrying away a large amount of the beam energy [15,16]. The laser backscattering mechanism will allow the NLC to operate in three different modes, e^+e^-, $e^-\gamma$, and $\gamma\gamma$, opening up an opportunity for a deeper search for compositeness.

We investigate here the possibility of the Next Linear Collider, to search for spin-$\frac{1}{2}$ and spin-$\frac{3}{2}$ excited leptons as resonances in the reaction $e^-\gamma \to e^-, L^-_{1/2(3/2)} \to e^-\gamma$. We point out the best strategy to disentangle the spin and the coupling chiral structure of the fermionic excitations, exploring the possibility of polarizing both the electron and laser beams. We establish discovery limits in the $M_{1/2(3/2)} \times \Lambda$ plane which demonstrate that an $e\gamma$ collider can probe up to $\Lambda \sim 200$ TeV for excited electron masses of the order of 400 GeV.

The cross section for the reaction $e^-\gamma \to X$ in an electron–positron linear collider, operating in the $e\gamma$ mode through the laser backscattering mechanism, is given by

$$d\sigma_{P_e\xi_2}(e^-\gamma \to X) = \int_{x_{\rm min}}^{x_{\rm max}} dx\; F(x,\zeta;P_p,P_l)\; d\hat{\sigma}_{P_e\xi_2}(e\gamma \to X)\,, \qquad (3)$$

where $F(x,\zeta;P_p,P_l)$ is the backscattered photon distribution function for polarized beams [16] which depends on the laser (P_l) and positron P_p polarization. $d\hat{\sigma}_{P_e\xi_2}$ is the polarized elementary cross section for the subprocess $e^-\gamma \to X$ written in terms of the longitudinal polarization of the electron beam (P_e) and the mean backscattered photon helicity (ξ_2). The results for these cross sections can be found elsewhere [17].

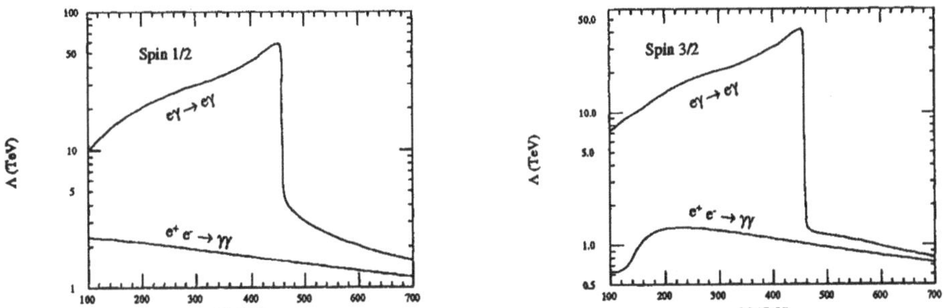

Figure 1: Comparison of the discovery regions for e^+e^- and $e\gamma$ modes with unpolarized beams, at $\sqrt{s} = 500$ GeV.

In order to quantify the potential of an $e\gamma$ collider to search for excited states, we defined the statistical significance of the signal ($\mathcal{S}$) by

$$\mathcal{S} \equiv \frac{|\sigma_{\rm exc} - \sigma_{\rm QED}|}{\sqrt{\sigma_{\rm QED}}}\sqrt{\mathcal{L}}\,, \qquad (4)$$

where $\sigma_{\rm exc}$ ($\sigma_{\rm QED}$) is the cross section associated to the excited lepton (QED) contributions and $\mathcal{L}$ is the integrated luminosity of the machine.

For the sake of comparison, we present in Fig. 1 the discovery contour in the $M_{1/2(3/2)} \times \Lambda$ plane for excited fermions of spin $\frac{1}{2}$ and $\frac{3}{2}$ at the NLC operating in the e^+e^- and $e\gamma$ modes. We required a 3σ effect in the total cross section and made a cut of $5°$ on the polar angle. From this figure, it is quite clear that $e\gamma$ reactions is the most suitable way to probe compositeness at the NLC, even for masses above the kinematical reach of the machine.

The reach of the machine can be further improved with the use of polarized beams, and integrating over a bin of 5 GeV around the excited lepton mass. This can be seen from Fig. 2 where we show the discovery contour for different polarization configurations of the initial particles. We can learn from this figure that electron–photon collisions are able to extend considerably the limits available from the current experiments and that polarization plays an important role to enlarge the accessible region. For an excited spin–$\frac{1}{2}$ lepton with mass around 400 GeV we can probe up to $\Lambda \sim 200$ TeV with this kind of machine.

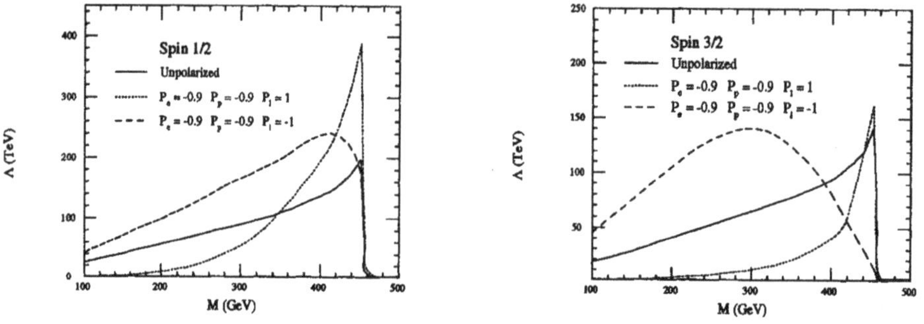

Figure 2: Discovery contour for LH excited states for $\sqrt{s} = 500$ GeV and $\mathcal{L} = 10$ fb^{-1} for different polarizations, with an angular cut of $5° < \theta < 175°$.

Once the existence of excited electrons is established, it is important to obtain some information about its spin and couplings. We can see that now the use of polarized beams becomes an essential tool. In order to discriminate between LH and RH excited electron coupling, we can analyse the cross section (3), using just polarized electron beam. In Fig. 3(a) we exhibit the deviation

$$\delta_e = \frac{\sigma_{e0} - \sigma_{00}}{\sigma_{00}} \tag{5}$$

of the polarized cross section (σ_{e0}) with respect to the unpolarized one (σ_{00}), as function of the excited state mass. We have assumed a degree of longitudinal polarization of 90% and required a 3σ effect in a bin of 5 GeV around the resonance. It becomes evident that the measurement of δ_e is able to clearly discriminate RH from LH couplings. However, no information about the the spin of the excited fermions can be extracted from this measurement. This can be accomplished by studying this same process, now using just polarized photon beams.

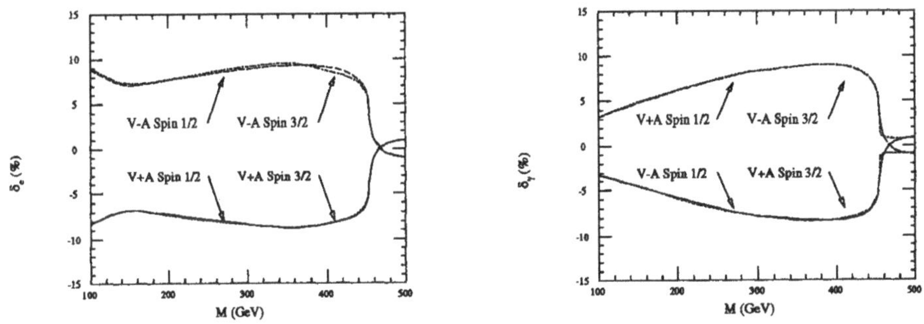

Figure 3: (a) Deviation δ_e [Eq. (5)]; (b) Deviation δ_γ [Eq. (6)].

We show in Fig. 3(b) the deviation

$$\delta_\gamma = \frac{\sigma_{0\gamma} - \sigma_{00}}{\sigma_{00}} , \tag{6}$$

where σ_{00} is the unpolarized cross section and $\sigma_{0\gamma}$ is the one with polarized photons. The polarized backscattered photons is obtained assuming that the positron beam has a 90% degree of polarization and that the laser beam is not polarized, which enables that the polarization of the backscattered photon (ξ_2) has the same sign of the parent positron polarization in the whole spectrum. We have also required that the total number of events of the polarized case differs by 3σ from the unpolarized one.

We can witness that RH spin-$\frac{1}{2}$ and LH spin-$\frac{3}{2}$ fermions furnish $\delta_\gamma > 0$, whereas LH spin-$\frac{1}{2}$ and RH spin-$\frac{3}{2}$ fermions yields $\delta_\gamma < 0$. Therefore, once the chirality of the fermion coupling is established through the measurement of δ_e, the spin of this state can be easily obtained through the analyses of δ_γ.

In conclusion, we have shown that the $e\gamma$ mode of the NLC is the most promising way to probe compositeness at these colliders. Furthermore, the use of polarized beams in $e\gamma$ collisions is able to determine whether the spin of the excited state is $\frac{1}{2}$ or $\frac{3}{2}$ and also to discriminate the chiral structure of its coupling to photons and usual fermions.

Acknowledgments

One of us (S.F.N.) would like to thank the Instituto de Física Corpuscular - IFIC/C.S.I.C. of Universidade de València for its kind hospitality. This work was partially supported by Generalitat Valenciana (Spain), Conselho Nacional de Desenvolvimento Científico e Tecnológico (Brazil) and by Fundação de Amparo à Pesquisa do Estado de São Paulo (Brazil).

References

1. See, for instance: H. Terazawa, M. Yasue, K. Akama and M. Hayashi, Phys. Lett. **B112**, 387, 1982; H. Harari, Phys. Reports, **104**, 159 (1984); W. Buchmüller, Acta Phys. Austriaca, Suppl. XXVII, 517 (1985); M. E. Peskin, Proceedings of the 1985 International Symposium on Lepton and Photon Interactions at High Energies, Kyoto, (Kyoto Univ., Kyoto, 1985), edited by M. Konuma and K. Takahashi, p. 714.
2. F. M. Renard, Phys. Lett. **B116**, 264 (1982).
3. F. E. Low, Phys. Rev. Lett. **14**, 238 (1965); N. Cabibbo, L. Maiani, and Y. Srivastava, Phys. Lett. **B139**, 459 (1984).
4. W. Rarita and J. Schwinger, Phys. Rev. **60**, 61 (1941); Ph. Salin, Nuovo Cimento **32**, 521 (1964); J. Leite Lopes, J. A. M. Simões and D. Spehler, Phys. Lett. **B94**, 367 (1980); Phys. Rev. **D23**, 797 (1981); Phys. Rev. **D25**, 1854 (1982); D. Spehler, O. J. P. Éboli, G. C. Marques, A. A. Natale and S. F. Novaes, Phys. Rev. **D36**, 1358 (1987).
5. J. Kühn and P. Zerwas, Phys. Lett. **B147**, 189 (1984).
6. F. M. Renard, Phys. Lett. **B139**, 449 (1984); M. Suzuki, Phys. Lett. **B143**, 237 (1984).
7. K. Hagiwara, S. Komamiya, and D. Zeppenfeld, Z. Phys. **C29**, 115 (1985).
8. K. Enqvist and J. Maalampi, Phys. Lett. **B135**, 329 (1984).
9. N. Cabibbo, L. Maiani, and Y. Srivastava, Phys. Lett. **B139**, 459 (1984); F. A. Berends and P. H. Daverveldt, Nucl. Phys. **B272**, 131 (1986); A. Feldmaier, H. Salecker, and F. C. Simm, Phys. Lett. **B223**, 234 (1989); M. Martinez, R. Miquel, and C. Mana, Z. Phys. **C46**, 637 (1990); F. Boudjema and A. Djouadi, Phys. Lett. **B240**, 485 (1990); Maria Bardadin–Otwinowska, Z. Phys. **C55**, 163 (1992); J. C. Montero and V. Pleitez, Phys. Lett. **B321**, 267 (1994).
10. F. Boudjema, A. Djouadi, and J. L. Kneur, Z. Phys. **C57**, 425 (1993).
11. I. F. Ginzburg and D. Yu. Ivanov, Phys. Lett. **B276**, 214 (1992); T. Kon, I. Ito and Y. Chikashige, Phys. Lett. **B287**, 277 (1992); E. Boos,

A. Pukhov, and A. Beliaev, Phys. Lett. **B296**, 452 (1992).

12. ALEPH Collaboration, D. Decamp *et al.*, Phys. Rep. **216**, 253 (1992); L3 Collaboration, O. Adriani *et al.*, Phys. Rep. **236**, 1 (1993);
13. H1 Collaboration, I. Abt *et al.*, Nucl. Phys. bf B396, 3 (1993); ZEUS Collaboration, M. Derrick *et al.*, Phys. Lett. **B316**, 207 (1993).
14. R. B. Palmer, Ann. Rev. Nucl. Part. Sci. **40**, 529 (1990).
15. F. R. Arutyunian and V. A. Tumanian, Phys. Lett. **4**, 176 (1963); R. H. Milburn, Phys. Rev. Lett. **10**, 75 (1963); I. F. Ginzburg, G. L. Kotkin, V. G. Serbo, and V. I. Telnov, Nucl. Instrum. Methods **205**, 47 (1983); V. I. Telnov, Nucl. Instrum. Methods **A294**, 72 (1990).
16. I. F. Ginzburg, G. L. Kotkin, V. G. Serbo, and V. I. Telnov, Nucl. Instrum. Methods **219**, 5 (1984).
17. O. J. P. Éboli, E. M. Gregores, J. C. Montero, S. F. Novaes, and D. Spehler, *Excited Leptonic States in Polarized $e^-\gamma$ and e^+e^- Collisions*, Preprint IFT–P.018/95 and IFUSP–P 1145 (1995).

TRILINEAR VECTOR-BOSON COUPLINGS PRESENT AND FUTURE

CH. PETRIDOU

Division of Nuclear and Particle Physics,
Aristotle University of Thessaloniki,
Thessaloniki 54006, Greece
and INFN Trieste Padriciano 99,
Trieste 34100, Italy

The current status of trilinear vector boson couplings from hadron colliders is reviewed. The prospects in the near and far future on the accuracy in the measurement in both hadron and e^+e^- colliders is discussed. The size of possible three-boson anomalous coupling is compared to the experimental sensitivity

1 Introduction

From the present precision measurements at LEP and SLC the gauge boson couplings to fermions are known with impressive accuracy [1]. Theory and experiment agree to $0.1 - 1\%$ and the results confirm the gauge nature of the W and Z particles. However, the non-abelian self couplings of the gauge bosons, which is also the most striking consequence of the $SU(2)_L \times U(1)_Y$ gauge symmetry, remains poorly measured to date. If the self interaction of the gauge bosons are not constrained by $SU(2) \times U(1)$ this would imply that the W and Z are composite particles and therefore we expect anomalous couplings in the $WW\gamma$ and WWZ vertices. Indirectly the electroweak precision data imply restrictions on non-standard boson self-couplings. Nevertheless, direct and precise measurement of these couplings will either confirm our theory of electroweak interactions or, in case of deviations, will provide indications for new physics. A first direct measurement of the $WW\gamma$ coupling was performed by UA2 at the $Sp\bar{p}S$ collider at CERN [2] from the reaction $p\bar{p} \rightarrow e\nu\gamma + X$. Current measurements at the TeVatron by CDF and D0 are using the pair production processes: $p\bar{p} \rightarrow W^+W^-$, WZ, $W\gamma$, $Z\gamma$. The same channels will be also studied at LHC. Similarly at LEP200 the channels: $e^+e^- \rightarrow W^+W^-, Z\gamma$ can be accessed. (The $e^+e^- \rightarrow Z\gamma \rightarrow \nu\bar{\nu}\gamma$ has been looked at by L3 [3]).
In order to study the effects of the anomalous couplings for the WWV ver-

tex, where V is Z or γ, the most general effective Lagrangian is used, which respects $U(1)$ and Lorentz invariance. After imposing C and P conservation becomes [4 5]:

$$\begin{aligned} L_{eff} &= -ie(A_\mu + g_{ZWW} Z_\mu)(W^{-\mu\nu} W^+_\mu - W^{+\mu\nu} W^-_\nu) \\ & -ie(\kappa_\gamma F^{\mu\nu} + \kappa_Z g_{ZWW} Z_{\mu\nu}) W^{+\mu} W^{-\nu} \\ & + \frac{ie}{M_W^2}(\lambda_\gamma F^{\nu\lambda} + \lambda_Z cot\theta_W Z^{\nu\lambda}) W^-_{\lambda\nu} W^{+\mu}_\nu, \end{aligned} \quad (1)$$

where the field strength tensors $F_{\mu\nu}$, $Z_{\mu\nu}$ and $W^\pm_{\mu\nu}$ are defined from the corresponding vector fields V_μ (A_μ, Z_μ, $W^\pm_\mu$ respectively) as $(\vartheta_\mu V_\nu - \vartheta_\nu V_\mu)$. The above effective Lagrangian has five free parameters: g_{ZWW} which in the Standard Model (SM) has the value of $cot\theta_W$, (θ_W is the Weinberg angle), κ_γ, κ_Z which in the SM are one and λ_γ and λ_Z with SM values zero. The terms in the $WW\gamma$ effective Lagrangian with C and P invariance, correspond to the lowest order terms in a multipole expansion of the W-photon interactions; the charge Q_W, the magnetic dipole moment μ_W and the electric quadrupole moment [7] q_W of the W^+ , which are:

$$\begin{aligned} Q_W &= e \\ \mu_W &= \frac{e}{2M_W}(1 + \kappa_\gamma + \lambda_\gamma), \\ q_W &= -\frac{e}{M_W^2}(\kappa_\gamma - \lambda_\gamma). \end{aligned} \quad (2)$$

In the following the conventions of [6] are used to express deviations from the $SU(2) \times U(1)$ standard values:

$$\begin{aligned} \delta g_Z &= g_{ZWW} - cot\theta_W \\ \chi_\gamma &= \kappa_\gamma - 1 \\ \chi_Z &= (\kappa_Z - 1)(cot\theta_W + \delta g_Z) \\ y_\gamma &= \lambda_\gamma \\ y_Z &= \lambda_Z cot\theta_W \end{aligned} \quad (3)$$

The above parameters are all zero in the SM. On the Lagrangian, equation 1, it is imposed that the terms containing deviations from SM should respect the $SU(2) \times U(1)$ symmetry. We rewrite then the effective Lagrangian as:

$$L = L_0 + \Sigma L_i, \quad (4)$$

where L_0 is the piece of the Lagrangian already tested at the hadron and e^+e^- colliders ($Sp\bar{p}S$, TeVatron, SLC, LEP100). The pieces L_i have all possible

Table 1: The free and constrained parameters in the five commonly used models together with the Lagrangian components that contribute to each as in [6].

Model	Free Parameters	Parameters constrained to SM values	Other parameter relations	Physical content
1	δg_Z	$\kappa_\gamma, \kappa_Z, \lambda_\gamma, \lambda_Z$		$L_{W\Phi} - L_{B\Phi}$
2	κ_γ	$\lambda_\gamma, \lambda_Z$	$\chi_Z = -\chi_\gamma \tan\theta_W$	$L_{W\Phi}$
3	λ_γ	$\delta g_Z, \kappa_\gamma, \kappa_Z$	$\lambda_Z = \lambda_\gamma$	L_W
4	$\kappa_\gamma, \delta g_Z$	$\lambda_\gamma, \lambda_Z$	$\chi_Z = -\chi_\gamma \tan\theta_W$	$L_{W\Phi}$, $L_{B\Phi}$
5	κ_γ, κ_Z		$\chi_Z = -\chi_\gamma \tan\theta_W$ $\delta g_Z = \frac{\chi_\gamma}{\sin\theta_W \cos\theta_W}$ $\lambda_Z = \lambda_\gamma$	$L_{W\Phi}, L_W$

deviations from SM and are mainly operators of dimmension-6 which respect $SU(2) \times U(1)$ symmetry [8]. In equation 4 the additional terms which satisfy the gauge symmetry and can give rise to non-standard triple gauge boson couplings are essentially three: $L_{W\Phi}$, $L_{B\Phi}$, L_ϕ, defined in [6]. Their presence in the Lagrangian give rise to deviations from SM of order $O(\frac{M_W^2}{\Lambda^2})$ where Λ is the scale at which new physics may appear [8]. Analogous to the WWV vertex it is possible to parametrize anomalous $Z\gamma V$ couplings. These originate from dimension-8 operators or higher [9] and are thus expected to be of order $O(\frac{M_Z^4}{\Lambda^4})$. At present five models are commonly used as shown in table 1.

2 Measuring $WW\gamma$ and WWZ at the Hadron Colliders

Following the measurement of UA2 [2] significant improvement has been made recently in the accuracy of the measurement of anomalous couplings at the TeVatron in both CDF and D0 experiments. For a complete report see [10]. At hadron colliders the best channel to probe the WWV and $Z\gamma V$ vertices is di-boson production. From the Feynman diagrams contributing to di-boson production the $pp, p\bar{p} \to W^+W^-$ is sensitive to both $WW\gamma, WWZ$ couplings, while in $pp, p\bar{p} \to W^\pm\gamma(W^\pm Z)$ only the $WW\gamma(WWZ)$ is tested. Anomalous couplings contribute only via the s-channel of W, Z or γ exchange. Therefore the transverse momentum distribution of the corresponding vector boson is expected to be particularly sensitive to anomalous couplings. The $Z\gamma V$ couplings are probed in $pp, p\bar{p} \to Z\gamma$. In order to reduce QCD backgrounds at

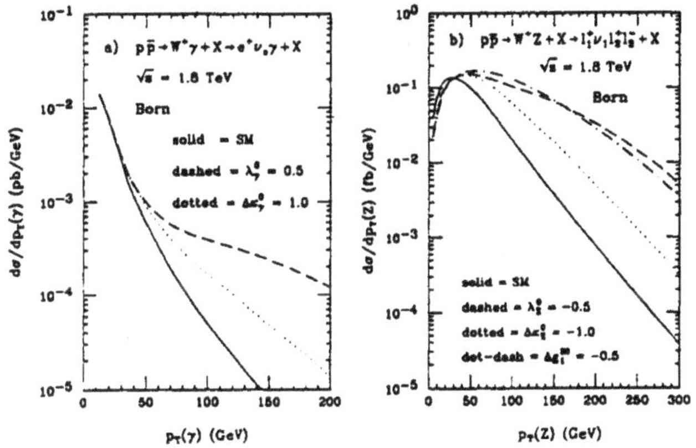

Figure 1: The differential cross-section for the transverse momentum of a)the photon in $p\bar{p}\to W\gamma$ and b)the Z boson in $p\bar{p}\to WZ$ for various anomalous WWV couplings.

least one of the W and/or Z bosons are required to decay leptonically. Typically the channels $pp, p\bar{p}\to W^{\pm}\gamma, W^{\pm}Z$ are used while the $pp, p\bar{p}\to W^{+}W^{-}$ is more difficult due to backgrounds from $t\bar{t}$ production.

2.1 Present: TeVatron

Recently CDF [11] and D0 [12] at Fermilab have improved the direct limits for $WW\gamma$ in the channel $p\bar{p}\to e\nu\gamma+X$.
Both experiments have also looked for anomalous couplings in: $p\bar{p}\to WZ$ [13], $p\bar{p}\to Z\gamma$ [15 14], $p\bar{p}\to WW$ [13 16]. Figure 1 shows the effect of anomalous coupling (only one parameter was changed at a time) on the p_T distribution of the produced γ or Z with appropriate selection criteria for the two channels: $W\gamma$ and WZ. The $p\bar{p}\to W\gamma$ and $Z\gamma$ channels are searched in the inclusive e/μ channels of the W and Z samples. Look for a high p_T lepton and missing E_T (in the case of $W\gamma$) or two high p_T leptons (in the case of $Z\gamma$ with the di-lepton mass consistent with a Z and an isolated high p_T photon. Main source of backgrounds is $W/Z+jets$ and QCD background. In particular, it has to be determined how often a jet can fake an isolated high p_T photon. By increasing the cut on the p_T of the photon this background can be substantially reduced. In order to select the $p\bar{p}\to WZ$, WW events from the data it is required one W/Z to decay leptonically and the other in two jets. Search for high p_T lepton and missing energy (if WW) or equivalently for two high p_T leptons with mass near the Z mass (for the WZ channel). The second boson should

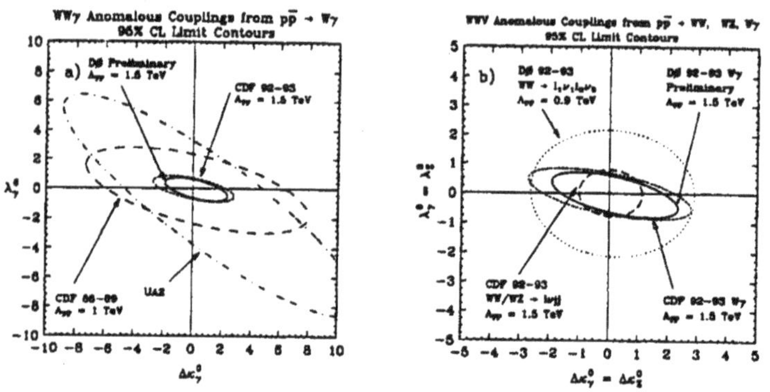

Figure 2: Summary of the present limits on anomalous WWV couplings. $(\lambda_\gamma, \chi_\gamma = \Delta\kappa^0_\gamma = \chi_Z = \Delta\kappa^0_Z)$

decay hadronically, therefore the existence of two high p_T jets in the boson's mass region is required. Finally the WW channel it is also searched in the leptonic channel, that is: missing energy and two high p_T leptons. The main source of background in the WZ, WW case is again $W/Z + jets$. By requiring the system of two high p_T jets to have a mass consistent with W/Z and lasrge p_T most of the background is eliminated.
The experimental limits on the $W\gamma$ and $Z\gamma$ were obtained via binned maximum likelihood fits in the p_T distributions of the photon, while the limits obtained from the WW and WZ samples were either via direct comparison of observed events to the expected signal after cuts (CDF), or by comparing the experimentally measured cross-sections to the predicted ones for various values of anomalous couplings (D0). The current limits on λ_γ and χ_γ for CDF and D0 are shown in figure 2. The limits are based on the 1992 and 1993 data. On the same figure the UA2 limits and the CDF results from the 1988-1989 runs are also shown. Table 2 summarizes the current best limits from CDF and D0.

2.2 Future: TeVatron, LHC

In table 3 the limits expected from the TeVatron for a total integrated luminosity of $1fb^{-1}$ and $10fb^{-1}$ are shown. At LHC for the channels $WW\gamma$, WWZ the high p_T spectra of the γ and Z respectively will be studied [17]. On the WW and WZ channels only the leptonic decays can be used while the semihadronic suffer from $t\bar{t}$ background. Anomalous couplings will manifest themselves as hight p_T tails in the p_T^γ and p_T^Z distributions. With $10^5 pb^{-1}$ an

Table 2: 95% CL limits from CDF and D0 on anomalous WWV ($V = \gamma,\ Z$) couplings.

Experiment	Channel	Limit
CDF	$p\bar{p} \to W^{\pm}\gamma \to l^{\pm}\nu\gamma$ $l = e, \mu$	$-2.3 < \chi_\gamma < 2.2$ $-0.7 < \lambda_\gamma < 0.7$
D0	$p\bar{p} \to W^{\pm}\gamma \to l^{\pm}\nu\gamma$ $l = e, \mu$	$-1.6 < \chi_\gamma < 1.8$ $-0.6 < \lambda_\gamma < 0.6$
CDF	$p\bar{p} \to W^{\pm}Z \to l^{+}l^{-}jj$ $l = e, \mu$	$-8.6 < \chi_Z < 9.0$ $-1.7 < \lambda_Z < 1.7$
CDF	$p\bar{p} \to W^{+}W^{-}, W^{\pm}Z \to l^{\pm}\nu jj$ $l = e, \mu$	$-1. < \chi_Z < 1.1$ $-0.8 < \lambda_Z < 0.8$
D0	$p\bar{p} \to W^{+}W^{-} \to l_1\nu_1 l_2\nu_2$ $l = e, \mu$	$-2.6 < \chi_Z < 2.8$ $-2.28 < \lambda_Z < 2.2$

Table 3: Expected 95% CL limits on anomalous WWV couplings from the future runs at Fermilab.

Channel	$1fb^{-1}$	$10fb^{-1}$
$p\bar{p} \to W^{\pm}\gamma \to e^{\pm}\nu\gamma$	$-0.38 < \chi_\gamma < 0.38$ $-0.12 < \lambda_\gamma < 0.12$	$-0.21 < \chi_\gamma < 0.21$ $-0.057 < \lambda_\gamma < 0.057$
$p\bar{p} \to W^{+}W^{-}$, $W^{\pm}Z \to l^{\pm}\nu jj, l^{+}l^{-}jj$	$-0.31 < \chi_\gamma < 0.41$ $-0.19 < \lambda_\gamma < 0.19$	$-0.17 < \chi_\gamma < 0.24$ $-0.1 < \lambda_\gamma < 0.11$
$p\bar{p} \to W^{\pm}Z \to l^{\pm}\nu l_1^{+}l_2^{-}$	$-0.26 < \chi_\gamma < 0.70$ $-0.24 < \lambda_\gamma < 0.32$	$-0.09 < \chi_\gamma < 0.32$ $-0.1 < \lambda_\gamma < 0.13$

accuracy of about 8% on χ_γ and 0.5% on λ_γ is expected, that is about a factor of 4 (10) in χ_γ (λ_γ) from the TeVatron at $10fb^{-1}$.

3 Measuring anomalous couplings at the e^+e^- Colliders.

W-pair production at LEP200 ($\sqrt{s} = 176 - 190GeV$) provides the possibility to study the WWV vertex. The reaction: $e^+e^- \to WW \to l\nu jj$, contrary to the $p\bar{p} \to WW \to l\nu jj$, at hadron colliders does not suffer from large backgrounds. Moreover the reconstruction of the leptonically decaying W's is possible since the center of mass energy of the interaction is precisely known, and the reconstruction of the W-mass from the two jets can be done with good accuracy. At hadron colliders limits on anomalous couplings can be derived either from transverse momentum distributions of one of the vector bosons, or simply from comparison of the measured production cross section of the process to the predicted one for various values of the anomalous couplings, as discussed in subsection 2.1. At LEP200 on the other hand, the angular distribution of the W and its decay products can be studied. Anomalous couplings contribute to the helicity amplitudes and therefore affect the angular distributions of the W and its decay products [6 4]. The measurement of the anomalous couplings is made by using the five characteristic angles of each event. These are: the W production angle θ and the decay angles (θ_1, ϕ_1) and (θ_2, ϕ_2) of the W^- and W^+ respectively. The production angle is defined as the one between the direction of the incoming e^- and the outgoing W^- in the overall center of mass frame. For the decay angles the convention of [6] is used where for W^- the polar and azimuthal angles of the outgoing fermion are taken in the W^- rest frame (a right handed coordinate system is used) with the $z-axis$ along the direction of the W^- and the $y-axis$ along the direction of $e^- \times W^-$. For the W^+ the angles of the outgoing antifermion are taken with in the W^+ rest frame with the $z-axis$ along the W^+ direction and the $y-axis$ along the $e^- \times W^+$. The differential cross-section can be expressed as a sum of products of production and decay helicity amplitudes. The expressions for these amplitudes can be found in [6 4]. A study of the accuracy expected at LEP200 can be found in [18] where it is assumed that the two W's are produced on-mass shell, in the narrow width approximation without initial state radiation (ISR). In a more complete study [20] more diagrams are used which contain: WW production, $Wl\nu$, $Wq\bar{q}$ (single W-production) and non-resonant channels $l\nu l\nu, l\nu q\bar{q}$, and $q\bar{q}q\bar{q}$. A treatment in the finite W-width is underway. The analysis in [18] is making use of all the five angles and the angular data are fitted with an unbinned maximum likelihood technique. It should be noted however that in any analysis of the reactions: $l\nu jj$ and $jjjj$ not all of the five angles are known without ambiguity. As long as the jet charge is not known, the hadronically decaying W's are known with a twofold ambiguity: $cos\theta^* \rightarrow -cos\theta^*$ and $\phi \rightarrow \phi + \pi$. In the reaction where both W's decay hadronically this ambiguity holds for both pairs of quarks, while the cosine of the production angle is also known wthin a sign. The expected accuracy at 176 GeV and 192 GeV for the one-parameter models, in the $l\nu jj$ and $jjjj$ channels for $500pb^-1$ can be found in table 2 of reference [18]. In the semileptonic channel only the e, μ were considered. An efficiency of about 70% was used for the semileptonic channel and about 50% for the hadronic. For example the error on κ_γ for the semileptonic channel at 190 GeV, when the three variables of the leptonic decay were used unfolded and the other two (from the hadronic channel) folded is expected to be 0.022 and on λ_γ 0.032. In the $jjjj$ channel with all five variables

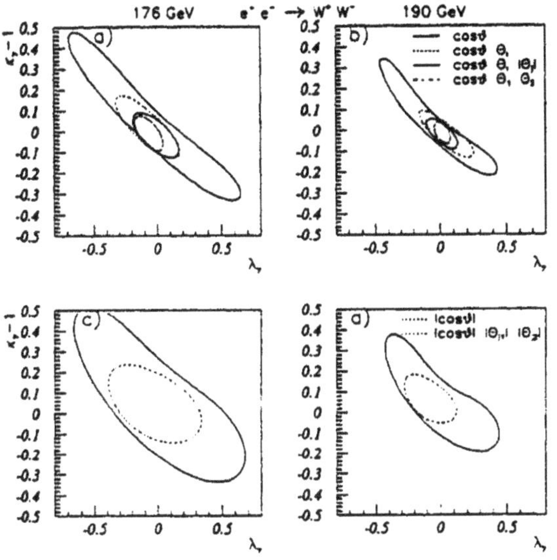

Figure 3: 95% CL contours in the $\lambda_\gamma, \chi_\gamma = \kappa_\gamma - 1$ for: a), b) semileptonic channels in 176 GeV and 190 GeV respectively, and c), d) hadronic channels for the same $\sqrt{s}$ correspondingly. The legend for the various contours of a) and b) are shown on b) and for c) and d) on d). In the legend $cos\theta$ and $|cos\theta|$ represent the production angle without and with ambiguity, while Θ and $|\Theta|$ the decay angles $cos\theta, \phi$ without and with ambiguity (see text).

folded (at 190 GeV) the error on κ_γ and λ_γ will be a factor of two worse. Finally in Figure 3 the results from the two-parameter fits are shown for model 5. The 95% CL contours are shown for the semileptonic and hadronic channels when all or part of the angular information is used. The experimental expectation at LEP200 is in most cases better than for Fermilab at the highest possible luminosity of $10fb^{-1}$. However the accuracy improves significantly in the Next Linear Collider (NLC) where a factor of 10 improvement is expected [19] with respect to LEP200.

4 Conclusions

The measurement of the trilinear couplings of the vector bosons is of interest in itself in order to complete the picture of the Standard Model. Possible deviations will provide indications for new physics. However, the high presicion measurements from SLC and LEP provide severe constraints to the size of such deviations (of the order of a few %). The current measurements from the TeVatron have at best a 50% accuracy. LEP200, in the best cases, will reach an accuracy of 3% which only in the very high luminosity at Fermilab may be achieved. Accuracy in the trilinear boson couplings at the 10^{-3} can only be obtained at LHC or at a Next Linear Collider (NLC).

Acknowledgments

It is my pleasure to thank the organizers and in particular Prof. J.W.F. Valle for the interesting scientific program, the ambience and the hospitality.

References

1. D.Schaile, CERN-PPE/94-162, Oct.94; LEP-Collaborations, CERN-PPE/94-187, Nov.94.
2. Ch.Petridou, Lepton Photon Conference, Geveva 1991 (World Scientific, Singapore, 1991); J.Alitti et al, *Phys. Lett.* B **277**, 194 (1992).
3. M.Acciari et al, *Phys. Lett.* B **376**, 190 (1995).
4. K.Hagiwara et al, *Nucl. Phys.* B **282**, 253 (1987)
5. K.Gaemers and G.Gounaris, *Z. Phys.* C **1**, 259 (1979).
6. M.Bilenky et al, *Nucl. Phys.* B **409**, 22 (1993).
7. H.Aronson, *Phys. Rev.* **186**, 1434 (1969); K.J.Kim and Y.S.Tsai, *Phys. Rev.* D **7**, 3710 (1973).
8. A.DeRujula, *Nucl. Phys.* B **384**, 3 (1992).
9. K.Hagiwara et al, *Phys. Lett.* B **283**, 353 (1992); K.Hagiwara et al, *Phys. Rev.* D **48**, 2182 (1993).
10. H.Aihara et al, Fermilab-Pub-95/031-E.
11. F.Abe et al, *Phys. Rev. Lett.* **74**, 1936 (1995).
12. J.Ellison, Fermilab-Conf-94/329.
13. F.Abe etal, *Phys. Rev. Lett.* **74**, 1941 (1995).
14. S.Abachi et al, Fermilab-Pub-95/042-E.
15. T.A.Fuess, Fermilab-Conf-94/283-E; F.Abe et al, Fermilab-Pub-95/036.
16. S.Abachi et al, Fermilab-Pub-95/044-E.
17. ATLAS Technical Proposal, LHC-2, Dec. 1994.
18. R.Sekulin, *Phys. Lett.* B **338**, 369 (1994).
19. M.Bilenky et al, *Nucl. Phys.* B **419**, 240 (1994).
20. C.Papadopoulos, CERN-TH/95-46.

Chapter 2

OBSERVATION OF THE TOP QUARK AT THE TEVATRON

J.M. BENLLOCH

Instituto de Física Corpuscular,
Avda. Dr. Moliner 50, 46100 Valencia, SPAIN

This report is a summary of the observation for top quark production in $p\bar{p}$ collisions at a center of mass energy of 1.8 TeV reported by the CDF and DØ Collaborations in February 1995. During Run Ia (1992-93) and Run Ib (1994-95), the CDF and DØ detectors collected a total luminosity of 67 pb^{-1} and 50 pb^{-1}, respectively. CDF reported the observation of a signal consistent with $t\bar{t}$ decay to $WWb\bar{b}$, but inconsistent with the background prediction by 4.8σ. CDF measured the top quark mass to be 176 ± 8(stat.) ± 10(syst.) GeV/c^2, and the $t\bar{t}$ production cross section to be $6.8^{+3.6}_{-2.4}$ pb. DØ also reported an excess over signal of 4.6σ consistent with top quark decay and measured its mass to be 199 $^{+19}_{-21}$(stat.) ± 22(syst.) GeV/c^2, and its production cross section to be 6.4±2.2pb.

1 Introduction

The CDF Collaboration reported [1] evidence for top quark production in April 1994 with Run Ia data (19.3 pb^{-1}). The DØ detector took data for the first time during Run Ia collecting 15 pb^{-1} and found [2] no significant excess of signal over background, establishing a lower limit on the top quark mass of 131 GeV. The Fermilab Tevatron improved its performace during the 1994-95 run (Run Ib) and in February 1995, CDF and DØ had already collected a total of 67 pb^{-1} and 50 pb^{-1}, respectively, from Run Ia and Run Ib. Both CDF [3] and DØ [4] confirmed the previous CDF evidence.

In the Standard Model, the top quark is the weak isospin partner of the b quark. The existence of the top quark was inferred by several measurements, such as the forward backward asymmetry in $Z \to b\bar{b}$ at LEP and the absence of flavor-changing neutral currents. An indirect value of the top quark mass of approximately 150-210 GeV/c^2 has been estimated[5] by comparing the precision measurements of the Z boson resonance at LEP with electroweak calculations including radiative corrections, assuming that no new physics is present.

The top quark is produced predominantly in pairs at the Tevatron energies. The cross section is a steeply falling function of the top quark mass varying from 30 pb at 120 GeV to 5 pb at 190 GeV. The top quark decays into a W boson and a b-quark assuming no deviations from the Standard Model (see Figure 1). So, the produced top quark pairs have four possible decay channels, giving different search methods:

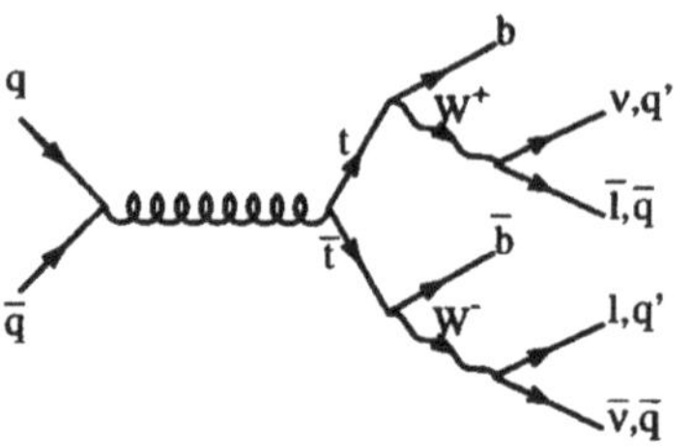

Figure 1: Top production and standard model decay chain. The W may decay into either a lepton and a neutrino or into a quark-antiquark pair.

1. The single lepton plus three or four jets (the b-quark jets and the two jets coming from the hadronic decay of a W) mode, having a branching ratio of 24/81.

2. The dilepton (electrons or muons) and two jets (the b-quark jets) mode. This is the cleanest channel but has the lowest branching ratio (4/81).

3. The 6-jets mode, when both W's decay into quark-antiquark pairs. This channel has the highest branching ratio (36/81) but due to the huge QCD background its study has only started recently.

4. Modes including τ's, with a branching ratio of 21%. Since τ's are tagged with very low efficiency, this mode has not achieved a sufficiently good signal to noise ratio so far.

2 Lepton+Jets Channel at CDF

The lepton+jets analysis starts with the high-P_T inclusive lepton data sample which requires at least one isolated electron ($E_T \geq 20$ GeV) or muon ($P_T \geq$ 20 GeV) in the central region ($|\eta| < 1.0$). From that sample, about 70000 W events were selected by requiring a good electron or muon and a missing transverse energy in the event $\not{E}_T \geq 20$ GeV. Events containing Z bosons, with an ee or $\mu\mu$ invariant mass between 70 and 110 GeV/c^2, were removed. The dominant background in the lepton+jets top search is the direct production of W+jets.

Approximately 75% of the $t\bar{t}$ signal for M_{top} = 160 GeV/c^2 is kept by requiring $N_{jet} \geq$ 3 (jets were requiered to have $E_T \geq 15GeV$ and $|\eta| < 2.0$), whilst only 203 events with $N_{jet} \geq 3$ were observed in the W sample. In the $W+ \geq$3-jet sample, about 30 $t\bar{t}$ events are expected for $M_{top} = 175\ GeV/c^2$. Additional background rejection is needed to isolate a possible $t\bar{t}$ signal.

A way to significantly reduce the QCD background is to ask for tagging a b-quark jet in the event. The presence of a b-quark jet can be tagged either by a secondary vertex or by a semileptonic decay.

b-quark Tagging with the SVX.

Because of nonzero b lifetime, tracks from b decays may have a measurable displacement from the primary vertex. The CDF Sillicon Vertex Detector[8] (SVX) is an $r - \phi$ tracker that consists of 4 layers situated concentrically right after the beam pipe. The SVX provides a transverse impact parameter resolution of:

$$\sigma_d \sim 16\,(1 + (0.8/P_T)^2)\ \mu m$$

Displaced tracks measured with the SVX were used to reconstruct secondary vertices in a jet (see Figure 2). The secondary vertex algorithm first searches for verteces with at least three good tracks with $P_T \geq 1.5\ GeV/c$, associated with a jet (cos(jetaxis − track) > 0.8) and with impact parameter significance $d/\sigma_d \geq 2.5$. The secondary vertex is calculated using a three-dimensional fit, with the jet's tracks constrained to originate from a common vertex. This procedure takes advantage of the fact that 3 o more tracks have much lower background than 2-track verteces. If this fails, the secondary vertex algorithm searches for 2-track verteces using more stringent track and vertex quality criteria. The error on the transverse decay length is typically $\sigma_{L_{xy}} \approx 130\ \mu$m. Tracks consistent with K_s or Λ are removed.

Jets that have a secondary vertex displaced in the direction of the jet, with significance $L_{xy}/\sigma_{L_{xy}} \geq 3.0$ were positively tagged ($+L_{xy}$), and those with a secondary vertex behind the primary, with significance $L_{xy}/\sigma_{L_{xy}} \leq 3.0$ were negatively tagged ($-L_{xy}$). The largest contributions to the negative tags are resolution smearing and mistakes in the tracking reconstruction.

A control sample of inclusive electrons, which are enriched in b-decays, is used to measure the efficiency for tagging a semileptonic b jet. The transverse decay length, L_{xy}, is converted into an estimate of the effective proper decay length ($c\tau_{eff}$) using the expression $c\tau_{eff} = L_{xy}\frac{M}{P_T}F$, where M is the invariant mass of the tracks associated with the displaced vertex, P_T is their total vector transverse momentum, and F is a scale factor, determined from a b Monte Carlo sample. The scale factor F accounts for b-hadron decay products that are not attached to the secondary vertex. There is good agreement for the

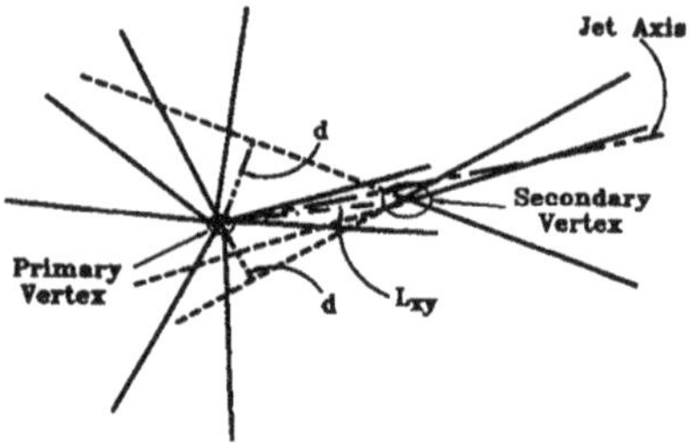

Figure 2: Simplified view of an event containing a secondary vertex shown in the transverse plane. The primary vertex is the point where the $\bar{p}p$ interaction occurs. The secondary vertex is the point of decay for a long-lived particle originated at the primary vertex. The impact parameter, d, is shown for representative tracks. Also shown is L_{xy}, the two-dimensional decay distance to the secondary vertex, measured in the transverse plane.

effective proper decay length $c\tau_{eff}$ between the inclusive electron CDF data and the tags in a $b\bar{b}$ ISAJET [6] + CLEO [7] Monte Carlo sample.

The radiation-soft SVX used in Run Ia was replaced by a radiation-hard SVX' in Run Ib. Due to the lower level of noise of SVX' and an improvement in the secondary vertex algorithm, the b-quark tagging efficiency with the SVX' for tagging at least one b jet in a $t\bar{t}$ event with three or more observed jets increased to $42 \pm 5\%$. The number of expected SVX-tagged $t\bar{t}$ events with $N_{jet} \geq 3$ is shown in table 1.

M_{top}	160 GeV/c^2	170 GeV/c^2	180 GeV/c^2
$\sigma_{t\bar{t}}^{Theor}$ (pb)	8.2	5.8	4.2
$N_{expected}(SVX)$	20	16	10
$N_{expected}(SLT)$	20	16	10

Table 1: The theoretical cross section in the $W + jets$ channel with b-tag and the expected event yield for $t\bar{t}$ in $67pb^{-1}$.

Figure 3 shows the number of events before SVX tagging, the number of SVX tags observed and the estimated background, as a function of jet multiplicity. Based on the excess number of tags in events with $\geq$3 jets, we expect an additional 0.5 and 5 tags from $t\bar{t}$ decay in the 1 and 2 jet bins respectively. The effective proper decay length $c\tau_{eff}$ for all the jets tagged in the $W+ \geq$3-jet sample is also shown the Figure.

The largest contribution of the background in the top signal region ($W+ \geq$3-jets) is expected to arise from W production in association with heavy flavor.

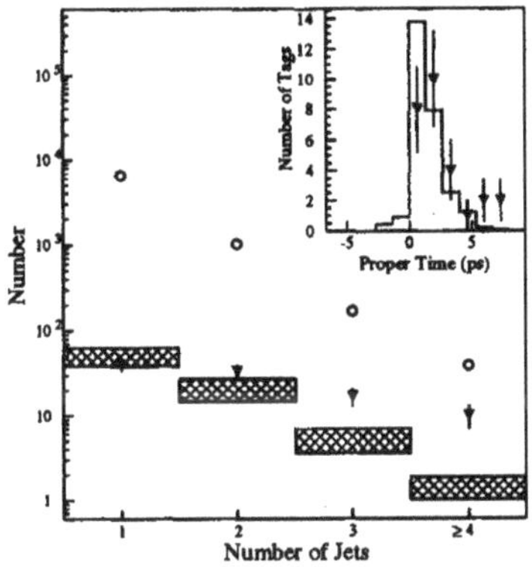

Figure 3: The number of events observed in the W+jets data sample before SVX tagging (circles) and the number of tags observed (solid triangles) versus jet multiplicity. The shaded area is the background estimates with its uncertainty. The 3-jet and $\geq$ 4-jet bins are the $t\bar{t}$ signal region. The inset shows the secondary vertex proper time distribution for the 27 tagged jets in the signal region compared to the expectation for b quark jets from $t\bar{t}$ decay.

A method to estimate directly from the data the number of tags expected in the 203-event sample is to assume that the heavy-quark (b and c) content of jets in W+jets background events is the same as in the inclusive-jet sample. This assumption is too conservative and is not used, since heavy quarks are produced in W+jets events mostly through gluon splitting[9] whilst the inclusive-jet sample contains direct production contributions as well. Instead, the heavy-quark background is estimated using Monte Carlo calculations of the $Wb\bar{b}$, $Wc\bar{c}$ processes in W+jets events. Fake-tags due to track mismeasurement were estimated from jet data (tracking errors generate with equal probability tags with positive and negative decay length L_{xy}, whilst tags from heavy quark decays populate the positive L_{xy} region). The negative tag rates measured in the inclusive-jet sample, parameterized by the E_T, track multiplicity and η of each jet, were applied to the jets in the 203 events to yield the total expected number of SVX-tagged events due to fake tags.

The small contributions from Wc, WW, WZ production, and $Z \to \tau\tau$ are estimated with Monte Carlo.

The backgrounds from non-W sources (direct $b\bar{b}$ production and hadrons misidentified as leptons) are determined from the data by studying the isolation of the lepton candidates in the low $\not{E}_T$($\not{E}_T$< 15 GeV/c) and high $\not{E}_T$($\not{E}_T$> 20 GeV/c) regions[10].

There are 27 tags in 21 events in the W+ $\geq$3-jets sample. The total estimated background to SVX tags in the signal region is 6.7 ± 2.1 events. The 6 events with 2 tagged jets can be compared with 4 expected for the top+background hypothesis and less than 1 for background alone.

Cross Section

From the number of events SVX tagged in the $W+ \geq 3$-jets channel, the estimated background, the calculated $t\bar{t}$ acceptance, and the integrated luminosity of the data sample, the $t\bar{t}$ production cross section is calculated to be:

$$\sigma_{t\bar{t}}(M_{top} = 176\ GeV/c^2) = 6.8\ ^{+3.6}_{-2.4}\ pb$$

b-quark Tagging with Soft Leptons.

b-quark jets can also be tagged identifying the leptons of the decay $b \to l$ ($l = e$ or μ) or $b \to c \to l$. For this purpose, leptons with $P_T > 2$ GeV/c were considered. The leptons from b decay have a much softer spectrum ($< P_T > \approx 4\ GeV/c$ for $M_{top} \approx 120\ GeV/c^2$) than leptons from W decays, and are therefore called "soft lepton tags", or SLT tags.

Electrons are identified by matching extrapolated CTC tracks with an electromagnetic cluster in the calorimeter, consistent in size, shape and position with expectations for electron showers. The efficiency of the soft electron algorithm, excluding isolation cuts, was measured through a photon conversion sample, where the first electron was identified in the calorimeter and the second, unbiased, electron was selected using a track-pairing algorithm. Combined with electron isolation efficiencies determined from $t\bar{t}$ Monte Carlo, the total efficiencies are (53±3)% and (23±3)% (statistical uncertainties only) for electrons from b and sequential c decays respectively.

To identify muons, track segments in the muon chambers are matched to tracks in the CTC. The efficiency for reconstructing track segments in the muon chambers is measured to be 96% using a $J/\psi \to \mu^+\mu^-$ sample. This number was combined with the P_T-dependent efficiency of the track-matching requirements to give an overall efficiency of approximately 85% for muons from a combination of both b and c decays.

The acceptance of the SLT analysis for $t\bar{t}$ events was calculated using the ISAJET + CLEO Monte Carlo programs. The efficiency for tagging at least one jet in a $t\bar{t}$ event by detecting an additional lepton with $P_T > 2\ GeV/c$ is $\epsilon_{tag} = 20 \pm 2\%$, approximately independent of M_{top}. The efficiency, ϵ_{SLT}, for inclusive $t\bar{t}$ events to pass the lepton-identification, kinematic, and SLT b-tag requirements is shown in Table 1 along with the number of expected SLT-tagged $t\bar{t}$ events.

The main backgrounds to the SLT search are hadrons misidentified as leptons, and $Wb\bar{b}$, $Wc\bar{c}$ production. As in the SVX analysis, the background from hadrons misidentified as leptons is estimated from the inclusive-jet sample. By studying tracks in such events, the probability of misidentifying a hadron as an electron or muon, or of tagging a true semileptonic decay is determined. In the $W+\geq 3$ jet sample there are 23 tags in 22 events, with 15.4±2.0 tags expected from the total background. Six events contain both an SVX and SLT tag, compared to the expected 4 for top+background and 1 for background alone.

3 Dilepton Channel at CDF

3.1 Dilepton Event Selection

The dilepton data sample also starts with the single lepton data sample, with the same lepton identification requirements used in the Lepton and Jets analysis. An additional high P_T ($P_T > 20\ GeV$) lepton (electron or muon) with opposite charge is required in this channel. This second lepton may pass looser cuts.

An invariant mass cut is applied to remove $Z's$ in ee, $\mu\mu$: $75\ GeV/c^2 < M_{ll} < 105\ GeV/c^2$. For $M_{top} = 160\ GeV/c^2$, 80 % of dielectron and dimuon events pass the invariant mass cut.

Due to the presence of two high P_T neutrinos in the top dilepton channel a missing E_T cut is applied on its magnitude: $\not{E}_T > 25\ GeV$. The missing E_T is corrected to account for jet energy mismeasurement. At this point, the background is dominated by Drell-Yan in the ee, $\mu\mu$ channels and by $Z \to \tau\tau$ in the $e\mu$ channel. For events with $\not{E}_T < 50\ GeV$ it is also required that the azimuthal angle between the $\not{E}_T$ direction and the direction of the closest jet be $\Delta\phi(\not{E}_T, jet) > 20^0$ to reject the Drell-Yan continuum background, which would pass the missing $\not{E}_T$ magnitude cut due to mis-measured jet energies from cracks, etc. A similar cut is applied to the direction of $\not{E}_T$ with respect to leptons to reject the $Z \to \tau\tau$ background ($\Delta\phi(\not{E}_T, l) > 20^0$ when $\not{E}_T < 50\ GeV/c^2$), since the missing $\not{E}_T$ produced by neutrinos from $\tau's$ goes primarily along the lepton direction. For $M_{top} = 160\ GeV/c^2$, 76 % of the dilepton events pass all the missing $\not{E}_T$ cuts.

For a high mass top ($M_{top} > 120\ GeV/c^2$) the two b-quarks will form high E_T jets with high efficiency. So, two or more central ($|\eta| < 2.0$) jets are required with tranverse energy greater than $10\ GeV$. The two-jet cut reduces the background by a factor 4 while being 84 % efficient for $M_{top} = 160\ GeV/c^2$.

After all these cuts, there are 5 $e\mu$ events, 2 dimuon events and no dielectron events in the signal region. One of the dimuon events contains an energetic photon with a $\mu\mu\gamma$ invariant mass of 86 GeV/c^2. Although we have estimated the expected background from radiative Z decay to be small, we have removed that event from the final sample, to be conservative. Three of these events contain a total of 5 b-tags, compared with an expected 0.5 if the events are background. We would expect 3.6 tags if the events are from $t\bar{t}$ decay.

Dilepton Acceptance

The dilepton acceptance has been studied mainly with the ISAJET Monte Carlo program. The predicted central value of the $t\bar{t}$ production cross section from the NNLO theoretical cross section[11] and the number of expected events in 67 pb^{-1} is given in Table 2. After all the cuts, the contribution from the different channels is: $e\mu$: 59%; ee : 21%; $\mu\mu$: 20%.

The dilepton acceptance uncertainty is dominated by the two-jet cut which varies from 36% to 3% (M_{top} 100 − 180 GeV/c^2). The other cuts have uncertainties which do not depend strongly on the value of the top mass.

Table 2: The predicted central value of $t\bar{t}$ production cross section and the number of events expected in 67 pb^{-1}, as functions of top mass.

M_{top} GeV/c^2	$\sigma_{t\bar{t}}$ pb	$N_{e\mu,ee,\mu\mu}$ events
160	8.2	4.4
170	5.8	3.0
180	4.2	2.4

Dilepton Backgrounds

The most important backgrounds have been estimated to be WW, Drell-Yan ($\gamma/Z \rightarrow ee$, $\mu\mu$), $Z \rightarrow \tau\tau$, $b\bar{b}$ processes and QCD or W+jets processes with at least one misidentified lepton.

The WW background is estimated using the ISAJET Monte Carlo program, normalized to a total cross section of 9.5 pb, and the CDF detector simulation. The error is dominated by the theoretical uncertainty in the cross section. Only 13% of the events pass the two-jet cut. The effects of initial state radiation for the two-jet cut are checked using the Drell-Yan process. The total WW background after the two-jet cut is predicted to be 0.21 ± 0.07 events.

The $\tau\tau$ background is determined using the hybrid technique of taking the $Z \to ee$ events and substituting Monte Carlo generated taus in place of the electrons. After the two-jet cut 0.38 ± 0.07 events are expected.

The $b\bar{b}$ background is estimated using the ISAJET Monte Carlo for the production and the CLEO Monte Carlo[7] for b decays. To check ISAJET, the lepton P_T cut is lowered to 15 GeV/c and 5 GeV/c (for the first and second lepton, respectively), where heavy flavor is dominant, and the distributions found to be in good agreement with the $e\mu$ data. 0.03 ± 0.02 events are predicted from this background.

The lepton misidentification background (QCD or W+jets processes with at least one fake lepton) is determined using jet data samples to assign a probability per track to pass the electron and muon identification cuts. These fake probabilities are applied to events with a good lepton plus additional tracks to estimate a total background of 0.23 ± 0.15 events.

The non-resonant Drell-Yan backgrounds for ee, $\mu\mu$ are determined using Z data to get scale factors for cuts (0.1% of Z events pass all the cuts, based on one $Z \to \mu\mu$ event), and apply the same scale factors to the number of non-resonant ee and $\mu\mu$ events. 0.44 ± 0.28 events are expected from this background.

The total expected background in the dilepton channel is 1.3 ± 0.3 events.

4 Significance of the CDF Result and Consistency Checks

Each of the channels presented above shows an excess of events over expected background. The lepton+jets b-tag analysis identifies 37 events: 27 SVX tags with an expected background of 6.7 ± 2.1, and 23 SLT tags with an estimated background of 15.4 ± 2.0. The dilepton analysis observes 6 events with an expected background of 1.3 ± 0.3 events. For each of these results the probability P that the estimated background has fluctuated up to the number of candidate events seen or greater is calculated to be: $P_{SVX} = 2 \times 10^{-5}$, $P_{SLT} = 6 \times 10^{-2}$, and $P_{DIL} = 3 \times 10^{-3}$.

After understanding the SVX-SLT correlations by correctly predicting the number of double tagged jets and events in the inclusive-jet sample, the combined probability is calculated. The result is 1×10^{-6}, which is equivalent to a 4.8σ excess from background in a Gaussian distribution.

Two features of the data were found in reference [1] which did not support the interpretation of the excess as $t\bar{t}$ production and now are no longer present:

1. The Z+multijet sample is, in principle, a good cross check of the heavy flavor content of the W+multijet sample. The standard model production mechanisms are similar in Ws and Zs, but there should be no top in

the Z sample. In practice, however, $\sigma \times BR$ for $Z \rightarrow ee, \mu\mu$ is an order of magnitude less than for $W \rightarrow e\nu, \mu\nu$. Backgrounds in the Z+jets sample are calculated using the same method used in the W+jets background calculation, but no backgrounds from non-Z sources are included ($t\bar{t}$, $b\bar{b}$, etc.). The number of tags observed in the Z+1-jet, 2-jet, and $\geq$3-jet samples is 15, 3, and 2, respectively, compared to the background predictions of 17.5, 4.2, and 1.5. The excess over background in the $Z+\geq$ 3-jet region that was seen in reference [1] has disapeared.

2. There is no discrepancy between the measured and predicted $W+\geq$4 jets background, in contrast to a deficit described in reference [1].

5 Direct Determination of the Top Mass at CDF

W events with 4 or more jets can be kinematically reconstructed to the $t\bar{t} \rightarrow WbW\bar{b}$ hypothesis, yielding for each event an estimate of the top quark mass. There are multiple solutions due to the assigments of the highest 4 jets to the parent $W's$ and $b's$ and the quadratic ambiguity in determining the longitudinal momentum of the neutrino. For each event, the solution with the lowest fit χ^2 is chosen. The E_T threshold of the fourth jet is lowered to 8 GeV and its angle allowed to be within $|\eta| < 2.4$. Figure 4 shows the mass distribution for the 88 $W+\geq$ 4 jets events giving a good fit. The distribution is consistent with the predicted mix of approximately 30% $t\bar{t}$ signal and 70% W+jets background. The VECBOS[12] Monte Carlo program is used to make an estimate of the shape of the direct W+jets background. The Monte Carlo background shape agrees well with that measured in a limited-statistics sample of Z+4-jet events as well as in a QCD sample selected to approximate non-W background.

The mass distribution for the 19 events tagged, of which $6.9^{+2.5}_{-1.9}$ are expected to be background, is shown in Figure 5. For these events, only solutions in which the tagged jet is assigned to one of the b-quarks are considered.

To find the most likely top mass, the mass distribution is fit to a sum of the expected distributions from the W+jets background and a top quark of mass M_{top}. The -ln(likelihood) distribution from the fit is shown in the Figure 4 inset. The best fit mass is:

$$M_{top} = 176 \pm 8(stat.) \pm 10(syst.)\ GeV/c^2$$

The shape of the mass peak in Figure 4 provides additional evidence for top quark production, since the observed b-tags cluster around the 175 GeV/c^2 mass region and the number of observed b-tags is independent of the observed

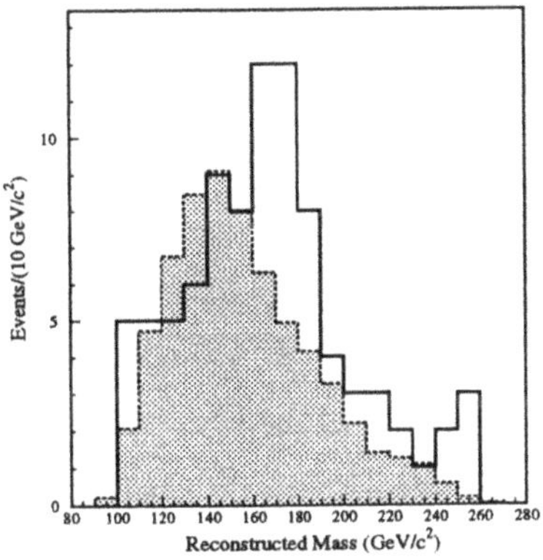

Figure 4: Top mass distribution for the data (solid histogram), the W+jets background (dots), and the sum of background and Monte Carlo $t\bar{t}$ for $M_{top} = 175GeV/c^2$ (dashed). The background distribution has been normalized to the 1.4 background events expected in the mass-fit sample. The inset shows the ln(likehood) fit used to determine the top mass.

mass distribution. After including systematic effects in the predicted background shape, a probability of 2×10^{-2} is found that the observed mass distribution is consistent with the background.

6 Observation of the Top Quark at D0

In the dilepton channel DØ has performed a similar search than CDF, though the lepton coverage of the DØ detector is better. DØ found 2 $e\mu$ events and 1 $\mu\mu$ event all of them with two jets above 15 GeV.

In the W+jets channel DØ did not applied secondary vertex b-tagging due to the lack of a central magnetic field and a microstrip sillicon vertex detector. DØ based its analysis on its superior jet resolution that allows to perform better kinematic tests. Four jets or more with $P_T > 15\ GeV$ and event shape cuts were required. Events coming from $t\bar{t}$ production are more spherical and aplanar than the ones coming from the direct W+jets background. Also, the scalar sum of the E_T of the jets, a quantity called H_T, is a powerful discriminator between background and high mass top quark production. So, the event structure cuts are the following: aplanarity of the W+jets system larger than 0.05 and H_T larger than 200 GeV. Five and three events are observed in the e+jets and μ+jets modes, respectively.

In the case of muon-tagged channel, only three jets are requested and looser cuts are applied to the event structure. Six events are observed in the

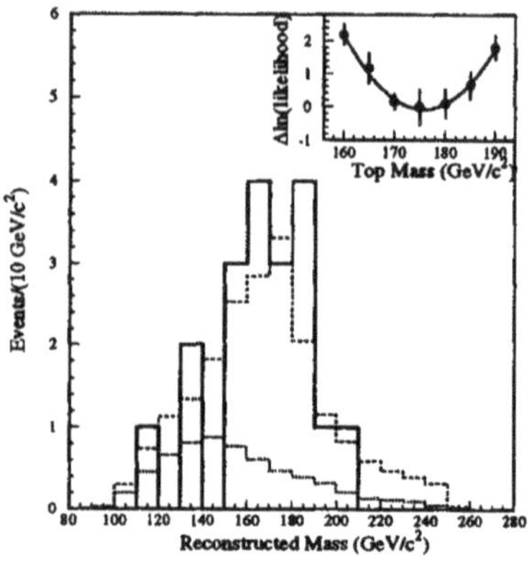

Figure 5: Top mass distribution for the data (solid histogram), the W+jets background (dots), and the sum of background and Monte Carlo $t\bar{t}$ for M_{top} = 175 GeV/c^2 (dashed). The background distribution has been normalized to the 1.4 background events expected in the mass-fit sample. The inset shows the ln(likehood) fit used to determine the top mass.

W+jets channel where one of the jets is muon-tagged.

In total, 17 events are observed with an expected background of 3.8±0.6 events. The probability of an upward fluctuation of the background to the observed number of events is 2×10^{-6}, which corresponds to 4.6 standard deviations for a Gaussian probability distribution. Assuming a top quark mass of 200 GeV/c^2, the production cross section is 6.3±2.2 pb.

Of 27 lepton+four-jet events (removing the H_T requierement to avoid any bias), 24 were fitted successfully to the $t\bar{t}$ hypothesis. A likelihood fit to the mass distribution resulted in a top quark mass of:

$$M_{top}(D0) = 199^{+19}_{-21}(stat.) \pm 22(sys.)\ GeV/c^2$$

where the systematic error is dominated by the jet energy scale.

7 Conclusions

The top quark has been discovered by both the CDF and DØ Collaborations. The value of the top quark mass measured by CDF is: M_{top}(CDF) = 176 ± 8(stat.) ± 10(sys.) GeV/c^2. The $t\bar{t}$ production cross section measured by CDF is $6.8^{+3.6}_{-2.4}$ pb. The top quark mass measured by DØ is M_{top}(D0) = 199^{+19}_{-21}(stat.) ± 22(sys.) GeV/c^2 and its production cross section is 6.4 ± 2.2 pb.

The Tevatron is still running and has already delivered a total luminosity in excess of 100 pb^{-1} by the end of 1995. The Main Injector, scheduled to turn on in year 2000, should provide enough luminosity (about 1 fb^{-1} per year) to allow studies in detail of the top quark properties.

References

1. CDF Collaboration, F. Abe et al, *Phys. Rev. Lett.* **73**, 225 (1994); *Phys. Rev.* D **50**, 2966 (1994).
2. DØ Collaboration, S. Abachi et al, *Phys. Rev. Lett.* **72**, 2138 (1994).
3. CDF Collaboration, F. Abe et al, *Observation of Top Quark Production in $\bar{p}p$ Collisions with the CDF Detector at Fermilab*, FERMILAB-PUB-95-022-E, February 1995.
4. DØ Collaboration, S. Abachi et al, *Observation of the Top Quark*, FERMILAB-PUB-95-028-E, February 1995.
5. D. Schaile, CERN-PPE/94-162, presented at the 27th International Conference on High Energy Physics, Glasgow, July 1994 (unpublished).
6. F. Paige and S.D. Protopopescu, BNL Report No. 38034 (1986).
7. P. Avery, K. Read, and G. Trahern, Cornell Internal Note CSN-212 (1985).
8. P. Azzi et al., FERMILAB-CONF-94/205-E.
9. M.L. Mangano, *Nucl. Phys.* B **405**, 536 (1993).
10. CDF Collaboration, F. Abe et al, *Phys. Rev.* D **44**, 29 (1991).
11. E. Laenen, J. Smith and W.L. Van Neerven, *Phys. Lett.* B **321**, 254 (()1994).
12. F.A. Berends, W.T. Giele, H. Kuijf and B. Tausk, *Nucl. Phys.* B **357**, 32 (1991).

Status of SUSY Searches at DØ

Douglas Norman[a]

Texas A&M University / Fermilab, M.S. 352, P.O. Box 500, Batavia, IL 60510

DØ is very active in searches for SUSY. The current status of SUSY searches at DØ will be presented in this paper. Results on searches for squarks and gluinos through multijet plus missing transverse energy signatures will be covered. Also current results on a search for light scalar SUSY top will be presented. Finally, the search for SUSY gauginos through leptonic signatures will be discussed. Preliminary results on excluded mass spaces and cross sections are given.

1 New Phenomena at DØ

DØ is involved with many searches for new phenomena. Non-SUSY searches include searches for leptoquarks and extended gauge bosons such as W' bosons, Z' bosons, and right handed W bosons. We also look for excited quarks, evidence of a composite structure of quarks through a total event energy signature, fourth generation b quarks and leptons, and bosonic Higgs which couples to two photons.

For SUSY we do direct searches for squarks, gluinos, the lightest charged gaugino, and the next lightest neutral gaugino. These particles decay to final states containing quarks, leptons, and for R-parity conserving SUSY theories lightest neutral gauginos. R-parity is the SUSY quantum number (-1 for SUSY particles and +1 for Standard Model (SM) particles) which if conserved provides for a lightest SUSY particle (LSP) that is stable (Cold Dark Matter candidate). Conservation of R-parity also requires that SUSY particles be produced in pairs. Most of the SUSY searches at DØ, including those discussed in this paper, assume R-parity conservation. This report will give an overview of analyses and results for: 1) searches for squarks and gluinos (excluding the scalar top) through hadronic signatures, 2) a separate search in a hadronic channel for scalar tops which could be much lighter than the other squarks due to mass splitting, and 3) searches for gauginos through leptonic signatures. These analyses are based on approximately 13 pb^{-1} of data taken in 1992-1993.

---

[a]representing the DØ Collaboration

2 Searches for Squarks and Gluinos

Searches for squarks and gluinos have been done in the hadronic channels. We assume for these analyses that pair produced squarks and gluinos cascade decay through intermediate gauginos to final state quarks, gluons, and LSP's. We also assume that the squark masses are degenerate and that the slepton masses are much heavier than the masses of the squarks. This last assumption is necessary in order to have the branching fractions to hadronic channels less sensitive to the choice of SUSY parameters. The expected signature is multijets and large $\not{E}_T$. There are two search analyses for hadronic signature of squarks and gluinos: a three jet analysis [1] where three jets with $E_T > 25$ GeV and $\not{E}_T > 75$ Gev are required and a four jet analysis [2] where four jets with $E_T > 20$ GeV and $\not{E}_T > 65$ GeV are required. In both analyses we require there be only one vertex (single interaction) in the event because $\not{E}_T$ is not well defined for multi-interaction events. We also make jet-$\not{E}_T$ angular correlation cuts and require there be no high E_T leptons, since they are not part of the squark/gluino multijet signature. The single interaction cut reduces the integrated luminosity to an effective 7.4 pb^{-1}. The jet azimuthal correlation cuts are used to remove multijet events where the jet E_T of one or more jets is mismeasured producing large $\not{E}_T$. For example if the E_T of the leading jet is overestimated, the $\not{E}_T$ would point opposite the jet. With these cuts we have 14 candidate events (after three events were eliminated by eye scan) in the three jet search and 5 candidate events in the four jet search.

The backgrounds for squark and gluino signatures are primarily vector boson plus jet events where the lepton from the vector boson is missed, or the leptons are neutrinos from the decay of the Z^0 boson. Other less significant backgrounds come from multijet processes. The total number of vector boson backgrounds is estimated to be 14.2±4.4 events for the three jet search and 5.2±2.2 events for the four jet search. The number of multijet background is estimated to be 0.4±0.4 and 1.6±0.9 for the three and four jet searches. These estimates are determined from studies of the data and Monte Carlo. The numbers for the three jet search are an update of the number of expected background events previously reported [1].

There is no excess above expected background in either search. Subtracting these backgrounds (the multijet background for the four jet search is not subtracted), the 95% confidence level limit excluded contours given in Fig. 1 are determined. The signal efficiencies for the cuts used in the analyses are determined from ISAJET [3] Monte Carlo with a detailed detector simulation. The selection of SUSY parameters used in the Minimum Supersymmetric Standard Model (MSSM) to generate the signal events are:

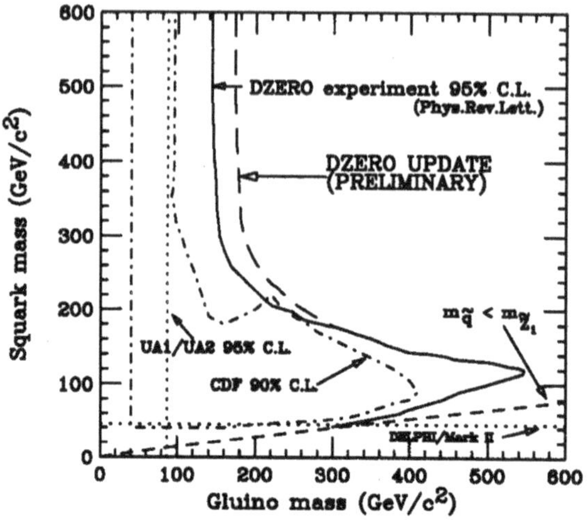

Figure 1: The DØ published and new preliminary updated 95% CL excluded squark versus gluino mass regions.

$\tan\beta = 2.0, \mu = -250GeV, m_{H^+} = 500GeV$ with various values of the squark and gluino mass. Given in Fig. 1 are the published results from the three jet analysis and an updated limit derived from the combined three and four jet analyses. The new preliminary limit is 173 GeV/c^2 for the gluino mass for heavy squarks and 229 GeV/c^2 for the mass of the squark and gluino if they are equal.

3 Search for a Scalar Top

The SUSY scalar top is handled separately from the other squarks because the mass of the scalar top (stop) may be lighter than the mass of the other squarks due to mass splitting induced by the heaviness of the top quark and its effect on the Yukawa coupling to the Higgs. The signature for the stop squark depends in the mass of the stop. If the mass of the stop is heavy enough, then $\tilde{t} \rightarrow b\tilde{W}_1$ where $\tilde{W}_1$ denotes the lightest chargino. Otherwise, $\tilde{W}_1$ becomes virtual and the stop can have the three body decays $\tilde{t} \rightarrow bl\tilde{\nu}$ or $\tilde{t} \rightarrow b\nu\tilde{l}$ unless the sleptons, $\tilde{\nu}\tilde{l}$, are heavier that the stop. In this case the stop will primarily decay to an LSP and c quark, and the event signature for pair produced stops is large $\not{E}_T$ and acollinear jets. The search for this signature [4] from a low mass stop is reported here.

The analysis uses the following cuts: Single interactions, $\not{E}_T > 40$ GeV, two jets with $E_T > 30$ GeV, $90° < \Delta\phi(j_1, j_2) < 165°$, $10° < \Delta\phi(j_1, \not{E}_T) < 125°$, $10° < \Delta\phi(j_{3,4}, \not{E}_T)$, and no leptons with $E_T > 10$ GeV. The single interaction

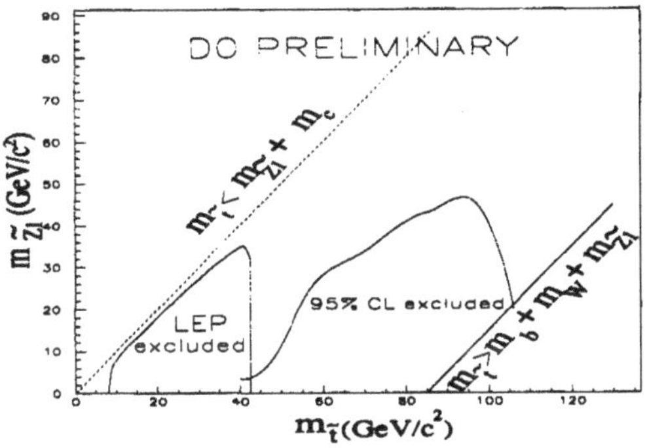

Figure 2: The DØ preliminary 95% CL excluded LSP versus stop mass region.

requirement reduces the integrated luminosity to an effective 7.4 pb^{-1}. The $\Delta\phi$ cuts reduce the backgrounds from QCD dijet events and from false $\not{E}_T$ coming from mismeasured jets. After these cuts there are two candidate events remaining.

The sources of background are the same as those for the squark/gluino search. The total number of background from vector bosons is 2.86 ± 0.93. The amount of background from multijet sources is negligible. The total expected background indicates that there is no excess above background for the stop signature. A 95% CL excluded LSP (denoted by $\tilde{Z}_1$) versus stop mass region is given in Fig. 2. The efficiency for the stop signal is determined from ISAJET Monte Carlo. The gap between the DØ limit and the LEP limit is due to the high $\not{E}_T$ trigger threshold used in taking the data used in this analysis. In the current 1994-95 Tevatron collider run, we have a trigger utilising a much lower $\not{E}_T$ threshold, and we hope to cover this region.

4 Search for Gauginos

The signature for gaugino pair production that we have investigated [5] is that of three charged leptons (electrons and muons) and $\not{E}_T$ from associated pairs of the lightest chargino and the second lightest neutralino (denoted as $\tilde{Z}_2$). These are produced by the decay of an off-shell W boson from a $q\bar{q}$ scattering. We assume the gaugino pairs decay to off-shell W and Z bosons and LSP's with subsequent decay of the W and Z bosons to leptons. Since the signature involves three charged leptons and no jets, it is relatively free of Standard Model backgrounds. The only physics background is W and Z boson associated pair production; this background contributes negligibly. The major background is from Z boson plus fake electron where the sources of the fake electrons are photons, neutral pions, and jets.

The values of MSSM parameters used in this analysis are: μ = -500 GeV, $\tan\beta = 2$, $A_t = 0$ GeV, $m_{H_A} = 500$ GeV, $m_{top} = 170$ GeV, $m_{\tilde{q}} = 1000$ GeV, and $m_{\tilde{l}} = 200$ Gev but $m_{\tilde{\nu}} = 1000$ GeV. The mass of the charginos is varied by changing the mass of the gluino. One should note that the kinematics of the Monte Carlo samples for a given mass are insensitive to the choice of MSSM parameters which result in that mass. However, the parameters effect the branching fraction to leptons. From the decay of the gauginos one expects moderate lepton E_T's and moderate $\not\!\!E_T$. The third highest E_T lepton has, however, a rather soft E_T spectrum, so it is important that the threshold on the third lowest E_T lepton remain as low as possible in order to preserve the acceptance for the signal.

About 12 pb^{-1} of data is used in this analysis. The triggers require one or two leptons as follows: one muon with $p_T >$ 15 GeV/c, two muons with $p_T >$ 3 GeV/c, one muon with $p_T >$ 5 GeV/c plus one electron with $E_T > 7$ GeV, a single electron with $E_T > 20$ GeV, or two electrons with $E_T > 10$ GeV. Offline, threshold cuts 2 GeV above the trigger are required for the leading leptons. For the remaining leptons in the events the thresholds are set at 5 GeV. We require that the leptons pass quality criteria, be isolated, and have no angular correlations with other leptons that one expects in SM backgrounds. We require a $\not\!\!E_T$ cut of 10 GeV in the *eee* channel. We also require that there be no more than one electron in the forward detector. This cut reduces the background from photon conversions where the conversion rate in the forward detector approached 40% . With these cuts we see no candidates in any of the four channels containing three muons or electrons. The expected backgrounds range between 0.2 and 0.8 events. The efficiency for the signal is determined by a study of ISAJET Monte Carlo. The resulting no-background subtracted zero event 95% CL limit on the cross section as a function of the lightest chargino mass is given in Fig. 3. This limit is largely independent of the MSSM parameters assumed.

5 Conclusion

DØ has searched for SUSY and has excluded a significant amount of SUSY parameter space. In the near future we expect to have six to seven times the data used in the current analysis. With this amount of data we can expect to be able to probe a much larger portion of the parameter space, excluding it or discovering SUSY!

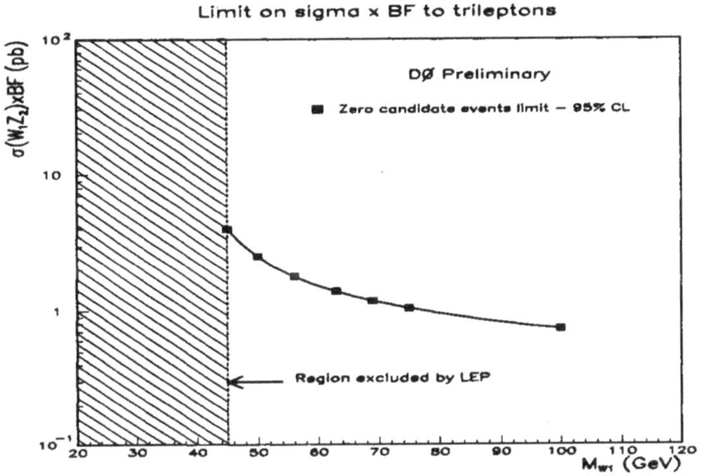

Figure 3: The DØ preliminary 95% CL upper limit on the cross section times the branching fraction to three leptons as a function of the lightest chargino mass.

Acknowledgments

We thank the Fermilab Accelerator, Computing, and Research Divisions, and the support staffs at the collaborating institutions for their contributions to the success of this work. We also acknowledge the support of the U.S. Department of Energy, the U.S. National Science Foundation, the Commissariat à L'Energie Atomique in France, the Ministry for Atomic Energy and the Ministry of Science and Technology Policy in Russia, CNPq in Brazil, the Departments of Atomic Energy and Science and Education in India, Colciencias in Colombia, CONACyT in Mexico, the Ministry of Education, Research Foundation and KOSEF in Korea and the A.P. Sloan Foundation.

References

1. S. Abachi *et al.* Search for Squarks and Gluinos in $p\bar{p}$ Collisions at $\sqrt{s}$ = 1.8 TeV. Submitted to Phys. Rev. Lett. FERMILAB-PUB-95/057-E.
2. M. Goforth, DØ internal Note (1995) (unpublished).
3. F. Paige and S. D. Protopopescu, H. Baer, X. Tata, BNL Report No. 38304, 1986 (unpublished).
4. D. Claes, DØ internal Note 2546 (1995) (unpublished).
5. L. Sawyer, M. Sosebee, A. White, J. T. White, DØ internal Note 2520 (1995).

Top Quark Physics at the Tevatron

C.-P. Yuan

Department of Physics and Astronomy
East Lansing, MI 48824 , USA

We discuss physics of the top quark at the Fermilab Tevatron. By the year 2000, many properties of the top quark can be measured at the Tevatron.

1 Discovery of the Top Quark

The discovery of the top quark is one of the most important discoveries at the Fermilab Tevatron which is currently the only collider that produces top quark on its mass-shell. The mass of the top quark has been measured to be $m_t = 176 \pm 8\,(\text{stat.}) \pm 10(\text{sys.})\,\text{GeV}$ by CDF group and $m_t = 199^{+29}_{-21}\,(\text{stat.}) \pm 22(\text{sys.})\,\text{GeV}$ by DØ group, by detecting the $t\bar{t}$ events. The standard $t\bar{t}$ event selection is based on the expected Standard Model (SM) decay chain $t\bar{t} \rightarrow (W^+b)(W^-\bar{b})$ and the subsequent decays of the W's into fermion pairs. At least one W is tagged in the mode $W \rightarrow \ell + \nu$ by requiring an isolated high p_T (transverse momentum) lepton ($\ell = e$ or μ) and large $\not{E}_T$ (missing transverse energy). In the "dilepton" analysis the leptonic decay of the other W is identified with a loose lepton selection; this mode has small backgrounds but small branching ratio of just $4/81 \simeq 5\%$. In the case of the "lepton+jets" mode, the second W decays to quark pairs, giving larger branching ratio of $24/81 \simeq 30\%$. The final state of $(\ell\nu b)(jj\bar{b})$ is separated from the primary background, $W+$jets, by requiring a large multiplicity of high p_T jets and also evidence of a b-jet, using either secondary vertex (silicon detector) or soft lepton ($b \rightarrow c\ell\nu X$) identification. For more detailed discussions on the event selection and the detector configuration which determines the acceptance and the detection efficiency of the events, we refer the readers to Ref. [1].

Assuming the SM decay mode of the top quark, the cross section $\sigma_{t\bar{t}}$ for the QCD production processes $q\bar{q}, gg \rightarrow t\bar{t}$ was measured to be $7.6^{+2.4}_{-2.0}\,\text{pb}$ (by CDF) and $6.4 \pm 2.2\,\text{pb}$ (by DØ). For comparison, the SM result for $m_t = 175\,\text{GeV}$ at $\sqrt{s} = 1.8\,\text{TeV}$ is $\sigma_{t\bar{t}} = 5.52^{+0.07}_{-0.45}$ pb quoted from Ref. [2] in which the effects of multiple soft-gluon emissions have been properly resummed. Since the measurement of the cross section (obtained from the "counting" experiments, i.e. counting the observed total $t\bar{t}$ event numbers in various decay modes) and the measurement of the mass of the top quark (obtained from reconstructing the invariant mass of the top quark) are two independent measurements, one can combine these results to find the best fitted values for m_t and $\sigma_{t\bar{t}}$ [3]. We find that $m_t = 174 \pm 4\,\text{GeV}$ and $\sigma_{t\bar{t}} = 5.9 \pm 0.7\,\text{pb}$ [4]. If the $t\bar{t}$ event rate is indeed given by the SM prediction for $\sigma_{t\bar{t}}$, then the results of the fit described above conclude the branching ratio (BR) of $t \to X$ for $X \neq bW$ has to be less than $\sim 10\%$ [4].

2 Is Top Quark just another heavy quark?

m_t ($\simeq v/\sqrt{2} = 174\,\text{GeV}$) is about the scale ($v$) of the electroweak symmetry breaking (EWSB). Studying top quark might provide clues to generation of the fermion masses which could be closely related to the EWSB. Furthermore, effects of new physics originating from the EWSB would be more apparent in the top quark sector than any other light sector of the electroweak theory. Hence, the top quark system not only serves as a stage for testing the SM but also provides a window to new physics beyond the SM.

A few examples are discussed in Ref. [5] to illustrate that different models of EWSB mechanism will induce different interactions among the top quark and the W- and Z-bosons. These interactions may strongly modify the production and/or the decay of the top quark. In some models (e.g., the TopColor model [6]) observable flavor-changing neutral current (FCNC) processes (e.g., $t \to cZ$, cg, $c\gamma \ldots$) can be mediated by new underlying dynamics, and some new resonances can strongly couple to $t\bar{t}$ (e.g., a degenerate, massive color octet of "colorons" and a singlet heavy Z') or $t\bar{b}$ (e.g., a triplet of "top-pions") system. With all these new effects possibly appearing in the top quark system, we conclude that top quark is likely not just another heavy quark.

Is it a SM top quark? What do we know about the interaction of the top quark? Can we learn about that from the radiative effects to the precision LEP/SLC data (physics at the Z-pole)? A few analyses for studying the top quark couplings to the gauge bosons show that current low energy data still allow rooms for new physics [5, 7], and that only the direct measurements (with on-shell top quark produced) of these couplings can be conclusive.

3 Top Quark Physics for the Tevatron at Run-II and Beyond

To determine how well an observable can be measured at the Tevatron in the Main Injector Era (Run-II and beyond), we need to set up a reference for top quark event rates. For this purpose, we consider a $\bar{p}p$ collider with $\sqrt{S} = 2\,\text{TeV}$ and an integrated luminosity of $2\,\text{fb}^{-1}$ (or, $1\,\text{fb}^{-1}$ for each experimental group). In the SM, for a 175 GeV top quark, there will be about 1.4×10^4 $t\bar{t}$

pairs and 5×10^3 single-t or single-$\bar{t}$ events produced. After taking into account the b-tagging efficiency and the detection efficiency [1], there are about 1000 single-b-tagged $t\bar{t}$ pairs in the $\ell+$ jets sample (among those 600 are also double-b-tagged), 100 in the dilepton sample, and 250 single-t or single-$\bar{t}$ events (in the $\ell+$ jets sample) available for testing various properties of the top quark. In the following, we discuss the relevant observables and show that with a $2\,\mathrm{fb}^{-1}$ luminosity, many *first* measurements can already be done to a good accuracy at the Tevatron. With a $(10-100)\,\mathrm{fb}^{-1}$ integrated luminosity (beyond Run-II), many further improvements are expected.

3.1 $t\bar{t}$ production rate $\sigma_{t\bar{t}}$

At the Tevatron, the dominant $t\bar{t}$ pair production mechanism is $q\bar{q} \to t\bar{t}$ not $gg \to t\bar{t}$, the former contributes about 90% of the rate because the quark luminosities are larger than the gluon luminosities for large x (i.e. for producing heavy top quarks). To test QCD, we need an accurate measurement for $\sigma_{t\bar{t}}$ which is experimentally limited by the systematic uncertainties. It is concluded in Ref. [1] that the experimental error in $\sigma_{t\bar{t}}$ is $\delta\sigma_{t\bar{t}} \simeq 10\%$ which is about the same as the theoretical error for calculating $\sigma_{t\bar{t}}$ [2]. (Note that counting the dilepton event rate is a direct measurement of $\sigma_{t\bar{t}}$, and the statistical error for the dilepton sample of $t\bar{t}$ events is about $1/\sqrt{100} = 10\%$.)

3.2 Mass of the top quark m_t

With enough $t\bar{t}$ pair events, the accuracy in measuring m_t will be determined by the systematic uncertainty which is dominated by the error in measuring the jet energy scale due to the imperfections in the calorimetry and the effects of initial and final state gluon radiations. The determination on the jet energy scale can be greatly improved by studying the $Z+1$ jet and $\gamma+1$ jet events [1]. It is expected that m_t can be measured to within a couple percent. So, the uncertainty in measuring m_t is $\delta m_t \simeq (2-4)\,\mathrm{GeV}$.

3.3 Distributions of invariant mass $M_{t\bar{t}}$ and transverse momentum $p_T(t)$

If a heavy new resonance (V) can be produced in $\bar{p}p$ collision and can strongly couple to $t\bar{t}$ [8], then the observed distributions of $M_{t\bar{t}}$ and $p_T(t)$ will be different from the SM predictions. (The event rates can either increase or decrease.) By carefully comparing these distributions with the predictions of a given theory, one can then either approve or exclude that theory. Since $M_{t\bar{t}}$ can be reconstructed on an event-by-event basis by requiring that there are two top quarks observed in the event, the shape and the magnitude of this distribution can therefore set a model-independent limit on possible new physics coupled to $t\bar{t}$ pairs. Demanding a resonance to be observed at the 5σ level (i.e. $S/\sqrt{B} \gtrsim 5$) over the $t\bar{t}$ continuum in the $M_{t\bar{t}}$ spectrum, one can set the minimum bound on $\sigma(\bar{p}p \to V) * (V \to t\bar{t})$ to about $(0.4-0.8)$ pb and $(0.1-0.2)$ pb for M_V equal to 500 GeV and 800 GeV, respectively [1].

3.4 Top quark decays and FCNC decay modes of top

Because top quark is heavy, it will decay via weak interaction before it feels non-perturbative strong interaction. This is the first opportunity we have for studying the properties of a bare quark. In the SM, the total decay width of a SM top quark is $\Gamma_t \simeq 1.6\,\mathrm{GeV}(m_t/180\,\mathrm{GeV})^3$, and the branching ratio of the weak two body decay $t \to bW^+$ is about one hundred percent. In this decay mode the top quark will analyze its own polarization [9]. If t is found dominantly decaying to bW^+, the second top ($\bar{t}$) in each $t\bar{t}$ event should be carefully studied as a window for small non-SM decay modes of top quarks. We call this t-tagging.

In the SM, the branching ratios for the FCNC decay modes were found to be too small to be detected, but they can be observable in some other models. Consider the MSSM with light chargino and top-squark. One of the t in the $t\bar{t}$ event decays to $\tilde{t}_1\tilde{\chi}_1^0$, and $\tilde{t}_1$ subsequently decays to $c\tilde{\chi}_1^0$. The signature of this event is $W+2\,\mathrm{jet}+\not{E}_T$ which is not included in the counting experiments that only count events with $W+\geq 3\,\mathrm{jets}$. A careful study on this signature can approve the MSSM or set limit on the MSSM parameters [4]. Other studies [1] show that a $2\,\mathrm{fb}^{-1}$ luminosity can be sensitive to $\mathrm{BR}(t \to cZ) \sim 2\%$ (from $3\ell+2\,\mathrm{jets}$ or $2\ell+4\,\mathrm{jets}$ sample) and $\mathrm{BR}(t \to c\gamma) \sim 0.3\%$ (from $\ell+\gamma+2\,\mathrm{jets}$ or $\gamma+4\,\mathrm{jets}$ sample).

3.5 Ratios of branching ratios: $\mathcal{R}_\ell$ and $\mathcal{R}_b$

Define $\mathcal{R}_\ell$ to be the ratio of the $t\bar{t}$ cross sections measured using dilepton events to that measured using $\ell+$ jets events. If $\mathcal{R}_\ell$ differs from the SM prediction, then it implies new physics that would allow t decay without a W boson in the final state, such as charged Higgs ($t \to bH^+$ [10]) and top-squark ($t \to \tilde{t}_1\tilde{\chi}_1^0$). Hence, $\mathcal{R}_\ell$ measures $\mathrm{BR}(t \to bW)$. With $2\,\mathrm{fb}^{-1}$, the error on $\mathrm{BR}(t \to bW)$ is about 10% [1]. Another useful ratio $\mathcal{R}_b$ is the ratio of the $t\bar{t}$ cross sections measured using double-b-tagged events to that measured using single-b-tagged events. This determines the upper limit on the branching ratio of $t \to X$ where X does not contain any b-quark. (This can be applied to both $\ell+$ jets and dilepton samples, from a known b-tagging efficiency.) With $2\,\mathrm{fb}^{-1}$, this upper limit can be set to about $(3-5)\%$ [1]. This result can be interpreted as the error on measuring BR($t \to bW$) if a W boson is confirmed in t decay. (Generally a small BR($t \to X$) is better measured from direct search of the rare decay mode than from the measurements of $\mathcal{R}_\ell$ and $\mathcal{R}_b$.)

3.6 Partial decay width $\Gamma(t \to bW)$ and the lifetime of top

The total decay width Γ_t of a SM top quark cannot be measured from the invariant mass of t reconstructed in the $t\bar{t}$ events [11]. An elegant way to determine the lifetime (the inverse of the total decay width) of the top quark is to measure the partial decay width $\Gamma(t \to bW)$ and the branching ratio BR($t \to bW$), since $\Gamma_t = \Gamma(t \to bW)/\text{BR}(t \to bW)$. As shown in Ref. [11], $\Gamma(t \to bW)$ can be *model-independently* measured by counting the production rate of single-top events produced from the W-gluon fusion process (which is *equivalent* to the W-b fusion process), because the production rate of $Wb \to t$ is directly proportional to the decay rate of $t \to bW$. The cross section for the W-gluon fusion process is known to about $(15-20)\%$ [11], the lifetime of the top quark can therefore be determined to about $(20-30)\%$.

Before concluding this section, we note that measuring the single-top event rate from the W-gluon fusion process is an *inclusive* method for detecting effects of new physics which might produce large modifications to the interactions of the top quark [11]. This provides the *first* hint of possible large new physics effects in the production of the single-top events at the Tevatron, and a further detailed test of the t-b-W couplings can be achieved by studying the decay of top quarks in both the $t\bar{t}$ pairs and the single-top events.

3.7 Form factors of t-b-W, V_{tb} and m_t

To describe the decay of $t \to bW(\to \ell\nu)$, the most general form factors contain $f_1^{L,R}$ and $f_2^{L,R}$ [9]. Using the invariant mass ($m_{b\ell}$) of b and ℓ, one can determine the polarization of the W boson which depends on the values of these form factors. A study [12] shows that the errors $\delta f_1^L \sim (2-3)\%$ and $\delta f_1^R \sim 0.2$, if $f_1^L \approx 1$ and f_1^R and $f_2^{L,R}$ are almost zero (SM values). The measurement of f_1^L can be interpreted as the measurement of the CKM element V_{tb} which is therefore known to within 3%. (We note that this result is more accurate than that obtained from measuring the $qq' \to W^* \to t\bar{b}$ production rate which yields a 10% error in measuring V_{tb} [1].) Furthermore, the fraction (F_L) of longitudinal W's from top decays is equal to $\frac{1}{2}\frac{m_t^2}{M_W^2}$, and is independent of $f_1^{L,R}$, it is therefore a good tool for measuring m_t [9]. (The quadratic dependence of F_L on m_t helps in δm_t by a factor of 2.) A 2% accuracy in determining F_L yields a $\sim 2\,\text{GeV}$ error in measuring m_t.

3.8 Exotic production mechanism and testing CP violation in top

If new physics strongly modifies the coupling of t-c-g, then the production rate of t-c pair from $q\bar{q} \to g \to t\bar{c}$ can be largely enhanced. By measuring its production rate in high M_{tc}, one can set the minimum energy scale Λ_{tcg} at which new physics must set in. It is found that $\Lambda_{tcg} \gtrsim 2\,\text{TeV}$ [13].

Besides all the potential physics discussed above, the Tevatron, as a $\bar{p}p$ collider, is unique for being able to test CP violation by measuring the production rates of single-top events. A nonvanishing asymmetry ($\mathcal{A}_t$) in the inclusive production rates of the single-t events and the single-$\bar{t}$ events signals CP violation [14]. With $2\,\text{fb}^{-1}$, it is possible to observe this asymmetry for $\mathcal{A}_t \gtrsim 20\%$ [14].

This work is supported in part by the NSF under grant no. PHY-9309902. We thank our colleagues who contribute to the studies in Ref. [1].

References

1. Dan Amidei and Chip Brock, "Report of the *TeV2000* Study Group on Future ElectroWeak Physics at the Tevatron", 1995; and references therein.
2. E.L. Berger and H. Contopanagos, ANL-HEP-PR-95-31, May 1995.
3. D. Soper, talk at the QCD session of the XXX Rencontre de Moriond, Les Arcs, France, March 1995.
4. S. Mrenna and C.–P. Yuan, in preparation.
5. Ehab Malkawi and C.–P. Yuan, *Phys. Rev.* **D50**, 4462 (1994).
6. C.T. Hill, *Phys. Lett.* **B345**, 483 (1995).
7. J. Feliciano, F. Larios, R. Martinez and M.A. Perez, CINVESTAV-FIS-09-95, 1995;
   T. Han, R.D. Peccei and X. Zhang, Fermilab-pub-95/160-T, 1995.
8. C.T. Hill and S. Parke, *Phys. Rev.* **D49**, 4454 (1994);
   E. Eichten and K. Lane, *Phys. Lett.* **B327**, 129 (1994).
9. G.L. Kane, G.A. Ladinsky and C.–P. Yuan, *Phys. Rev.* **D45**, 124 (1992).
10. J. Guasch, R. Jimenez and J. Sola, UAB-FT-370, July 1995.
11. D.O. Carlson and C.–P. Yuan, MSUHEP-50823, August 1995.
12. D.O. Carlson, Ph.D. thesis, Michigan State University, MSUHEP-050727, August 1995.
13. E. Malkawi and T. Tait, private communication.
14. C.–P. Yuan, *Mod. Phys. Lett.* **A10**, 627 (1995).

On Gluon Radiation in $t\bar{t}$ Production and Decay

Lynne H. Orr

Department of Physics, University of Rochester
Rochester, NY 14627-0171, USA

T. Stelzer

Department of Physics, University of Durham
Durham DH1 3LE, England

W.J. Stirling

Department of Physics and Mathematical Sciences, University of Durham
Durham DH1 3LE, England

Understanding the pattern of gluon radiation in $t\bar{t}$ production processes is important for making an accurate determination of the top mass from the measurement of its decay products. In a recent paper we showed that the exact matrix element and parton shower (HERWIG) calculations gave very different results for the distribution of gluon jets in $t\bar{t}$ production at the Tevatron $p\bar{p}$ collider. By repeating the calculation for the simpler $e^+e^- \to b\bar{b}W^+W^-g$ process, we reveal even more dramatic differences between the two approaches. We conclude that there are significant differences in gluon radiation between HERWIG and the matrix element calculation in regions of phase space where one would expect agreement.

Having established the existence of the top quark through its production in $p\bar{p}$ collisions at the Fermilab Tevatron [1,2], the CDF and D0 collaborations must now accurately measure the top mass. To do so requires understanding gluon radiation in top events. Because top is observed only through its decays, measuring its mass requires reconstructing the momenta of its decay products and comparing measured invariant mass distributions with those predicted by theory. And because gluons can be radiated in top production *and* decay, top events often contain extra jets which may or may not need to be counted among the decay products. While this cannot be decided on an event-by-event basis, in principle the problem of extra jets can be dealt with as long as their distributions are properly understood.

In previous work [3,4] we studied the distributions of extra jets in $t\bar{t}$ production and decay at the Tevatron $p\bar{p}$ collider. In [4] we performed a complete tree-level calculation of the process $p\bar{p} \to bW^+\bar{b}W^- +$ jet $+X$, in which we included all contributions from gluons emitted in the production and decay stages as well as all top width effects and spin correlations. We compared our exact matrix element results to those obtained using the parton-shower Monte Carlo program HERWIG [5], which is widely used in the experimental analyses. We found a significant discrepancy between the two calculations. Although we were unable to identify the reason for the discrepancy, the differences in jet distributions seemed to indicate a relative lack of gluons radiated in top decay compared to top production in HERWIG.

In this paper we pursue further the comparison of matrix element (ME) and parton shower (PS) calculations of gluon radiation in top production and decay. In light of the subtle issues associated with making ME–PS comparisons, and of the complications associated with hadronic top production, it is necessary to investigate the differences in a systematic way. Therefore we focus here on the simpler case of e^+e^- collisions, where many of the complications of $p\bar{p}$ collisions, such as initial state radiation, are absent, and where a cleaner comparison can be performed. Then, armed with our e^+e^- results, we return to a discussion of the hadronic case.

Comparing a fixed order calculation with a full parton shower program presents several challenges. Most notable is that the ME calculation produces only partons directly from the hard process, whereas in the PS calculation, any number of partons can be present as a result of showers, and multiple gluon effects can be important. In order to develop a meaningful method for comparison, we begin by studying the well-understood process $e^+e^- \to q\bar{q}g$, where q is a light quark. Next, we make the same comparison but including a non-zero quark mass ($e^+e^- \to Q\bar{Q}g$). From these processes a useful method of comparison is found, and the regions of applicability for the different calculations are studied. Having understood the simpler processes, we then consider $e^+e^- \to bW^+\bar{b}W^- + g$ via top pair production. Cuts are kept to the minimum necessary to avoid singularities.

In each case, the final state hadrons in the ME calculation are two quarks and a single extra gluon (we ignore the W's from top decay; they can be considered to decay leptonically). In the PS calculation the final state contains two quarks plus any number of showered partons. In order to make the ME–PS comparison, we must use some algorithm to combine showered partons in the PS calculation so that we are left with at most three jets.[a] We use the Durham successive combination algorithm [6], which has the advantages [7] that it reproduces LEP data well and lends itself to direct comparison with ME calculations.

[a]The PS results were obtained using HERWIG [5] (v5.8), with hadronization and b quark decays turned off.

We apply the Durham algorithm as follows. For each pair of partons (or jets) i and j, we compute the quantity

$$y_{ij} \equiv \frac{2\min(E_i^2, E_j^2)(1 - \cos\theta_{ij})}{s}, \tag{1}$$

where $E_{i(j)}$ is the energy of parton $i(j)$, θ_{ij} is the angle between them, and s is the process center-of-mass energy. The pair with the smallest y_{ij} are combined into a single jet with four-momentum $p = p_i + p_j$. The jet replaces the partons in the y_{ij} computations and we repeat the process until no more than three jets remain. Note that for small angles, the numerator of y_{ij} is simply the k_T^2 of the less energetic parton with respect to the more energetic one. Hence the algorithm causes the partons with the smallest relative transverse momentum to be combined. In addition, the dimensionless variable y is somewhat like an angular variable, so that this method is analogous to the angular cone algorithm used in our comparisons for $p\bar{p}$ [4].

After using the jet algorithm, we have three-'jet' events from both PS and ME calculations, and they can now be compared. To do so, we make further use of the Durham variable: for each event, we plot the distribution as a function of the smallest value of y ($\equiv y_{\rm min}$) for all pairings of the remaining three jets. Because large values of $y_{\rm min}$ correspond to large relative k_T, *i.e.*, large angles and energies, the ME and PS results should agree well in this region. Small values of $y_{\rm min}$ bring us into the soft/collinear regime, where the ME result diverges as $y_{\rm min} \to 0$. In this region multiple gluon effects become important, and we expect the PS result, which takes these into account via resummation, to remain finite.

This behavior is illustrated in Fig. 1(a), which shows the $y_{\rm min}$ distribution for $e^+e^- \to q\bar{q}g$ (massless quarks) with a center-of-mass energy of 100 GeV. Since we are primarily interested in shapes of distributions, the normalization of the ME calculation is chosen (via the α_s scale) so that the distributions for $e^+e^- \to q\bar{q}g$ agree at large values of $y_{\rm min}$. We see, then, that ME and PS agree quite well for large $y_{\rm min}$, not only in normalization, which we have fixed, but in shape, which we are not free to manipulate. As $y_{\rm min}$ decreases, the ME distribution rises, diverging as $\alpha_s \ln^2(1/y_{\rm min})$, but we see the multigluon effects begin to show in the PS curve, which turns over and remains finite.

The effect of giving the final quark a mass is shown in Fig. 1(b) for the process $e^+e^- \to b\bar{b}g$. Note that the b mass has weakened the ME divergence, and the distribution for all $y_{\rm min}$ is less steep than in the massless case. The PS calculation again reproduces the ME result (including the rounding due to the mass) quite well in the large $y_{\rm min}$ region, and we see the two results begin to disagree as multigluon effects come in for smaller $y_{\rm min}$.

Satisfied that we understand the above simple cases, we turn to top production. The ME and PS calculations performed here for e^+e^- production of top are similar to those performed in [4]. We take $m_t = 175$ GeV. To obtain the ME results we have performed an exact parton-level calculation of $e^+e^- \to bW^+\bar{b}W^- + g$, including top width effects, using helicity amplitudes generated by the MadGraph [8] package. The PS results were obtained using HERWIG [5] (v5.8), with hadronization and b quark decays switched off. As in the case of hadronic top production, we do not consider radiation off the W decay products, and the W's can be considered to decay leptonically.

We proceed with the jet algorithm as above, and compare the ME and PS $y_{\rm min}$ distributions for center-of-mass energy 420 GeV in Fig. 2.[b] The difference is quite dramatic. The ME curve displays further rounding as one might expect from a higher mass quark or from the lack of correlation between the final state b quarks and gluons radiated from the t or $\bar{t}$. The behavior of the PS curve is unexpected. We see the usual turn-over at small values of $y_{\rm min}$, but the large $y_{\rm min}$ behavior is surprising. Not only does PS not reproduce ME; its shape is so different that there is no shift in normalization that would cause the two to agree.

Given that the matrix element calculation is of fixed order in perturbation theory, one expects that its region of validity would start at large values of $y_{\rm min}$ and extend to the left to the region where we expect multiple gluon emissions to become important. Since the parton shower has these multiple emissions, but uses an angular ordering algorithm, we expect its region of validity to be from small values of $y_{\rm min}$ and extend to the right. The fact that there is almost no region of overlap is very disturbing.

We have been unable to pin down the reason for the above discrepancy. Although it is in principle difficult to compare fixed order (ME) with approximate all-orders (PS) calculations, the fact that the comparison works well for *light* quarks suggests that this is not the explanation for the difference. Furthermore, there is no reason why a matrix element calculation which works well for light quarks should fail for heavy quarks, provided of course that appropriate cuts are used to avoid the soft and collinear regions.

We are therefore forced to conclude that the implementation of very heavy quark production and/or decay in HERWIG may not be correct. One can perhaps shed further light on this by comparing the two calculations for *stable* top production, thus testing the 'production' part of the jet cross section in each case. The result of such a comparison is shown in Fig. 3, where the distribution in $y_{\rm min}$ is shown for stable $m_t = 175$ GeV $t\bar{t}g$ production in e^+e^- collisions

---

[b]This energy is chosen to approximate the average subprocess center-of-mass energy for $t\bar{t}$ production in $p\bar{p}$ collisions at the Tevatron.

at $\sqrt{s} = 420$ GeV. In contrast to Fig. 1 (essentially the same distribution but for light quarks), there is a marked difference between the two calculations in the medium-small $y_{\rm min}$ region. Note that in this case there is no collinear singularity and so the divergence as $y_{\rm min} \to 0$ is weaker ($\sim \alpha_s \ln y_{\rm min}$) in the ME distribution. It is difficult to see how higher-order multi-gluon contributions could produce such a pronounced *increase* at $y_{\rm min} \sim 10^{-4}$ in the PS calculation. By comparing Figs. 2 and 3, we see that the difference in the 'production' part can also be seen in the full 'production + decay' distribution. It is difficult to estimate the effect of this difference on top quark analyses, and in particular on the measurement of the top mass, but clearly every effort should be made to ensure that the event simulators used to study the top quark do indeed constitute a good approximation to the underlying physics.

Acknowledgments

Two of us (TS,WJS) are grateful to the UK PPARC for financial support. Useful discussions with Nigel Glover, Michelangelo Mangano, Stephen Parke and Bryan Webber are acknowledged. This work was supported in part by the U.S. Department of Energy, under grant DE-FG02-91ER40685 and by the EU Programme "Human Capital and Mobility", Network "Physics at High Energy Colliders", contract CHRX-CT93-0537 (DG 12 COMA).

1. CDF collaboration: F. Abe *et al.*, Phys. Rev. Lett. **74** (1995) 2626.
2. D0 collaboration: S. Abachi *et al.*, Phys. Rev. Lett. **74** (1995) 2632.
3. L.H. Orr and W.J. Stirling, Phys. Rev. **D51** (1995) 1077.
4. L.H. Orr, T. Stelzer and W.J. Stirling, Durham University preprint DTP/94/112 (1994), hep-ph/9412294, to be published in Phys. Rev. **D**.
5. G. Marchesini and B.R. Webber, Nucl. Phys. **B310** (1988) 461.
   G. Marchesini, B.R. Webber, G. Abbiendi, I.G. Knowles, M.H. Seymour and L. Stanco, Comp. Phys. Commun. **67** (1992) 465.
6. Yu.L. Dokshitzer, contribution to the *Workshop on Jets at LEP and HERA*, Durham, England, December (1990); see summary by W.J. Stirling, J. Phys. G **17** (1991) 1572.
   S. Catani, Yu.L. Dokshitzer, M. Olsson, G. Turnock and B.R. Webber, Phys. Lett. **B269** (1991) 432.
7. S. Bethke, Z. Kunszt, D.E. Soper, and W.J. Stirling, Nucl. Phys. **B370** (1992) 310.
8. T. Stelzer, W.F. Long, Comp. Phys. Commun. **81** (1994) 357.
9. S. Frixione, M.L. Mangano, P. Nason and G. Ridolfi, CERN preprint CERN-TH-95-52 (1995).

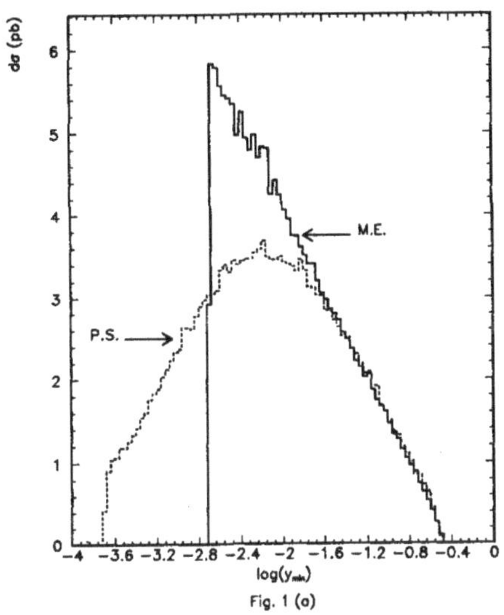

e+e- -> qq at sqrt(s)=100 GeV. PS (HERWIG) vs ME as a function of y_min.

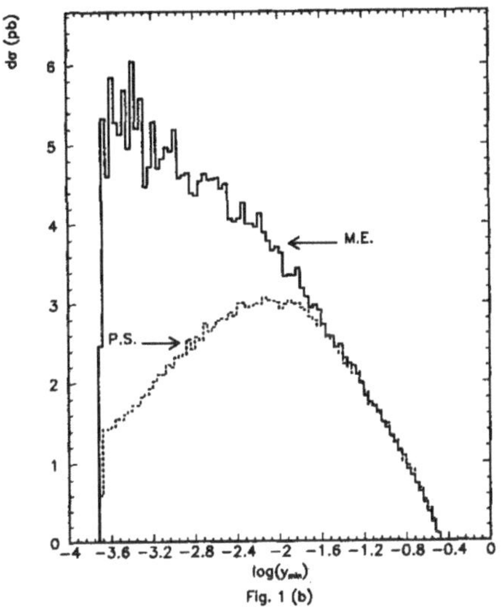

e+e- -> bb at sqrt(s)=100 GeV. PS (HERWIG) vs ME as a function of y_min.

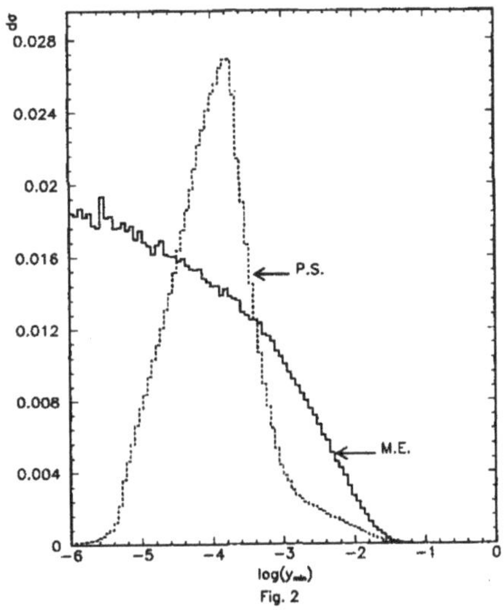

e+e- ->ttg at sqrt(s)=420 GeV.
PS (HERWIG) vs ME as a function of y_min.

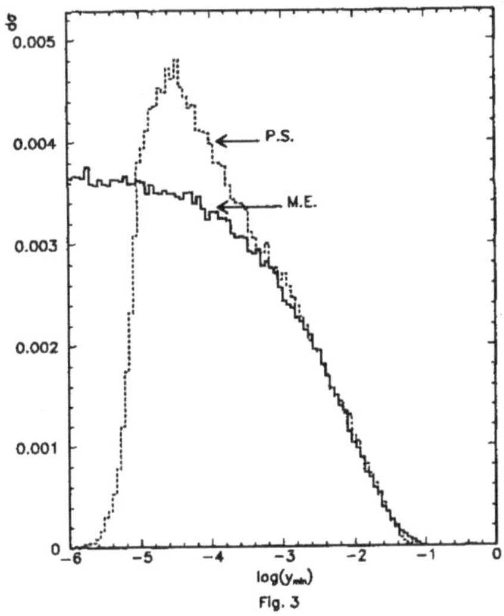

e+e- ->WWbb at sqrt(s)=420 GeV.
M_top=175 and no top decay.
PS (HERWIG) vs ME as a function of y_min.

THE CMS DETECTOR AND PHYSICS AT THE LHC

D. Denegri,
CE SACLAY, DAPNIA/SPP AND CERN/PPE

ABSTRACT

We discuss the physics issues at the LHC and the CMS detector, one of the two general purpose detectors planned at the LHC, the expected physics performance of LHC detectors in the search for SM and SUSY Higgs bosons, some SUSY particle searches and some possibilities in B and heavy ion physics.

1. INTRODUCTION

The main motivation for building the LHC is to investigate the mechanism responsible for electroweak symmetry breaking. The Standard Model (SM) Higgs mechanism is such a possibility [1]. The Minimal Supersymmetric Standard Model (MSSM) is a possible extention, but alternative symmetry breaking schemes with strongly interacting W_L, Z_L in the TeV mass range have been proposed too [2]. The present mass constraints on the SM Higgs are loose ($m_H < 1$ TeV) [1]. The design and optimisation of an LHC detector to be able to explore the entire expected SM Higgs mass range taking into account the the variety of decay modes and expected experimental signatures is an adequate design criterion. It provides enough flexibility to allow an investigation of the ew symmetry breaking mechanism, even if there were no elementary Higgs bosons. The ATLAS and CMS collaborations propose to build detectors designed for such investigations at the highest luminosities available ($L \approx 10^{34}$ cm^{-2}s^{-1}) in $\sqrt{s} = 14$ TeV proton–proton collisions at the LHC.

The CMS detector (Fig.1) is built around a large, 13m long, 6m diameter, high-field superconducting solenoid (4 T) leading to a compact design for the muon spectrometer, hence the name Compact Muon Solenoid (CMS) [3]. The hadronic and electromagnetic calorimeters are located inside the coil. The innermost part of the detector is occupied by a 6m long, 1.3m radius central tracker. To detect signatures of new physics efficiently, identification and precise measurement of muons, photons and electrons is emphasised in the design of CMS. The goal of this experiment is to measure these particles with an energy resolution of about 1% over a large momentum range.

The central feature of the ATLAS detector (Fig.2) is a large air-core toroidal magnet muon spectrometer consisting of an 8-coils barrel toroid with 0.8 T average field, each coil being 26m long and 4.7m wide, and two endcap toroids [4]. This system would allow precise muon measurements at highest luminosities using the external muon system alone. The measurements are done in the air behind calorimeters that have absorbed all the hadrons. The electromagnetic calorimeter is a liquid argon one with a particular "accordion" geometry. A thin 2T solenoid is placed in front of the em calorimeter to allow momentum measurements in the inner tracker of similar dimension to the CMS one. Although the goals of these two general purpose detectors are very much the same, the experimental techniques are different and complementary, in particular in the B-field configuration and calorimeter choices.

With the investigation of the origine of particle masses, the second, and not unrelated, most important task for these detectors will be the search for the supersymmetric partners of Standard Model particles. The production cross sections for the strongly interacting squarks and gluinos are large at hadron colliders. At the LHC this allows to search for these particles up to a mass of ~ 2 TeV ie over the range where supersymmetry could be relevant to ew symmetry breaking. A significant discovery potential also exists for sleptons and charginos/neutralinos for masses up to ~ 0.2 TeV [5]. Particularly important would be the (indirect) detection of the lightest neutralino as the LSP as it is the best cold (or mixed) dark matter candidate. Introduction of supersymmetry cures the mass divergence problems of the SM Higgs, but at the expense of introducing several Higgs bosons [6]. In the MSSM there are five such states and they can be investigated over a large portion of parameter space in ATLAS and CMS. Detector requirements for the search for the MSSM Higgs bosons are similar as for the SM Higgs. Searches for squarks and gluinos depend crucially on the missing momentum measurements. Both detectors are thus designed to be highly hermetic. Missing momentum resolution is important in detecting MSSM Higgs modes with τ's in the final state, $H^{\pm} \rightarrow \tau\nu$ and h, H, $A \rightarrow \tau\tau$.

Although high luminosity is essential to cover the entire range of mechanisms of electroweak symmetry breaking and explore a significant fraction of SUSY parameter space, the LHC will start at a significantly lower luminosity. Both the ATLAS and CMS detectors are designed to take full benefit of the lower luminosities ($L \leq 10^{33}$ cm^{-2}s^{-1}) to study large cross-section phenomena such as beauty and top production. Such studies require powerful inner tracking and microvertexing systems. The recent observation by CDF of a number of exclusive modes $B \rightarrow J/\psi K^0_s$, $J/\psi K^{\pm}$, $J/\psi K^*$, $J/\psi\phi$ shows that at least some exclusive B channels can be extracted with a good signal to noise ratio[7]. The 10^{12} to 10^{13} $b\bar{b}$ produced per year at the LHC should allow the study of CP violation in the B sector, and of some rare B decays, such as $B \rightarrow \mu\mu$, testing physics beyond the Standard Model. B-factories at 10^{33}cm^{-2}s^{-1} are limited to few times 10^7 $b\bar{b}$/year, whilst CDF/D0 or HERA/B to $\sim 10^9$ $b\bar{b}$/year. A dedicated B-physics detector, LHC-B, is also under active study [8]. It would be a single arm spectrometer covering the angular range from ~ 10 mrad to ~ 400 mrad to the beam line, optimised for data the 10^{32} to $\sim 10^{33}$ cm^{-2}s^{-1} range.

An important part of the LHC physics programme will be devoted to heavy ion collisions. The expected luminosities go from 10^{27}cm^{-2}s^{-1} in Pb-Pb collisions to 3×10^{31}cm^{-2}s^{-1} in O-O collisions. The nucleon-nucleon center of mass energy would be 5.4 TeV. Heavy ion beams at LHC should provide collision energy densities well above the expected threshold for formation of a quark-gluon plasma. A dedicated detector, ALICE [9], is being considered for studies of a broad range of signatures of quark-gluon plasma formation. The emphasis in this detector is on efficient detection of soft particles and on particle identification; a limited solid angle calorimetric coverage is foreseen and a possibility to adjoin a muon spectrometer is also contemplated. Particular signatures of quark-gluon plasma could be looked for in the general purpose detectors too. Colour screening effects on heavy quark bound states is such a possibility [10]. The CMS collaboration plans to measure the $\mu^+\mu^-$ rates in the Υ family and study the suppression of Υ(2S) and Υ(3S) relative to Υ(1S) with different ion species, and relative to pp collisions [3]. The muon arm in ALICE could allow to perform these studies too.

In the following we discuss in more detail the CMS detector as an example of a general purpose detector at LHC, then we discuss the expected physics performance and reach for SM and SUSY Higgs boson searches, make some remarks on SUSY particle searches and finally we give some examples of B and heavy ion physics possibilities at the LHC.

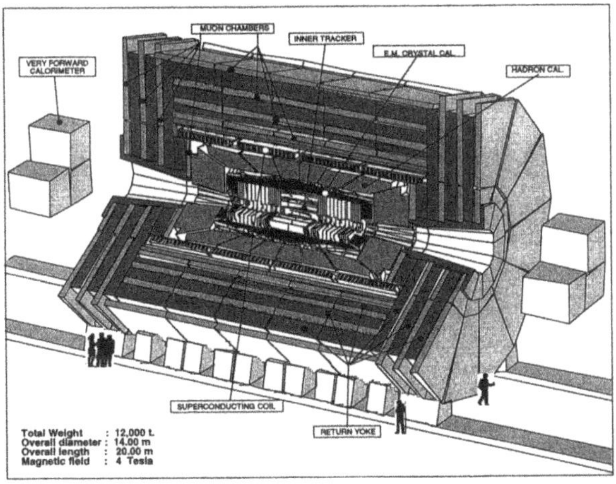

Fig. 1 : Three-dimensional view of CMS

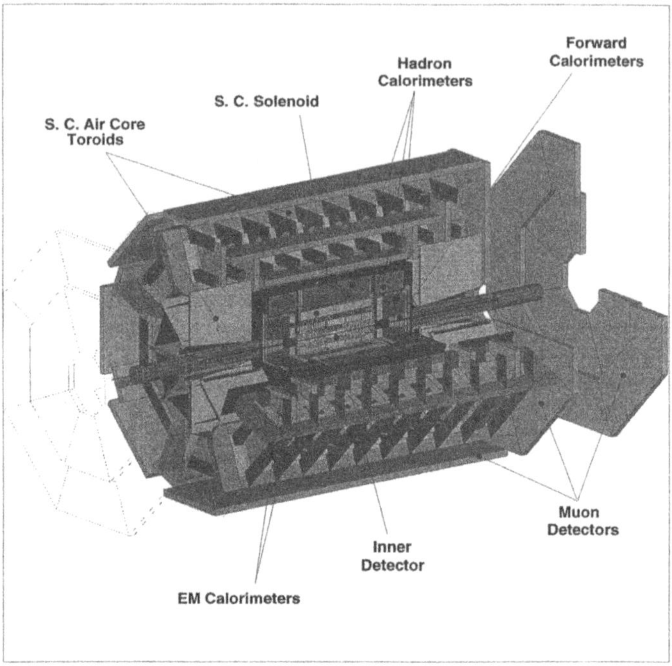

Fig. 2 Three-dimensional view of the ATLAS detector. The overall length is 40m and the diameter 20m.

2. THE CMS DETECTOR

Magnet

The central piece of CMS is a long solenoid (L = 13m) with an inner radius of 2.95 m generating a uniform magnetic field of 4 T [1]. The magnetic flux is returned through a 1.8 m thick saturated iron yoke (1.8 T) instrumented with muon chambers. The integral bending power of the solenoid is 17 Tm up to rapidity of 1.5 and decreases to ≈ 6 Tm at $|\eta| = 2.5$. A single magnet thus provides the necessary bending power for precise inner and muon tracking, and efficient muon detection and measurement up to rapidities of 2.5. Measurements within the iron yoke provide the muon stand-alone capability. Figure 1 shows a 3-dimensional view of CMS. The overall dimensions of the detector are: a length of about 22 m, a diameter of 14.6 m and a total weight of 14500 tons.

Muon system

The 4 T solenoidal field leads to an excellent momentum resolution and facilitates sharp muon momentum thresholds at trigger level. Starting from the primary vertex, muons are first measured in the inner tracker, then traverse the calorimeters, the coil and a return yoke (Fig. 3). They are identified and measured in four muon stations inserted in the return yoke in the barrel and endcap regions. The four stations provide redundancy and optimize geometrical acceptance. The four muon stations also include triggering planes that identify the bunch crossing and enable a cut on the muon transverse momentum at the first trigger level. This arrangement of chambers (Fig. 3) allows low p_t (~ 4 GeV) muon triggers required for CP violation studies and for the study of $\Upsilon(\Upsilon',\Upsilon'') \rightarrow \mu^+\mu^-$ production in heavy ion collisions.

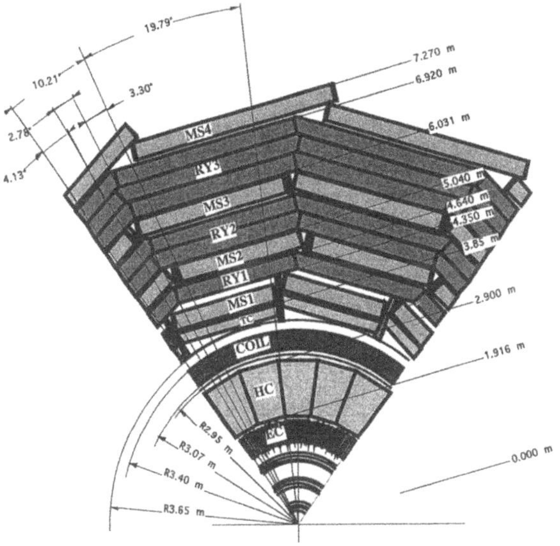

Fig. 3: Transverse view of CMS in the barrel region with the four muon stations MS1 to MS4.

In a barrel muon station, about 40 cm deep, two groups of 4 layers of aluminium DTBX chambers (drift tubes with bunch crossing capability) are used in the bending plane to measure precisely the rϕ coordinate and the local trajectory slope, and to define the bunch crossing using a mean-timer technique [11]. They are supplemented by 4 DTBX layers for the measurement of the other (z) coordinate. There are about 2×10^5 drift tube channels in the barrel. The goal is to achieve a space resolution of better than $\approx$ 200 μm per layer giving a position accuracy of $\approx$ 100 μm and angular accuracy of $\approx$ 1mrad per station. At trigger level the DTBX system provides a local trajectory slope with ~ 15 mrad accuracy. A time resolution of much better than 25ns is needed to identify the bunch crossing. Six layers of resistive plate chambers (RPC), two in MS1 and MS2 and one in MS3 and MS4, with an excellent time resolution (σ ~2nsec) and sufficient spatial resolution are also used in a first-level muon trigger [12]. The four endcap muon stations which must work in a magnetic field use six-layer CSC (cathode strip chambers) for precision measurement of muon position and momentum [13]. The precision of CSC's is ~ 100μm per layer. The CSC's are also used for triggering and they are supplemented, as the DTBX's in the barrel region, by the RPC trigger system which extends up to $|\eta| = 2.1$. This double triggering scheme insures robustness of muon triggering.

In CMS the lowest values of the muon thresholds (determined by muon penetration through calorimeters) for 90% trigger efficiency in various rapidity ranges are the following:

$$p_t^{\mu} > 4.3\,\mathrm{GeV} \quad \text{for: } 0.0 < |\eta| \le 1.5:$$

$$p_t^{\mu} > 3.4\,\mathrm{GeV} \quad \text{for: } 1.5 < |\eta| \le 2.0:$$

$$p_t^{\mu} > 2.4\,\mathrm{GeV} \quad \text{for: } 2.0 < |\eta| \le 2.5:$$

These thresholds will be useful for multi-muon final states at ~$10^{33}\mathrm{cm}^{-2}\mathrm{s}^{-1}$, in B physics studies or in heavy ion runs in particular. Two-muon final states from Z or Higgs decays are less demanding, requiring $p_t^{\mu} > 7$-10 GeV. At $10^{34}\mathrm{cm}^{-2}\mathrm{s}^{-1}$ the di-muon trigger would have a threshold at ~10 GeV and the more demanding, rate limited, single muon trigger with a threshold at 20 GeV would result in a 6kHz first-level trigger output rate [3]. If the magnetic field were reduced to 3T the same trigger threshold would lead to a 10 kHz first-level trigger rate.

Tracking

The goal of the inner tracking system of CMS is to reconstruct high p_t muons and isolated electrons in $|\eta| < 2.5$ with a momentum resolution of $\Delta p_T/p_T \approx 0.15 p_T \oplus 0.5\%$ (p_T in TeV), as well as hadrons down to low transverse momenta (≥ 2 GeV). Lepton and photon isolation is a very important criterion in signal selection for a number of important signals such as H_{SM}, susy $\rightarrow ZZ^* \rightarrow 2l^+2l^-$, $H_{SM,SUSY} \rightarrow \gamma\gamma$ or in sparticle searches, in particular for sleptons, charginos, neutralinos leading to leptons in the final state [5,3]. To identify isolated leptons or photons it is thus important to reconstruct all high p_T tracks ($p_T > 2$ GeV) including hadrons in the central rapidity region. Track measurement, counting and isolation is also the main requirement in selecting τ candidates. An excellent momentum resolution (≤1 % below 100 GeV) is needed for electromagnetic calorimeter calibration through E/p matching using $Z \rightarrow e^+e^-$ and $W \rightarrow e\nu$ decays, and is desirable for a good signal to background ratio for narrow states such as $J/\Psi \rightarrow \mu^+\mu^-$, $\Upsilon(\Upsilon',\Upsilon'') \rightarrow \mu^+\mu^-$, $B^0_d \rightarrow \pi^+\pi^-$, $B^0_s \rightarrow \mu^+\mu^-$ or $Z \rightarrow l^+l^-$, $H_{SM} \rightarrow ZZ^* \rightarrow 2l^+2l^-$, $H_{SUSY} \rightarrow ZZ^*$, $ZZ \rightarrow 2l^+2l$, and h, H, $A \rightarrow \mu^+\mu^-$.

The main problem in tracking is that of pattern recognition. At a luminosity of 10^{34}cm^{-2}s^{-1}, interesting events will be superimposed on a background of about 500 soft charged tracks within the rapidity range considered from ~15 minimum bias events occurring in the same bunch crossing. Their vertices are distributed along the beam direction (z-axis) with a r.m.s. of 5.3 cm. To solve the pattern recognition problem at high luminosity detectors with small cell sizes are required. In CMS silicon and gas microstrip detectors provide the required granularity and precision. Strip lengths of the order of 10 cm are necessary to maintain the cell occupancies below 1%. This leads to a large number of detection channels ($\approx 10^7$).

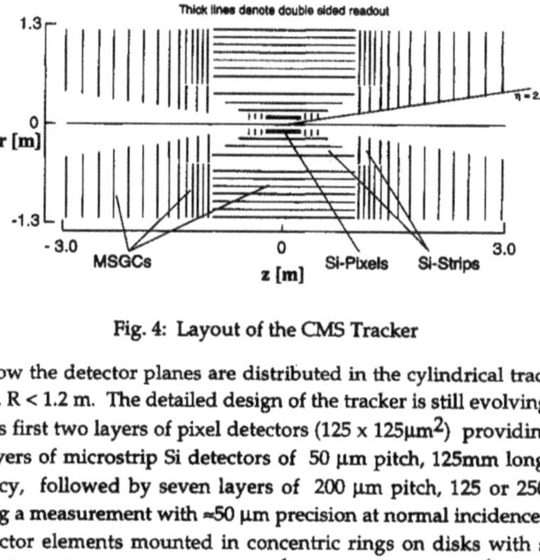

Fig. 4: Layout of the CMS Tracker

Figure 4 shows how the detector planes are distributed in the cylindrical tracking volume of CMS with dimensions $|z| < 3.0$ m, $R < 1.2$ m. The detailed design of the tracker is still evolving. A track in the barrel part of the tracker encounters first two layers of pixel detectors (125 x 125μm^2) providing a measurement accuracy of 15 μm, then three layers of microstrip Si detectors of 50 μm pitch, 125mm long providing high precision points of 15 μm accuracy, followed by seven layers of 200 μm pitch, 125 or 250 mm long gas microstrip chambers (MSGC) giving a measurement with ≈50 μm precision at normal incidence[14]. The forward tracking system is made of detector elements mounted in concentric rings on disks with silicon strip detectors and MSGC detectors. The total number of channels is 11x10^6 for MSCG's and 3x10^6 for silicon detectors. High track finding efficiencies are expected for isolated tracks, as well as for tracks in jets, the mean efficiency in $|\eta| < 2.5$ is about 95% even inside jets [3, 15]. Figure 5 shows the expected track momentum resolutions in the CMS tracker alone. For high momentum muons the combination of tracker and muon chamber measurements improves very much the resolution: $\Delta p_T/p_T \approx 0.06$ for a p =1 TeV muon in $|\eta| <1.6$ [3]. The relative alignement precision required for the inner tracker and external muon system is ~ 100 μm.

The two pixel layers at a radial distance of 7.5 and 11 cm from the beam line (Fig. 4) with the three forward pixel discs, altogether 80x10^6 channels, insure precise impact parameter measurements, with an asymptotic (high momentum) accuracy of σ_{IP} = 23 μm in the transverse plane, Fig. 5 [3]. The possibility to bring the inner layer to a ~4cm radius for initial low luminosity running is also being considered. Impact parameter measurements play an essential role in B-physics (CP violation, B^0 oscillations) and in tagging b-jets in high p_t

events, such as $t\bar{t}$ production, associated $b\bar{b}H$ production, in the search for $H \to b\bar{b}$ decays, as well as for modes involving τ's ($H^{\pm} \to \tau\nu$ and h, H, A $\to \tau\tau$). Microvertex b-tagging played a key role in the recent discovery of top. The pixels are also very useful for pattern recognition.

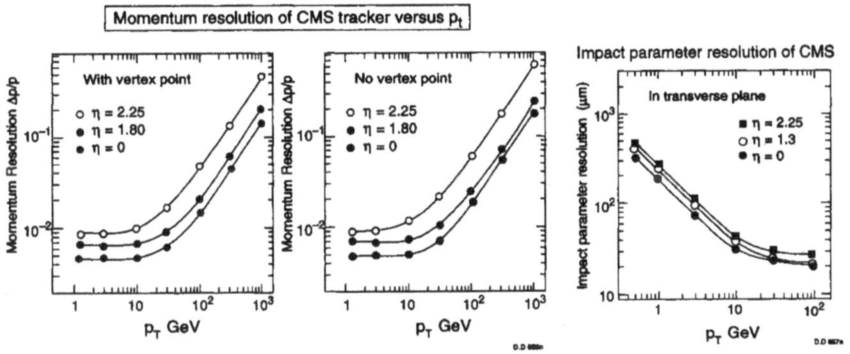

Fig. 5: Momentum and impact parameter resolutions as a function of p_t at various rapidities.

Calorimetry

The primary function of the electromagnetic (EM) calorimeter is to measure precisely electrons and photons. In conjunction with the hadron calorimeter it measures also jets. The calorimeter system of CMS, is made of a high resolution lead-tungstate ($PbWO_4$) crystal EM calorimeter and the hadron calorimeter behind it. In the endcap region the EM calorimetry extends up to rapidity 2.6 (possibly up to 3.0) and the hadron calorimeter up to $|\eta| = 3.0$. This central calorimetric system is complemented in the forward region $3.0 < |\eta| < 5.0$ by 'very forward calorimeters' (Fig.1). Their function is to insure detector hermeticity for good missing transverse energy resolution, and to measure forward 'tagging' jets signing Higgs production through WW or ZZ fusion. Hermeticity is particularly important for processes where the physical missing E_t is on the order of few tens of GeV as in h, H, A $\to \tau\tau$, $W \to l\nu$, $t \to l\nu\, b$, $t \to H^{\pm}b \to \tau\nu\, b$ and for sparticle searches connecting the LEP2 and Fermilab with the LHC search ranges.

The desired performance, choice of detection technique and design of the EM calorimeter is to a large extent determined by the requirements imposed by the $H \to \gamma\gamma$ channel. This is the most appropriate channel to search at a hadron collider the SM Higgs boson or the lightest MSSM Higgs boson h in the $\approx$ 80 - 130 GeV mass range. The natural width of the Higgs in this mass range is very small (<<1 GeV) thus the observed signal width is entirely determined by the experimental $\gamma\gamma$ effective mass resolution. This resolution and the level of the large and irreducible $\gamma\gamma$ background ultimately determine the signal significance. To achieve a high mass resolution requires first an excellent electromagnetic energy resolution σ_E/E. However the $\gamma\gamma$ mass resolution depends also on the two-photon angular separation θ:

$$\sigma_M/M = 1/2(\, \sigma_{E1}/E_1 \oplus \sigma_{E2}/E_2 \oplus \sigma_\theta/tg\theta/2) \qquad (\oplus \text{ are quadratic sums})$$

At $\sim 10^{33}$ $cm^{-2}s^{-1}$ the event vertex is known or is effectively designated by a hard track in the event [16] and the angular term has only a minor effect on the $\gamma\gamma$ mass resolution (a contribution $\delta m_{\gamma\gamma} < 200$ MeV). At 10^{34} $cm^{-2}s^{-1}$ there are, however, on average 15 minimum bias events superimposed on the triggered $\gamma\gamma$ event. If the mean longitudinal vertex position were used in calculating the $\gamma\gamma$ effective mass, too large a contribution to the measured signal width would be introduced [17]. It is thus necessary to foresee a photon direction measurement capability for full luminosity running (would be needed at > $6\text{x}10^{33}$ $cm^{-2}s^{-1}$[16]). A directional precision of $\approx$ 8mrads is sufficient to keep this contribution to the resolution to < 500 MeV. This can be obtained from the shower position measurements in the calorimeter and a preshower detector located in front, Fig. 6 [3]. The goal is to have an overall resolution $\sigma_M \approx 800$ MeV for $m_H = 100$ GeV at 10^{34} $cm^{-2}s^{-1}$.

CMS has chosen a $PbWO_4$ crystal calorimeter. The main reasons for choosing this crystal are its short radiation length (9mm) and small Moliere radius (2.0cm) leading to a compact calorimeter, the short scintillation decay time constant matched to the 25 nsec bunch spacing, and good radiation hardness (with Nb doping) [18]. The low light-yield of this crystal can be overcome using Si avalanche photodiodes (or more radiation hard phototetrodes in the end-caps) as readout elements [19]. The arrangement of crystals is shown in Fig. 6. In the barrel the crystals have a length of 23 cm($25X_0$ deep) and the lateral granularity is ≈ 2cm x 2cm (front face) i.e. $\Delta\eta \times \Delta\phi = 0.014 \times 0.014$. The preshower, $3X_0$ deep, would cover in the barrel $|\eta| < 1.1$. The total volume of crystals is about 11 m^3 and the total number of crystals (channels) is ≈ 1.1×10^5. With such a calorimeter an energy resolution of $\sigma_E/E \approx 0.6\%$ can be expected for electrons (photons) of $E_t = 120$ GeV. This is at present obtained (0.7%) with prototypes in a test beam [20]. Calibrating and monitoring the response of the calorimeter at the required level is a major task and requires in-situ calibration. This is possible with E/p matching using the better than 1% precision momentum measurements of isolated electrons from W and Z decays [3, 21].

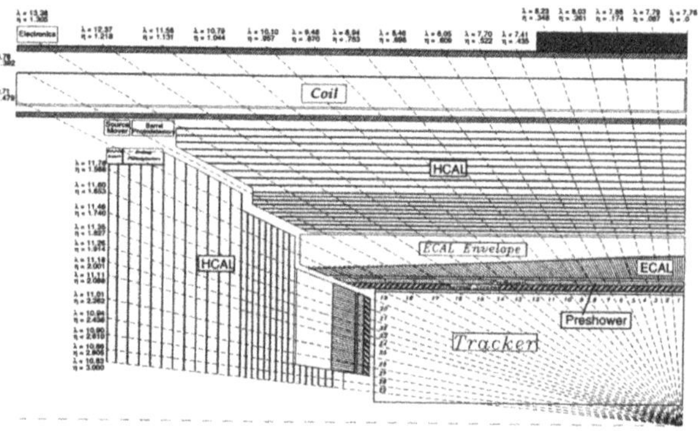

Fig. 6: Layout of the electromagnetic calorimeter, preshower and hadron calorimeter of CMS. The tower structure of the hadron calorimeter is indicated.

Hadron calorimetry with large geometrical coverage for measurement of multi-jet final states and missing transverse energy is essential in both squark and gluino searches, and for detection of the Higgs at m_H > 700 GeV. Hadron calorimetry plays also an essential role in t-quark physics, QCD jet studies and in channels involving τ's in the final state.

In CMS the hadron calorimeter is made of copper absorber plates interleaved with scintillator tiles read-out with embedded wavelength shifting fibers. The expected hadronic resolution is $\sigma_E/E = 70\%/\sqrt{E} \oplus 5\%$. The tiles are organised in towers (Fig. 6) giving a lateral segmentation of $\Delta\eta \times \Delta\phi \approx 0.09 \times 0.09$, adequate for good di-jet separation and mass resolution. Particularly demanding in this respect is the reconstruction of highly boosted W or Z decaying to two jets, produced in decays of a ~ 1 TeV mass Higgs. The main concern for the very forward calorimeters are the very high radiation levels expected in the region $3.0 < |\eta| < 5.0$. The two techniques considered at present are a sampling calorimeter with Fe plates and PPC's (parallel plate chambers)[22], or a quartz-fiber calorimeter [23].

Trigger and data acquisition

The task of the trigger system is to reduce the input rate of ~10^9 events/sec at high luminosity to <100Hz, the maximum rate at which events can be written on the permanent storage device. The average event size is ~ 1MByte, most of the information being provided by the tracker. The needed data reduction is done in two steps. The level-1 trigger system, using only calorimetric and muon system information in hardware processors operating at 40MHz input rate reduces the data rate to < 100kHz. To remain well within this cumulative first-level trigger output rate at 10^{34} $cm^{-2}s^{-1}$, typical trigger thresholds would be: for single electron (γ) $E_t > 30$ GeV, di-electrons (γs) $E_t > 15$ GeV on each, for inclusive jets $E_t > 140$ GeV, for single muons $p_t > 20$ GeV, for dimuons $p_t > 10$ GeV, all within $|\eta| < 2.5$, and $E_t^{miss} > 150$ GeV [3].

For the higher-level triggers, a high bandwidth ($\approx$ 500 Gbit/s) readout network (event builder) treating the $\approx$ 1000 front-end readout units with ~ 1000 Bytes/event and a high processing power (10^6-10^7 MIPS) event filter is needed [3]. The event filter is implemented in an on-line processor farm (about 1000 units) where full detector analysis and event reconstruction is performed, reducing the event rate by a factor ~ 1000 for writing on mass storage. The total data production will be on the order of 1TByte/day.

Collaborations, costs and schedule

At present all subdetectors in CMS have been chosen, except for the very-forward calorimeters. The evaluation of the cost of the major components of the detector is the following: the coil 74 MSF, the tracker 90 MSF, the EM calorimeter 80 MSF, the hadron calorimeter 42 MSF, the muon system 63 MSF, the trigger/DAQ system 47MSF and the iron yoke about 42 MSF. The overall cost of the CMS detector is estimated at 475 MSF [3]. The cost of the ATLAS detector is comparable [4].

The two collaborations ATLAS and CMS are also of comparable size, with at present about 1350 physicists and engineers from about 130 institutions in about 30 countries in each of them. About 50% of collaborators in CMS and about 40% in ATLAS are from CERN non-member states. The number of physicists from the US approaches 300 in each collaboration, with a somewhat smaller but comparable number from Russia and other Dubna-member states [3,4]. At present the number of physiciste and engineers in ALICE is about 300 and the cost of the detector is of the order of 120 MSF, whilst for LHCB it is ~150 participants and ~ 85 MSF.

With the LHC approved in December 1994 and a possible first-stage approval of CMS and ATLAS by the end of 1995, taking into account prototype construction and testing, detector construction, assembly and tests, civil engineering and installation, these detectors could be completed in 2004 [3,4].

3. PHYSICS AT THE LHC

Extensive studies have been made of the expected performance of CMS and ATLAS [3.4]. We discuss expectations for the SM Higgs boson and some of the MSSM Higgs bosons, and comment on the gluino and squark mass reaches, some CP-violation measurements and possible signals for QCD deconfinement via the relative suppressions within the Y family.

Signals have been evaluated for $\sqrt{s}$ = 14 TeV using in most cases PYTHIA, ISAJET is used for SUSY signals. For channels that are particularly sensitive rather complete GEANT simulations of detector responses are performed, but often GEANT simulations of momentum resolution, radiative losses, isolation efficiencies have been parameterised and implemented with Gaussian smearings [3,4].

3.1 Standard Model Higgs Boson

Fits of radiative corrections to electroweak data provide indirect information on m_H. The dependence is quadratic on m_{top}, but is only logarithmic on the Higgs mass. Fits to data available before summer 1995, including the combined CDF and D0 value m_{top} = 181 ± 12 GeV [24], the LEP value of R_b= 0.2204 ± 0.0020 and the new SLAC-95 value of A_{LR} giving $\sin^2\theta_{eff}(A_{LR})$ = 0.2305 ± 0.0005 (combined 1994 + 95), are shown in Fig. 7a [25]. The minimum χ^2 is obtained for m_H ~ 100 GeV, but it is shallow and the upper limit of m_H is not stringent. More exactly, this analysis gives $m_H = 76^{+152}_{-50}$ GeV and the 95% C.L. upper bound is $m_H \leq$ 0.7 TeV. A reanalysis including LEP data from summer 1995 does not change significantly the best value of m_H, although there is a significant degradation of the best χ^2 due to R_b, R_c [25]. If errors on R_b, R_c are arbitrarily increased by a factor 3, the optimal m_H increases by $\approx$ 45 GeV, well within the uncertainty of the fits. Very similar fits are obtained by the LEP electroweak working group [26]. A best χ^2 is consistently obtained for m_H ~ 100 GeV and a $\Delta\chi^2 = \chi^2 - \chi^2_{min} < 4$ limit for $m_H < 0.85$ TeV. Thus a broad mass range must be explored in future experiments. The present lower bound from LEP is $m_H \geq 65$ GeV [27] and LEP2 should allow to extend the search to ≈ 90 GeV. Figure 7b shows the allowed regions for m_{top} and m_H [28], the contours indicating the scale up to which the SM is supposed to be valid; the region to the right is excluded by the top rad. corrections destabilizing the Higgs potential. If there is no new physics before the GUT scale, m_H is limited to ~ 200 GeV; for a lower breakdown scale m_H could be as large as ~ 0.5 TeV. From these bounds and the (high) measured value of m_{top}, if a Higgs signal were found at LEP2, it is unlikely to be the SM one, i.e. new physics would be at a TeV scale, possibly supersymmetry, a most interesting prospect for the LHC.

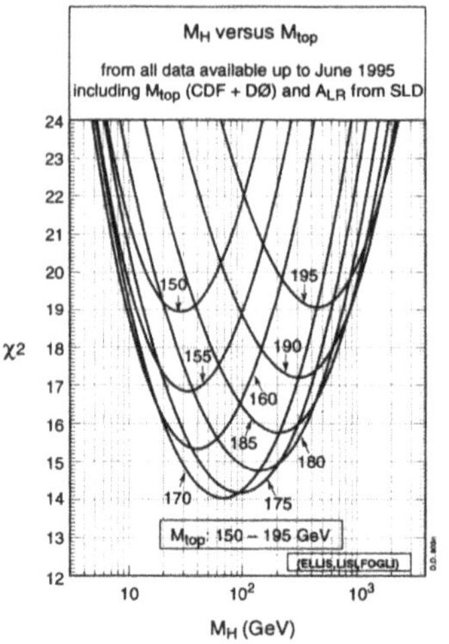

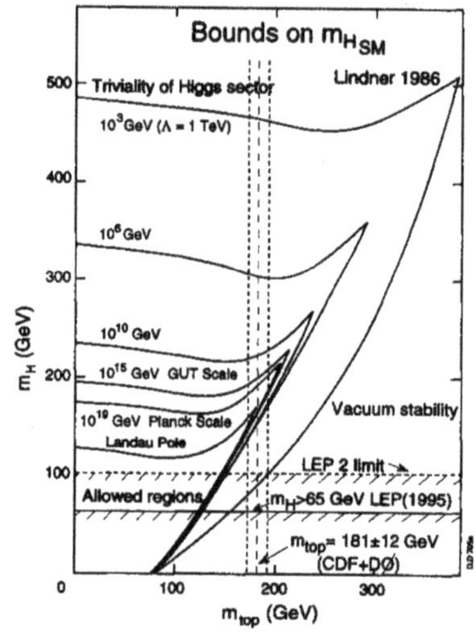

Fig. 7a : χ^2 as a function of m_H for various m_{top} values [25].

Fig. 7b: Bounds on the top and SM Higgs masses from ref. [28]; the allowed region extends from the lower left corner up to the indicated scale of assumed SM break-down.

ATLAS and CMS have been designed to allow investigations over the entire mass range up to ~1TeV. For the $80 < m_H < 140$ GeV domain the $H \to \gamma\gamma$ decay mode provides the best chances. Supporting evidence could be obtained from $H \to b\bar{b}$ for $m_H < 100$ GeV [4]. This mass range is most important as it connects to the LEP domain and, as discussed later, if the lightest MSSM Higgs boson exists, it must be within $m_h < 130$ GeV. For $130 \leq m_H \leq 700$ GeV the channel $H \to ZZ^*, ZZ \to 2l^+2l^-$ is the most appropriate one. For $m_H > 650$ GeV the difficulties increase as the Higgs becomes broader and the cross-section decreases. We discuss in more detail these points, in particular the rad. correction favored $m_H < 300$ GeV range which is also more demanding in terms of detector performance and versatility.

$H \to \gamma\gamma$

For $H \to \gamma\gamma$ let us use the CMS study [17]. The cross section $\sigma \cdot B(H \to \gamma\gamma)$ for $\sqrt{s} = 14$ TeV is 76 (68) fb for $m_H = 110$ (130) GeV. The photon acceptance is $|\eta| < 2.5$, and the transverse momentum cuts $p_t^{\gamma_1} > 40$ GeV, and $p_t^{\gamma_2} > 25$ GeV. The background to $H \to \gamma\gamma$ is made of: i) prompt di-photon production from quark annihilation and gluon fusion, ii) prompt di-photon production from bremsstrahlung from the quark line, and iii) reducible background from jets or from jets + 1 prompt photon, with a leading π^0 (or η) decaying to $\gamma\gamma$. At $m_{\gamma\gamma} = 110$ GeV for example, the cross-sections $d\sigma/dm_{\gamma\gamma}$ are 84 fb/GeV for isolated bremsstrahlung, 61 fb/GeV for quark annihilation and 72 fb/GeV for gluon fusion. The jet and bremsstrahlung backgrounds can be reduced by a photon isolation cut. Photons from π^0s in the $p_t \sim 50$ GeV range are separated in the calorimeter by ~1 cm. The presence of two impacts can be detected using a lateral shower profile in the calorimeter, or the preshower. The jet-jet and jet-γ backgrounds are reduced to < 15% of the irreducible $\gamma\gamma$ background by photon isolation and π^0 rejection cuts [3]. An efficiency of 64% is assumed for the reconstruction of each photon; this accounts for conversion, fiducial volume, isolation and π^0 rejection losses. Subsequent studies have shown that about 50% of conversions could be recovered thus an efficiency of 74% would at present seem more appropriate [29].

Figure 8a shows a background-subtracted $\gamma\gamma$ effective mass plot for an integrated luminosity of 10^5 pb^{-1} taken at high luminosity with expected signals at 90, 110, 130 and 150 GeV. The crystal calorimeter is assumed to have an energy resolution of $\Delta E/E = 2\%/\sqrt{E} \oplus 0.5\% \oplus 0.200/E$ in the barrel and of $5\%/\sqrt{E} \oplus 0.5\% \oplus 0.200/E$ in the barrel and endcap regions where there is a preshower detector. The $\gamma\gamma$ mass resolution at $m_{\gamma\gamma}$ = 110 GeV is 870 MeV. Figure 8b shows the level of σ.B required to have a specified signal significances$\left(N_S/\sqrt{N_B}\right)$ as a function of mass with the performance and selection cuts of CMS. The expected σ.B for the SM Higgsis also shown; the uncertainty on this prediction is ~ ± 15% due to structure functions alone and the uncertainty due to 'k factors' is comparable. From Fig. 8b we conclude that with 10^5 pb^{-1} the SM Higgs could be discovered with > 5σ signal significance across the range 90 to 150 GeV. A significant advantage of the potentially excellent resolution is that in the 10^{33} cm^{-2}s^{-1} regime where the resolution is best (known vertex), with a reduced luminosity of 3×10^4 pb^{-1} the mass range ≈ 95 to 140 GeV could be covered.

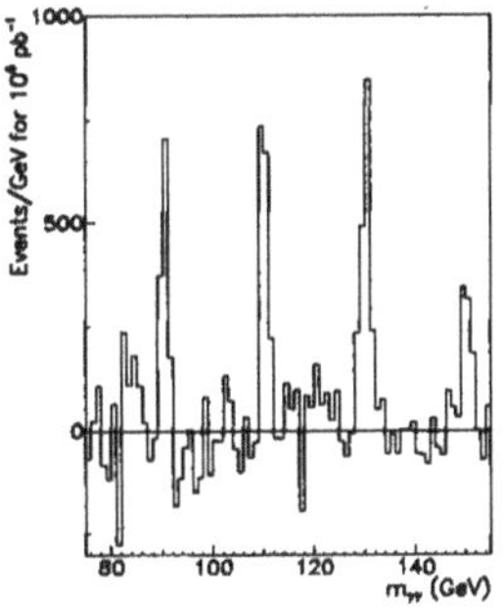

Fig. 8a: Background-subtracted 2γ mass for 10^5 pb^{-1} with signals at m_H = 90, 110, 130 and 150 GeV in the $PbWO_4$ calorimeter.

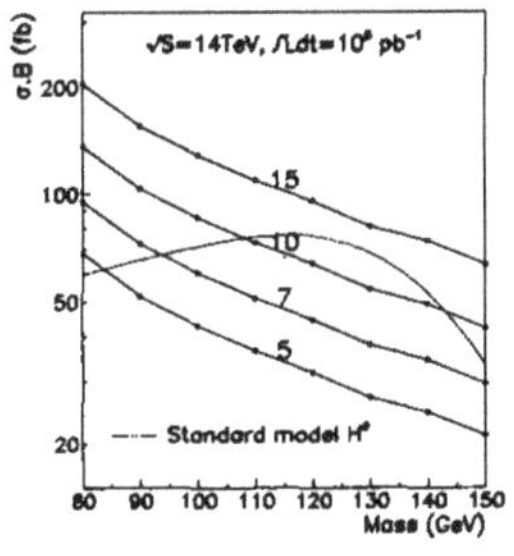

Fig. 8b: Signal significance contours for 10^5 pb^{-1} taken at high luminosity.

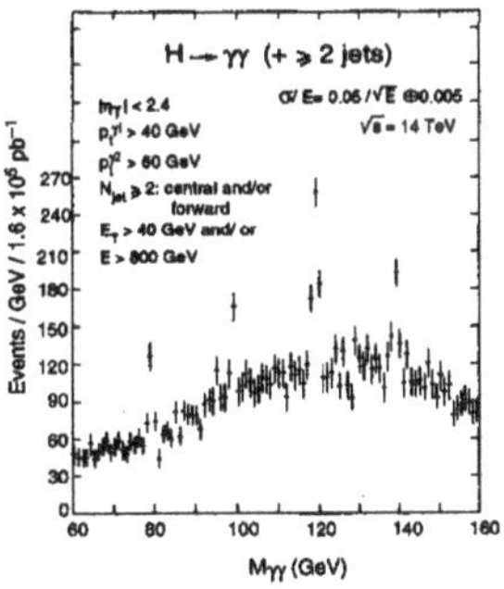

Fig. 8c: Reconstructed $H \rightarrow \gamma\gamma$ signal in a crystal calorimeter for H produced in association with hard jets for 1.6×10^5 pb^{-1}.

In the inclusive $H \rightarrow \gamma\gamma$ search the signal to background ratio is ~ 1/10. This can be improved using Higgs production in association with hard jets or isolated leptons. This is more demanding in terms of statistics, but is significantly less demanding in terms of calorimeter performance, intrinsic energy resolution or calibration and there is no need for a preshower. Sources of such events are radiative corrections to $gg \rightarrow H$ with hard real gluons, associated production of $t\bar{t}H$, WH, ZH and WW (ZZ) fusion. Figure 8c shows the signal for m_H = 80, 100, 120, 140 GeV superimposed on the background in a search where at least two hard jets are demanded [30]. The number of signal events is about 100 for 1.6×10^5 pb^{-1}, an order of magnitude less than in the inclusive $\gamma\gamma$ search, but the signal-to-background ratio is ~1. Signal significance is ≈ 7 to 14σ depending on m_H. A similar analysis is possible requiring a hard and isolated lepton [3,4]. The hard jets or lepton present in the event indicate the interaction vertex at any luminosity and could dispense us of a use of a preshower. The drawback is the uncertainty in Higgs production at high p_t and even more so for the backgrounds [30].

$H \rightarrow ZZ^*$, $ZZ \rightarrow 4$ Charged Leptons

The four-lepton channel allows to search the m_H ≈ 130 to ~ 800 GeV range. For $m_H < 2m_Z$ one of the Z's is off-mass-shell and the backgrounds are $t\bar{t}$, $Zb\bar{b}$ and ZZ^*. The ZZ^* background is irreducible, $Zb\bar{b}$ can be suppressed by lepton isolation and lepton impact parameter cuts and $t\bar{t}$ by a Z mass cut and lepton isolation or impact parameter cuts [31]. The signal cross sections are $\sigma \cdot B$ = 2.9, 5.3 and 1.4 fb for m_H = 130, 150 and 170 GeV respectively. For m_H < 200 GeV the width of the Higgs is small, Γ_H < 1 GeV, and the signal significance is sensitive to the mass resolution. In CMS the $Z \to \mu^+\mu^-$ resolution is σ_Z = 1.6 GeV and the $H \to 2\mu^+2\mu^-$ resolution σ_H = 0.9 GeV for m_H < 200 GeV [32]. Internal bremsstrahlung effects are included and detailed GEANT simulations are performed to investigate the effects of external bremsstrahlung for electrons [21]. With a 5 × 7 crystal matrix to reconstruct electrons, the mass resolution for $Z \to e^+ e^-$ is 2.2 GeV, and the mass resolution for $H \to ZZ^* \to 2\,e^+\,2\,e^-$ is 2.0 GeV.

The selection cuts for electrons are: one with p_t > 20 GeV, one with p_t > 15 GeV, and the remaining two with p_t > 10 GeV; for muons the p_t cuts are 20, 10 and 5 GeV. There is a certain flexibility in the ways to suppress $t\bar{t}$ and $Zb\bar{b}$ backgrounds using the Z mass cut, lepton isolation or lepton impact parameter cuts according to the m_H range investigated, detector performance and instantaneous luminosity [31]. Figure 9a shows the reconstructed Higgs signals at 130, 150 and 170 GeV for 10^5pb^{-1}. Figure 9b shows the σ.B required to obtain a given signal significance with CMS performance and cuts, compared to the expected σ.B in the SM. The mass range covered goes from ≈ 120 GeV up to $2m_Z$. Observation of the signal in the vicinity of m_H = 165 GeV (opening of the $H \to WW$ decay channel) is difficult and requires 10^5 pb^{-1} for a 5σ significance signal. Figure 10 shows the results of a similar study made for the ATLAS detector [4].

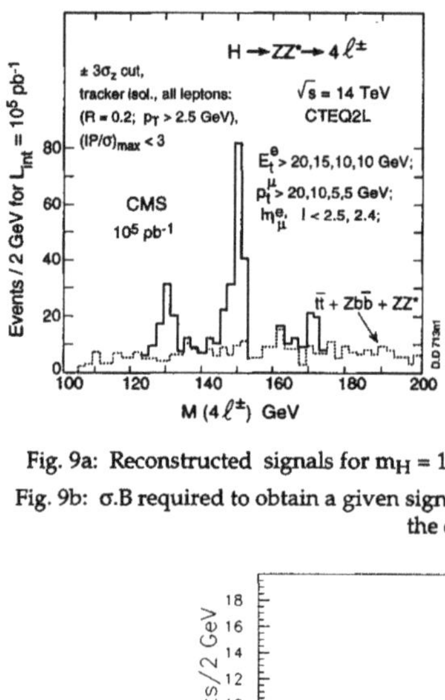

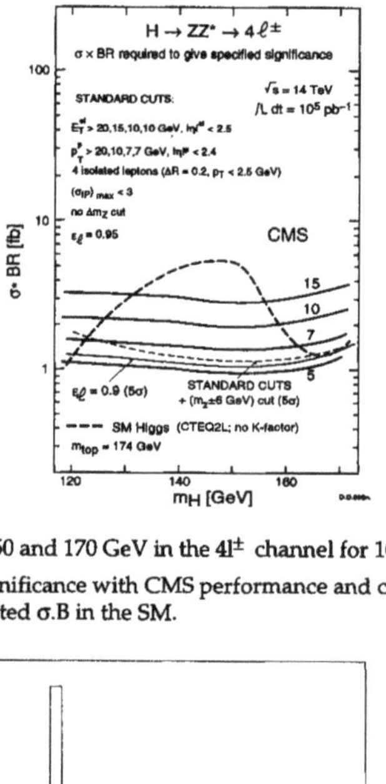

Fig. 9a: Reconstructed signals for m_H = 130, 150 and 170 GeV in the $4l^{\pm}$ channel for 10^5 pb^{-1} in CMS.
Fig. 9b: σ.B required to obtain a given signal significance with CMS performance and cuts, compared to the expected σ.B in the SM.

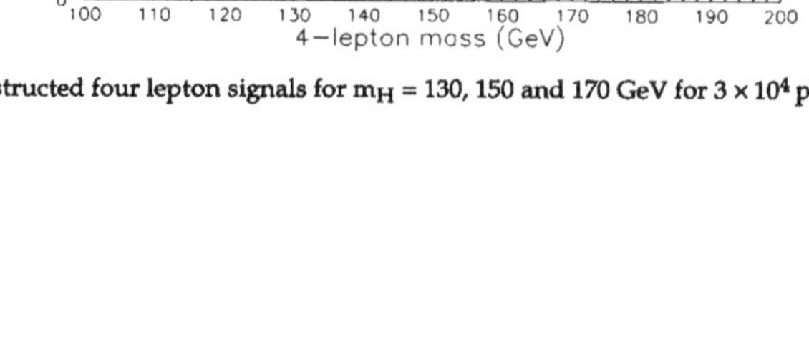

Fig. 10: Reconstructed four lepton signals for m_H = 130, 150 and 170 GeV for 3×10^4 pb^{-1} in ATLAS.

For $m_H > 2m_Z$ the main background is non-resonant ZZ production. The p_t and η selection cuts are the same as for the $H \to ZZ^*$ channel and the effects of internal and external bremsstrahlung are treated the same way [21]. Two e^+e^- or $\mu^+\mu^-$ pairs consistent with a Z mass are required. The Z mass window can be large since there are no backgrounds which are critically affected by the two-lepton mass cut. Lepton isolation cuts are not needed as the $t\bar{t}$ background is negligible. Figure 11 shows the expected 4-lepton mass spectrum for several m_H values. Figure 12a summarizes the expected $H \to ZZ^*$, ZZ signal significance for several integrated luminosities. With 10^4 pb^{-1}, the 5σ discovery region extends from $\approx$ 135 to 155 GeV and from $2m_Z$ up to $m_H \approx$ 400 GeV. Figure 12a also shows the effect of varying the input structure functions. With 10^5 pb^{-1} the upper mass reach is $\approx$ 650 GeV, but it can be extended to $\approx$ 800 GeV with a more carefull choice of selection criteria exploiting the scalar nature of the Higgs and the longitudinal polarisation of decay Z's, Figs. 11c and 12b [33].

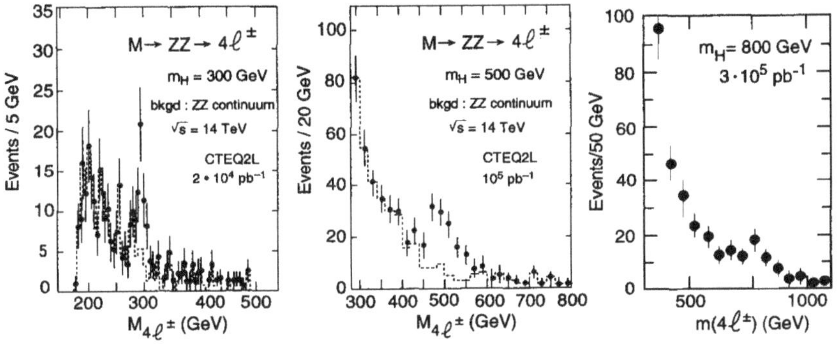

Fig. 11: The four-lepton mass distributions from $H \to ZZ \to 4l^\pm$ s and the ZZ continuum background for m_H = 300 and 2×10^4 pb^{-1}, m_H = 500 GeV with 10^5 pb^{-1} and m_H = 800 GeV with 3×10^5 pb^{-1} in CMS.

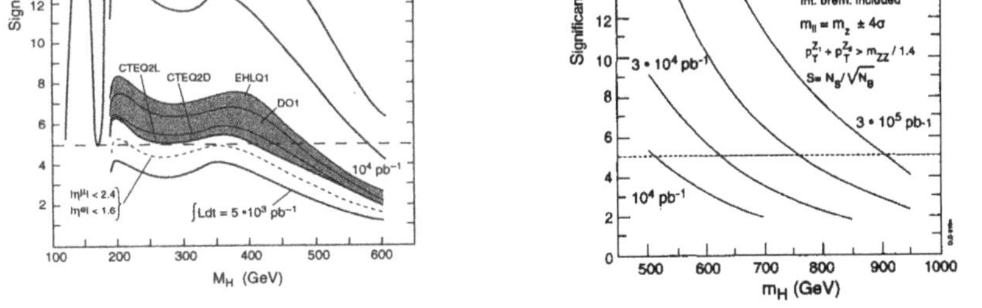

Fig. 12a: Significance of $H \to ZZ^*$, $ZZ \to 4l^\pm$ signals as a function of m_H for integrated luminosities from 5×10^3 to 3×10^5 pb^{-1}; Fig. 12b is for cuts optimized for the high mass range.

H → llνν

To explore the m_H ~ 1TeV region decay modes with larger branching ratio must be used. The channel H → llνν has a six times larger branching ratio than H → $4l^{\pm}$ and a distinct signature: two high p_t leptons from the Z decay and high E_t^{miss}. For m_H > 500 GeV, the signal is a broad Jacobian peak in the two-lepton transverse momentum distribution. The main background channels are ZZ, ZW, $t\bar{t}$ and Z + jets. For masses beyond ~700 GeV the signal becomes less distinct, as the Higgs width increases rapidly. Only a precise knowledge of background would allow the unambiguous observation of such a signal. To suppress backgrounds forward jets from WW (ZZ) fusion production can be used as an additional signature. Details on such an analysis can be found in [34]. Figure 13 shows the expected signal for m_H = 800 GeV and 10^5 pb^{-1}. This channel would allow discovery of a SM Higgs up to 1 TeV with an integrated luminosity of 10^5 pb^{-1}. Figure 14 shows the results of a similar analysis done for the ATLAS detector [34].

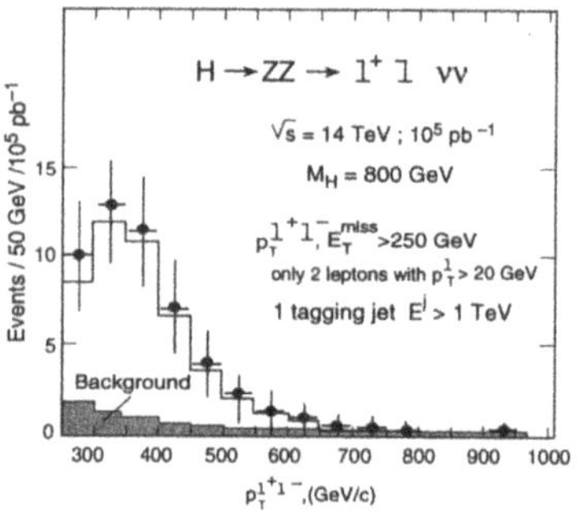

Fig. 13: H → $l^+l^-\nu\nu$ signal for m_H = 800 GeV and 10^5 pb^{-1} in CMS. One tagged jet with E > 1 TeV is assumed.

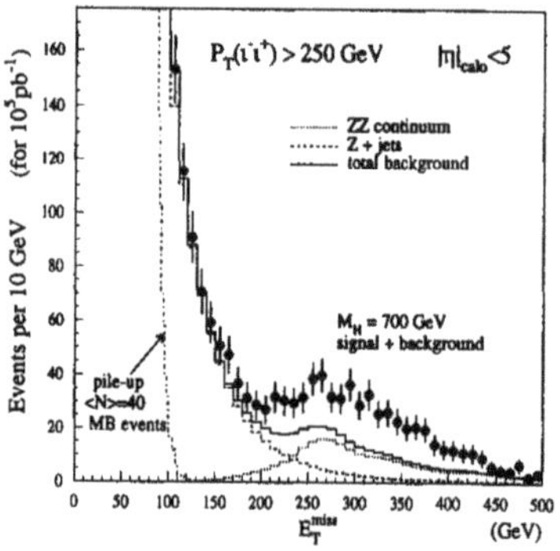

Fig. 14: H → $l^+l^-\nu\nu$ signal for m_H = 700 GeV and 10^5 pb^{-1} in ATLAS, the various background components are shown separately.

H → WW → lνjj and H → ZZ → lljj

These decay channels are needed to search the $m_H \approx 1$ TeV region exploiting the large W, Z → $q\bar{q}$ branching ratios. The signature is a high p_t lepton pair with $m_{ll} \approx m_Z$, for H → ZZ, or a high p_t lepton plus large E_t^{miss} for H → WW. In addition, hard central jets from the hadronic decays of Z/W, with $m_{jj} \approx m_{W/Z}$ are required. The main backgrounds are Z + jets, ZW, ZZ for H → ZZ, and W + jets, $t\bar{t}$, WW, WZ for H → WW. In this mass range the Higgs is very broad ($\Gamma_H \approx$ 0.5 TeV for $m_H \approx 1$ TeV) and indistinct in effective mass distributions. The forward 'tagging jets' can improve the signal to background ratio as WW/ZZ fusion represents about 50% of the cross-section. A set of appropriate cuts can be found in [34]. Figure 15 shows the expected H → WW → $l\nu_t$jj signal for 3×10^4 pb^{-1}; one expects ≈ 22 signal events over a background of ≈ 8 events for m_H = 1 TeV, giving a 4σ signal significance. Simultaneous observation in the lljj, lνjj and llνν channels would give a ≈ 5.5σ signal for 3×10^4 pb^{-1}.

Overview of SM Higgs detection possibilities

Figure 16 summarizes the mass range each mode would allow to investigate. The entire SM Higgs mass range can be covered with some safety margin. For the 80-200 GeV range high performance and high integrated luminosity are needed. Most demanding is the 80-130 GeV range, the detection of H → γγ justifying the $PbWO_4$ calorimeter. The calorimeter performance is critical, resolution, calibration and monitoring will be of decisive importance. If a γγ resolution better than 1 GeV is achieved, few× 10^4 pb^{-1} can be enough for discovery in the inclusive γγ search. In case of insufficient performance, say effective resolution ~2GeV, we must resort to the γγ +lepton, or γγ + multijets subchannels instrumentally less demanding, but requiring in excess of $10^5 pb^{-1}$. For the ~130-200 GeV domain muon and electron resolution and acceptance will be critical. Relaxing the m_Z cut and compensating with lepton isolation and impact parameter cuts, masses down to 120 GeV are accessible. There is a window at $m_H \approx 165 \pm 5$ GeV where $10^5\ pb^{-1}$ are mandatory for 5σ signal significance. For $m_H >$ 650 GeV the difficulties increase and excellent knowledge and control of backgrounds will be of decisive importance. The performance of the hadron calorimeter and detector hermeticity are critical. An integrated luminosity of $\approx 10^5\ pb^{-1}$ is needed to cover the upper m_H range. If the SM Higgs conforms to present expectations it should not escape detection at the LHC.

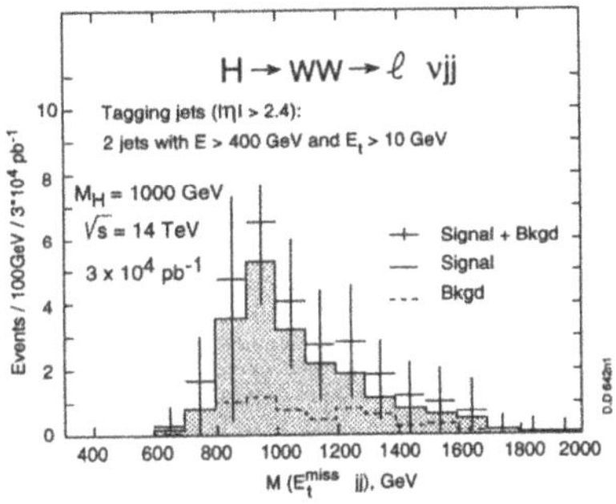

Fig.15: Higgs signals in $H \to WW \to l\nu jj$ for $m_H = 1$ TeV and $3 \times 10^4\ pb^{-1}$.

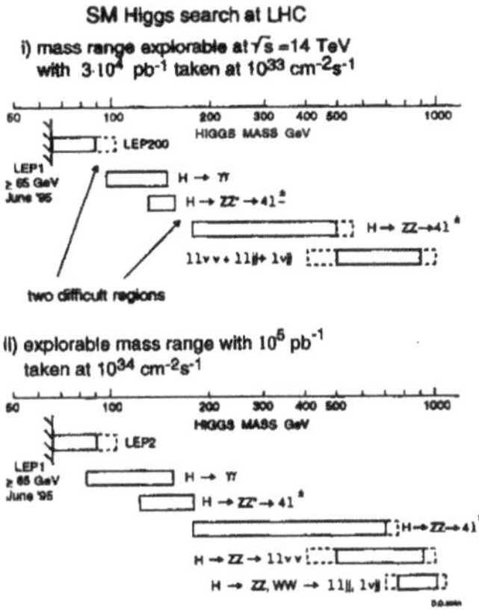

Fig.16: Overview of Standard Model Higgs observability at LHC versus luminosity

3.2 Searches for MSSM Higgs bosons

The difficulties of the SM Higgs scenario are alleviated in SUSY theories at the expense of introducing several Higgs bosons. The Minimal Supersymmetric Standard Model (MSSM) contains one charged ($H^{\pm}$) two CP-even (h, H) and one CP-odd (A) state. The specific scenario that has been investigated by ATLAS and CMS is the one of ref. [6] with a heavy mass scale at 1 TeV. The decay channels of the MSSM neutral Higgs bosons are similar to the S M Higgs, but the production rates are significantly modified by MSSM couplings. Particularly interesting is the the light scalar h whose mass is bounded by $m_h < m_Z$ at lowest order; radiative corrections modify this upper bound to $m_h < 130$ GeV [6, 35]. Thus the h could be outside the reach of LEP2 limited to $m_h < 90$ to 100 GeV. This puts additional emphasis on the $h \to \gamma\gamma$, $b\bar{b}$ and ZZ^* searches at LHC to test this clear prediction of the MSSM. The present LEP experimental mass limit on m_h are > 65 GeV for $tg\beta = 1$, >45.5 GeV for moderate $tg\beta$ (10 to 20) and > 55 GeV for large $tg\beta$ [27].

$h,H \to \gamma\gamma$

The MSSM Higgs bosons h,H are extremely narrow in the kinematic regions for which the $\gamma\gamma$ decay has a useful branching ratio. The experimental requirements and expected backgrounds are the same as in the SM $H \to \gamma\gamma$ search. From the Higgs mass and $\sigma \cdot B$ as a function of m_A and $\tan\beta$, the significance of a possible $h,H \to \gamma\gamma$ signal can be extracted from Fig. 8b. Figure 17 shows several contours in the (m_A, $\tan\beta$) plane, a 5σ discovery contour for $10^5 pb^{-1}$ taken at high luminosity, a 3σ exclusion contour for same conditions, and a 5σ discovery contour for $3\times10^4 pb^{-1}$ taken at $10^{33} cm^{-2}s^{-1}$. The region of (m_A, $\tan\beta$) parameter space which can be explored with LEP2 at $\sqrt{s}$ = 190 GeV and 500 pb^{-1} (assuming the heavy mass scale at 1 TeV) is also shown. The two machines, LHC and LEP2, are complementary in their possibilities, LHC being sensitive at large $tg\beta$, large m_A and LEP2 at low $tg\beta$ or low m_A.

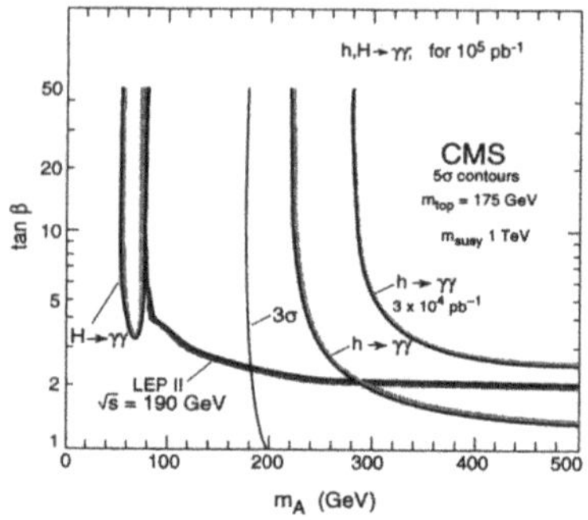

Fig. 17: Significance contours in the MSSM parameter space for h,H → γγ signals in CMS. The region explorable at LEP2 is also shown. The explorable regions are inside the shaded boundaries.

H → ZZ*, ZZ and h → ZZ*

The H and h bosons couple to W and Z pairs and can be searched in h, H → ZZ*, ZZ → $4l^{\pm}$. Radiative corrections have a large effect on the upper bound of m_h, and the opening of the h → ZZ* channel is possible only if m_h exceeds ≈120 GeV. Representative values of σ·B for H → ZZ*, ZZ→ $4l^{\pm}$ and h → ZZ* → $4l^{\pm}$ are given in Table 1. The simulation procedure and experimental criteria are the same as for the SM H → ZZ*, ZZ. Both the h and H are narrow, with widths well below 1 GeV in the accessible parameter range. The regions of MSSM parameter space in which the h and H can be looked for via ZZ*/ZZ are disconnected, a low tanβ region for H and a high tanβ region for h. The 5σ discovery limits in these two channels are shown in Fig. 18a for 10^5pb^{-1}. The limits of the h → ZZ* discovery region are very sensitive to changes in m_h, the contour in Fig. 18a is for m_{top} = 174 GeV and m_h calculated with one-loop radiative corrections giving m_h < 128 GeV [36].

Table 1. σ·B (fb) for H → ZZ*, ZZ, h → ZZ* for different MSSM parameter values.

	m_A = 80 GeV	m_A = 150 GeV	m_A = 200 GeV	m_A = 300 GeV
H → ZZ (tanβ = 2)	2.2	1.6	1.8	1.2
h → ZZ* (tanβ = 10)	0.01	1.3	1.7	1.6

h, H, A → ττ → $l^{\pm}$ + $h^{\pm}$ + X

The ττ final states can be looked for in lepton + hadron or e + μ channels. As there are several neutrinos in the final state, mass peaks are difficult to reconstruct and evidence for a signal comes mainly from an excess of events over expected SM backgrounds. Backgrounds are due to Z, γ* → ττ, $t\bar{t}$ → ττ + X, τ + X and $b\bar{b}$ → ττ + X, τX, and to events with one hard lepton and jets with a jet misidentified as a τ. Event selection algorithms have been developed requiring : i) one isolated lepton, ii) one τ-jet candidate and iii) no other significant jet activity in the event [3,4,37]. The τ → charged hadron decay is selected by requiring a collimated high E_t jet in the calorimeters and exactly one high p_t charged track within ΔR < 0.1 of the calorimeter jet axis. The small Q-value of the τ decay allows reconstruction of m_A, if A has a transverse boost, by taking the neutrinos parallel with the hadron and the lepton. The reconstructed mass distribution has a mass resolution of ~ 15 GeV for m_A = 100 GeV [4]. The 5σ discovery contours in the (m_A, tanβ) plane for A, H, h → ττ → $l^{\pm}$ + $h^{\pm}$ and A, H, h → ττ → e + μ are shown in Fig. 18a for 10^4pb^{-1}. The ττ decay modes allow the exploration of a substantial region of MSSM parameter space.

Charged Higgs $H^{\pm}$ in t → $H^{\pm}$ b, $H^{\pm}$→ $\tau\nu_\tau$

In the MSSM the top quark can decay to a charged Higgs, t → H^+b. The t → H^+b branching ratio is complementary to the t → Wb one, it is large at low and high tanβ values, having a minimum at tanβ ≈ 6. The H^+ has two main decay modes, $H^+ \to c\bar{s}$ and $H^+ \to \tau\nu_\tau$. The $H^+ \to \tau\nu_\tau$ branching ratio is large, ≈ 98%, for tanβ >

2, and only slightly dependent on tanβ. Studies have been made [3,4,38] of the observability of the $H^{\pm}$ signal from $t\bar{t}$ events where one top decays leptonically and the other to a charged Higgs followed by $H^{+} \rightarrow \tau\nu$ using the one-prong τ signature. The existence of H^{+} in the data can only be inferred from an excess of τ production over what is expected from the SM backgrounds. Irreducible ones involve real τ's as in $t\bar{t}$ with $t \rightarrow Wb$, $W \rightarrow \tau\nu_{\tau}$, reducible ones are from $t\bar{t}$, where a W decay jet fakes a τ, or a b-jet gives a real or a fake τ candidate, and from $b\bar{b}$. The main selection criteria are i) an isolated high p_t lepton and ii) one jet with $E_t > 40$ GeV fulfiling the τ selection criteria. To suppress the dominant W + jet background at least one tagged b-jet is required in the event [38,39]. One contour in Fig. 18a shows the (m_A, tanβ) region that can be explored with $t \rightarrow H^{\pm}b$ at 5 σ significance with 10^4 pb^{-1} integrated luminosity.

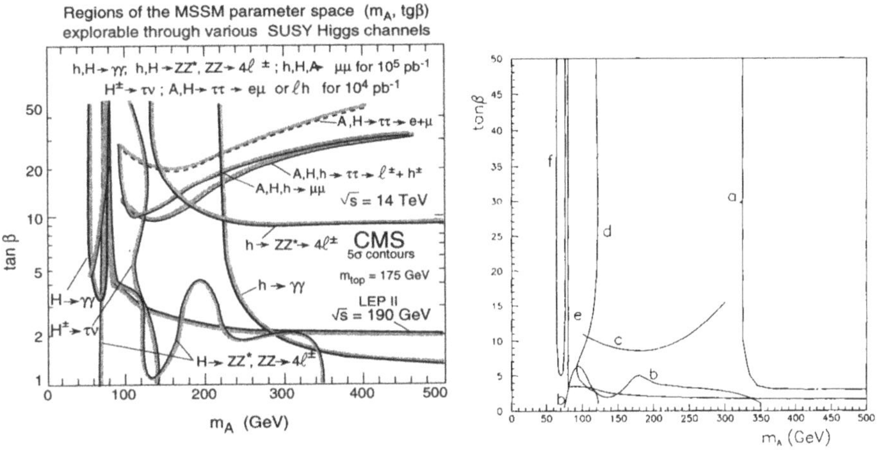

Fig. 18a: Significance contours (5σ) in the MSSM parameter space for different Higgs boson decays accessible to CMS. The region explorable at LEP2 with 500 pb^{-1} is also shown. The explorable regions are inside the shaded boundaries. Fig. 18b: Same analysis in ATLAS: contours a) $h \rightarrow \gamma\gamma$, b) $H \rightarrow ZZ \rightarrow 4l^{\pm}$, c) $A \rightarrow \tau\tau$, d) $t \rightarrow H^{+}b$, e) LEP2 sensitivity and f) $H \rightarrow \gamma\gamma$,; c) and d) are for 10^4pb^{-1} a), b) and f) are for 10^5pb^{-1}.

h, H, A → μμ

In both the SM and MSSM the branching ratio of $H \rightarrow \mu\mu$ is small, about 3×10^{-4}. For the SM Higgs the μμ channel is overwhelmed by the large Drell-Yan background. In the MSSM, however, the cross-sections are enhanced relative to the SM at large tg. An appropriate set of signal selection cuts is: i) two muons with $p_t^{\mu} \geq 10$ GeV and ii) no more than one jet with $E_t \geq 40$ GeV within $|\eta| \leq 2.4$ [40]. Figure 19 shows the signal for two values of m_A and large tgβ. The h and A mass peaks remain unresolved when $m_A \leq 130$ GeV, and similarly the A and H peaks cannot be separated for $m_A \geq 130$ GeV. The peaks become resolvable at tanβ values ≈ 10 – 15. Close to the Z-peak the signal is difficult to observe as it sits on the shoulder of the much larger Z-peak, but the signal-to-background ratio can be improved by b-tagging [39]. Whilst $gg \rightarrow b\bar{b}H$ associated production is negligible for the SM Higgs compared to $t\bar{t}H$, in the MSSM the rate of $gg \rightarrow b\bar{b}H_{SUSY}$ is~50% of the total production rate for $m_A \sim 100$ and increases with m_A (tanβ = 10 – 30). B-taggingthus favours $b\bar{b}$ H_{SUSY} production relative to $Z b\bar{b}$ which amounts to only few percent of Z + jets [3]. The 5σ discovery contour for H, h, $A \rightarrow \mu\mu$ for an integrated luminosity of 10^5 pb^{-1} is shown in Fig. 18a. It is close to the H, h, $A \rightarrow \tau\tau$ contour for 10^4 pb^{-1}, but the μμ channel provides a much better signal identification and mass resolution.

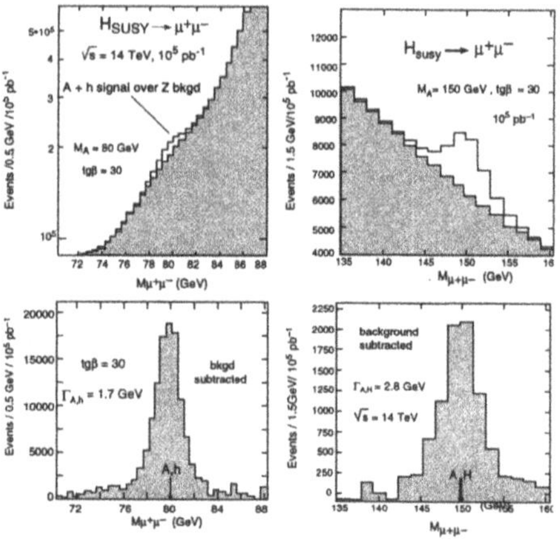

Fig. 19: Inclusive $H_{SUSY} \to \mu\mu$ at $\tan\beta$ = 30 for m_A = 80, 150 GeV and 10^5 pb^{-1}. The h, A and H, A masses are almost degenerate.

Conclusions on SUSY Higgs Searches

Figure 18a summarises investigations done by CMS on the regions of MSSM parameter space that can be explored using various channels with luminosities ranging from 10^4pb^{-1} to 10^5pb^{-1}. Figure 18b shows the results of a similar study done by ATLAS [4]. There are regions of parameter space that are explorable through several channels. This may eventually allow clarification as to which of the various SUSY models is correct, if any. Figures 18 also show the region of parameter space which can be explored with LEP2 at $\sqrt{s}$ = 190 GeV. LEP2 and LHC clearly complement each other. With the channels investigated in detail up to now and with 10^5pb^{-1} there remains, however, a region inaccessible to both machines given approximately by 110 < m_A < 220 GeV, 2.5 < $\tan\beta$ < 10. A channel which could at LHC shed light on this region is h, H $\to b\bar{b}$, from Wh, Zh and $t\bar{t}$h final states [41]. The pixel detectors included in both ATLAS and CMS tracking system may allow this, but more work is needed to evaluate how realistic is this possibility. Another way to cover most of this region would be to accumulate > 3x 10^5pb^{-1} [4]. The LEP2 contour could also be much more favourable if the energy could be increased to ≈ 205 GeV as then the horizontal branch of the LEP2 coverage would reach up to tgβ ≈ 8 with 300pb^{-1}, but this may be financially unafordable. If the stop-quark mass, or more generally the heavy mass scale of MSSM were lower than the 1 TeV assumed here, the LEP2 coverage would improve in the same way [6].

3.3 SUSY searches

Physics performance studies for signals other than the MSSM Higgs bosons have not been performed to the same depth. A general squark/gluino signal detection capability has been tested, but a systematic exploration of model parameter space has not yet been done, partly due to the great variety of scenarios to be investigated, but mainly because it is sensitive to detector E_t^{miss} resolution which is a global variable requiring a realistic simulation of the entire calorimetric system. The tail of the E_t^{miss} distribution is affected by geometrical coverage, cracks and gaps due to mechanical structures, cable and cooling system passages etc, thus a reliable E_t^{miss} response of a detector can come only late in the detector design phase.

The present lower mass bound on squark and gluino masses from Tevatron experiments is 212 GeV if $m_{squark} = m_{gluino}$, and m_{gluino} > 179 GeV if $m_{squark} >> m_{gluino}$ [42]. CDF and D0 should be able to explore masses up to ~ 0.4 TeV, beyond that is the LHC domain. Studies done in ATLAS and CMS on the benchmark jets + E_t^{miss} channel concern the squark/gluino upper mass reach which, depending on the scenario, is ~2 TeV for 10^5pb^{-1} [3,4]. This channel is fed, for example, by squark or gluino pair production followed by $\tilde{q}_R \to q\chi_1^0$

or $\tilde{g} \rightarrow q\bar{q}\chi_1^0$ decays. No mass peaks can be expected in these searches and evidence for a signal depends on an excess of events over expected backgrounds. For high gluino/squark masses the E_t^{miss} due to signal neutralinos and to neutrinos from SM backrounds $t\bar{t}$, W($\rightarrow$ lν)+jets, Z($\rightarrow$ νν)+jets, dominates over the instrumental E_t^{miss} from detector response to jets. The lower mass reach, at the junction between the FERMILAB and LHC search regions, has not been so reliably explored yet, as QCD multijet events are here the dominant backround [3]. The production cross sections are large, however, and significant searches will be possible already with 10^3pb^{-1}[3,4,43]. Figure 20 shows expected squark and gluino signals, with masses 0.5 to 1.5 TeV, in the jets + E_t^{miss} final states. Selection criteria are at least three hard jets of E_t > 200 GeV and high circularity C > 0.2 [43]. The expected instrumental (jets) and $t\bar{t}$, W, Z backgrounds are shown separately. Another experimental signature that has been studied is leptons + E_t^{miss} + jets. These final states are fed by cascade decays of squarks or gluinos through $\chi^{\pm}_i$ and χ^0_i decaying to leptons. Studies of gluino pair production have been performed for final states with one, two and three high p_t muons. Two same-sign isolated muons are a particularly clean signature. These channels allow to explore the ~0.3 to 1.5 TeV gluino mass range [3,44]. Through the internal SUSY mass relations [5,35] the exploration of the ~TeV squark/gluino mass range would provide also a very significant constraint on, or in case of discovery provide indirect evidence for the cosmologically relevant LSP-lightest neutralino for a mass up to ~ 0.2 TeV.

A comprehensive phenomenological study of the explorable regions of supergravity constrained MSSM parameter space exploiting the E_t^{miss}+ multijets, leptons + E_t^{miss} + jets final states from gluino and squark production, and the two and three-leptons + E_t^{miss} final states from slepton and chargino, neutralino production has been performed in ref. [5] assuming a generic LHC detector response. The results are shown in Fig. 21. They seem to indicate that with 10^4pb^{-1} most of parameter space of this particular model could be explored. These very encouraging results have to be eventually substantiated with more detailed and realistic detector response simulations.

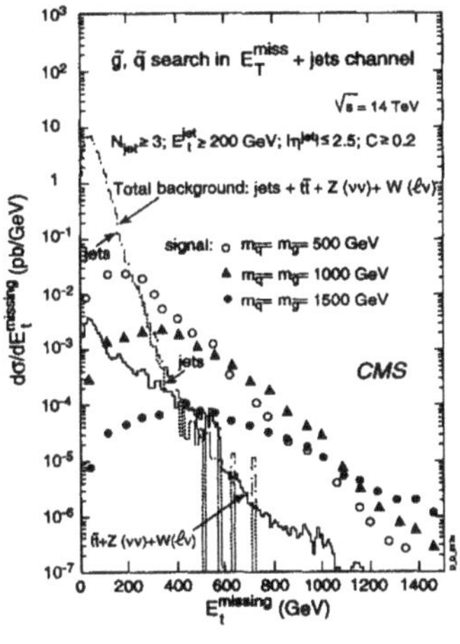

Fig. 20: E_t^{miss} distributions for squark and gluino signals and expected backgrounds in CMS.

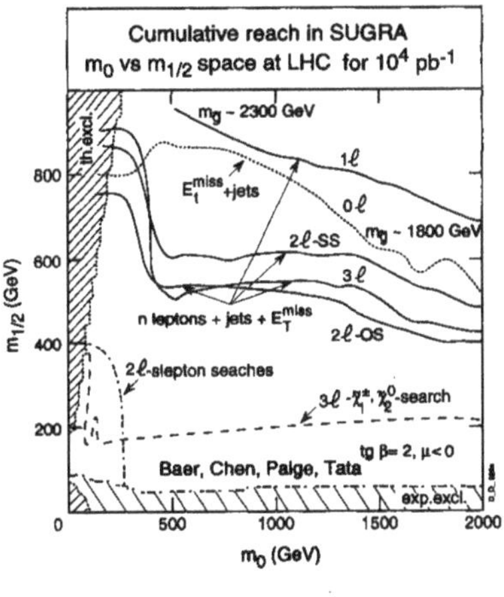

Fig. 21: SUSY parameter space explorable through different channels, from ref. [5].

3.4 B - Physics at the LHC

The main issue in B-physics is the observation of CP-violation. For such a measurement high statistics are vital as the interesting decay modes have small branching ratios (~10^{-5}) and the large B-production rate of a hadron collider is essential. At LHC ~10^{13} $b\bar{b}$ will be produced per year – the problem is to trigger on and select the interesting modes. ATLAS and CMS could be competitive for some particular channels, such as $B_d^0 \rightarrow J/\psi K_s^0$ and $B_d^0 \rightarrow \pi^+\pi^-$.

$B^0_d \to J/\psi\, K^0_s$

The decay $B^0_d \to J/\psi\, K^0_s$, with $J/\psi \to \mu^+\mu^-$ and $K^0_s \to \pi^+\pi^-$, is the most appropriate channel to measure the angle β of the unitarity triangle. A way to tag the B^0_d is through the associated B hadron requiring it to decay into lepton + X. The time-integrated asymmetry A is:

$$A = \frac{N^+ - N^-}{N^+ + N^-} = D \cdot \frac{x_d}{1 + x_d^2} \cdot \sin 2\beta,$$

where N^+ and N^- are the number of events with positively and negatively charged tagging leptons, D is the tagging dilution factor and $x_d/(1+x_d^2)$ is the time-integration factor. Detailed simulations of these final states have been performed. In CMS a two-muon trigger would be used. The main selection criteria are : i) two muons and ii) two charged hadrons with $p_t > 0.7$ GeV in $|\eta| < 2.4$ for K^0_s reconstruction. Figure 22a shows the $\mu^+\mu^-\pi^+\pi^-$ invariant mass after all selection cuts. To simulate the regime at $10^{33}\,cm^{-2}s^{-1}$, two minimum-bias events have been sûperimposed on signal and background events. The number of reconstructed signal events is 5500 for $p_t^{\mu_{tag}} > 5$ GeV and $10^4 pb^{-1}$. The measured asymmetry A is affected by dilution effects, the most important being mistagging of muons due to oscillations and cascade decays of b ($b \to c \to \mu$) [45]. The overall dilution factor in this channel is D = 0.49 for $p_t^{\mu_{tag}} > 5$ GeV and the sensitivity to $\sin 2\beta$ ie the expected measurement precision is: $\delta(\sin 2\beta) = 0.05 \pm 0.014$. A time dependent analysis is of similar sensitivity. The sensitivity can be improved using electrons for tagging and in J/ψ decays as done by ATLAS, and is investigated at present in CMS.

$B^0_d \to \pi^+\pi^-$

$B^0_d \to \pi^+\pi^-$ is a promising channel to measure the angle α of the unitarity triangle. Even without particle identification CMS can be expected to perform well thanks to its excellent mass resolution. A trigger is provided by the semileptonic decay of the associated b-hadron which is used to tag the flavour. The main selection criteria are: i) a muon with $p_t^{\mu} > 9$ GeV, ii) two opposite-sign hadrons with $p_t^h > 5$ GeV within $|\eta| < 2$, iii) B^0_d isolation, $I < 0.3$ ($I = \Sigma p_t$ of hadrons within $\Delta R < 1$ around the B^0_d $/p_t(B^0_d)$) and iv) impact parameter significance > 3 for each pion. Without particle identification the two-body decays of B-hadrons generate fake mass peaks when the charged hadrons are assigned a pion mass as visible in Fig. 22b. The background to signal ratio, B/(S + B), is 45% in a mass window of $\pm 1\sigma$ (σ_B = 27 MeV). The combinatorial background gives a flat ~10% background contribution. The number of $B^0_d \to \pi^+\pi^-$ events within a $\pm 1\sigma$ mass-bin around the nominal B^0_d peak after all cuts is 4300 [3,45]. Including dilution effects due to oscillations, mistagging and background, the sensitivity to the unitarity triangle angle α is estimated to be: $\delta(\sin 2\alpha) = 0.057^{+0.018}_{-0.014}$. A significant improvement in sensitivity can be obtained using electrons to trigger and tag, but triggering at an electron threshold of E_t ~10 GeV at $10^{33}\,cm^{-2}s^{-1}$ is still under investigation.

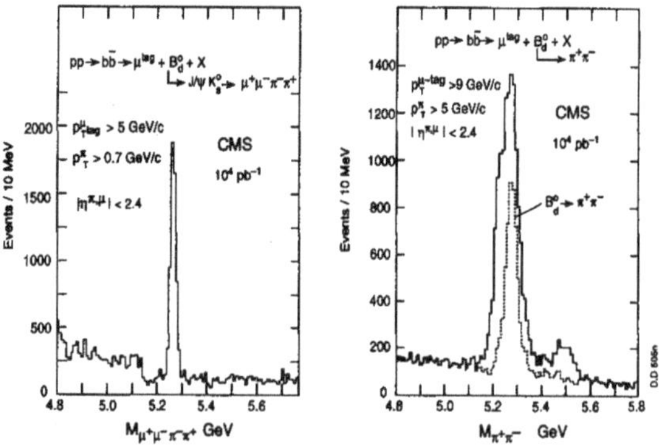

Figs. 22a, b: $\mu^+\mu^-\pi^+\pi^-$ and $\pi^+\pi^-$ invariant masses after all selection cuts.

Perspectives on CP violation measurements

Figure 23 shows the expected sensitivities in the measurements of the unitarity triangle angles α and β in the various experiments over the coming years. A reasonable guess is made for the start-up of the machines and detectors or upgrades as indicated [46, 47, 8, 48]. A data taking and analysis period of at least one year is assumed preceeding the first quoted sensitivity point. For CMS and ATLAS $10^4 pb^{-1}$ is assumed for the first point. Subsequent improvement is due to increased statistics ($1/\sqrt{N}$ behaviour) or to significant detector upgrades as in case of CDF [48]. For each of the experiments there are difficulties. HERA-B is very difficult due to smallness of the B cross section and unfavorable S/B, thus pattern recognition and radiation damage difficulties; for B-factories the machine is critical , if luminosities in excess of $10^{33}cm^{-2}s^{-1}$ are not obtained the statistics will be insufficient for significant results; for CDF and D0 the uncertainties are with the performance upgrades, ultimately they should have comparable reach as ATLAS or CMS as collider experiments are limited by tagging efficiency/purity; LHCB could be the best, but whilst CDF shows that central collider B-physics is feasible, not much can be said up to now on forward B experiments. In any case, around AD 2000 the competition for the first observation of CP violation in the B system may be fierce. The present theoretical lower bound is $\sin 2\beta > 0.17$ [49]. LHC experiments will be the natural higher-sensitivity continuation of these studies.

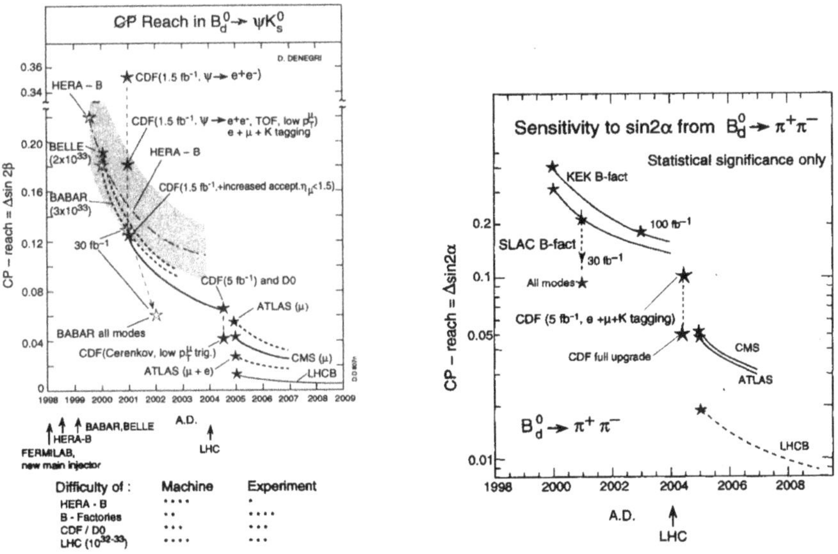

Fig. 23: Expected sensitivities to the measurement of the angles α and β for various experiments over the coming years. The shaded area illustrates the effect of the uncertainty on the B cross section at HERA.

Search for $B_s^0 \to \mu\mu$

$B_s^0 \to \mu^+\mu^-$ is a prototype of a rare decay for which only a hadron collider has a sufficient $b\bar{b}$ production rate to possibly allow its study. The expected SM branching ratio is $\approx 4 \times 10^{-9}$ [50]. The two-muon trigger should be used for $B_s^0 \to \mu\mu$. The background is due to muons from $b\bar{b}$ and a reduction factor of at least 10^7 is needed to suppress it. Appropriate selection criteria are: i) transverse momentum $p_t^{\mu\mu} > 12$ GeV, ii) $\mu^+\mu^-$ pair isolation, iii) a dimuon mass cut, iv) a fight-path cut $d > 3\sigma_{sv}$ and v) $\delta < 0.04$ rad where δ is the angle between the vectors $\vec{p}_t^{\mu\mu}$ and flight-path $\vec{d}$ [45,51]. Critical cuts are the secondary vertex one and the dimuon isolation. In CMS the secondary vertex resolution in the transverse plane is $\sigma_{sv} \approx 70$ μm. After all cuts, 11 signal and less than 64 background events (at 90% C.L.) are expected for $10^4 pb^{-1}$. The upper limit on the branching ratio which can be set at a 90% C.L. for 10^4 pb^{-1} is 4 .5x 10^{-9}.

3.5 Heavy ion collisions

Preliminary studies of $\Upsilon(\Upsilon',\Upsilon'') \to \mu^+\mu^-$ detection in CMS were performed with ^{16}O, ^{40}Ca, ^{97}Nb and ^{208}Pb ion beam collisions. The trigger is provided by the two muons penetrating to the first two muon stations with an effective threshold of $p_t^{\mu} \approx 4$ GeV. The critical aspect is one of pattern recognition for the muons in the midst of the many soft particles [52]. For central Pb–Pb collisions expectations for particle densities go from $dn^{\pm}/dy = 3000$ to as high as $dn^{\pm}/dy = 8000$. Using this latter value in the calculation of occupancy and track reconstruction efficiency in the tracker, the occupancy in the outer four MSGC layers is ~10% (with 25cm long strips) which is at the limit of the acceptable. The muon track reconstruction efficiency in this case is 75% [52].

The cross-sections for Υ production in A-A collisions are obtained multiplying the cross-sections for pp reactions at the same c.m.s. energy by a factor $A^{2\alpha}$, with $\alpha = 0.95$. The pp cross-section at LHC energies was estimated by extrapolating existing experimental data assuming a linear increase with $\sqrt{s}$, giving $B(d\sigma/dy)_0 = 1.5 \times 10^{-33}$ cm^2 at $\sqrt{s_{pp}} \approx 6$ TeV. The earlier estimates were based on the saturating cross section indicated CR in Fig. 23 showing also the recent extrapolations with MRSD structure functions. These would give a factor~2 larger cross section still [53]. The main dimuon background is due to uncorrelated muon pairs from π,K decays, the contributions from b and c decays are less significant [3].

Dimuon mass spectra expected from a typical experimental run with Pb, Ni and Ca beams assuming the high particle density option in each case and no colour supression are shown in Fig.24. The mass resolution is 85 MeV (FWHM). The signal-to-background ratios indicated for Pb collisions have been calculated in the mass band $M(\Upsilon) \pm 50$ MeV. For lighter nuclei the S/B ratio improves, as well as the overall production rate, Fig.24. The plan is to measure the production ratios of $\Upsilon(2S)$ and $\Upsilon(3S)$ relative to $\Upsilon(1S)$ with different ion species, possibly at different energies, and relative to pp collisions, and correlate it with the centrality of the collisions measured by the calorimetric global or transverse energy flows. Colour screening resulting in the suppression of all heavy quark bound states except for $\Upsilon(1S)$ is the signature of quark-gluon plasma formation that would be investigated [10]. Another probe of QGP formation is jet production, where jet quenching is expected [54]. The CMS 4π calorimetry could allow to test this signature too [55].

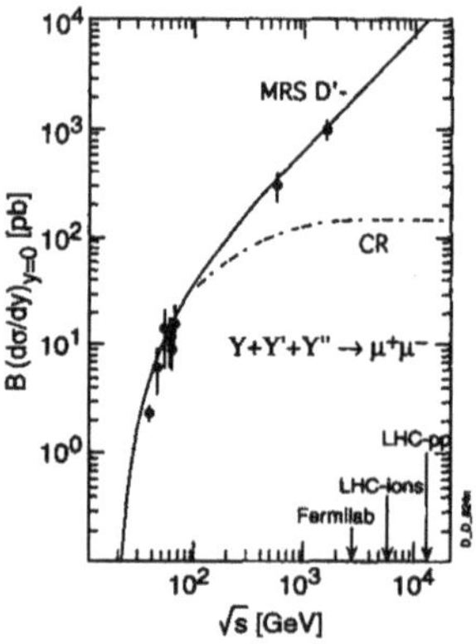

Fig. 23 Energy dependence of Y production in pN

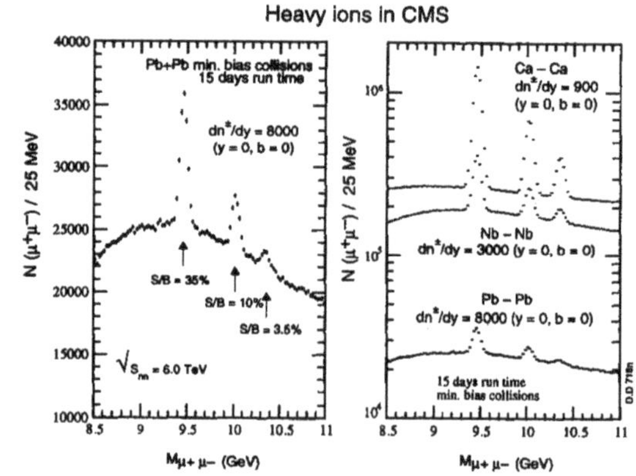

Fig. 24: Mass distributions of opposite sign muon pairs for Pb-Pb, Ni-Ni and Ca-Ca collisions in CMS.

4. CONCLUSIONS

The main task at the LHC will be to shed light on the mechanism responsible for electroweak symmetry breaking. The simplest scenario is the SM Higgs one, the MSSM one is its simplest supersymmetric extention, but no-elementary-Higgs and non-resonant schemes are possible too [2]. CMS and ATLAS are powerful and well balanced general pourpose detectors able to thoroughly explore these possibilities through a variety of experimental signatures. The entire expected SM Higgs mass range can be explored, and with few$\times 10^{5}pb^{-1}$ the MSSM definitively tested. Even if ew symmetry breaking is not realised through a Higgs mechanism, the detector designs are flexible enough that they would allow to explore a number of alternative symmetry breaking schemes [1,2]. A hadron collider is also the most appropriate machine to look for squarks and gluinos; the LHC will allow to cover the mass range where supersymmetry could be of relevance for ew symmetry breaking. The lightest supersymmetric particle, if a neutralino, is of key cosmological importance being a prime candidate for dark matter. A mass range up to ~0.2 TeV can be explored indirectly through squark, gluino, chargino, neutralino or slepton searches. Very exciting possibilities exist too for CP violation studies in the B system and for uncovering and studying QCD deconfinement.

5. REFERENCES

[1] G. Altarelli, Proc. of LHC Workshop, Aachen, CERN 90-10, 1990 and references therein;
M. Lindner, Z. Phys C31 (1986) 295;
M. Chanowitz, Ann. Rev. Nucl. Part. Sci. 38 (1988) 323.

[2] S. Dimopoulos and M. Lindner, Proc. of LHC Workshop, Aachen, CERN 90-10, 1990.

[3] Technical Proposal, CMS Collaboration, CERN/LHCC 94-38, LHCC/P1, December 1994;
Letter of Intent, CMS Collaboration, CERN/LHCC 92-3.

[4] Technical Proposal, ATLAS Collaboration, CERN/LHCC/94-43,LHCC/P2, December 1994.

[5] H. Baer et al. Low Energy Supersymmetry Phenomenology, CERN-PPE/95-45, April 1995;
H. Baer et al., Phys. Rev. D50 (1994) 4508;
R. Barbieri et al. , NP B367 (1993) 28.

[6] Z. Kunszt and F. Zwirner, Nucl. Phys. B385 (1992) 3.

[7] P. Sphicas, CDF Collaboration, "Four Seas Conference", Trieste, June 1995 and Proc. SLAC Summer Institute 1995, July 1995;
J.A. Mueller, 'Beauty 94', M. St. Michel, France, May 1994, NIM A 351 (1994) 59.

[8] Letter of Intent, LHC-B Collaboration, CERN/LHCC 95-5, LHCC/I8, August 1995.

[9] Letter of Intent, ALICE Collaboration, CERN/LHCC93-16, March 1993.

[10] T. Matsui and H. Satz, Phys. Lett. B178 (1986) 416.

[11] F. Gasparini et al., Nucl. Instrum. Methods A336 (1993) 91.

[12] R. Cardarelli et al., Nucl. Instrum. Methods A340 (1994) 466;

[13] D. Carlsmith, CMS Cathode Strip Chambers in Endcap Magnetic Field, CMS TN/94-217 (1994);
Y. Fisyak and J. Rowe, CMS TN/94-160 (1994).

[14] F. Angelini et al., Nucl. Instrum. Methods A343 (1994) 441;

[15] V. Karimaki, Fast tracker response simulation, CMS TN/94-275 and references therein;
A. Khanov and N. Stepanov, Fast track finder algorithm, CMS TN/94-309;

[16] D.J. Graham, An algorithm using tracks to locate the two-photon vertex at high luminosity, CMS TN/95-115.

[17] C. Seez and T.S. Virdee, The Higgs to two photon Decay in CMS, an update, CMS TN/94-289 and references therein.

[18] P. Lecoq et al., Lead tungstate scintillators for LHC em calorimetry, CMS TN/94-308 (1994);

[19] E. Lorenz et al., Max Planck Institute, Munich, MPI-PHE-93-23 (1993);

[20] PbWO4 test beam results, CMS working group meetings 1995;
J.P. Peigneux et al. , Results from tests on matrices of lead tungstate crystals using high energy beams, CMS TN/95-166.

[21] A. Nikitenko, P. Verrecchia and D. Bomestar, GEANT reconstruction of $H \rightarrow ZZ^{*} \rightarrow 4e$, CMS TN/95-019 (1995);

[22] A. Ferrando et al., CMS TN/92-19 (1992);

[23] O. Ganel and R. Wigmans, SSC note SDC-93/575 (1993);

[24] F. Abe et al., CDF collaboration, Phys. Rev. Lett. 74 (1995) 2676.
S. Abachi et al., D0 Collaboration, Phys. Rev. Lett. 74 (1995) 2632.

[25] J. Ellis, G.L. Fogli and E. Lisi, The Higgs Boson Mass from Precision Electroweak Data, CERN TH/95-202 (1995) and CERN–TH 7261/94; J. Ellis, New fits to m_{Higgs}, communication at the CMS Physics Meeting, October 1995.

[26] LEP Electroweak Working Group, CERN Report no. LEPEWWG/95-01,

[27] J.F. Grivaz, "New Particle Searches", LAL 95-83, rapporteur talk at EPS Brussels Conferece, July 1995;
F. Richard, contribution to International Conf. on High Energy Physics, 94- Glasgow.

[28] M. Lindner, Z. Phys C31 (1986) 295; M. Sher, Phys. Rep179 (1989) 273;
G. Altarelli, Proc. of LHC Workshop, Aachen, eds G. Jarlskog and D. Rein, CERN 90-10, 1990;
see also R. Barbieri, talk at this conference and ref.[22].

[29] K. Lassila, $H \to \gamma\gamma$ conversion recovery, mass resolution, communications in CMS working groups.

[30] S. Abdullin et al., 'Possibilities to improve the observability of SM light Higgs in the $\gamma\gamma$ channel', CMS TN/94–247 (1994).

[31] I. Iashvili et al., 'Study of $H \to ZZ^* \to 4l^{\pm}$ inCMS', CMS /TN 95-059.

[32] D. Bomestar et al., Study of $H \to ZZ \to 4l^{\pm}$ with full GEANT simulation of CMS, CMS TN/95-018.

[33] M. Dzelalija et al., 'Study of Heavy $H \to ZZ \to 4l^{\pm}$ inCMS, 'CMS TN/95-076.

[34] S. Abdullin and N. Stepanov, CMS TN/94–179 (1994) and CMS TN/94–178 (1994).

[35] G. Kane et al., 'Study of Constrained Minimal Supersymmetry', UM- TH-93–24 (1993);
J.A. Casas et al., ' The Lightest Higgs Bosons Mass in the Minimal Supersymmetric Standard Model', CERN–TH.7334/94 (1994).

[36] We are highly indebted to Z. Kunszt and F. Zwirner, who provided us with their programs RADCORR and BBSUSY.

[37] R. Kinnunen, J. Tuominiemi, D, Denegri, CMS TN/93–98 (1993) and TN/93–103 (1993).

[38] R. Kinnunen et al., CMS TN/94–233 (1994).

[39] D. Denegri, et al., 'Justifications for a pixel detector in CMS', CMS TN/94–258.

[40] N. Stepanov, CMS TN/94–182 (1994);
C. Kao and N. Stepanov, Discovering MMSM Higgs Bosons with Muons,UR-1407,ER-40685-854,1995.

[41] J. Dai, J. F. Gunion and R. Vega, UCD-93–20, SLAC PUB-6277.

[42] J. M. Benlloch, CDF Collaboration, XXX th Rencontre de Moriond, QCD and High Energy Hadronic Interactions, March 1995; D. Claes, presentation at the 10th Topical Workshop on Collider Physics, FERMILAB, May 1995.

[43] V. Genchev and M. Greiter, Backgrounds for the SUSY Searches, CMS TN/95-066;
V. Genchev, Gluino and squark searches at LHC, "Four Seas Conference", Trieste, June 1995.

[44] L. Rurua, CMS TN/94–207 (1994).

[45] D. Denegri et al., Int. Journal of Mod. Phys A,Vol 9, No 24 (1994) 4211; D. Denegri, et al., in 'Beauty 94', M. St. Michel, France, May 1994, NIM A 351 (1994) 95.

[46] HERA-B Proposal, DESY-PRC 94/02, May 1994; P. Krizan et al., in 'Beauty 94', NIM A 351 (1994) 111.

[47] G. Wormser, 'Beauty 94', M. St. Michel, France, May 1994, NIM A 351 (1994) 54; G. Wormser LAL preprint 95-85; BABAR Technical Design Report, SLAC-R-95-457, March 1995.

[48] N.S. Lockyer, B physics at CDF and D0, Present and Future, CDF/DOC/BOTTOM/PUBLIC/3039;
F. DeJongh, B physics with CDF RunII upgrade, "Four Seas Conference", Trieste, June 1995.

[49] A. Ali and D. Landon, CERN–TH 7398/94.

[50] B. Buchalla and A. Buras, Nucl Phys. B400 (1993) 225;
A. Ali, C. Greub and T. Mannel, 'Rare B Decays in Standard Model', DESY 93–016 (1993).

[51] A. Nikitenko and A. Starodumov, CMS TN/94–186.

[52] O. Kodolova and M. Bedjidian, Pattern recognition and track reconstruction in heavy ion collisions, CMS TN/95-124.

[53] R. Gavai et al. Int. Journal of Mod. Phys. A,Vol 10 (1995) 3043.

[54] H. Satz, ECFA LHC Workshop Proceedings, CERN 90–10,

[55] R. Kvatadze and R. Shanidze, CERN CMS TN/94–270 and V. Korotkikh, et al., CERN CMS TN/94–244.

Production of Stop and Sbottom at LEP2 and at the LHC*

A. Bartl, W. Porod
Institut für Theoretische Physik, Universität Wien
A-1090 Vienna, Austria

H. Eberl, S. Kraml, W. Majerotto
Institut für Hochenergiephysik
Österreichische Akademie der Wissenschaften
A-1050 Vienna, Austria

Abstract

This talk deals with the question how sbottoms and stops can be discovered at LEP2 or at the LHC. We present numerical predictions within the Minimal Supersymmetric Standard Model for production cross sections and decay rates of these particles.

1 Introduction

The search for supersymmetric (SUSY) particles plays an essential rôle at the present colliders LEP1 [1] and TEVATRON [2] and will become even more important at LEP2 [3] and the LHC [4]. The stop and the sbottom are particularly interesting candidates for SUSY searches. Since they are the partners of the quarks of the third generation, they could be produced in e^+e^- annihilation and in pp ($p\bar{p}$) collisions. Moreover, they could be the lightest visible SUSY particles. The stop can be much lighter than the other squarks due to its large Yukawa coupling [5]. If the parameter $\tan\beta \gtrsim 10$, the bottom Yukawa terms are also non-negligible, and the scalar bottom quark could turn out to be the lightest charged SUSY particle [6].

Theoretical studies of stop production in e^+e^- annihilation were carried out in [7,8]. Stop production at the TEVATRON and at the LHC was recently studied in

*presented by A. Bartl at the International Workshop on Elementary Particle Physics: Present and Future, Valencia, Spain, 5 - 9 June, 1995.

[9,10]. In this talk we shall present our results for the production of stops and sbottoms at LEP2 and at the LHC. We shall also discuss the decay properties of these particles.

We shall work in the Minimal Supersymmetric Standard Model (MSSM) [11]. The main parameters which determine the phenomenology of the stop and sbottom system are $M_{\tilde{Q}}$, $M_{\tilde{U}}$, $M_{\tilde{D}}$, A_t, A_b, μ, and $\tan\beta = v_2/v_1$. For the description of the charginos and neutralinos also the parameters M and M' are necessary, with $M'/M = \frac{5}{3}\tan^2\Theta_W \approx 0.5$.

Concerning the present experimental lower mass bounds for stops, LEP obtained $M_{\tilde{t}} \gtrsim 45$ GeV [1]. If the stop mixing angle is near the special value $\Theta_{\tilde{t}} = 0.98$, the bound is only $M_{\tilde{t}_1} \gtrsim 42$ GeV. The TEVATRON experiment D0 excludes the mass range 40 GeV $\lesssim M_{\tilde{t}_1} \lesssim 100$ GeV, provided $M_{\tilde{t}_1} - m_{\tilde{\chi}_1^0} \gtrsim 30$ GeV [2].

2 Production and Decay of Stops and Sbottoms at LEP2

The mass terms of the stop and sbottom systems can most conveniently be written in terms of the 2×2 mass mixing matrices. The diagonal elements are $M^2_{\tilde{t}_L} = M^2_{\tilde{Q}} + m_Z^2 \cos 2\beta(\frac{1}{2} - \frac{2}{3}\sin^2\Theta_W) + m_t^2$, $M^2_{\tilde{t}_R} = M^2_{\tilde{U}} + \frac{2}{3}m_Z^2 \cos 2\beta \sin^2\Theta_W + m_t^2$, and $M^2_{\tilde{b}_L} = M^2_{\tilde{Q}} - m_Z^2 \cos 2\beta(\frac{1}{2} - \frac{1}{3}\sin^2\Theta_W) + m_b^2$, $M^2_{\tilde{b}_R} = M^2_{\tilde{D}} - \frac{1}{3}m_Z^2 \cos 2\beta \sin^2\Theta_W + m_b^2$, respectively. The off-diagonal elements are $m_t(A_t - \mu\cot\beta)$ and $m_b(A_b - \mu\tan\beta)$ for top and bottom squarks, respectively. For scalar top quarks there can be an appreciable amount of $\tilde{t}_L$-$\tilde{t}_R$ mixing due to the large top quark mass. Similarly, for the scalar bottom quarks $\tilde{b}_L$-$\tilde{b}_R$ mixing is also non-negligible if $\tan\beta \gtrsim 10$. The mass eigenvalues for the top and bottom squarks are denoted by $M_{\tilde{t}_{1,2}}$ and $M_{\tilde{b}_{1,2}}$ where $\tilde{t}_1$ and $\tilde{b}_1$ denote the lighter eigenstates.

We show in Fig. 1 the contour lines in the $M_{\tilde{t}_1} - \cos\Theta_{\tilde{t}}$ plane for the total cross section of $e^+ + e^- \to \tilde{t}_1 + \bar{\tilde{t}}_1$ in fb, for $\sqrt{s} = 190$ GeV. This cross section can go up to 1.5 pb. Assuming an integrated luminosity of $500pb^{-1}$, the mass reach for $\tilde{t}_1$ is about 85 to 90 GeV at this CMS energy. Analogously, for other CMS energies the mass reach for stops is about $E_{beam} - 10$ GeV. Similarly, Fig. 2 is a contour plot of the total cross section of $e^+ + e^- \to \tilde{b}_1 + \bar{\tilde{b}}_1$ in the $M_{\tilde{b}_1} - \cos\Theta_{\tilde{b}}$ plane for $\sqrt{s} = 190$ GeV. The values of this cross section can go up to 1.2 pb. The expected mass reach for $\tilde{b}_1$ is slightly less than that for $\tilde{t}_1$, i. e. about 80 to 90 GeV, depending on the mixing angle $\Theta_{\tilde{b}}$.

The cross sections for $e^+e^- \to \tilde{t}_1\bar{\tilde{t}}_1$ and $e^+e^- \to \tilde{b}_1\bar{\tilde{b}}_1$ show a characteristic dependence on the squark mixing angle, due to the fact that the $Z^0\tilde{t}_1\bar{\tilde{t}}_1$ coupling vanishes at $\Theta_{\tilde{t}} = 0.98$, and the $Z^0\tilde{b}_1\bar{\tilde{b}}_1$ coupling vanishes at $\Theta_{\tilde{b}} = 1.17$. The cross sections have their minima at the mixing angles $\cos^2\Theta_{\tilde{t}} = \frac{4}{3}\sin^2\Theta_W \times F(s)$ and $\cos^2\Theta_{\tilde{b}} = \frac{2}{3}\sin^2\Theta_W \times F(s)$, with $F(s) = 1 + 2\frac{s-m_Z^2}{s}\frac{\cos^2\Theta_W(1-4\sin^2\Theta_W)}{1-4\sin^2\Theta_W+8\sin^4\Theta_W} \approx 1 - 0.32\frac{s-m_Z^2}{s}$.

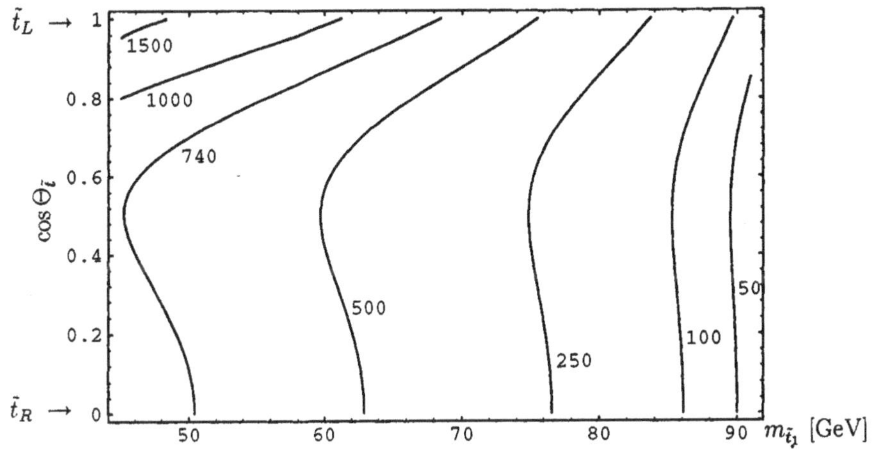

Fig.1: Contour lines for the total cross section (in fb) of $e^+e^- \rightarrow \tilde{t}_1\bar{\tilde{t}}_1$ as a function of $m_{\tilde{t}_1}$ and $\cos\theta_{\tilde{t}}$ for $\sqrt{s} = 190$ GeV. ISR corrections and QCD radiative corrections are included.

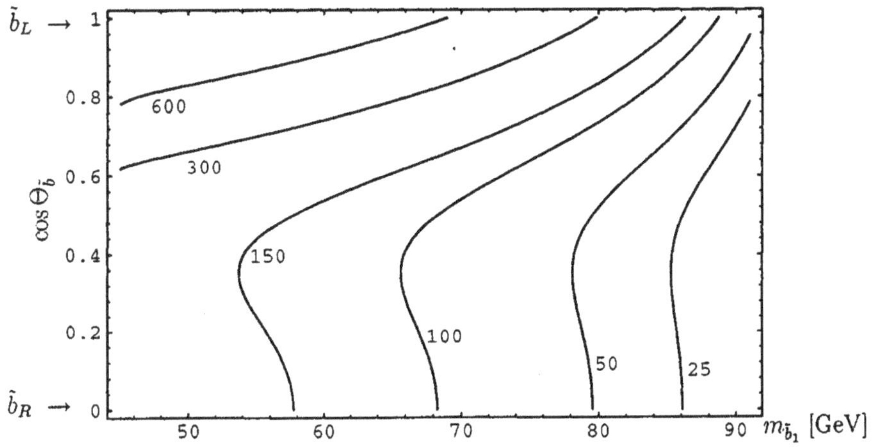

Fig.2: Contour lines for the total cross section (in fb) of $e^+e^- \rightarrow \tilde{b}_1\bar{\tilde{b}}_1$ as a function of $m_{\tilde{b}_1}$ and $\cos\theta_{\tilde{b}}$ for $\sqrt{s} = 190$ GeV. ISR corrections and QCD radiative corrections are included.

$M_{\tilde{q}}$ [GeV]	200	300	450
$\sigma(pp \to \tilde{u}_L\bar{\tilde{u}}_L X)$ [pb]	28	4.4	0.6
$\sigma(pp \to \tilde{t}_1\bar{\tilde{t}}_1 X)$ [pb]	24	3.7	0.5

Table 1:
Total cross sections for $pp \to \tilde{u}_L\bar{\tilde{u}}_L X$ and $pp \to \tilde{t}_1\bar{\tilde{t}}_1 X$ at $\sqrt{s} = 14$ TeV, for $M_{\tilde{q}} =$ 200 GeV, 300 GeV, and 450 GeV, taking $m_{\tilde{g}} = M_{\tilde{q}}$.

For $\sqrt{s} = 190$ GeV the minima of the cross sections are at $\cos\Theta_{\tilde{t}} = 0.55$ and $\cos\Theta_{\tilde{b}} = 0.39$.

In the calculations of the cross sections shown in Figs. 1 and 2 we have included the QCD corrections in the approximation of [8], and the corrections due to initial state radiation.

The possible decay modes of $\tilde{t}_1$ and $\tilde{b}_1$ at LEP2 are $\tilde{t}_1 \to c + \tilde{\chi}^0_1$, $\tilde{t}_1 \to b + \tilde{\chi}^+_1$, and $\tilde{b}_1 \to b + \tilde{\chi}^0_1$, $\tilde{b}_1 \to b + \tilde{\chi}^0_2$. In the parameter domain $M_{\tilde{t}_1} \lesssim M \lesssim 2M_{\tilde{t}_1}$ and $|\mu| \gtrsim M_{\tilde{t}_1}$ the decay $\tilde{t}_1 \to c\tilde{\chi}^0_1$ has a branching ratio of 100 percent. Outside this parameter domain the decay $\tilde{t}_1 \to b\tilde{\chi}^+_1$ occurs with practically 100 percent branching ratio. In a very restricted parameter region the decay $\tilde{t}_1 \to c\tilde{\chi}^0_2$ may also play a rôle. As for the $\tilde{b}_1$ decays, for $M_{\tilde{b}_1} \lesssim M \lesssim 2M_{\tilde{b}_1}$ and $|\mu| \gtrsim M_{\tilde{b}_1}$ the $b\tilde{\chi}^0_1$ decay mode has 100 percent branching ratio. If $M \lesssim M_{\tilde{b}_1}$ or $|\mu| \lesssim M_{\tilde{b}_1}$, then the branching ratios for the $b\tilde{\chi}^0_1$ and the $b\tilde{\chi}^0_2$ decay modes depend on $\cos\Theta_{\tilde{b}}$ and on M, μ, $\tan\beta$. For example, for $M_{\tilde{b}_1} = 80$ GeV, $M = \mu = 60$ GeV, $\tan\beta = 30$, $BR(\tilde{b}_1 \to b\tilde{\chi}^0_2)$ is about 10 percent for $\cos\Theta_{\tilde{b}} \approx 0$ and goes up to 80 percent for $|\cos\Theta_{\tilde{b}}| \approx 1$.

3 Direct Production of Stops and Sbottoms at the LHC

In this talk we shall concentrate on the direct production of stop and sbottom pairs at the LHC, i. e. on the reactions $p + p \to \tilde{t}_1 + \bar{\tilde{t}}_1 + X$ and $p + p \to \tilde{b}_1 + \bar{\tilde{b}}_1 + X$. We show in Tab. 1 the values of the total cross section $\sigma(pp \to \tilde{t}_1\bar{\tilde{t}}_1 X)$ [12] for three typical values of stop masses at $\sqrt{s} = 14$ TeV, as obtained with ISASUSY, taking $m_{\tilde{g}} = M_{\tilde{t}_1}$. $pp \to \tilde{b}_1\bar{\tilde{b}}_1 X$ has the same cross section as $pp \to \tilde{t}_1\bar{\tilde{t}}_1 X$. For comparison we also show $\sigma(pp \to \tilde{u}_L\bar{\tilde{u}}_L)$. At $\sqrt{s} = 10$ TeV the cross sections are about a factor 2 smaller.

In the mass region accessible at the LHC the stops and sbottoms decay according to $\tilde{t}_i \to t + \tilde{\chi}^0_k$, $b + \tilde{\chi}^+_k$, and $\tilde{b}_i \to b + \tilde{\chi}^0_k$, $t + \tilde{\chi}^-_k$. If $M_{\tilde{t}_1} < m_{\tilde{\chi}^0_1} + m_t$ as well as $M_{\tilde{t}_1} < m_{\tilde{\chi}^\pm_1} + m_b$, then the three-body decay $\tilde{t}_1 \to b + W^+ + \tilde{\chi}^0_1$ and the higher-order decay $\tilde{t}_1 \to c + \tilde{\chi}^0_1$ are possible [7]. We show in Fig. 3a and b the allowed parameter domains in the $M - \mu$ plane for the various $\tilde{t}_1$ and $\tilde{b}_1$ decay modes, for

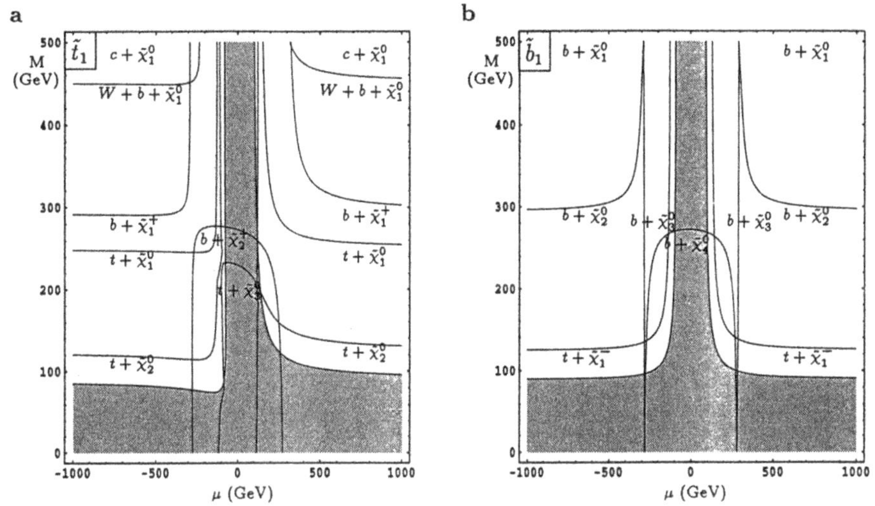

Fig.3: Allowed parameter domains in the M–μ plane for the decays of (**a**) $\tilde{t}_1$ for $\tan\beta = 2$, $m_{\tilde{t}_1} = 300$ GeV, and (b) $\tilde{b}_1$ for $\tan\beta = 30$, $m_{\tilde{b}_1} = 300$ GeV. We have taken $m_t = 175$ GeV, $m_b = 5$ GeV. The grey area is covered by LEP2.

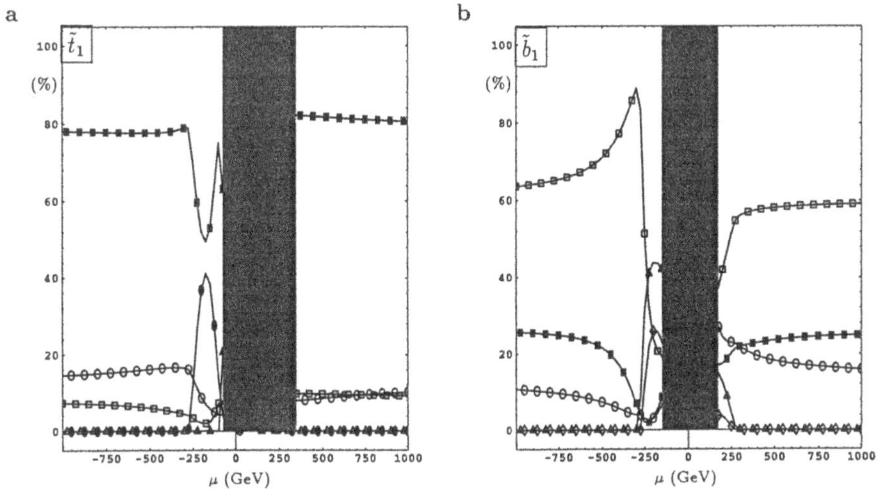

Fig.4: Branching ratios of the decays (**a**) $\tilde{t}_1 \to b\tilde{\chi}_1^+$ (■), $\tilde{t}_1 \to b\tilde{\chi}_2^+$ (•), $\tilde{t}_1 \to t\tilde{\chi}_1^0$ (○), $\tilde{t}_1 \to t\tilde{\chi}_2^0$ (□), $\tilde{t}_1 \to t\tilde{\chi}_3^0$ (△), $\tan\beta = 2$, $m_{\tilde{t}_1} = 300$ GeV, $\cos\theta_{\tilde{t}} = 0.71$, and (**b**) $\tilde{b}_1 \to b\tilde{\chi}_1^0$ (○), $\tilde{b}_1 \to b\tilde{\chi}_2^0$ (□), $\tilde{b}_1 \to b\tilde{\chi}_3^0$ (△), $\tilde{b}_1 \to b\tilde{\chi}_4^0$ (◇), $\tilde{b}_1 \to t\tilde{\chi}_1^-$ (■), $\tan\beta = 30$, $m_{\tilde{b}_1} = 300$ GeV, $\cos\theta_{\tilde{b}} = 0.69$, as a function of μ with $m_t = 175$ GeV, $m_b = 5$ GeV, $M = 110$ GeV. The grey area is covered by LEP2.

$M_{\tilde{t}_1} = M_{\tilde{b}_1} = 300$ GeV, $\tan\beta = 2$ and $\tan\beta = 30$, respectively, taking $m_t = 175$ GeV and $m_b = 5$ GeV. In Fig. 4a and b we show the branching ratios for the decays of $\tilde{b}_1$ and $\tilde{t}_1$ as a function of μ, for $M_{\tilde{t}_1} = M_{\tilde{b}_1} = 300$ GeV, $M = 110$ GeV, $\tan\beta = 2$, $\cos\Theta_{\tilde{t}} = 0.71$, and $\tan\beta = 30$, $\cos\Theta_{\tilde{b}} = 0.69$, respectively, with $m_t = 175$ GeV, $m_b = 5$ GeV. As can be seen in Figs. 4a and b, for this choice of the parameters the main decay modes are $\tilde{t}_1 \to b + \tilde{\chi}_1^+$ and $\tilde{b}_1 \to b + \tilde{\chi}_2^0$.

Acknowledgements

A. B. is very grateful to Prof. J. W. F. Valle and the organizers for inviting him to this interesting and inspiring workshop, and for the kind hospitality extended to him.

References

[1] F. Richard, these Proceedings

[2] A. White, Proc. of the Int. Workshop SUSY-95, Palaiseau, France, 15-19 May 1995; D. Normann, these Proceedings

[3] D. Treille, these Proceedings

[4] D. Denegri, these Proceedings

[5] J. Ellis, S. Rudaz, Phys. Lett. B128 (1983) 248; G. Altarelli, R. Rückl, Phys. Lett. B144 (1984) 126; I. Bigi, S. Rudaz, Phys. Lett. B153 (1985) 335; M. Fukugita, H. Murayama, M. Yamaguchi, T. Yanagida, Phys. Rev. Lett. 72 (1994) 3009

[6] A. Bartl, W. Majerotto, W. Porod, Z. Phys. C64 (1994) 499

[7] K. I. Hikasa, M. Kobayashi, Phys. Rev. D36 (1987) 724

[8] M. Drees, K. I. Hikasa, Phys. Lett. B252 (1990) 127

[9] H. Baer, J. Sender, X. Tata, Phys. Rev. D50 (1994) 4517

[10] H. Baer, M. Drees, C. Kao, M. Nojiri, X. Tata, Phys. Rev. D50 (1994) 2148; T. Kon, T. Nonaka, Phys. Rev. D50 (1994) 6005

[11] H.E. Haber, G.L. Kane, Phys. Rep. 117 (1985) 75

[12] H. Rohringer, private communication

Searching for B_c mesons in the ATLAS experiment at LHC

M. Galdón, M. A. Sanchis-Lozano [a]

Instituto de Física Corpuscular (IFIC)
Departamento de Física Teórica
Dr. Moliner 50, E-46100 Burjassot, Valencia (Spain)

R. Pérez Ochoa, J.A. Valls

Instituto de Física Corpuscular (IFIC)
Departamento de Física Atómica, Molecular y Nuclear
Dr. Moliner 50, E-46100 Burjassot, Valencia (Spain)

We discuss the feasibility of the observation of the signal from B_c mesons in the ATLAS experiment at the LHC. In particular, we address the decay mode $B_c \to J/\psi\pi$ followed by the leptonic decay $J/\psi \to \mu^+\mu^-$, which should permit an accurate measurement of the B_c mass. The semileptonic decay $B_c \to J/\psi\, \mu^+ \nu_\mu$ is also considered for a possible extraction of $|V_{cb}|$.

1 Introduction

There is a general consensus in the scientific community [1] that the scope of a future high-luminosity, high-energy hadron collider like LHC should not be restricted to the hunting of the standard model Higgs and its extensions, or the search for supersymmetry. Other topics requiring lower luminosities like top and beauty physics deserve in their own right a close attention. In particular, a lot of work has been recently devoted to the observation of the B_c meson regarding both the hadroproduction [2] and decay [3 4] (see references therein). Specifically, we shall focus on the feasibility of its detection in the ATLAS experiment at the LHC through some decay modes.

2 B_c signal

At the center-of-mass energy $\sqrt{s} = 14$ TeV, the cross-section for beauty production is assumed to be 500 μb leading to 5×10^{12} $b\bar{b}$ pairs per year-run (10^7 s)

[a]presented by M.A. Sanchis-Lozano; E-mail: mas@evalvx.ific.uv.es

at a luminosity of $\mathcal{L} \approx 10^{33}$ cm^{-2} s^{-1}, corresponding to an integrated luminosity of $\sim$ 10 fb^{-1}. The number of bottom pairs reduces, however, to 2.3×10^{10} by requiring events with a triggering muon coming from either a b or a $\bar{b}$ under the kinematic cuts $p_\perp > 6$ GeV/c and $|\eta| < 2.2$ [5].

On the other hand, assuming that the b-quark fragmentation yields a B_c or a B_c^* with probability of the order of 10^{-3} [2], the yield of B_c mesons (not yet triggered) per year of running would be roughly $\simeq 10^{10}$.

$B_c \rightarrow J/\psi\ \pi$ channel

This exclusive channel followed by the leptonic decay of the J/ψ resonance into a pair of oppositely charged muons offers several important advantages. First of all, it allows for the mass reconstruction of the B_c meson. Observe also that anyone of the two muons can trigger the decay. Besides, it is very clean topologically with a common secondary vertex for all three charged particles, two of them (the muons) with the additional constraint of their invariant mass compatible with a J/ψ. Furthermore, the expected branching fraction is not too small, about 0.2% [6], which combined with the branching fraction of the leptonic decay of the resonance $BR(J/\psi \rightarrow \mu^+\mu^-) \simeq 6\%$ [7], yields an overall branching fraction for the signal of 10^{-4}. Thus, the number of such events turns out to be $\simeq 10^6$ per year of running.

A study has been performed in order to estimate the signal detection efficiency and background for the $B_c \rightarrow J/\psi(\rightarrow \mu^+\mu^-)\ \pi$ channel. The Monte Carlo employed for the signal simulation corresponds to a sample of $B_c \rightarrow J/\psi\pi$ events while for the background we have used a sample of inclusive b muon decays generated in all the cases with PYTHIA 5.7 .

Two types of background were considered: [b]

a) Combinatorial background due to muons from semileptonic decays of $b\bar{b}$ pairs produced at the main interaction. Cascade contributions such as $b \rightarrow c \rightarrow \mu$ are included as well for random combinations with any other muon in the same event.

b) Contamination from prompt J/ψ's in combination with another charged hadron (interpreted as a pion) from the main vertex. (In fact data released by Tevatron on the J/ψ yield point out a production rate quite larger than initially expected [9].) Incorrect tracking may give rise to the

[b] Muons coming from semileptonic decays of long-lived particles such as pions or kaons contribute in a negligible amount to trigger rates. On the other hand B decays into J/ψ and a charged particle would give an invariant mass quite below the B_c mass so they are of no concern

reconstruction of a (fake) secondary vertex, becoming a potential source of a large amount of background.

In a first step, we imposed the following cuts on events based on kinematic constraints:

- $p_{\perp min}(trig.\mu) = 6$ GeV/c ; $|\eta_{max}(trig.\mu)| = 2.2$
- $p_{\perp min}(\mu) = 3$ GeV/c ; $|\eta_{max}(\mu)| = 2.5$
- $p_{\perp min}(\pi) = 1$ GeV/c ; $|\eta_{max}(\pi)| = 2.5$
- $M_{\mu^+\mu^-} = M_{J/\psi} \pm 50$ MeV

The two first cuts correspond to the requirement of the 1st-level B physics trigger leading in our case to an efficiency of $\sim 15\%$ in triggering one of the two muons from the J/ψ. We next take into account the detection efficiency for the signal after applying the rest of $p_\perp$ and η cuts which turns out to be $\sim 21\%$. Setting the efficiency for muon identification as 80% and the track reconstruction as 95% [5] we get a combined detection efficiency of $\sim 2\%$ leading to an observable signal of about 20,000 events per year of running.

The last of the cuts described above constrains the two muons invariant mass to be compatible (within two standard deviations [5]) with the nominal J/ψ mass, thus drastically reducing random combinations. However, background of class b) can potentially pass all the kinematic cuts by a large amount, so another type of rejection is required.

To this end, we adapted to our needs the vertex reconstruction (i.e. vertex finding and fitting) routines of the LEP experiment DELPHI at CERN [10]. The vertex fitting algorithm provides as output the coordinates of the secondary vertex, the track momenta re-evaluated with the vertex constraint and the goodness of the fit by means of the total χ^2 as well as the contribution of each single track to it. In particular, we employed for background rejection the three spatial coordinates and the χ^2 for each fitted secondary vertex formed by the two muons and the charged hadron (assumed to be a pion) satisfying the above kinematic constraints. The distance between the reconstructed vertex and the primary (pp) interaction point was thereby determined. We shall refer to it as the decay length even for background events of class b).

Hence, candidate (either signal or background) events were required to pass the following extra cuts:

- total $\chi^2 < \chi_0^2$
- $\chi_i^2 < \frac{\chi_0^2}{3}$ for each single track-i

• **decay length larger than L_0**

where χ_0^2, L_0 have to be optimized to remove the background as much as possible but with a good acceptance for the signal. In our analysis we found $\chi_0^2 = 8$ and $L_0 = 350$ μm.

Figure 1 shows the reconstructed $(\mu^+\mu^-)_{J/\psi}$ π mass distribution for the expected signal above the surviving background once all the cuts have been applied, for an integrated luminosity of 10 fb^{-1} [3].

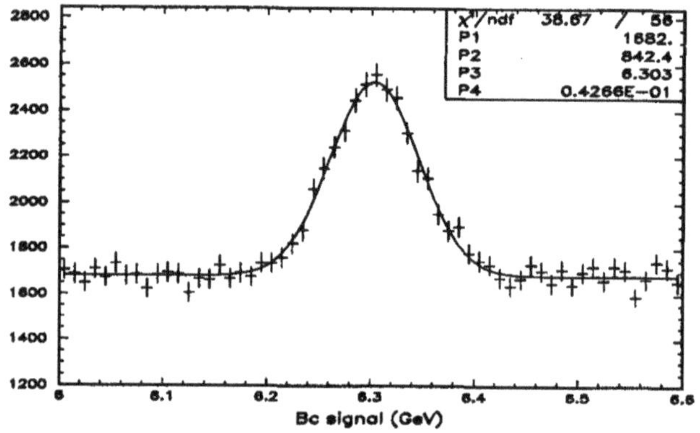

Figure 1: Reconstructed $(\mu^+\mu^-)_{J/\psi}$ π mass distribution after cuts. The nominal value for the B_c mass was set equal to 6.3 GeV. The solid line corresponds to a linear+Gaussian fit

In summary, we have found that the self-triggering weak decay $B_c \rightarrow J/\psi\ \pi$, followed by the leptonic decay of the J/ψ into two muons, could be clearly observed in the ATLAS detector at LHC. Under rather conservative assumptions, a total number of $\approx 10,000$ signal events could be fully reconstructed after one year run, corresponding to 10 fb^{-1} at "low" luminosity ($\approx 10^{33}$ cm^{-2} s^{-1}). This represents a signal to background ratio of about 0.5 with a statistical significance of ≈ 20 standard deviations above a nearby almost flat background. The foreseen mass resolution (standard deviation) of the B_c meson is about 40 MeV, representing an error of $\simeq$ MeV for the determination of the B_c mass.

$B_c \rightarrow J/\psi\ \mu^+\ \nu_\mu$ channel

In spite of the fact that this channel does not permit the measurement of the B_c mass, its signature would be quite clean experimentally when the J/ψ decays into a pair of muons, providing a three muon vertex. In the following we shall argue that in LHC experiments this decay mode could provide a reliable extraction of the mixing matrix element $|V_{cb}|$.

The expected semileptonic branching ratio $\simeq 2\%$ [6] together with the leptonic branching ratio for the J/ψ into two muons $\simeq 6\%$ [7], yields an overall branching fraction of order 10^{-3}. We next combine the last BF with the acceptance of the muon trigger $\simeq 15\%$ (where the three possible triggering muons per decay have been taken into account.) Besides, the detection of the two remaining particles (i.e. the non-triggering muons satisfying some $p_\perp$ and η cuts [8]) amounts to an acceptance for signal events of $\simeq 12\%$. Finally, we assume an identification efficiency for each muon of 0.8 [5] yielding a combined value of $0.8^3 \simeq 0.5$.

Thereby, the expected number of useful B_c decays [c] could reach several 10^5 per year at low luminosity. Such large statistics together with the foreseen precision in the muon momentum measurement [5] should permit the experimental access to the kinematic region near zero-recoil of charmonium (with respect to the B_c meson) with a good energy/momentum resolution.

Theory inputs and determination of $|V_{cb}|$

Next we examine in detail the possibility of extracting $|V_{cb}|$ from experimental data. Let us start by recalling that

$$\frac{d\Gamma(B_c \rightarrow J/\psi\ \mu^+\nu)}{dw} = \frac{1}{\tau}\ \frac{dBr(B_c \rightarrow J/\psi\ \mu^+\nu)}{dw} \tag{1}$$

The differential branching ratio dBr/dw can be written as

$$\frac{dBr(B_c \rightarrow J/\psi\ \mu^+\nu)}{dw} = \frac{1}{N(B_c)}\ \frac{dN(B_c \rightarrow J/\psi\ \mu^+\nu)}{dw} \tag{2}$$

where dN/dw represents the number of semileptonic decays per unit of w for a certain integrated luminosity, once corrected by detection efficiency; $N(B_c)$ is the corresponding total yield of B_c's produced for such integrated luminosity. (Notice that τ could be determined accurately from the same collected sample of semileptonic B_c decays.)

In order to get $N(B_c)$ it will be convenient to compare the B_c production rate to the prompt ψ' yield in pp collisions, for the same integrated luminosity. Therefore

$$N(B_c) = \frac{\sigma(pp \rightarrow B_c + X)}{\sigma(pp \rightarrow \psi' + X)} \times \frac{N(\psi' \rightarrow \mu^+\mu^-)}{\epsilon_{\psi'} \cdot BF(\psi' \rightarrow \mu^+\mu^-)} \tag{3}$$

[c] Semi-electronic $B_c \rightarrow J/\psi e^+\nu$ decays, followed by the J/ψ decay into $\mu^+\mu^-$ can be considered as well. In addition, if the muon from the semi-muonic decay (1) gives the trigger, the event can be selected with the mode $J/\psi \rightarrow e^+e^-$. The trigger acceptance for the former is $\simeq 12\%$ whereas for the latter is $\simeq 3\%$ [8]

where $N(\psi'\to\mu^+\mu^-)$ stands for the number of prompt ψ' to be experimentally detected through the muonic decay mode and $\epsilon_{\psi'}$ denotes its detection efficiency. Contamination from weak decays of B mesons into ψ' states should be efficiently removed by means of the vertex capability of the inner detector.

Therefore, we suggest normalizing the B_c sample with the aid of the ψ' yield through expression (3). The ψ' state is probably preferable to the J/ψ mainly because the former expectedly should not be fed down by higher charmonium states.[d] Instead, data released by Tevatron shows that the J/ψ indirect production through χ_c intermediate states may be also important. Let us also remark that the production of the ψ' resonance in $p\bar{p}$ collisions is more than one order of magnitude larger than initially expected [13 9].

According to the set of papers in [2], perturbative QCD can provide a reliable calculation for the fragmentation function of a high-$p_\perp$ parton into heavy quarkonia (charmonia, B_c states) with only few input parameters: the QCD running constant α_s, the charm and bottom masses, and the square of the radial wave function of heavy quarkonium at the origin $|R(0)|^2$. Note that the largest uncertainty for normalization purposes comes from the (third power of the) charm mass in the fragmentation function (see [2] for explicit expressions). However, we are only interested in the ratio

$$r = \frac{\sigma(pp\to B_c + X)}{\sigma(pp\to\psi' + X)} \tag{4}$$

where each cross section can be written as a convolution of the parton distribution functions of the colliding protons, the cross sections for the hard subprocesses leading to the fragmenting gluon or heavy quark and the respective fragmentation functions [2]. Thus notice that the fragmentation functions $D_{\bar{b}\to B_c}$, $D_{g\to B_c}$ and $D_{c\to\psi'}$, $D_{g\to\psi'}$ [14], which themselves are independent of the parton-level subprocesses, appear combined as a ratio. Therefore, those uncertainties coming from the common overall factor m_c^3 automatically cancel each other. Moreover, those uncertainties introduced by the parton distribution functions also should significantly diminish in r as well. Still one must evaluate the ratio

$$\frac{|R_{B_c}(0)|^2}{|R_{\psi'}(0)|^2} \tag{5}$$

In fact there is the possibility of expressing the above ratio (aside trivial factors)

[d] See Ref. [12] for an alternative explanation of the observed ψ' surplus found in Tevatron. However this mechanism would require an uncomfortable large radiative BF for higher resonances

according to the general factorization analysis of [15] in a more rigorous way as

$$\kappa_0 = \frac{<0|O_1^{B_c}(^1S_0)|0>}{<0|O_1^{\psi'}(^3S_1)|0>} \tag{6}$$

where O_n^X are local four fermion operators[e] and the matrix elements can be evaluated from NRQCD. A similar expression holds for the B_c^*. (Moreover, the denominator can be determined from the measured leptonic width of the ψ'.)

Recently, an additional color-octet fragmentation mechanism [16] has been suggested in order to reconcile the experimental results on inclusive ψ' production with theoretical predictions. This mechanism assumes the creation from gluon fragmentation of a $c\bar{c}$ pair in a color-octet state, in analogy to χ_c production [13]. Then a new nonperturbative parameter $H'_{8(\psi')}$ is required, appearing in r combined as the dimensionless factor

$$\kappa_{8(\psi')} = \frac{<0|O_8^{\psi'}(^3S_1)|0>}{<0|O_1^{\psi'}(^3S_1)|0>} \approx \frac{2\pi m_c^2 H'_{8(\psi')}}{|R_{\psi'}(0)|^2} \tag{7}$$

where we remark that it can be computed in terms of NRQCD matrix elements [15] [f].

Furthermore, if we do not neglect the contribution to the total B_c production of those orbital excitations like P-wave states, new fragmentation functions $D_{\bar{b}\to\bar{b}c(P)}$ come into play [19]. Accordingly, some new κ_i factors will appear which can be conveniently expressed, for instance in analogy to Eqs. (6) and (7), as:

$$\kappa_{1(B_c)} = \frac{1}{\hat{m}^2}\frac{<0|O_1^{B_c}(P_J)|0>}{<0|O_1^{B_c}(^1S_0)|0>} \propto \frac{8\pi\hat{m}^2 H_{1(\bar{b}c)}}{|R_{B_c}(0)|^2} \tag{8}$$

and

$$\kappa_{8(B_c)} = \frac{1}{\hat{m}^2}\frac{<0|O_8^{B_c}(^3S_1)|0>}{<0|O_1^{B_c}(^1S_0)|0>} \propto \frac{8\pi\hat{m}^2 H'_{8(\bar{b}c)}}{|R_{B_c}(0)|^2} \tag{9}$$

where $\hat{m}$ is the reduced mass of the b and c quarks. The long-distance H_1 parameter [18] is related to the square of the derivative of the color-singlet wave function at the origin.

In sum, the evaluation of such κ_i parameters would allow one to complete the computation of the ratio r and thereby to obtain from the experimental

[e]This implies the application of NRQCD to the evaluation of the long-distance part of the fragmentation function of a b quark into a $\bar{b}c$ bound state. Therefore four quark operators would contain b- and c-quark fields in this case[21].

[f]$H'_{8(\psi')}$ can be phenomenologically estimated from B or Υ decays as well [13 17 18].

measurement of the ψ' yield, the total number of B_c events for an integrated luminosity with the aid of expression (3).

On the other hand, the hadronic transition $B_c \to J/\psi$ involves two heavy-heavy systems whereby heavy quark spin symmetry should be still valid as a first order approach [4 21]. This permits the introduction of the analogue of the Isgur-Wise function for transitions involving doubly heavy hadrons, denoted by $\eta_{12}(v_1 \cdot v_2)$. At the non-recoil point $v_1 = v_2$, we shall write

$$< J/\psi | \, A^{\mu} \, | B_c \, > = \, 2 \, \eta_{12}(1) \, \sqrt{m_1 m_2} \, \varepsilon_2^{*\mu} \tag{10}$$

where $A^{\mu} = \bar{c}\gamma^{\mu}\gamma_5 b$ stands for the axial-vector current and ε_2^{μ} represents the four-vector polarization of the J/ψ.

Following similar steps as in the B decay into D^* [22], we write

$$\lim_{w \to 1} \frac{1}{\sqrt{w^2-1}} \frac{d\Gamma(B_c \to J/\psi \, \mu^+ \nu)}{dw} \; = \; \frac{G_F^2}{4\pi^3} \, (m_1 - m_2)^2 \, m_2^3 \, \eta_{12}^2(1) \, |V_{cb}|^2 \tag{11}$$

Note that $\eta_{12}(1)$ may be interpreted in first approximation as the overlap of the initial and final hadron wave functions [4 20]. Moreover, due to the intrinsic nonrelativistic nature of its heavy constituents, nonrelativistic QCD on the lattice or even refined potential based models of hadrons, could be applied in order to get η_{12} and its slope at zero recoil in a reliable way.

This fact in conjunction with the presumed accuracy of the J/ψ momentum measurement suggests determining $|V_{cb}|$ in a similar way as in the method proposed by Neubert [23] through the $\overline{B} \to D^* \ell \overline{\nu}$ decay.

For the purpose of illustration, we have depicted in figure 2 some expected typical points and error bars for the decay $B_c \to J/\psi(\to \mu^+ \mu^-) \, \mu^+ \nu$ for a definite prediction of the ratio of production cross sections r in Eq. (4).

Under this condition, let us remark that the uncertainty on w (i.e the horizontal error bars) is of systematic nature, mainly coming from the B_c flight-path reconstruction. With regard to the vertical coordinate $\eta_{12}(w) \cdot |V_{cb}|$, we have assumed that its uncertainty is essentially statistical [3].

It is important to stress that the quantity to be experimentally measured with a large accuracy from $B_c \to J/\psi \, \mu^+ \nu$ decays can be written as the linear combination:

$$r^{1/2} \times \eta_{12}(1) \times |V_{cb}|$$

In summary, we suggest to keep an open mind on the possibility of an alternative determination of $|V_{cb}|$ from semileptonic decays of B_c mesons at LHC. Even if at the time LHC will start to run, B factories would have already provided a (still) more precise value of $|V_{cb}|$ than at present via the semileptonic

B decay, an independent determination of it should bring a valuable cross-check. On the other hand, a lot of activity is being devoted to the analysis of inclusive production of prompt charmonium resonances at Fermilab. Therefore we conclude that if the fragmentation approach to describe the production of heavy quarkonia at the large transverse momentum domain is accurately tested (and tuned) from experimental data, B_c semileptonic decays could be competitive with the B ones.

Alternatively, one can turn the question round and consider $|V_{cb}|$ as a well-known parameter thus verifying QCD calculations, either at the production or at the weak decay level. Indeed, fragmentation into heavy quarkonia offers an interesting check of perturbative QCD and a deep insight into the nonperturbative dynamics in the hadronic formation. Besides, a precise knowledge of the B_c production rate is a necessary condition for the experimental measurement of the absolute branching fraction of any of its decay modes.

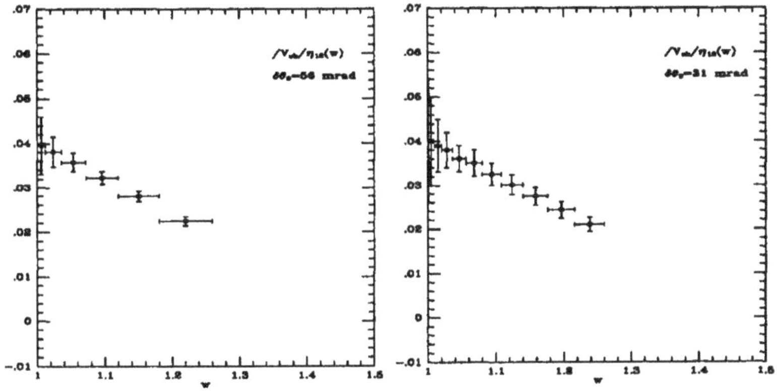

Figure 2: Hypothetical $\eta_{12}(w)\cdot|V_{cb}|$ distributions of the decay $B_c \rightarrow J/\psi \mu^+ \nu$ for a 500 events sample tentatively assigned to an integrated luminosity of 20 fb^{-1} (i.e. two years of running at low luminosity). It was assumed an spatial resolution for the B_c flight-path reconstruction of $\sigma \approx 100$ μm and its lifetime is allowed to vary in the interval $\tau = 0.5-1$ ps. Horizontal bars show systematic errors basically limiting the number of experimental points. Vertical bars take into account the statistical fluctuations of the sampled events in the bin (distributed according to the expected differential rate for a pseudoscalar to vector semileptonic decay[22]). Points line up with η'_{12} set equal to -2. For more details see Ref. [24]

Acknowledgments

Work partially supported by CICYT under contract AEN 93-0234 and by IVEI.

References

1. G. Carboni et al, Proceedings of the ECFA Large Hadron Collider Workshop, Aachen (1990)
2. K. Cheung and T.C. Yuan, Phys. Lett. **B325** (1994) 481 ; E. Braaten, K. Cheung and T.C. Yuan, Phys. Rev. **D48** (1993) 5049; **D48** (1993) R5049 ; K. Cheung, Phys. Rev. Lett. **71** (1993) 3413
3. F. Albiol et al. ATLAS internal note PHYS-NO-058 (1994); IFIC/95-24 (hep-ph/9506306)
4. M.A. Sanchis-Lozano, Nuc. Phys. **B440** (1995) 251
5. ATLAS Technical Proposal, CERN/LHCC 94-43
6. M. Lusignoli and M. Masetti, Z. Phys. **C51** (1991) 549
7. Particle Data Group, Phys. Rev. **D50** (1994)
8. M.A. Sanchis-Lozano et al, ATLAS internal note, in preparation
9. M. Cacciari and M. Greco, Phys. Rev. Lett. **73** (1994) 1586 ; E. Braaten, M.A. Doncheski, S. Fleming and M.L. Mangano, Phys. Lett. **B333** (1994) 548
10. P. Billoir and S. Qian, Nucl. Instr. and Meth. **A311** (1992) 139
11. C-H Chang and Y-Q Chen, Phys. Rev. **D49** (1994) 3399
12. F.E. Close, Phys. Lett. **B342** (1995) 369 ; D.P. Roy, K. Sridhar, Phys. Lett. **B345** (1995) 537
13. E. Braaten, T.C. Yuan, Phys. Rev. **D50** (1994) 3176 ; J.P. Ma, Phys. Lett. **B332** (1994) 398 (1994) 1586 ; E. Braaten et al., Phys. Lett. **B333** (1994) 548 CERN-TH/95-129 (hep-ph/9505379)
14. E. Braaten and T.C. Yuan, Phys. Rev. Lett. **71** (1993) 1673
15. G.T. Bodwin, E. Braaten and G.P. Lepage, Phys. Rev. **D51** (1995) 1125
16. E. Braaten and S. Fleming, Phys. Rev. Lett. **74** (1995) 3327 ; P. Cho and A. Leibovich, CALT-68-1988
17. H.D. Trottier, Phys. Lett. **B320** (1994) 145
18. G.T. Bodwin, E. Braaten, T.C. Yuan and G.P. Lepage, Phys. Rev. **D46** (1992) R3703 ; G.T. Bodwin, E. Braaten and G.P. Lepage, Phys. Rev. **D46** (1992) R1914
19. K. Cheung and T.C. Yuan, CPP-94-37 (hep-ph/9502250)
20. M.A. Sanchis, Phys. Lett. **B312** (1993) 333 ; Z. Phys. **C62** (1994) 271
21. T. Mannel and G. A. Schuler, CERN preprint CERN-TH.7468/94
22. M. Neubert, Phys. Rep. **245** (1994)
23. M. Neubert, Phys. Lett. **B264** (1991) 455 ; CERN-TH/95-107 (hep-ph/9505238)
24. M. Galdón and M.A. Sanchis-Lozano, IFIC/95-31 (hep-ph/9506307)

SPAGHETTI CALORIMETER SPACAL

JAN HLADKY

Institute of Physics of the Czech Academy of Sciences,Prague
H1 Collaboration

Abstract

The H1 Collaboration has built a Pb-scintillation fiber calorimeter.It is installed in the backward scattering region of the H1 detector at e-p collider HERA. The "spaghetti" calorimeter has both electromagnetic and hadronic part, where fibers of diameter 0.5 and 1.0 mm are embedded into thin lead sheets.The main construction parameters as well as main characteristics and the performance of the new calorimeter are presented.

Introduction

HERA is the only ep-collider.It enables a lot of new studies at the CMS energy about $sqrt(s) = 300$ GeV.For such a purpose it was necessary to construct a very sophisticated detector.My contribution is connected with one part of the H1-set up, the upgraded backward calorimeter. It consists of three main parts: - the motivation, - the design and construction of the new calorimeter and finally - the performance.

♡ * The motivation

HERA is dedicated to study parton structure.At the beginning the search was mostly extended for high Q^2 up to 20,000 GeV^2.Recently there has also arisen the interest to study the region of Q^2 close to 1 GeV^2 and Bjorken x down to 10^{-5}.It enables us to investigate the behaviour of a gluon content in partons [1].

To reach such conditions it was necessary to upgrade the backward part of the detector in which the scattered electron can be very precisely detected also with the hadronic final state and in which on can register single minimum ionizing particles.The so called "backward " calorimeter - see on Fig.1. [2] should improve:

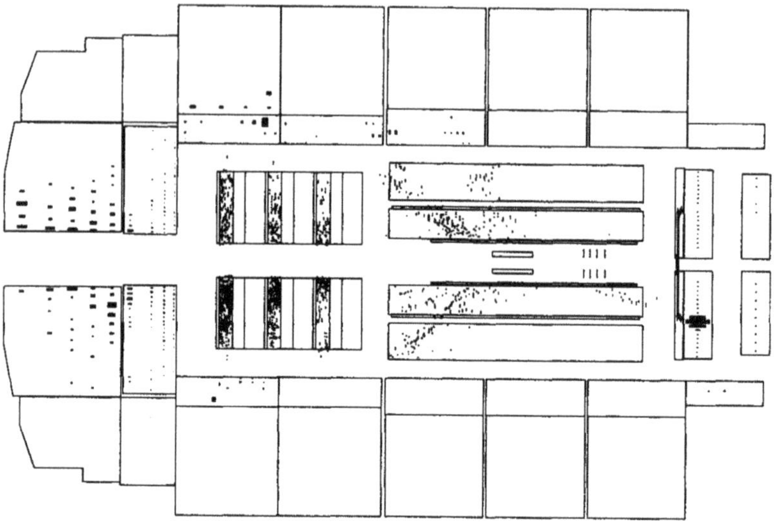

Figure 1: The two sections of the Spacal in the "backward"(right side) part of H1 detector.The electron made a shower in elmg.section, hadronic one is empty.

- geometrical acceptance enabling us to register scattering angles down to 3° relative to the beam axis,
- electron's energy resolution up to $7\%/sqrt(E) + 1\%$,
- position/angle resolution of electrons down to few mm / mrad.,
- time resolution for e/hadron/min.ioniz.particle better than 1 nsec,
- electron/hadron separation.

To fulfill the tasks the spaghetti calorimeter SPACAL consisting of lead sheets and scintillation fibers has been chosen.

♡ * The construction of the Spacal

The Spacal calorimeter has two parts - electromagnetic and hadronic [3].Each is composed of totally independent individual modules, containing mechanical, optical and electronic systems.The construction parameters are given in Table 1.

Table 1: Construction parameters of the Spacal.

Parameter	Electromagnetic part	Hadronic part
Fiber diameter	0.5 mm	1.0 mm
Lead/fiber ratio	2:1	4:1
Number of channels	1192	128
Size of standard cell	40.5 x 40.5 mm^2	120 x 120 mm^2
Radiation length	0.88 cm	0.74
Interaction length	25 cm	20 cm
Active length	25 cm	25 cm
Molière radius	2.55 cm	-

The mechanical part is represented mainly by thin lead profiled sheets made from pure lead. In the sheets for the electromagnetic section, grooves have been rolled on one side while for the hadronic one, grooves were pressed on both sides in a steel matrix using 200 ton pressure.The mechanical tolerances of the sheets were up to 20 resp. 100 microns for elmg. resp. hadronic part and represent the limit of todays production technology.Fig.2. illustrates a cut of both types of the sheets.The fibers are embedded into grooves and then the sheets are sandwiched into the modules using pressure.

The optical part is composed of the scintillation fibers,light collection system and photomultipliers.The fibers should have a long attenuation length, high light yield and good radiation hardness. After several tests the fibers Bicron type BCF-12 have been chosen. The mechanical and optical parameters of the fibers had to be permanently controlled and monitored [4]. The Fig.2. shows the changes of the fiber average cross section during the production time in various batches.The front side of the fibers has been mirrored with Al to increase the light yield.On the rear ends are glued together with black glue to kill a cladding light which has shorter attenuation length. The light collection system concentrates the light from the fiber bundles to a smaller area of the photomultipliers and compensates for the nonuniformity of their photocathodes. Light-mixers of the form of truncated pyramids have been used on the basis of simulation and experimental results.

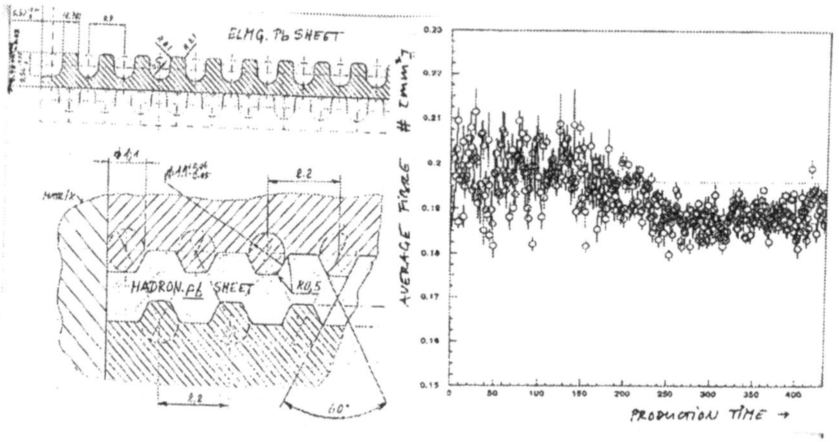

Figure 2: Left: The profiles of the Pb sheets. Right: The average cross section of the fibers.

Photomultipliers used in the Spacal had to have good longterm stability, gain greater than 10^4 in the strong magnetic field of the H1 detector (of about 1 Tesla) and low noise to be sensitive to minimum ionizing particles.Precise timing was also asked. Mesh-type Hamamatsu photomultipliers had such properties.The electromagnetic part contains type R5505 diameter of 1" and the hadronic one 2" type R2490 photomultipliers, all controlled and monitored during the production of the Spacal.

The signal obtained from the photomultipliers is integrated in the base of their housing tubes and then transported to the electronics system placed about 18 m outside of the beam area.The signal is distributed (see Fig.3.) via analog cards to the standard calorimeter readout and to the frontend card.This card splits the signal for the trigger branch, via Inclusive Electron Trigger card (selecting e-p interactions), and for veto, via ATOF card (to reject huge background owing to beam-gas and beam wall interactions).The TDC card provides on-line monitoring of timing and off-line time measurements.

A high voltage system provides the HV supply of the photomultipliers using 16-channel HV boards, where each channel has individual regulation.

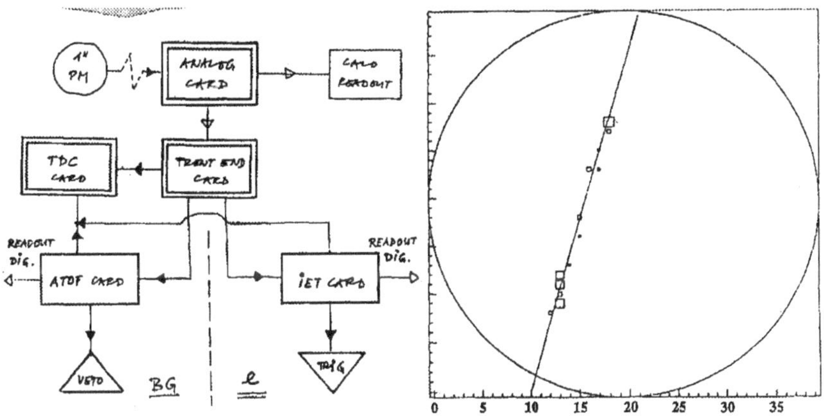

Figure 3: Left: Distribution of the signal in the electronics.Right: A cosmic ray particle in the elmg. part of the Spacal.

Calibration system.

The optical system, as well as the electronics chain and also the photomultiplier gain, needs to be continuously calibrated.The first two show only small variations in long periods.An absolute energy calibration and a longtime control is performed by a check of the position of the deep inelastic kinematic peak of the electron's energy.The gain of the photomultipliers can show the shorter-time variations to be calibrated by use of a reference light source.The light calibration system injects the light from two different light emission diodes into lightmixers. The stability of the LED's light output is controlled and monitored by photodiodes.

♡ * The performance

The Spacal modules have been tested in electron and pion beams at CERN and ITEP Moscow and in an electron beam at DESY.The results of the tests are shown in Table 2. Fig.4. illustrates the energy resolutions of the electrons in the electromagnetic and hadronic parts and of the pions in the hadronic part (having one λ).

The modules have been tested also by a radioactive β-source scanning during the production to monitor homogeneity.

Table 2: Main parameters of the Spacal.

Parameter	Electromagnetic part	Hadronic part
Linearity	better than 1%	few %
$\sigma_{E_e}/E_e(\%)$	$(7.1 \pm 0.2)E_e^{-1/2}GeV$ $+(1.0 \pm 0.1)$	$(13.3 \pm 0.9)E_e^{-1/2}GeV$ $+3.6$
$\sigma_{E_h}/E_h(\%)$	about 35	$(71 \pm 8)E_h^{-1/2}GeV + 20$
$\sigma_x(\mathrm{mm})$	$(3.8 \pm 0.4)E^{-1/2}GeV$ $+(0.3 \pm 0.2)$	
σ_{time}	better than 1 nsec	better than 1 nsec

After assembly of the electromagnetic and hadronic parts both sections were tested using cosmic rays-see Fig.4.The inclusion of the Spacal into the H1 setup at HERA (as illustrated on the Fig.1.)enabled first energy reconstruction in the backward scattering region on e - p beams.

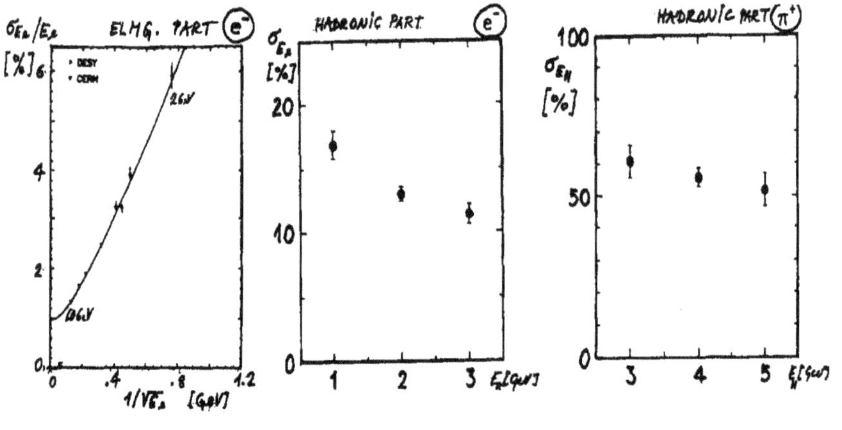

Figure 4: Left: The relative resolution of electron energy in the elmg.part. Middle: The resolution of electron energy in the hadronic part. Right: The resolution of pion energy in the hadronic part (having only one λ).

♠ * Summary.

The spaghetti calorimeter SPACAL consisting of the electromagnetic and hadronic part is installed in the H1 setup at the HERA collider. The resolutions of the energy, position and time as well as other important parameters reached the proposed values.First results on the e-p beams with Spacal are obtained.

Acknowledgements
I am indebted to thank to all my colleagues - physicists, engineers and technicians who participated on the Spacal project and thus enabled me to present the talk. I would like to thank the organizers of the Conference for their kind hospitality and a pleasant atmosphere they created during our stay at Valencia.

References

[1] J.Bartels and J.Feltesse in *QCD at low x,Proc.Phys.at HERA Workshop*,(DESY,Hamburg, 1991).

[2] H1 Collab. in *Techn. Proposal to Upgrade the Backward Scatt. Region of the H1 Detector*,(DESY PRC 93/02).

[3] U.Goerlach in *Proc.IV.Calo.in H.En.Ph.,La Biodola 1993,p.170*, S.Dagoret in *ibid,p.176*,(World Sci.,1994), G.Mueller in *Proc.V.Calo.in H.En.Ph.,Brookhaven 1994*, (World Sci.,in print), M.Weber in *Proc.1994 Beijing Calo.Symp.,p.107*, (Beijing 1995)-see also for other ref.

[4] J.Cvach et al. in *H1 Note No. H1-06/95-443*.

Chapter 3

Topics in String Unification

L.E.Ibáñez

Departamento de Física Teórica,
Universidad Autónoma de Madrid,
Cantoblanco, 28049 Madrid, Spain.

Abstract

I discuss several issues concerning the use of string models as unified theories of all interactions. Topics covered include gauge coupling unification, the construction of string-GUTs and the computation of SUSY-breaking soft terms from the dilaton/moduli sectors of the theory.

1 Strings and gauge coupling unification

Heterotic 4-D strings are considered today our best candidates for the construction of unified theories of all interactions including gravity.In 4-D heterotic strings the strength of both gauge and gravitational coupling constants is governed by the vev of the dilaton, $< ReS >= 4\pi/g^2$. Consider for example an hypothetical 4-D string whose gauge group contains the SM group. The corresponding SM couplings would be related as [1]

$$g_3^2 k_3 \;=\; g_2^2 k_2 \;=\; g_1^2 k_1 \;=\; G_{Newton} M_{String}^2 \;=\; \frac{4\pi}{ReS} \,. \qquad (1)$$

at the string scale $M_s \simeq 5.3 g 10^{17} GeV$ [3] . Here k_2, k_3 are positive integers (the "levels" of the $SU(2)$ and $SU(3)$ algebras) and k_1 is a rational number which gives the normalization of the $U(1)_Y$ weak hypercharge. If the corresponding 4-D string was constructed by a simple compactification from either of the two supersymmetric 10-D heterotic strings, one always has $k_2 = k_3 = 1$ whereas k_1 is a model-dependent rational number. In this way one has at the string scale $sin^2\theta_W = 1/(1+k_1)$ and $\alpha_s = (1+k_1)\alpha_e$, where $\alpha_{s,e}$ are the strong and electromagnetic fine-structure constants. The standard $SU(5)$ GUT predictions are recovered for the choice $k_1 = 5/3$. As a first try one can assume this value for k_1 and compute $sin^2\theta_W$ and α_s at the weak scale M_Z by running the renormalization group equations (r.g.e.) down to low energies. Assuming that the only massless particles charged under the SM are those of the minimal supersymmetric SM (MSSM) one finds [4, 5] $sin^2\theta_W = 0.218$ and $\alpha_s = 0.20$, several standard deviations away from the experimental values $sin^2\theta_W(M_Z) = 0.233 \pm 0.003$ and $\alpha_s(M_Z) = 0.11 \pm 0.01$. Is this a serious problem for the idea of a direct string unification of the SUSY standard model?

One may argue that, for an $SU(3) \times SU(2) \times U(1)$ string the value of k_1 should be considered as a free parameter [6, 7] . If we indeed take k_1 as a free parameter one can find much more successfull results for the coupling constants. In particular,for $k_1 = 1.4$ one gets $sin^2\theta_W(M_Z) = 0.235$ and $\alpha_s(M_Z) = 0.13$, in not unreasonable agreement with experiment [7] .

In fact it is amussing to note that, if the historical order of theoretical ideas would have been slightly different, the joining of coupling constants could have been considered an outstanding success of string theory! Indeed, imagine the stringers of the early seventies would have discovered the supersymmetric heterotic strings before GUTs (and SUSY-GUTs) would have been introduced. Georgi, Quinn and Weinberg would have told them how to extrapolate the couplings down to low energies. However they would not have had any prejudice concerning the value of k_1 and they would have taken it as a free parameter. Then they would have found that for $k_1 = 1.4$ one gets the above successfull results. This could had been then interpreted as a great succsess of string theory! (In fact even the results stated above for $k_1 = 5/3$ would have been considered rather successful given the experimental precission in the late seventies).

Let us forget now about virtual history and come back to the question whether direct string unification of the supersymmetric SM is or not a problem. Leaving k_1 as a free parameter is certainly a possibility. It is true however that up to now nobody has constructed any string model with $k_1/k_2 < 5/3$, which is what it seems to be required. For example, in the $k = 1$ orbifold models constructed up to now one has $k_1 \geq 5/3$. E_6-like models ((2,2) $E_8 \times E_8$ compactifications) have $k_1 = 5/3$ and often bigger results are obtained for (0,2) orbifold models. In fact, straightforward compactifications (e.g., level=1) of the heterotic strings are likely to yield always $k_1 \geq 5/3$ [2] , but this is not necesarily the case in all generality. In particular, in higher KM level models one can think of a value $k_1/k_2 \leq 5/3$ even though $k_1 > 5/3$. It would be very interesting to find model-independent constraints on k_1 or else find examples with $k_1/k_2 \leq 5/3$.

There are of course, other alternatives to understand the disagreement found for the joining of coupling constants. The infinite massive string states can give rise to substantial one-loop corrections [3, 8] to the gauge coupling constants (string threshold corrections). Indeed this possibility has been studied and the potential exists for such corrections to explain the discrepancy [4, 9]. Each coupling gets one-loop corrections as

$$\frac{4\pi}{g_i^2(M_s)} = k_i ReS + \Delta^i_{Th} \quad ,i = 1,2,3 \tag{2}$$

where $ReS = 4\pi/g^2$ and Δ^i_{Th} are the threshold corrections. There are two type of threshold corrections: field-independent and field-dependent. The first of these are expected to be small and indeed this smallness has been confirmed in explicit computations for some $(0,2)$ orbifold examples [3, 10] . The field-dependent threshold corrections may on the contrary be large depending on the values of the fields. The fields relevant in this case are those related to marginal deformations of the underlying CFT like the moduli T_a, U_a fields. These fields parametrize the size (R) and shape of the six-dimensional compactification variety (orbifold, Calabi-Yau manifold). The corresponding threshold functions $\Delta^i_{Th}(T_a, T_a^*, U_a, U_a^*)$ have been computed in a variety of 4-D strings [3, 8, 10, 11]. It seems to be a common feature that for large compactification radius R^2 one gets for all the examples studied a leading correction of the form [12]

$$\Delta^i = \frac{b_i'}{12} R^2 \tag{3}$$

where the b_i' are model-dependent constant coefficients [8]. Now one can see that for $k_1 = 5/3$ and not too large values of R (e.g. $R \simeq 2-4$) one can obtain good results for $sin^2\theta_W$ and α_s by apropriately chosing the b_i' coefficients [4, 9]. In fact this result is less trivial than it sounds since these b_i' coefficients are numbers which may be computed in specific (orbifold) models in terms of the quantum numbers and "modular weights" of the massless field of the theories. It was found in particular that only a quite limited class of orbifold models [9] could possibly have adequate threshold corrections. A random distribution of modular weights for the SM particles will in general lead to threshold corrections in the wrong direction. So even though indeed threshold corrections may be large there is no reason for them to conspire precisely in the way we want.

Another obvious alternative is to give up the idea of a direct string unification of the SM involving just the content of the MSSM, i.e., to consider the presence of extra massless particles in addition to those of the SM [5, 13]. After all, the explicit 4-D string models constructed up to now have always plenty of additional stuff! The problem with this is that we open a Pandora box of virtually unlimited possibilities in which we, rather than predicting the weak angle, are just adjusting it. A second problem with this is that explicit computations of the running couplings for different possible choices of extra matter fields have shown that there is no posible choice of extra matter fields which yields direct unification at the string scale *new intermediate scale thresholds are necessary* [5, 13, 2]. Thus we lose the beauty of direct string unification altogether. But, if we have to deal with extra intermediate scales, why not considering the possibility of GUTs themselves which naturally require a scale of order 10^{16} GeV?

Are there any motivations for constructing GUTs from strings?Standard SUSY-GUTs like $SU(5)$ and $SO(10)$ predict the unification of coupling constants at a scale M_X which is a free parameter. This allows for the computation of one coupling constant as a function of the others. One finds $\alpha_s(M_Z) = 0.12$ for $sin^2\theta_W(M_Z) = 0.233$ by chosing $M_X = 2\times10^{16}$ GeV, in good agreement with experiment. Given this success, in principle much better than the one obtained from direct string unification of the MSSM, it is natural to try and embed standard SUSY-GUTs into string theory.

This is obviously an important motivation. However, I must remark that in order for a SUSY-GUT to yield the above nice prediction for α_s, it is crucial the assumption that bellow the string scale the only particles present in the spectrum are those of the MSSM. So unification into a simple group like $SU(5)$ or $SO(10)$ is not enough, the breaking of those groups has to be such that the remaining low energy theory is the MSSM.

One could also consider as an argument in favour of GUTs the nice way in which the observed SM generations fit into representations of the $SU(5)$ or $SO(10)$ groups. A random 4-D string with $SU(3)\times SU(2)\times U(1)\times G$ group typically contains extra vectorlike heavy leptons or quarks. These extra particles are in fact chiral with respect to the extended group G and remain light as long as the extended symmetries are unbroken. Furthermore, these extra leptons and quarks are not guaranteed to have integer charges or to obey the usual charge quantization of the SM particles. In order to obtain (level=1) $SU(3) \times SU(2) \times U(1) \times G$ string models in which the light fermions are just ordinary SM generations with standard charge asignements the simplest way is to start with heterotic strings with some underlying (level=1)

$SU(5)$, $SO(10)$ or E_6 symmetry which is broken down to some smaller group *at the string scale* through the Hosotani-Witten flux-breaking mechanism. Notice that these models are *not* GUTs, because the symmetry breaking is not carried out through a Higgs mechanism (i.e., through an adjoint vev) and also because the GUT-like symmetries are never realized as GUT symmetries at any scale, they constitute just an intermediate step in the construction technique. The actual gauge symmetry is the one of the SM (or some simple extension) and hence they are just SM strings of a particular class. Anyway, if we need to have at some level some GUT-like structure (although not realized as a complete gauge symmetry) it is reasonable to try and study whether symmetries like $SU(5)$ and $SO(10)$ may be promoted to a complete symmetry of the massive spectrum. This would be the essence of string GUTs.

There are other features of SUSY-GUTs which their practitioners love, like the prediction for the m_b/m_τ ratio and other fermion mass relationships; predictions for proton-decay and lepton-number violating processes etc. These are more model dependent and may also be present in string models so they would not constitute by themselves motivations to construct string GUTs. On the other hand it would be usefull to construct GUTs from strings to check whether the dynamical assumptions that the GUT practitioners assume are or not natural within the context of strings. So one can try to extract some selection rules to constrain the rules of SUSY-GUT model-building. I will briefly discuss some of these in the next section.

2 SUSY-GUTs from strings

It is essential for SUSY-GUTs the existence in the spectrum of chiral fields (e.g., adjoints) appropriate to induce the breaking of the gauge symmetry down to the standard model. In the context of N=1, 4-D strings this is only possible if the affine Lie algebra associated to the GUT symmetry is realized at level $k \geq 2$. Straightforward compactifications of the supersymmetric heterotic strings have always $k = 1$ algebras which they inherit from the $k = 1$ $E_8 \times E_8$ or $Spin(32)$ $D = 10$ heterotic strings.

To obtain 4-D strings with higher level one has to go beyond simple compactifications of the heteorotic strings. At the begining it was thought that such higher level models would be very complicated to construct.

This is why in the early days of string model-building there were no attempts at the construction of string-GUTs. Only a few papers dealt with the explicit construction of 4-D strings with affine Lie algebras at higher levels [14, 15] .

In the last year there have been renewed attempts for the construction of string GUTs at $k = 2$ using orbifold [16] and free fermion techniques [17] . The first of these methods is relatively easy and is the one I am more familiar with. Furthermore world-sheet supersymmetry (which is quite a technical difficulty in fermionic models[18]) is guaranteed by construction. Here I will thus discuss mostly results obtained using the (symmetric) orbifold methods of ref. [16] , although many of the conclusions may be easily extended to other 4-D string constructions.

The general idea is the following [16]. One starts with a $(0,2)$ orbifold compactification of the 10-D heterotic string. It turns out that it is convenient to start with the $Spin(32)$ heterotic (instead of $E_8 \times E_8$). Models in which the gauge group has the structure $G_{GUT} \times G'$, where G' in turn contains as a subgroup a copy of G_{GUT} (i.e. $G'_{GUT} \in G'$) are searched for. We would just have at this point a usual level $k = 1$ (0,2) orbifold model with a particular gauge structure. Now we do some kind of modding or projection (to be specified below) in such a way that only gauge bosons corresponding to the *diagonal* G^D_{GUT} subgroup of $G_{GUT} \times G'_{GUT}$ survive in the massless spectrum. We are thus left with a structure of type $G^D_{GUT} \times G''$ where G^D_{GUT} is realized at $k = 2$.

In fact this scheme is quite general and may be implemented in other classes of $(0,2)$ 4-D string constructions. For example, it may be used starting with the class of $(0,2)$ models [19] obtained by adding gauge backgrounds and/or discrete torsion to Gepner and Kazama-Suzuki models . The final step leading to the $k = 2$ group may be achieved by embedding an order-two symmetry (γ in the notation of [19]) by a permutation of G_{GUT} with G'_{GUT}.

In the case of orbifolds, three methods in order to do the final step $G_{GUT} \times G'_{GUT} \to G^D_{GUT}$ were discussed in refs. [15]. In the first method (I) the underlying $k = 1$ model with the $G_{GUT} \times G'$ structure is obtained by embedding the twist action of the orbifold into the gauge degrees of freedom by means of an automorphism of the gauge lattice (instead of a shift). In $k = 1$ models of this type one can have "continuous Wilson-line" backgrounds [20, 21] which can be added in such a way that the symmetry is broken continuously to the diagonal subgroup G^D_{GUT}. In the second method (II) one does the final step by modding the original model by a Z_2 twist under which the two groups G_{GUT} and $G'_{GUT} \in G'$ are explicitly permuted. The third method (III) is field-theoretical. One explicitly breaks the original symmetry down to the diagonal subgroup by means of an ordinary Higgs mechanism. Although these three methods look in principle different, there are many $k = 2$ models wich can be built equivalently from more than one of the above methods.

Giving all the details of these constructions here would be pointless. Let me just explain a few features. There are a few of them which are quite general due to their, in some way, kinematical origin. Consider the mass formula for the left-moving (bosonic) string states from any 4-D toroidal orbifold model:

$$\frac{1}{8}M_L^2 \;=\; N_L \;+\; h_{KM} \;+\; E_0 \;-1\,. \tag{4}$$

Here N_L is the left-moving oscillator number, h_{KM} is the contribution of the KM gauge sector to the conformal weight of the particle and E_0 is the contribution of the internal (compactified) sector to the conformal weight. Let us consider the case of symmetric $(0,2)$ Abelian orbifolds. All Abelian Z_N and $Z_N \times Z_M$ orbifolds may be obtained by toroidal compactifications in which the 6 (left and right) compactified dimensions are twisted. There are just 13 possible orbifold twists [22] which can be characterized by a shift $v = (v_1, v_2, v_3)$, where $e^{2i\pi v_i}$ are the three twist eigenvalues in a complex basis. A consistent symmetric orbifold model is obtained by combining different twisted sectors in a modular invariant way. This procedure is well explained in the literature [22, 23] . To each possible twisted sector there corresponds a value for E_0 given by the general formula:

$$E_0 \;=\; \sum_{i=1}^{3} \frac{1}{2}|v_i|(1-|v_i|) \tag{5}$$

Notice also that $E_0 = 0$ for the untwisted sector which is always part of any orbifold model. In the case of asymmetric orbifolds [24], obtaining $N = 1$ unbroken SUSY allows the freedom of twisting the right-movers while leaving untouched the (compactified) left-movers.

In this case one can then have $E_0 = 0$ even in twisted sectors. Let us go now to the other relevant piece in eq. (4), namely the contribution h_{KM} of the KM sector to the conformal weight of the particle. A state in a representation $(R_1, R_2, \cdots)$ will have a general weight

$$h_{KM} \;=\; \sum_i \frac{C(R_i)}{k_i + \rho_i} \tag{6}$$

Here $C(R)$ is the quadratic Casimir of the representation R. $C(R)$ may be computed using $C(R)dim(R) = T(R)dimG$, where $T(R)$ is the index of R. Unless otherwise explicitly stated, we use the standard normalization in which $T = 1/2$ for the N-dimensional representation of $SU(N)$ and $T = 1$ for the vector representation of $SO(2N)$.

With this normalization, for simply-laced groups the Casimir of the adjoint satisfies $C(A) = \rho$. The contribution of a $U(1)$ factor to the total h_{KM} is instead given by Q^2/k, where Q is the $U(1)$ charge of the particle and k is the normalization of the $U(1)$ generator, abusing a bit it could be called the level of the $U(1)$ factor. Formula (6) is very powerful because the h_{KM} of particles can be computed without any detailed knowledge of the given 4-D string. This information is a practical guide in the search for models with some specific particle content.

Using eq. (4) and the values for E_0, h_{KM} computed through eq. (5), (6) we can learn, for instance, what $SU(5)$ or $SO(10)$ representations may appear in the massless spectrum of any possible twisted sector of any given Abelian orbifold. In the case of these groups we are interested in knowing which twisted sectors may contain 24-plets or 45 and 54-plets respectively.

For a 24-plet one has $h_{KM} = 5/7$; for $SO(10)$ 45-plets one has $h_{KM} = 4/5$ and, finally, for $SO(10)$ 54-plets one has $h_{KM} = 1$. From the condition $h_{KM} + E_0 \leq 1$ one draws the following conclusions:

i. All those representations may be present in the untwisted sector of any orbifold.

ii. 54s of $SO(10)$ ($k = 2$) can only be present in the untwisted sector of symmetric orbifolds.

iii. 45s of $SO(10)$ ($k = 2$) may only appear either in the untwisted sector or else in twisted sectors of the type $v = 1/4(0,1,1)$ or $v = 1/6(0,1,1)$. This is a very restrictive result since Abelian orbifolds containing these shifts are limited.

iv. 24-plets of $SU(5)$ can never appear in the twisted sectors of the Z_3, Z_4, Z_6' and Z_8 orbifolds.

From the above conclusions it transpires that looking for models with GUT-Higgs fields in the untwisted sector should be the simplest option, since they can always appear in *any* orbifold. This option has another positive aspect in that the multiplicity of a given representation in the untwisted sector is never very large, it is always less or equal than three in practically all orbifolds and is normally equal to one in the case of $(0,2)$ models. Proliferation of too many GUT-Higgs multiplets will then be avoided.

To be specific let us present a string $SO(10)$-GUT constructed from the first of the three orbifold methods mentioned above, i.e., the "continuous Wilson-line" method. In this method the orbifold twist θ is realized in the gauge degrees of freedom in terms of automorphisms Θ.

In the absence of Wilson line backgrounds L_i, the action of Θ can be described by an equivalent shift V. In the presence of L_i, the embedding is non-Abelian when ΘL_i does not give back L_i up to lattice vectors. When embedding by automorphisms, not all Cartan gauge currents are given by combinations of derivatives of the 16 bosonic coordinates ∂F_I since the lattice coordinates F_I are generically rotated by Θ and the unbroken gauge currents must be invariant under Θ. The Cartan sub-algebra, as well as the step currents, now arise from Θ invariant orbits of the $e^{iP\cdot F}$ operators of the form

$$|P\rangle + |\Theta P\rangle + \cdots + |\Theta^{N-1}P\rangle \tag{7}$$

where $|P\rangle \equiv e^{iP\cdot F}$ and $P^2 = 2$. After the continuous Wilson lines are turned on, states not satisfying $P.L_i = int$ drop out from the spectrum. This projection kills some Cartan generators thus forcing a reduction of the rank of the gauge group. This is a necessary condition to get a residual affine Lie algebra realized at higher level.

Consider [16] the simplest symmetric orbifold with order 2 symmetries, namely, $Z_2 \times Z_2$.

The internal six-dimensional twists θ and ω may be embedded into the gauge degrees of freedom by the order two automorphisms Θ and Ω defined by :

$$\begin{aligned}\Theta(F_1, F_2, \cdots, F_{16}) &= (-F_1, -F_2, \cdots, -F_{10}, -F_{11}, -F_{12}, F_{13}, F_{14}, F_{15}, F_{16}) \\ \Omega(F_1, F_2, \cdots, F_{16}) &= (-F_1, -F_2, \cdots, -F_{10}, F_{11}, F_{12}, -F_{13}, -F_{14}, F_{15}, F_{16})\end{aligned} \tag{8}$$

The unbroken gauge currents correspond to states $|P\rangle$ with P invariant plus the oscillators $\partial F_{15}, \partial F_{16}$. Also, from non-invariant P's we can form orbits invariant under both Θ and Ω. Altogether we find 200 currents that can be organized into an $SO(10) \times SO(18) \times U(1)^2$ algebra realized at level $k = 1$.

Next we turn on a Wilson line background L along, say, the compactified direction e_6. L has the form

$$L = (\lambda, \lambda, \lambda, \cdots, \lambda, 0, 0, 0, 0, 0, 0) \tag{9}$$

The parameter λ can take any real value since L is completely rotated by both Θ and Ω.

The gauge group is broken to $SO(10) \times SO(8) \times U(1)^2$. The currents associated to $SO(10)$ are given by

$$|\underline{+1, -1, 0, 0, \cdots, 0}, 0, 0, 0, 0, 0, 0\rangle + |\underline{-1, +1, 0, 0, \cdots, 0}, 0, 0, 0, 0, 0, 0\rangle \tag{10}$$

where underlining means that all possible permutations must be properly considered. One can check that $SO(10)$ is realized at level $k = 2$ whereas $SO(8)$ has $k = 1$.

In the untwisted matter sectors U_1, U_2 and U_3, the corresponding left-moving vertices transform under (Θ, Ω) with eigenvalues $(-1,1),(1,-1)$ and $(-1,-1)$ respectively. The momenta involved must also satisfy $P \cdot L = \; int$. In sectors U_1 and U_2 there are matter fields transforming as $(1,8)$ and with different $U(1)$ charges.

In the U_3 sector we find the states

$$\partial F_I \; , \; I = 1, \cdots, 10$$
$$|\underline{+1,-1,0,\cdots,0},0,0,0,0,0,0,0\rangle \; - \; |\underline{-1,+1,0,\cdots,0},0,0,0,0,0,0,0\rangle \qquad (11)$$

These states have no $U(1)^2$ charges and belong to a $(54,1) + (1,1)$ representation of $SO(10) \times SO(8)$. In U_3 we also find four extra singlets, charged under the $U(1)$s only. Altogether the spectrum of this GUT model is given in

Table 1. The charge Q is non-anomalous whereas Q_A is anomalous.

The gravitational, cubic and mixed gauge anomalies of Q_A are in the correct ratios in order to be cancelled by the 4-D version of the Green-Schwarz mechanism [25] . The degeneracies of the twisted sectors θ, ω and $\theta\omega$ depend on the way one realizes the $Z_2 \times Z_2$ twist in the compactifying cubic lattice (see ref.[16] for details). This model has 4 $SO(10)$ generations and two pairs of $16 + \overline{16}$ Higgs fields plus additional 10-plets.

There is an interesting feature which turns out to be quite generic in $SO(10)$ string GUTs obtained from this method (I). In the 0-picture the full emission vertex operator for the singlet in U_3 has the form

$$\partial X_3 \; \otimes \; \sum_{I=1}^{10} \partial F_I \qquad (12)$$

A Vev for this field precisely corresponds to the Wilson line background L in eq. (9).

The fact that this background may be varied continuously means that this singlet is a *string modulus*, a chiral field whose scalar potential is flat to all orders. Indeed, using the discrete Z_2 R-symmetries of the right-handed sector, it can be proven that its self-interactions vanish identically. The GUT Higgs contains the other 9 linear combinations of ∂F_I. These give the diagonal elements of the symmetric traceless matrix chosen to represent the 54-plet. the associated vertex operator is

$$\partial X_3 \; \otimes \; \sum_{I=1}^{10} c_I \partial F_I \;\; ; \;\; c_I \in R, \;\; \sum_I c_I = 0 \; . \qquad (13)$$

Sector	$SO(10) \times SO(8)$	Q	Q_A	A	B	C
gauginos	$(45,1)+(1,28)$	0	0	$3m_{3/2}^2$	$m_{3/2}^2$	0
U_1	(1,8)	1/2	1/2	$m_{3/2}^2$	$m_{3/2}^2$	$m_{3/2}^2$
	(1,8)	-1/2	-1/2	$m_{3/2}^2$	$m_{3/2}^2$	$m_{3/2}^2$
U_2	(1,8)	-1/2	1/2	$m_{3/2}^2$	$m_{3/2}^2$	$m_{3/2}^2$
	(1,8)	1/2	-1/2	$m_{3/2}^2$	$m_{3/2}^2$	$m_{3/2}^2$
U_3	(54,1)	0	0	$m_{3/2}^2$	$-m_{3/2}^2$	$-2m_{3/2}^2$
	(1,1)	0	0	$m_{3/2}^2$	$-m_{3/2}^2$	$-2m_{3/2}^2$
	(1,1)	0	1	$m_{3/2}^2$	$-m_{3/2}^2$	$-2m_{3/2}^2$
	(1,1)	1	0	$m_{3/2}^2$	$-m_{3/2}^2$	$-2m_{3/2}^2$
	(1,1)	-1	0	$m_{3/2}^2$	$-m_{3/2}^2$	$-2m_{3/2}^2$
	(1,1)	0	-1	$m_{3/2}^2$	$-m_{3/2}^2$	$-2m_{3/2}^2$
θ	$3(16,1)$	1/4	1/4	$m_{3/2}^2$	0	$-1/2m_{3/2}^2$
	$(\overline{16},1)$	-1/4	-1/4	$m_{3/2}^2$	0	$-1/2m_{3/2}^2$
ω	$3(16,1)$	-1/4	1/4	$m_{3/2}^2$	0	$-1/2m_{3/2}^2$
	$(\overline{16},1)$	1/4	-1/4	$m_{3/2}^2$	0	$-1/2m_{3/2}^2$
$\theta\omega$	$4(10,1)$	0	1/2	$m_{3/2}^2$	$m_{3/2}^2$	$m_{3/2}^2$
	$4(10,1)$	0	-1/2	$m_{3/2}^2$	$m_{3/2}^2$	$m_{3/2}^2$
	$3(1,8)$	0	1/2	$m_{3/2}^2$	$m_{3/2}^2$	$m_{3/2}^2$
	$(1,8)$	0	-1/2	$m_{3/2}^2$	$m_{3/2}^2$	$m_{3/2}^2$
	$8(1,1)$	1/2	0	$m_{3/2}^2$	$m_{3/2}^2$	$m_{3/2}^2$
	$8(1,1)$	-1/2	0	$m_{3/2}^2$	$m_{3/2}^2$	$m_{3/2}^2$

Table 1: Particle content and charges of the string-GUT example discussed in the text. The three rightmost columns desplay three examples of consistent soft masses from dilaton/moduli SUSY breaking.

Vevs for these nine components of the 54 would correspond to the presence of more general Wilson backgrounds of the form $L = (\lambda_1, \lambda_2, \cdots, \lambda_{10}, 0, 0, 0, 0, 0, 0)$ with $\sum_{I=1}^{10} \lambda_I = 0$.

These more general backgounds break the symmetry further to some $SO(10)$ subgroup like $SU(4) \times SU(2)_L \times SU(2)_R$. The fact that these other nine modes may be continuously varied means that they are also string moduli or, more generally, that the 54-plet of $SO(10)$ in this model is itself a string modulus! We find that this property of the GUT-Higgs behaving as a string modulus, on equal footing with the compactifying moduli T_i, is very remarkable.

This example belongs to a whole class of models obtained through continuous Wilson lines. A general characteristic is that they are $SO(10)$ models in which the GUT Higgs is a 54 multiplet.

Moreover, there is only one such GUT Higgs coming from the untwisted sector and behaving like a string modulus. On the other hand, the rest of the particle content is model dependent. This includes the number of generations, existence of Higgses 10s, $(16 + \overline{16})$s, hidden gauge group, etc. For instance, the number of generations can be changed by adding discrete Wilson lines to the original orbifold.

The second orbifold method (II), which is implemented by permutations, is more versatile [26].

In this case one may obtain models similar to the previous one both with either one 54 or one 45 in the untwisted sector. One can also find $SU(5)$ models with adjoints 24s in the untwisted sector (sometimes also in some twisted sectors). Instead of showing more examples it is perhaps more interesting to desplay some general properties and selection rules [26] which one can derive for this kind of string-GUTs. As will be clear, some of those will be more general and apply to any string-GUT constructed through any method.

General selection rules for any k=2 string-GUT

i) All superpotential terms have $dim \geq 4$ (i.e., no mass terms).

ii) At $k = 2$ the only reps. which may be present in the massless spectrum are: 5, 10, 15 and 24-plets for $SU(5)$; 10, 16, 45 and 54-plets for $SO(10)$.

iii) $SO(10)$: The rep. $\underline{54}$ for $k = 2$ is special. All its left-handed conformal weight comes from the KM sector. As a consequence: a) A $\underline{54}$ cannot be charged under any $U(1)$ nor any other gauge group. b) Couplings of the type $X(\underline{54})(\underline{54})$ or $X(\underline{54})(\underline{54}')$ (where X is some singlet) are not allowed.

In addition to these general rules one can prove several other ones for string-GUTs obtained from symmetric orbifolds. Some of these are as follows:

Selection rules for k=2 string-GUTs from symmetric orbifolds

i) There are no selfcouplings $\underline{54}^n$ for any n.

ii) There cannot be couplings of type $(\underline{54})(\underline{45})(\underline{45})$.

iii) There cannot be $SU(5)$ self-couplings of type $\underline{24}^3$.

In practice, when constructing explicit models in symmetric orbifolds, the constraints are even tighter. As we mentioned above, there is normally just one GUT-Higgs in $SO(10)$, either a $\underline{54}$ or a $\underline{45}$ in the untwisted sector. Being in the untwisted sectors, selfcouplings of the type $\underline{45}^n$ or couplings of type $X(\underline{45})(\underline{45})$ are also forbidden (see ref. [26] for a more detailed explanation of selection rules). Many of these couplings have been used in the past in SUSY-GUT model building in order to trigger GUT-symmetry breaking while obtaining as the low energy sector the MSSM. With the above type of constraints it seems it is very difficult (if not impossible) to construct models whose low energy sector is indeed the MSSM. The absence of some relevant GUT-Higgs selfcouplings cause extra chiral multiplets to remain massless. That will be the case of the GUT-partners of the Goldstone bosons of GUT-symmetry breaking. For example, upon symmetry breaking by an adjoint $\underline{24}$, twelve out of the 24 fields remain massless. They transform as

$$(8,1,0) \; + \; (1,3,0) \; + \; (1,1,0) \tag{14}$$

under $SU(3) \times SU(2) \times U(1)$. This seems quite a generic situation which I would expect to be present in more general 4-D string constructions like asymmetric orbifolds or models based on the fermionic construction. In these more general cases some of the above strict selection rules are relaxed in principle but not very much in practice. For example, if the $k = 2$ model comes from the diagonal sum of two $SO(10)$ $k = 1$ factors, the 45s or 54s obtained originate from a $(10,10)$ rep. of the original theory. Such reps. do not admit cubic selfcouplings and the same is expected for the 54 (due to its antisymmetry there are no cubic couplings for the 45 anyhow). Since the extra particles above will have masses only of the order of the weak scale, they will sizably contribute to the running of the gauge coupling constants.

One can perform a one-loop analysis of the running of the gauge coupling constants and check that, with the particle content of the minimal SUSY-SM plus the additional fields above, there is no appropriate gauge coupling unification in the vicinity of $10^{15} - 10^{17}$ GeV. In the case of $SO(10)$ an intermediate scale of symmetry breaking could improve the results. We thus see that gauge coupling unification is not particularly better in string-GUTs than in direct SM string unification, if the above analysis is correct.

The most severe problem of SUSY-GUTs is the infamous doublet-triplet splitting problem of finding a mechanism to understand why, for example, in the 5-plet Higgs of $SU(5)$ the Weinberg-Salam doublets remain light while their coloured triplet partners become heavy enough to avoid fast proton decay.

The most simple, but clearly unacceptable, way to achieve the splitting is to write a term in the $SU(5)$ superpotential

$$W_H = \lambda H \Phi_{24} \bar{H} + M H \bar{H} \tag{15}$$

and fine-tune λ and M so that the doublets turn light and the triplets heavy. Since there are no explicit mass terms in string theory this inelegant possibility is not even present. Another alternative suggested long time ago is the "missing partner" mechanism [27] . Formulated in $SU(5)$ it requires the presence of 50-plets in the massless sector which is only possible for level $k \geq 5$, a very unlikely possibility [15, 28] . A third mechanism, put forward in the early days of SUSY-GUTs, is the "sliding singlet" mechanism [29, 30] . This requires the existence of a singlet field X, with no self-interactions, entering in the mass term in eq. (15). W_H is then replaced by

$$W_X = \lambda H \Phi_{24} \bar{H} + X H \bar{H} \,. \tag{16}$$

The idea is that the vev of the 24 is fixed by other pieces in the potential but the vev of X is undetermined to start with, i.e. the vev "slides". Now, once the electroweak symmetry is broken by the vevs of $H, \bar{H}$, the minimization conditions give $\lambda\langle -\frac{3}{2}v\rangle + \langle X\rangle = 0$ where $\mathrm{diag}(\langle\Phi_{24}\rangle) = v(1,1,1,-3/2,-3/2)$. In this way X precisely acquires the vev needed for massless doublets. This is in principle a nice dynamical mechanism but it was soon realized that it is easily spoiled by quantum corrections [31, 32] .

Interestingly enough, one finds that in string GUTs, couplings of the "sliding singlet" type are frequent, the main difference now being that the GUT-Higgs field also "slides". In particular, this happens in models in which the GUT-Higgs is a modulus, as in the examples discussed above. Take for example the $SO(10)$ model discussed above whose massless spectrum is displayed in the table . The singlets in the U_3 sector $S^0 = (1,1)_{0,0}$, $S^+ = (1,1)_{0,1}$, $S^- = (1,1)_{0,-1}$ do also behave as moduli. Both these singlets and the 54 couple to the decuplets $H^+ = (10,1)_{0,1}$ and

$H^- = (10,1)_{0,-1}$. The sub-indices in all these fields refer to their Q and Q_A charges. It is easy to check that there are flat directions in this scalar moduli space in which the gauge symmetry is broken down to $SU(4) \times SU(2) \times SU(2)$ and some of the doublets remain light whereas the colour triplets remain heavy (the symmetry is broken down to the SM through the vevs of the $16 + \overline{16}$ pairs).

If the sliding-singlet argument were stable under quantum corrections, the regions in moduli-space in which there are light doublets would be energetically favoured.

As the above example shows, the appropriate language to describe the doublet-triplet splitting problem within the context of the above string-GUTs is in terms of the scalar moduli space of the model. At generic points in the moduli space there are no massless Higgs doublets at all, they are all massive. At some "multicritical" points in moduli space some Higgs fields become massless. This is very reminiscent of the behaviour of the moduli spaces of other well studied string moduli, those associated to the size and shape of the compact manifold usually denoted by T_i. It is well known that generically there are points in the T_i moduli space in which extra massless fields appear. This is also apparently the case of the moduli space associated to the dilaton complex field S. The problem of understanding the doublet-triplet splitting within this context would be equivalent to finding out why we are sitting on a region of moduli space in which massless doublets are obtained. It could well be that an appropriately modified version of the sliding-singlet mechanism is at work and that region of moduli space is energetically favoured.

To summarize this section, I believe that the doublet-triplet splitting problem is a crucial issue and should be addressed in any model before trying to extract any further phenomenological consequences such as fermion masses. It is also important to understand whether it is possible to build string GUTs in which the massless sector is just the MSSM, or else whether the existence of extra massless chiral fields is really generic. This would dictate the necessity of intermediate scales to attain coupling constant unification.

3 Soft SUSY-breaking terms from dilaton /moduli sectors

Let us turn now to a different subject. The idea is trying to extract some information about the structure of effective SUSY-breaking soft terms which are left out once supersymmetry is spontaneously broken.

This is a very important objective since by the year 2005 the spectrum of supersymmetric particles might be tested at LHC and this could be one of the few experimental windows we could have to test the theory. Since very little is known about the origin of supersymmetry-breaking in string theory, this aim looks impossible.

This is not as hopeless as it seems. We do not need to know *all* the details of a symmetry-breaking process in order to get important physical information. A well known example of this is the standard model itself.

We do not know yet how $SU(2) \times U(1)$ breaking takes place. But, assuming that somehow a (composite or elementary) operator with the quantum numbers of a doublet gets a vev, we get a lot of information.

The idea is to apply a similar philosophy for SUSY-breaking in string theory [33, 9].

We have to try and identify possible chiral fields ϕ_i such that their auxiliary fields F_i could get non-vanishing vev and break SUSY. In string models there are some natural candidates to do the job: the complex dilaton $S = 4\pi/g^2 + i\theta$ and the moduli fields T_i whose vevs determine the size and shape of the compact space. The field S is present in any 4-D strings and the T_i fields at least in any model obtained from compactification. Thus these singlet fields are generic in large classes of string models. An additional advantage is that these fields couple to matter only with non-renormalizable couplings supressed by powers of $1/M_{Planck}$. This is a condition which is required in supergravity models with supersymmetry breaking in a" hidden sector".

The scalar potential for S and T_i is flat order by order in perturbation theory and it is expected that non-perturbative effects will i) induce a non-trivial scalar potential for those fields yielding $< S > \neq 0$, $< T_i > \neq 0$ and ii) break supersymmetry spontaneously.

The crucial assumption here is to locate the origin of SUSY-breaking in the dilaton/moduli sector [33, 9, 34, 35]. It is perfectly conceivable that other fields in the theory, like charged matter fields, could contribute in a leading manner to supersymmetry breaking. If that is the case the structure of soft SUSY-breaking terms will be totally model-dependent and we would be able to make no model-independent statements at all about soft terms. On the contrary, assuming the seed of SUSY-breaking originates in the dilaton-moduli sectors will enable us to make some predictions which might be testable. We will thus make that assumption without further justification. Let us take the following parametrization [35, 36] for the dilaton/moduli auxiliary fields F^S and F^i vevs:

$$G_S^{S\,1/2} F^S \;=\; \sqrt{3} m_{3/2} sin\theta \;\;;\;\; G_i^{i\,1/2} F^i \;=\; \sqrt{3} m_{3/2} cos\theta \; \Theta_i \tag{17}$$

where $\sum_i \Theta_i^2 \;=\; 1$ and $m_{3/2}$ is the gravitino mass. The angle θ and the Θ_i just parametrize the direction of the goldstino in the S, T_i field space.

This parametrization has the virtue that when we plugg it in the general form of the supergravity scalar potential, its vev (the cosmological constant) vanishes by construction. We now need some information about the couplings of the dilaton/moduli fields. Those are given by the Kahler potential G and the gauge kinetic functions f^a. The latter are given at the tree level by $f^a \;=\; k^a S$, where k^a are the Kac-Moody levels, for any 4-D string. The tree-level kahler potential for S is $-log(S\,+\,S^*)$, also for any model. The kahler potential for the moduli are more complicated and model-dependent and we need to specify the class of models and moduli that we are considering. We will concentrate here on the three untwisted moduli T_1, T_2, T_3 always present in $(0,2)$ toroidal symmetric orbifold constructions (for particular examples there may be additional untwisted moduli and also complex structure structure fields U_1, U_2, U_3).

For this large class of models one can write for the kahler potential [38, 39]

$$K(S, T_i, C_r ..) \;=\; -log(S + S^*) \;-\; \sum_i log(T_i + T_i^*) \;+\; \sum_n |C_r|^2 \Pi_i (T_i + T_i^*)^{n_r^i} \tag{18}$$

where C_r are charged chiral fields and the n_r^i are fractional numbers called "modular weights" which depend on the given field C_r. The sum in i may be extended to all the three moduli T-fields and also to the complex-structure U-fields. Plugging this information into the supergravity lagrangian one finds the following results for the scalar masses, gaugino masses and soft trilinear coupling A_{rst} associated to a Yukawa coupling Y_{rst} [35, 36] :

$$\begin{aligned} m_r^2 &= m_{3/2}^2 (1 \;+\; 3cos^2\theta \; \vec{n_r}.\vec{\Theta}^2) \\ M^a &= \sqrt{3} m_{3/2} sin\theta \\ A_{rst} &= -\sqrt{3} m_{3/2} [sin\theta \;+\; cos\theta (\vec{\Theta}.(\vec{u} + \vec{n_r} + \vec{n_s} + \vec{n_t}))] \;. \end{aligned} \tag{19}$$

Here we have used vectorial notation in the space of the $T-U$ moduli and we define $\vec{u} = (1, 1, ...)$. (An additional term should be added to A_{rst} for Yukawa couplings Y_{rst} depending on the moduli).

Several observations are in order. Fist of all, in the case of dilaton dominance in the SUSY-breaking process ($sin\theta = 1$) one gets the remarkably simple universal[9] result [34, 35] :

$$-A_{rst} = M^a = \sqrt{3}m_{3/2} \tag{20}$$

$$m_r^2 = m_{3/2}^2. \tag{21}$$

This result in fact applies for any 4-D string (not only orbifolds) whenever the dilaton dominates. A second observation is that a similar structure of soft terms is obtained for orbifold models in which $\vec{n}_r.\vec{\Theta}^2 = -1/3$ and $\vec{n_r} + \vec{n_s} + \vec{n_t} = -\vec{u}$. This happens for any value of θ (i.e., not necessarily dilaton dominance) if one assumes $F^1 = F^2 = F^3$ (equal SUSY-breaking contribution from the three untwisted T-fields, $\vec{\Theta}^2 = (1/3, 1/3, 1/3)$) and that the charged fields C_r have overall modular weight $n_r = \vec{u}.\vec{n}_r = -1$ (this happens for all particles in the untwisted sector and particles in some types of twisted sectors).

A third important observation is that the $mass^2$ of the scalar fields is not positive definite and hence there are choices of $sin\theta$ and the $\vec{\Theta}^2$ for which tachyons may appear [33, 35] .

A simplified case in which only the dilaton S and the "overall modulus" T contribute to SUSY-breaking (i.e.$\vec{\Theta}^2 = (1/3, 1/3, 1/3)$) was studied in ref.[35] . In this case one finds that 1) the particles in both untwisted and twisted sectors with overall modular weights $n = -1$ never become tachyonic; 2) In order to avoid particles with smaller modular weights (i.e. $n = -2, -3..$) to become tachyionic one has to confine oneself to goldstino angles $cos^2\theta \leq 1/|n|$; 3) Due to these constraints one always has bigger gaugino than scalar masses ($M^2 \geq m_i^2$). There is only one situation in which the gauginos may become lighter than the scalars. This happens [35] when the chiral fields have all modular weights $n = -1$ and $sin\theta \to 0$, in which case both $M, m \to 0$ and including loop corrections to G and f^a can reverse the situation and yield gaugino masses smaller than scalar masses.

Considering the more general case [36, 37] in eq. with several moduli T_i modifies somewhat the general conclusions. One finds that 1) particles with overall modular weight $n = -1$ can also become tachyonic for some choices of the angles and 2) The gaugino masses may become lighter than scalar masses even at the tree level. As an example of the possibilities offered let me consider some consistent choices of soft mass terms for the string $SO(10)$ GUT discussed above. For simplicity I only consider the possibility of the S and the T_i $i = 1, 2, 3$, (and not the U_i) contributing to SUSY-breaking. Since it is a $Z_2 \times Z_2$ orbifold, the twisted modular weights are $\vec{n}_T = (\underline{\frac{-1}{2}, \frac{-1}{2}, 0})$, where the underlyning means permutations. The untwisted modular weights are as usual $\vec{n}_U = (\underline{-1, 0, 0})$. The three rightmost columns in the table show three consistent choices of soft masses: A) Dilaton dominated case ($sin\theta = 1$). All the scalars have the same mass, $\sqrt{3}$ times lighter than the gauginos. B) Case with $sin^2\theta = 1/3$ and $\vec{\Theta^2} = (0,0,1)$. In this case the gauginos and Higgs 10-plets have equal masses,the 16-plets have zero mass and the GUT-Higgs have negative $mass^2$. The latter property is interesting since it show us that SUSY-breaking may authomatically trigger GUT-symmetry breaking; C) Case with $sin\theta = 0$ and $\vec{\Theta^2} = (0,0,1)$. In this case the dilaton does not contribute to SUSY-breaking and the gauginos are massless at the tree-level. The Higgs 10-plets have positive $mass^2$ but the GUT-Higgs and the 16-plets get negative $mass^2$. The latter may also enforce that there are $16 + \bar{16}$ pairs getting a vev. Thus we see that a variety of phenomenological possibilities open up depending on what is the role of the different moduli in the process of SUSY-breaking.

It must be emphasized that, given a specific string model, there is only certain type of soft terms which can be added consistently with the assumptions of dilaton/moduli dominance in SUSY-breaking. Taking a random choice of soft terms would lead to inconsistencies. For example there is always a rule [36] which connects the soft masses of particles in the three untwisted sectors $i = 1,2,3$ with that of the gauginos, $m_1^2 + m_2^2 + m_3^2 = M^2$ (see the discussion below). The reader may check this constraint in the three examples in the table.

In fact a similar sum-rule is still correct for twisted particles with overall modular weight $n = -1$, and not only for untwisted fields. More details and examples can be found in a forthcoming publication [36] .

4 Conclussions and outlook

The above lines discussed several directions recently explored in trying to establish contact between the physics at the string scale and the physics at the weak scale, which is the one amenable to experimental test. It is important to realize that by the year 2005 the LHC should provide us with important experimental information about the origin of the weak scale. If low energy supersymmetry is correct, the spectrum of SUSY particles should be tested. We do not have at the moment a theory of supersymmetry breaking but we still have ten years ahead to find one! I certainly believe that it should be easier to find a theory of soft terms rather than a theory of fermion masses. At least, it seems that the former could have a more model-independent origin than the second.

References

[1] P. Ginsparg, Phys.Lett. B197 (1987) 139.

[2] K. Dienes and A. Faraggi, hep-th/9505018 ; hep-th/9505046.

[3] V. Kaplunovsky, Nucl.Phys. B307 (1988) 145 and erratum (1992).

[4] L.E. Ibáñez, D. Lüst and G.G. Ross, Phys.Lett. B272 (1991) 251.

[5] I. Antoniadis, J. Ellis, S. Kelley and D.V. Nanopoulos, Phys.Lett. B271 (1991) 31.

[6] A. Casas and C. Muñoz, *Phys.Lett.***B214** (1988) 543.

[7] L.E. Ibáñez, Phys.Lett. B318 (1993) 73.

[8] L. Dixon, V. Kaplunovsky and J. Louis, *Nucl.Phys.* **B355** (1991) 649; J.P. Derendinger, S. Ferrara, C. Kounnas and F. Zwirner, *Nucl.Phys.* **B372** (1992) 145; I. Antoniadis, K. Narain and T. Taylor, *Phys.Lett.* **B267** (1991) 37.

[9] L.E. Ibáñez and D. Lüst, Nucl.Phys. B382 (1992) 305.

[10] P. Mayr, H.P. Nilles and S. Stieberger, *Phys.Lett.* **B317**(1993) 53; *Phys.Lett.***B317** (1993) 53; P. Mayr and S. Stieberger, hep-th/9504129.

[11] V. Kaplunovsky and J. Louis, hep-th/9502077.

[12] L.E. Ibáñez and H.P. Nilles,*Phys.Lett.* **B169** (1986) 354.

[13] S. Martin and P. Ramond, hep-ph/9501244.

[14] D. Lewellen, Nucl. Phys. B337 (1990) 61.

[15] A. Font, L.E. Ibáñez and F. Quevedo, Nucl. Phys. B345 (1990) 389.

[16] G. Aldazabal, A. Font, L.E. Ibáñez and A.M. Uranga, Madrid preprint FTUAM-94-28 (1994); hep-th/9410206.

[17] S. Chaudhuri, S.-w. Chung and J.D. Lykken, "Fermion masses from superstring models with adjoint scalars", preprint Fermilab-Pub-94/137-T, hep-ph/9405374; S. Chaudhuri, S.-w. Chung, G. Hockney and J.D. Lykken, "String models for locally supersymmetric grand unification", preprint hep-th/9409151. G.B. Cleaver, "GUTs from adjoint Higgs fields from superstrings", preprint OHSTPY-HEP-T-94-007, hep-th/9409096.

[18] S. Chaudhuri, S. Chung, G. Hockney and J. Lykken, hep-th/9501361.

[19] A. Font, L.E. Ibáñez, F. Quevedo and A. Sierra, *Nucl.Phys.*B**337**(1990) 119.

[20] L.E. Ibáñez, H.P. Nilles and F. Quevedo, Phys. Lett. B192 (1987) 332.

[21] A. Font, L.E. Ibáñez, H.P. Nilles and F. Quevedo, Nucl. Phys. B307 (1988) 109.

[22] L. Dixon, J. Harvey, C. Vafa and E. Witten, Nucl. Phys. B274 (1986) 285.

[23] L.E. Ibáñez, H.P. Nilles and F. Quevedo, Phys.Lett. B187 (1987) 25; L.E. Ibáñez, J. Mas, H.P. Nilles and F. Quevedo, Nucl. Phys. B301 (1988) 137.

[24] K.S. Narain, M.H. Sarmadi and C. Vafa, Nucl. Phys. B288 (1987) 551; Nucl. Phys. B356 (1991) 163.

[25] M. Green and J. Schwarz, Phys.Lett. B149 (1984) 117.

[26] G. Aldazabal, A. Font, L.E. Ibáñez and A.M. Uranga, to appear.

[27] S. Dimopoulos and F. Wilczek, Sta.Barbara preprint, (1981); Proc. of the Erice summer school (1981); B. Grinstein, Nucl. Phys. B206 (1982) 387; A. Masiero, D.V. Nanopoulos, K.Tamvakis and T. Yanagida, Phys. Lett. 115B (1982) 380.

[28] J. Ellis, J. López and D.V. Nanopoulos, Phys.Lett. B245 (1990) 375.

[29] E. Witten, Phys. Lett. 105B (1981) 267.

[30] L.E. Ibáñez and G.G. Ross, Phys. Lett. 110B (1982) 227.

[31] See e.g. V. Kaplunovsky, Nucl.Phys.B233 (1984) 336.

[32] H.P. Nilles, M. Srednicki and D. Wyler, Phys. Lett. 124B (1983) 337.

[33] M. Cvetic, A. Font, L.E. Ibáñez, D. Lüst and F. Quevedo, *Nucl.Phys.* **B361** (1991) 194.

[34] V. Kaplunovsky and J. Louis, *Phys.Lett.* **B306** (1993) 269.

[35] A. Brignole, L.E. Ibáñez and C. Muñoz, *Nucl.Phys.*B**422**(1994) 125; erratum *Nucl.Phys.*B**436**(1995)747.

[36] A. Brignole, L.E. Ibáñez, C. Muñoz and C. Scheich, to appear.

[37] T. Kobayashi, D. Suematsu, K. Yamada and Y. Yamagishi, hep-ph/9408322; P. Brax and M. Chemtob, hep-th/9411022; D. Bailin, A. Love, W. Sabra and S. Thomas, *Mod.Phys.Lett.* **A9** (1994) 2543.

[38] E. Witten, *Phys.Lett.* **B155** (1985) 151.

[39] L. Dixon, V. Kaplunovsky and J. Louis, *Nucl.Phys.* **B329** (1990) 27.

[40] L.E. Ibáñez and C. Muñoz, unpublished (Dec. 1994); C. Muñoz at the Lake Tahoe Conference " Beyond the Standard Model IV", (December 13-18, 1994).

[41] I. Jack and D.R.T. Jones, Liverpool preprint LTH 345; hep-th-9501395 (1995).

[42] D.R.T. Jones, L. Mezincescu and Y.P. Yao, *Phys.Lett.* **B148**(1984)317.

[43] I. Jack and D.R.T. Jones, *Phys.Lett.* **B333** (1994) 372.

[44] A. Parkes and P. West, *Nucl.Phys. B222* (1983) 269.

[45] M. de Roo, *Nucl.Phys.***B255** (1985) 241.

[46] A. Font, L.E. Ibáñez, D. Lüst and F. Quevedo, *Phys.Lett.***B249**(1990) 35.

[47] J.H. Schwarz, hep-th/9503127; A. Sen, hep-th/9402002.

[48] N. Seiberg, *Nucl.Phys.***B435**(1995) 129; K. Intriligator and N. Seiberg, hep-th/9503179; D. Kutasov, hep-th/9503086; D. Kutasov and A. Schwimmer, hep-th/9505004.

[49] R.G. Leigh and M.J. Strassler, hep-th/9503121 .

[50] L.E. Ibáñez and F. Quevedo, unpublished.

[51] A. Font, L.E. Ibáñez, D. Lüst and F. Quevedo, *Phys.Lett.* **B245** (1990) 401; S. Ferrara, N. Magnoli, T. Taylor and G. Veneziano, *Phys.Lett.* **B245** (1990) 409; H.P. Nilles and M. Olechovsky, *Phys.Lett.* **B248** (1990) 268; P. Binetruy and M.K. Gaillard, *Phys.Lett.* **B253** (1991) 119.

Extending the symmetries of the Standard Model

Graham G. Ross[b*]

[a]Department of Physics, Theoretical Physics,
University of Oxford, 1 Keble Road, Oxford OX1 3NP

[b] Theory Division, CERN, CH-1211, Geneva, Switzerland

September 22, 1995

Abstract

We discuss whether extension of the symmetries of the Standard Model can lead to predictions for its parameters including the quark and lepton masses and mixing angles.

1 Introduction

The parameters of the Standard Model are not determined by the model itself and as such may be our only clue to what lies beyond. At first sight they seem to be very difficult to understand for parameters of a given dimension differ greatly in magnitude. However it has long been realised that such a characteristic hierarchy of scales may be obtained from an extension of the Standard Model posessing additional broken symmetries, the expansion in the symmetry breaking parameter providing the hierarchy of scales. This observation immediately leads us to consider supersymmetric extensions of the Standard Model. In these supersymmetry offers a solution to the original hierarchy problem, namely the problem of understanding why the ratio of the electroweak breaking scale to the Planck scale is so small, provided the scale of supersymmetry breaking is hierarchically smaller than the Planck scale. We shall discuss here whether a further extension of the Standard Model symmetries can similarly explain the hierarchical structure of the qark and lepton masses and mixing angles.

2 The Classic predictions

So far as determining the parameters of the Standard Model the minimal supersymmetric standard model (MSSM) improves on the Standard Model in only one respect; the quartic coupling of the Higgs fields are given in terms of the gauge coupling constants. However many new parameters are needed to describe the breaking of supersymmetry so the full theory contains even more unknown constants. Thus some further unification is necessary if these parameters are to be related.

One possibility is that there is a stage of Grand Unification. With the advent of precision measurements of the parameters of the Standard Model we have been able to

*SERC Senior Fellow, on leave from [a]

test quantitatively the possibility of such supersymmetric unification in several areas. In particular the couplings are found to come together at a scale $O(10^16GeV)$ the values corresponding to the $SU(5)$ normalisation of the $U(1)$ coupling with $sin^2(\theta_W) = 3/8$ (this is not the case in the Standard Model where the couplings are some six standard deviations apart)[1]. These analyses have been performed including the effects of a non-degenerate SUSY spectrum using the running masses at low energies and they are sensitive to the value of m_0 and $m_{1/2}$ (and other soft SUSY breaking terms) the common scalar mass and common gaugino mass, assuming that there is some unification of the soft SUSY breaking mass terms as well. It is found that, allowing for the range $\alpha_3(M_Z) = 0.102$-0.12, values for the supersymmetry mass scale M_{SUSY} in the range (10^2–10^5) GeV are allowed [I define M_{SUSY} to be $Min(m_0, m_{\frac{1}{2}}, \mu)$]. The unification scale is 3.10^{16}GeV.

The idea the soft SUSY breaking terms unify may be tested, too, because the Higgs mass as well as the squark and slepton mass are unified and deviate at low energy only through calculable radiative corrections. The exciting thing about soft mass unification is that it gives a reason why the pattern of symmetry breaking in the Standard Model is the way it is. It is found that due to the large top mass and associated Yukawa coupling the RG equations drive m_2^2 negative, triggering electroweak breaking. Now we can see why it is $SU(2)$ and not $SU(3)$ that is broken by this radiative breaking mechanism for, although the RG equations for both the Higgs scalars and the top (and bottom) squarks have destabilizing terms due to the top Yukawa coupling, only the squarks have large stabilizing terms due to QCD interactions [2]. Thus it is the Higgs $(mass)^2$ that is driven negative, and it is $SU(2)$ that is broken by the resultant Higgs vev. In addition the coefficient of the quartic term is determined in the supersymmetric theory in terms of gauge couplings.Thus one knows the full Higgs potential in supersymmetry. Indeed one finds that the initial value of the top Yukawa coupling, h, may be chosen to give the correct value of M_Z. It is found that the requirement of acceptable radiative electroweak breaking together with a heavy top quark leads to a very satisfactory explanation of both the existence of electroweak breaking and its magnitude for a wide range of supersymmetry breaking parameters consistent with the values obtained for gauge unification.

In specific GUTs there are relations between fermion masses, the best-known being the $SU(5)$ relation $m_b = m_\tau$ [4]. As with the relation amongst gauge couplings, this relation applies at the unification scale and must be radiatively corrected. These corrections offer a further tantalizing piece of circumstantial evidence in favour of supersymmetric unification for, starting with the relation $m_b(M_X) = m_\tau(M_X)$, they can bring the prediction for m_b/m_τ into agreement with experiment using the same value for M_X found in the analysis of gauge couplings. This cannot be done in the non-supersymmetric case without adding new interactions [5]. Agreement in the supersymmetric case is possible only for a restricted range of parameters, further tying down the range of supersymmetric masses. In particular a running b quark mass, $m_b(m_b) = 4.2$, can only result if the top Yukawa coupling is large, $h_t \approx 1.25$. For $h_t > h_b$ the top quark is heavy, $m_t = 170 \pm 20$ GeV, and the μ parameter is also large, $\mu \approx 2m_0$.

The final "classic" prediction is for supersymmetric dark matter. In the MSSM the interactions respect an R-parity under which the new SUSY states are odd while the states of Standard Model are even. As a result the SUSY states may only be pair-produced and the lightest supersymmetric state (LSP) is stable. The existence of the LSP is a prediction of the MSSM and, since its couplings are determined, the relic abundance of the LSP is also a prediction. The nature of the LSP and its abundance is determined in terms of the SUSY-breaking parameters introduced above and as we have seen there is only a small region allowed in this parameter space. Thus the range of possible relic abundance is strongly constrained and it is clearly of interest to see how this compares to the cosmological bounds on dark matter. It is found that almost all of the allowed range has relic abundance less than the closure density and is therefore not ruled out. Perhaps more interestingly, over a substantial part of the allowed region, the LSP abundance is large enough to be the candidate for dark matter! The sceptic may view this as a coincidence but, to me, the remarkable overlap between these abundances provides further circumstantial evidence in favour of the need for supersymmetry.

3 Fermion Masses and mixing angles

Is there any further evidence of additional symmetries relating the quark and lepton masses and mixing angles? The simple relations $m_s = m_\mu$ and $m_d = m_e$, plus radiative corrections, are grossly in disagreement with experiment. However it has been observed that experiment is consistent with the appearance of "texture" zeros in the fermion mass matrices [8, 9], possibly signalling the appearance of new symmetries enforcing these zeros. To see how texture zeros may be required consider the 2×2 submatrix, M_d, the down quark mass matrix, *assuming* $\bar{\psi}_L M_d \psi_R$ is (L-R) symmetric and has a texture zero in the (1,1) position

$$M_d = \begin{pmatrix} 0 & f \\ f & e \end{pmatrix} \tag{1}$$

This contains just 2 real parameters (the phases may be absorbed in a redefinition of the quark fields) yet describes 2 masses and a mixing angle. Thus it gives a prediction for the mixing angle, θ_d (the angle that diagonalises it) which is readily found to be $\theta_d \approx \sqrt{(m_d/m_s)}$. The Cabibbo angle is given by $\theta_d - \theta_u$ where θ_u is the equivalent angle coming from the up quark mass matrix. This illustrates the type of prediction that results from a mass matrix with "texture" zeros in some positions.

The major difficulty in finding the set of texture zeros is that laboratory measurements only determine the masses, i.e. the diagonal mass matrix M_d^{Diag}, M_u^{Diag}, and the CKM mass matrix, not the full mass matrices, M_d, M_u; i.e. we have

$$\begin{aligned} M_u^{Diag} &= V_u^L.M + M_u.V_u^{R\dagger},\ M_d^{Diag} = V_d^L.M_d.V_d^{R\dagger} \\ V_{CKM} &= V_u^L.V_d^{L\dagger} \end{aligned} \tag{2}$$

If we are to determine $V_{u,d}^{L,R}$, and hence the mass matrices separately, a simplifying assumption is needed. Since we are interested in those solutions showing maximum symmetry (perhaps indicating additional unification) it is interesting to ask whether texture zeros appear assuming *symmetric* mass matrices. In this case we can analyze the most general set of five or six texture zeros consistent with the measured masses and mixing angles [10]. The results of this general analysis are that there are no solutions with 6 texture zeros and just five with 5. An example of the texture is given in Table 1. As may be seen the mass matrices display an hierarchical structure characterised by the appearance of high powers in the expansion parameter λ. The question immediately arises whether we can understand this structure as resulting from a new symmetry beyon that of the Standard Model in which the role of λ is played by the magnitude of the symmetry breaking parameter?

4 A theory of texture zeros.

It turns out to be remarkably easy to construct a model with the form of Table 1 through the introductionof an abelian gauge symmetry, $U(1)_X$ (such additional symmetries abound in string theories). The simplest example has charge assignment of the

	U	D
2	$\begin{pmatrix} 0 & \lambda^6 & 0 \\ \lambda^6 & 0 & \lambda^2 \\ 0 & \lambda^2 & 1 \end{pmatrix}$	$\begin{pmatrix} 0 & 2\lambda^4 & 0 \\ 2\lambda^4 & 2\lambda^3 & 2\lambda^3 \\ 0 & 2\lambda^3 & 1 \end{pmatrix}$

Table 1: An example of a symmetric texture. The parameter λ is the CKM matrix element $\mathcal{U}_{12}^{CKM}$.

	Q_i	u_i^c	d_i^c	L_i	e_i^c	ν_i^c	H_2	H_1
$U(1)_{FD}$	α_i	α_i	α_i	a_i	a_i	a_i	$-2\alpha_1$	$-2\alpha_1$

Table 2: $U(1)_{FD}$ symmetries.

Standard Model states given in Table 2. The need to preserve $SU(2)_L$ invariance requires (left-handed) up and down quarks (leptons) to have the same charge. This plus the requirement of symmetric matrices then requires that all quarks (leptons) of the same i-th generation transform with the same charge $\alpha_i(a_i)$. If the light Higgs, H_2, H_1, responsible for the up and down quark masses respectively have $U(1)$ charge so that only the (3,3) renormalisable Yukawa coupling to H_2, H_1 is allowed, only the (3,3) element of the associated mass matrix will be non-zero as desired. The remaining entries are generated when the $U(1)$ symmetry is broken. We assume this breaking is spontaneous via Standard Model singlet fields, θ, $\bar{\theta}$, with $U(1)_{FD}$ charge -1, +1 respectively, which acquire approximately equal vacuum expectation values (vevs) along a "D-flat" direction[1]. After this breaking all entries in the mass matrix become non-zero. For example, the (3,2) entry in the up quark mass matrix appears at $O(\epsilon^{|\alpha_2-\alpha_1|})$ because U(1) charge conservation allows only a coupling $c^ctH_2(\theta/M_2)^{\alpha_2-\alpha_1}$, $\alpha_2>\alpha_1$ or $c^ctH_2(\bar{\theta}/M_2)^{\alpha_1-\alpha_2}$, $\alpha_1>\alpha_2$ and we have defined $\epsilon=(<\theta>/M_2)$ where M_2 is the unification mass scale which governs the higher dimension operators. As discussed in reference[11] one may expect a different scale, M_1, for the down quark mass matrices (it corresponds to mixing in the H_2, H_1 sector with M_2, M_1 the masses of heavy H_2, H_1 fields). Thus we arrive at mass matrices of the form

$$\frac{M_u}{m_t}\approx\begin{pmatrix} h_{11}\rho_{11}\epsilon^{|2+6a|} & h_{12}\rho_{12}\epsilon^{|3a|} & h_{13}\rho_{13}\epsilon^{|1+3a|} \\ h_{21}\rho_{21}\epsilon^{|3a|} & h_{22}\rho_{22}\epsilon^{2} & h_{23}\rho_{23}\epsilon^{1} \\ h_{31}\rho_{31}\epsilon^{|1+3a|} & h_{32}\rho_{32}\epsilon^{1} & h_{33} \end{pmatrix} \quad (3)$$

$$\frac{M_d}{m_b}\approx\begin{pmatrix} k_{11}\sigma_{11}\bar{\epsilon}^{|2+6a|} & k_{12}\sigma_{12}\bar{\epsilon}^{|3a|} & k_{13}\sigma_{13}\bar{\epsilon}^{|1+3a|} \\ k_{21}\sigma_{21}\bar{\epsilon}^{|3a|} & k_{22}\sigma_{22}\bar{\epsilon}^{2} & k_{23}\sigma_{23}\bar{\epsilon}^{1} \\ k_{31}\sigma_{31}\bar{\epsilon}^{|1+3a|} & k_{32}\sigma_{32}\bar{\epsilon}^{1} & k_{33} \end{pmatrix} \quad (4)$$

where $\bar{\epsilon}=(\frac{<\theta>}{M_1})^{|\alpha_2-\alpha_1|}$, $\epsilon=(\frac{<\theta>}{M_2})^{|\alpha_2-\alpha_1|}$ and $a=\alpha_1/(\alpha_2-\alpha_1)$. The light Higgs states are given by $H_{33}^2+\sum\rho_{ij}H_{ij}^2\epsilon^{n_{ij}}$ and $H_{33}^1+\sum\sigma_{ij}H_{ij}^1\bar{\epsilon}^{n_{ij}}$ where the powers n_{ij} are those appearing in eq(4) and ρ, σ are related to Yukawa couplings in the Higgs sector in a manner discussed below. These and the Yukawa couplings h_{ij}, k_{ij} are all assumed to be of $O(1)$. For $a>0$ there are two approximate texture zeros in the (1,1) and (1,3), (3,1) positions in the sense that the entries are so small they do not significantly affect the eigenvalues and mixing angles. . These give rise to excellent predictions for two combinations of the CKM matrix duplicating the phenomenologically successful form of table 1. Choosing $a=1$ the remaining non-zero entries of the mass matrixhave magnitude

[1]The spontaneous breaking of non-anomalous gauge symmetries at high scales in supersymmetric theories must proceed along such flat directions to avoid large vacuum energy contributions from D-terms, giving $<\theta>=<\bar{\theta}>$.

in excellent agreement with the measured values (cf Table 1). To a good approximation we have the relation [11]

$$\epsilon = \bar{\epsilon}^2 \tag{5}$$

which also implies that $M_2 > M_1$.

The charged lepton mass matrix may similarly be determined. In [11] we restricted the lepton charges by requiring the good relation $m_b = m_\tau$ at the unification scale. This requires $\alpha_1 = a_1$ giving

$$\frac{M_l}{m_\tau} \approx \begin{pmatrix} l_{11}\sigma_{11}\bar{\epsilon}^{|2+6a+2b|} & l_{12}\sigma_{12}\bar{\epsilon}^{|3a|} & l_{13}\sigma_{13}\bar{\epsilon}^{|1+3a+b|} \\ l_{21}\sigma_{21}\bar{\epsilon}^{|3a|} & l_{22}\sigma_{22}\bar{\epsilon}^{|2(1+b)|} & l_{23}\sigma_{23}\bar{\epsilon}^{|1+b|} \\ l_{31}\sigma_{31}\bar{\epsilon}^{|1+3a+b|} & l_{32}\sigma_{32}\bar{\epsilon}^{|1+b|} & l_{33} \end{pmatrix} \tag{6}$$

where $b = (a_2 - \alpha_2)/(\alpha_2 - \alpha_1)$ and again the Yukawa couplings, l_{ij}, are assumed of $O(1)$. At this stage b is not determined but for any value of $b > 0$ there are texture zeros in the $(1,1)$ and $(1,3)$ positions giving rise to the phenomenologically successful prediction $Det(M_l) \approx Det(M_d)$. To proceed further a value of b must be chosen and two viable choices were explored. For $b = 0$ the lepton charges are the same as the down quark sector, and so the structure of the down quark and lepton mass matrices are identical. In order to explain the detailed difference between down quark and lepton masses it is necessary in this case to assume that the constants of proportionality determined by Yukawa couplings which we have so far taken to be equal (and of O(1)) differ sightly for the lepton case. A factor 3 in the (2,2) entry is sufficient to give excellent charged lepton masses. An alternative which does not rely on different Yukawa couplings is to choose b half integral. For $a = 1$, $b = 1/2$ we found excellent agreement for the charged lepton masses (in this case there is a Z_2 symmetry forcing the $(1,3)$, $(3,1)$, $(2,3)$, $(3,2)$ matrix elements to vanish).

To complete the discussion we turn to a consideration of the anomaly structure of the model. Although the choice of charges for the quarks and letons given above has no $SU(3)^2U(1)_{FD}$, $SU(3)^2U(1)_{FD}$ or $U(1)^2U(1)_{FD}$ anomalies it is clear that the Higgs charges are not anomaly free. However in theories derived from a string theory there is a significant new possibility for a non-vanishing anomaly associated with a new $U(1)$ gauge factor can be cancelled by the Green-Schwarz (GS) anomaly cancellation mechanism[12, 13].

In the 4-D version of the GS mechanism one cancels the anomalies of a single $U(1)$ by an appropriate shift of the axion present in the dilaton multiplet of four-dimensional strings. This happens because such an axion has a direct coupling to $F\tilde{F}$. For the GS mechanism to be possible, the coefficients A_i, $i = 3,2,1$ of the mixed anomalies of the $U(1)$ with $SU(3)$, $SU(2)$ and $U(1)_Y$ have to be in the ratio $A_3 : A_2 : A_1 = k_3 : k_2 : k_1$. Here k_i are the Kac-Moody levels of the corresponding gauge factors and they determine the boundary condition of the gauge couplings at the string scale by the well-known equation $g_3^2 k_3 = g_2^2 k_2 = g_1^2 k_1^2$. The usual (e.g. GUT) canonical values for these normalization factors (corresponding to the successful result $sin^2(\theta_W) = 3/8$) yield $k_3 : k_2 : k_1 = 1 : 1 : 5/3$ and hence the GS mechanism can only work in this case if the mixed anomalies of the $U(1)$ with the SM gauge factors are in the ratio $A_3 : A_2 : A_1 = 1 : 1 : 5/3$[13].

One can easily convince oneself that there are only two $U(1)$ symmetries with anomalies consistent with this ratio of gauge coupling constants, namely $U(1)_X$ and $U(1)_{XX}$ given in Table 1.

Thus, in a supersymmetric SM coming from a string the most general family-independent anomaly-free $U(1)$ consistent with canonical gauge coupling unification is given by :

$$U(1)_{FI} \;=\; z\, U(1)_H \;+\; x\, U(1)_X \;+\; y\, U(1)_{XX} \,. \tag{7}$$

	Q	u	d	L	e	H_2	H_1
$U(1)_H$	0	0	0	0	0	1	-1
$U(1)_{XX}$	0	0	1	1	0	0	0
$U(1)_X$	1	1	0	0	1	0	0

Table 3: Anomaly-free $U(1)_{FI}$ symmetries.

$U(1)$	$\alpha_i + x$	$\alpha_i + x$	$\alpha_i + y$	$a_i + y$	$a_i + x$	z-$2\alpha_1$	-z+$w\alpha_1$

Table 4: Anomaly-free $U(1)$ symmetries.

Allowing for a combination of $U(1)_{FD}$ and $U(1)_{FI}$ the full charges under this group may now be determined using Tables 2 and 1 and give the charges shown in Table 4.

One can easily check that for $z = -2x$ one gets the results of Table 1 for the u-quark mass matrix. If one further has $3x + y = -4\alpha_1$ (and $w = -2$) one gets the results of Table 1 for the d-quark masses. (Note, however, the choice of the flavour-independent component allows for further possibilities for the d-quark matrices[11]). Thus we conclude that the generic problem raised by anomaly cancellation may naturally be solved in the context of string based models.

It is important to recall that the $U(1)$s whose anomalies are cancelled through a GS mechanism are *necessarily spontaneously broken not far below the string scale.* The reason for this is that the piece in the Lagrangian cancelling the anomalies has a supersymmetric counterpart which is a sort of field-dependent Fayet-Iliopoulos term for the $U(1)$. This term forces $U(1)$ symmetry breaking in a natural way at a scale of order $1/\sqrt{192}\pi M_{string}$ [13]. Thus the present scheme also explains why the extra $U(1)$ symmetry required to generate the fermion mass patterns does not survive down to low energies.

So far we have assumed that $sin^2(\theta_W) = 3/8$ at the string unification scale. In fact an acceptable pattern of fermion masses actually *requires* this value! To see this let us compute the mixed anomalies A_i for the $U(1)_{FD}$ symmetry of table 3 with $SU(3), SU(2)$ and $U(1)_Y$. One finds respectively:

$$\begin{aligned} A_3 &= 2\sum_{i=1}^{3}\alpha_i \\ A_2 &= \frac{3}{2}\sum_{i=1}^{3}\alpha_i + \frac{1}{2}\sum_{i=1}^{3}a_i + \frac{\alpha_1}{2}(w-2) \\ A_1 &= \frac{11}{6}\sum_{i=1}^{3}\alpha_i + \frac{3}{2}\sum_{i=1}^{3}a_i + \alpha_1(w-2) \end{aligned} \tag{8}$$

where, as we are working with the full $U(1)$ charges, we no longer have $\alpha_3 = -(\alpha_1 + \alpha_2)$ and $a_3 = -(a_1 + a_2)$ as the $U(1)_{FI}$ piece adds a family independent term to α_i and also to a_i. However, to maintain the result $m_b = m_\tau$, this term is the same for α_i and for a_i and so

$$\sum_{i=1}^{3}\alpha_i = \sum_{i=1}^{3}a_i \tag{9}$$

Now, in order that $A_3 = A_2$ so that the $SU(3)$ and $SU(2)$ couplings are unified eq(8)

$U(1)_F$	Q	u	d	L	e	H_1	H_2	θ	θ
3rd generation	0	0	0	0	0	0	0	1	-1
2nd generation	1	1	1	-1/2	-1/2				
1st generation	-4	-4	-4	-5/2	-5/2				

Table 5: Anomaly-free $U(1)$ gauge symmetry giving rise to the textures in eqs(A,B,C)

gives

$$w = 2 \quad or\ \alpha_1 = 0 \tag{10}$$

$$and\ A_3 : A_2 : A_1 = 1 : 1 : \frac{5}{3} \tag{11}$$

The second equation requiring $sin^2(\theta_W) = 3/8$ may be seen as a *consequence* of an acceptable mass matrix structure. The first condition in eq(11) requires $\alpha_1 = 0$ since $w = 2$ leads to very bad results for the d-quark masses.

Thus we see that a simple $U(1)_{FD}$ extension of the standard model generates much of the quark and lepton mass matrix structure through the constraints of anomaly cancellation, but that the full $U(1)$ extension including the Higgs doublets needed for the MSSM is only anomaly free in the context of string theory via a Green Schwarz term. The charges of each individual particle with respect to this anomaly-free $U(1)$ are shown in Table 4.

The simplicity of the assignments is remarkable. It is also worth to emphasize that this $U(1)$ symmetry may be made anomaly free through the GS mechanism *if and only if the normalization of the coupling constants is the canonical one* $g_3^2 = g_2^2 = 5/3g_1^2$ yielding the succesfull prediction $sin^2\theta_W = 3/8$. Thus the present scheme not only predicts a succesfull pattern of fermion masses and mixings but also predicts $sin^2\theta_W = 3/8$ even without any grand unification group.

5 Infra-red fixed point structure

Although the $U(1)_X$ symmetry generates all the structure of one of the symmetric mass matrices found to be consistent with the data it leaves some questions unanswered. In particular what are the Yukawa couplings and mixing angles of eq(3) and (4)? These we have taken to be of O(1) and determined by the underlying (string) symmetry. However there is another possibility, namely that the Yukawa couplings are determined by the infra red stable fixed point (IRSFP) structure of the theory lying beyond the Standard Model. In fact it seems to me quite likely that IRSFP are likely to play a role in the theory beyond the Standard Model. We suppose there is some extension of the Standard Model at a scale M. It is plausible (but not essential - see later) to identify this with the scale, M_X, at which the gauge couplings unify. However as there is only a small gap between the gauge unification scale $O(10^{16}GeV)$ and the compactification or Planck scale there seems little room for logarithmic corrections to drive couplings to their fixed points. However, as the example presented below shows, it is very likely that the extension of the Standard Model will have additional matter which causes the couplings to run so rapidly that they closely approach their fixed points even running over a very small energy range[14].

There are several attractive features of such a proposal. In the first place the predictions rely only on a knowledge of the dynamics of the theory, which follow from the symmetries, and do not require knowledge of the initial conditions. As a result the uncertainties associated with the final "Theory of Everything"[2] are avoided or reduced. Secondly the flavour changing problems associated with a family symmetry are under control in a class of theories with a family independent fixed point structure[14]. Indeed we may take this as a strong hint that the IRSFP structure plays an important role for it seems likely that there *must* be new family dependent interactions if the pattern of fermion masses is to be understood and then it is essential to explain why flavour changing processes are small. Finally the known structure of the quark and lepton masses and mixing angles may be used to infer the symmetry structure needed for the theory beyond the Standard Model. Given the symmetries the dynamics of the theory and thus the renormalisaton group (RG) equations are determined and so one may try to develop a systematic and *quantitative* investigation of the idea that IRSFP structure determines the masses and mixing angles.

5.1 The $SU(3) \otimes SU(2) \otimes U(1) \otimes U(1)_X$ model

To illustrate this possibility we will construct the model that realizes this (a=1,b=0) solution for the fermion mass structure[17]. This will illustrate how a knowledge of the symmetries of the theory largely determine the infra red structure of the theory. In order to implement the structure of eq(3) it is necessary to introduce at least 5 additional Higgs doublets H_i^2 with the appropriate $U(1)_X$ charges, i, to allow them to couple to the quarks in the $(2,3),(2,2),(1,3),(1,2)$ and $(1,1)$ directions respectively. These Higgs doublets will belong to massive supermultiplets pairing up with their partners $\bar{H}_i^2$ in the conjugate representation through a term in the superpotential $r_i^2\Phi^2 H_i^2 \bar{H}_i^2$ where Φ^2 is a singlet chiral superfield whose scalar component acquires a vacuum expectation value, V_2, giving the Higgs mass $r_i^2 V_2$. We also need Standard Model singlet fields Θ^2, $\bar{\Theta}^2$ as discussed above to break $U(1)_X$. When $\bar{\Theta}^2$ acquires a vev it causes mixing between the massive Higgs fields and the original $H_{2,0}$ field coupling in the $(3,3)$ position. For example the $U(1)_X$ symmetry allows the term $s_0^2 H_0^2 \bar{H}_{-1}^2 \bar{\Theta}^2$ giving the massive state proportional to $H_{-1}^2 + s_0^2 < \bar{\Theta}^2 > /(r_1^2 V_2) H_2^2$ leaving the massless state $H_0^2 - < \bar{\Theta}^2 > s_0^2/(r_1^2 V_2) H_{-1}^2$. This generates $\rho_{23} = s_0^2/r_1^2$ in eq(3) and $\epsilon =< \Theta^2 > /V_2$. While this multiplet structure may appear complicated it is of the type that appears in compactified string theories where in addition to the three generations of states left massless at the compactification scale there are typically a large number of additional states in vectorlike representations. In what follows we will assume that the five copies of H_i^2 plus $\bar{H}_i^2$ appear together with five copies of D_i^{2c} plus $\bar{D}_i^{2c}$ where D_i are additional chiral superfields in a down quark representation of the Standard Model. This will ensure the successful gauge unification of the MSSM persists in this model at one loop order for these new multiplets fill out complete five dimensional representations of $SU(5)$ and thus do not change the relative evolution of the couplings.. Given that the families too fill out complete representations of $SU(5)$ it is perhaps not so surprising that the vectorlike matter should do so too and presumably just reflects the underlying GUT structure of the string even though the low energy theory has a reduced (non-GUT) gauge group.

The remainder of the spectrum is chosen to complete the structure needed to generate the mass matrices of eq(4) and (6) in the manner just discussed. Thus we add five further representations $H_i^1 + \bar{H}_i^1 + D_i^{1c} + \bar{D}_i^{1c}$. We also introduce further Standard Model singlet fields Φ^1, Θ^1, $\bar{\Theta}^1$ which acquire vevs generating masses and mixing for these vectorlike representations in the manner just discussed. In particular we have $\sigma_{23} = s_0^1/r_1^1$ in eq(3) and $\bar{\epsilon} =< \Theta^1 > /V_1$ etc.

This completes the multiplet structure of the model. A summary of the multiplet

[2]For example the uncertainties associated with the moduli vevs in string theories.

	$(Q,\ u^c,\ d^c,\ L,\ e^c)_i$		$(H^{1,2},\ D^{1,2\,c})_j$	$(\bar{H}^{1,2},\ \bar{D}^{1,2\,c})_j$	$H_0^{1,2}$	$\Phi^{1,2}$	$\Theta^{1,2}$	$\bar{\Theta}^{1,2}$
i=1	0	j=-2	-2	2	0	0	1	-1
i=2	1	j=-1	-1	1				
i=3	-4	j=3	3	-3				
		j=4	4	-4				
		j=8	8	-8				

Table 6: The $U(1)_X$ charges of the chiral supermultiplets.

content and their $U(1)_X$ charges is given in Table 6. To complete the model the couplings must be specified. We will allow all trilinear couplings consistent with the symmetries of the theory. Apart from the gauge symmetries as usual there must be discrete symmetries to inhibit nucleon decay. We choose the simplest possibility, matter parity, under which the three generations of quarks and lepton supermultiplets are odd and the Higgs and the additional D quarks are even. With this symmetry the allowed couplings are just the normal Yukawa couplings plus the singlet couplings discussed above together with further singlet couplings of the singlets to D^c and $\bar{D}^c$ which will give all these states mass at the scale $V_{1,2}$. Note that the normal quarks cannot mix with the D^c quarks because of the R symmetry. In addition we assume there is a symmetry preventing the coupling of the H^1 to the H^2 fields in order to solve the μ problem i.e. such a coupling occurs only after supersymmetry breaking[15]. Again a simple Z_2 symmetry suffices. With this the allowed couplings are given in eq(12).

$$\begin{aligned} L_{Yuk} &= h_{ijk}Q_i u_j^c H_k^2 + k_{ijk}Q_i d_j^c H_k^1 + l_{ijk}L_i l_j^c H_k^2 \\ &+\sum_{l,j}(r_l^j \Phi^j H_l^j \bar{H}_l^j + s_l^j \Theta H_l^j \bar{H}_{l+1}^j + \bar{s}_l^j \bar{\theta} H_l^j \bar{H}_{l-1}^j) \\ &+\sum_{l,j}(t_l^j \Phi^j D_l^{j\,c} \bar{D}_l^{j\,c} + u_l^j \Theta D_l^{j\,c} \bar{D}_{l+1}^{j\,c} + \bar{u}_l^j \bar{\Theta} D_l^{j\,c} \bar{D}_{l-1}^{j\,c}) \end{aligned} \tag{12}$$

Having specified the couplings the renormalisation group equations are now specifed. If they turn out to determine the Yukawa couplings and soft masses of the theory we will have a self contained low-energy theory (low relative to the compactification scale!) in which all the parameters are determined independent of the underlying string theory. To investigate this possibility we turn to a discussion of the renormalisation group equations for the gauge and Yukawa couplings.

The renormalisation group equations for the gauge couplings $\tilde{\alpha}_i \equiv g_i^2/(4\pi)^2$ are

$$\frac{d\tilde{\alpha}_i}{dt} = -b_i \tilde{\alpha}_i^2 \tag{13}$$

where $b_i = -3 + n_V$, $1{+}n_V$, $11{+}n_V$, and 1197 for the group factors of $SU(3) \otimes SU(2) \otimes U(1) \otimes U(1)_X$ respectively. Here n_V is the number of vectorlike representations making up complete 5 dimensional representations of $SU(5)$; for the model just discussed $n_V = 10$. The reason the $U(1)_X$ factor is so large is mainly due to the fact that the colour and $SU(2)$ multiplicity factors are large. Note that none of the gauge couplings are asymptotically free due to the profusion of matter fields. We may see that the evolution of all the gauge couplings is fast, which is why the fixed point structure dominates the infra-red behaviour. Moreover, assuming rough equality at the compactification scale the relative size of $b_{U(1)''}$ ensures the Abelian coupling will be quite negligible at the gauge unification scale. This means that family violating effects coming from the family gauge interaction become negligible, an important feature for a viable model where small variations in the scalar mass spectrum can lead to large flavour changing neutral currents at low energies [16].

Turning to the renormalisation group equations for the Yukawa couplings we find they have (in the approximation of neglecting the small differences between the gauge couplings close to the unification scale) infra-red-stable fixed points (IRSFP) relating the Yukawa couplings to the gauge couplings. For the couplings of the quarks and leptons to the Higgs we find the maximally symmetric IRSFP solution

$$(h_2,\ k_2,\ l_2) \ = \ (1.79,\ 0.76,\ 0.73)\begin{pmatrix} 2 & 1 & 1 \\ 1 & 2 & 1 \\ 1 & 1 & 2 \end{pmatrix} \tag{14}$$

Note that the fixed points are relating lepton and quark couplings even though there is no underlying symmetry; another GUT type relation " without "Grand Unification"! It may appear that there are no free parameters left undetermined but this is not the case for what is determined is the modulus squared of the Yukawa couplings, their phase is not determined and will depend on the intial values i.e. they depend on the underlying (string) theory.

Before we can confront these forms with experiment we must determine the Yukawa couplings associated with the $SU(3)\otimes SU(2)\otimes U(1)$ singlet couplings for they in turn determine the composition of the light Higgs which, as discussed above, generates the quark and lepton masses. In fact the alert reader will have questioned why it was not necessary to include these when determining the IRSFP values for the Yukawa couplings just discussed. The answer to this is that the solution to the IRSFP for the quark and lepton Yukawa couplings requires the contribution to $N_I^{H_{1,2}}$ from these couplings to be independent of i. Thus the RG equations for the IRSFP factor into two parts, one determining the quark and lepton Yukawa couplings and one determining the Yukawa couplings involving the singlet fields. There are two IRSFP structures depending on the initial relative values of the Yukawa couplings relative to the gauge couplings. For large initial values the fixed point structure gives

$$\begin{aligned} (h_8,\ h_{-2},\ h_{-1}) &= \frac{1}{22}(4,\ 2,\ 1) \\ (s_3,\ s_{-2}) &= \frac{1}{44}(9,\ 4) \\ (\bar{s}_4,\ \bar{s}_0,\ \bar{s}_{-1}) &= \frac{1}{44}(7,\ 4,\ 4) \end{aligned} \tag{15}$$

where, as above, we have given the square of the couplings. The couplings h_4 and h_3 are driven smaller than these at the fixed point. Using these couplings we may determine the light Higgs eigenstates to be

$$\begin{aligned} H^1_{light} &= H^1_0 + H^1_{-1}\epsilon + \frac{1}{\sqrt{2}}H^1_{-2}\epsilon^2 + \sigma_{22}H^1_{1,2}\epsilon^3 \\ H^2_{light} &= H^2_0 + H^2_{-1}\bar{\epsilon} + \frac{1}{\sqrt{2}}H^2_{-2}\bar{\epsilon}^2 + \rho_{22}H^2_{1,2}\epsilon^3 \end{aligned} \tag{16}$$

where we have absorbed a factor of $\sqrt{2}$ in the original definition of ϵ, $\bar{\epsilon}$. Note that the IRSFP structure does not determine σ_{22} or ρ_{22} because we have not included light vectorlike states coupling $H_0^{1,2}$ to $H_3^{1,2}$. These terms must be driven by higher dimension operators or we must modify the Standard Model singlet Higgs sector to include some such coupling. While we have constructed schemes to implement this we will not pursue them here but concentrate on a discussion of the phenomenological implications of eq(16)as it stands. Using it in eqs(3), (4) and (6) determines the quark and lepton masses and mixing angles. Consider first the implications for quark and lepton masses. We find for the third generation at the scale M

$$\begin{aligned} \frac{m_b}{m_\tau} &= 1.02 \\ \frac{h_t}{h_b} &= 1.09 \\ h_t &= 1.34\, g \end{aligned} \tag{17}$$

After including radiative correctins the $b-\tau$ ratio is in good agreement with experiment. The initial value of the top Yukawa coupling implies the top mass at low energies will be close to the quasi fixed point $m_t \approx 195 Gev$ where the number follows since from eq(17) the bottom Yukawa coupling is large and $tan\beta \approx 40$.

For the lighter quarks and leptons we have the results

$$\begin{aligned} m_d m_s &= 1.04 m_\mu m_e \\ \frac{m_c}{m_t} &= \frac{1}{2}\sqrt{3-2\sqrt{2}cos\theta_u}\epsilon^2 \\ \frac{m_s}{m_b} &= \frac{1}{2}\sqrt{3-2\sqrt{2}cos\theta_d}\bar{\epsilon}^2 \\ \frac{m_\mu}{m_\tau} &= \frac{1}{2}\sqrt{3-2\sqrt{2}cos\theta_e}\bar{\epsilon}^2 \end{aligned} \tag{18}$$

where the angles $\theta_{u,d,e}$ come from the undetermined phases of the Yukawa couplings. After including the radiative corrections which increase the quark masses at low energies relative to the lepton masses by a factor approximately 3 the first relation is in excellent agreement with the experimental measurement. For arbitrary angles the last two relations imply $0.18 < m_\mu/m_\tau < 5.4$. Finally the CKM matrix elements are given by

$$\begin{aligned} \mid V_{us} \mid &= \mid V_{su} \mid = \left(\frac{m_d}{m_s} + \frac{m_u}{m_c} + 2\sqrt{\frac{m_d m_u}{m_s m_c}}\cos\phi\right)^{\frac{1}{2}} \\ \mid \frac{V_{ub}}{V_{cb}} \mid &= \sqrt{\frac{m_u}{m_c}} \\ \mid \frac{V_{bu}}{V_{ub}} \mid &= \sqrt{\frac{m_u m_s}{m_c m_d}} \\ \mid V_{cb} \mid &= \mid V_{cb} \mid = \left(\frac{m_s}{m_b'} + \frac{m_c}{m_t'} + 2\sqrt{\frac{m_s m_c}{m_b' m_t'}}\cos\phi'\right)^{\frac{1}{2}} \end{aligned} \tag{19}$$

where

$$\begin{aligned} m_b' &= m_b\sqrt{3-2\sqrt{2}cos\theta_d} \\ m_t' &= m_t\sqrt{3-2\sqrt{2}cos\theta_u} \end{aligned} \tag{20}$$

and the angles ϕ, ϕ' come from the undetermined phases. All but the last result of eq(19) comes entirely from the texture zero structure and are in excellent agreement with experiment (cf Table 3). The second result is in agreement with the experimental measurement only if there is a strong cancellation between the terms. This, together with the requirement that m_b/m_τ should be small, requires extremum values for the top and lepton phases, $\phi' \approx \pi$, $\theta_u \approx 0$, $\theta_e \approx \pi$. Further we need $\theta_d \approx \pi/4$ for maximum CP violation. Together this gives $\frac{m_\mu}{m_s} = 2.4$ at the scale M. Allowing for radiative corrections this gives a strange quark mass approximately $125 MeV$ and $\mid V_{cb} \mid = 0.05$

In summary we have examined the possibility that the Yukawa couplings determining the quark and lepton masses and mixing angles are given by symmetries beyond the

$\frac{m_d}{m_s}$	$\frac{m_u}{m_c}$	$\frac{m_s}{m_b}$	V_{cb}	$\frac{V_{ub}}{V_{cb}}$
0.04-0.067	0.003-0.005	0.03-0.07	0.02-0.05	0.05-0.13

Table 7: Experimental limits on quark masses and mixing angles

Standard Model. The idea was illustrated by a very simple extension of the Standard Model which had a single abelian family symmetry which immediately generated texture zeros in the correct positions and accounted for the hierarchy of quark and lepton masses and mixing angles. In addition the weak mixing angle is determined by the requirement of anomaly cancellation even without a stage of Grand Unification. Although the model still has unknown Yukawa couplings of $O(1)$ we found that many of these couplings were fixed by the IRSFP structure of the theory. Further the IRSFP structure increases the symmetry of the low energy effective theory leading to family independence in the squark and slepton mass matrices consistent with the phenomenological requirements. The resultant structure for the masses and mixing angles agrees well with the observed masses and mixing angles. This model demonstrates the possibility that the IRSFP structure is responsible for all the quark and lepton masses and mixing angles. If so the only relics of any underlying (string?) unification in the effective low-energy theory'may be the light multiplet structure, some of the symmetries of the theory and the gauge couplings; the Yukawa couplings themselves may be determined by the dynamics of the effective theory itself.

References

1 J.Ellis, S.Kelley, D.V.Nanopoulos,Phys.Lett.260B(1991)131;
U.Amaldi,W.deBoer,H.Furstenau,Phys.Lett.260B(1991)447;
P.Langacker,M.Luo,Univ. of Pennsylvania preprint,UPR-0466T(1991).

2 L. E. Ibanez, G. G. Ross, Phys. Lett., 110B (1982) 227
K. Inoue et al., Prog. Theor. Phys. 68, 927 (1982)
L. Alvarez-Gaume, J.Polchinski, M. Wise, Nucl. Phys. B221, 495 (1983)
J. Ellis, J. Hagelin, D.V. Nanopoulos, K. Tamvakis, Phys. Lett., B125, 275 (1983)
L. Ibanez, C. Lopez, Phys. Lett. B126, 54 (1983); Nucl. Phys. B233, 545 (1984)
C. Kounnas, A. Lahanas, D.V. Nanopoulos, M. Quiros, Phys. Lett. B132, 95 (1983)

3 G.G.Ross and R.G.Roberts, Nucl. Phys. B377(1992)571.
See also rapporteur talk presented at 1991 Joint International Lepton-Photon Symposium and EPS conference, Geneva, Switzerland, World Scientific, ed. S. Hegarty, K. Potter and E. Quercigh(1992)

4 A. Buras, J.Ellis, M.K.Gaillard and D.V.Nanopoulos, Nucl.Phys.B135(1978)66;
D.V.Nanopoulos and D.A.Ross, Nucl.Phys.B157(1970)273.

5 H.Arason, D.J.Castano, B.E.Keszthelyi, S.Mikaelian, E.J.Piard, P.Ramond and B.D.Wright, Phys.Rev.Lett.67(1991)2933.

6 R.G.Roberts and L. Roszkowski, Rutherford Appleton preprint RAL-93-003(1993)

7 for recent works see e.g. G. L. Kane, C. Kolda, L. Roszkowski and J. D. Wells, Phys. Rev. **D49** (1994) 49; M. Carena, M. Olechowski, S. Pokorski and C.E.M. Wagner, MPI preprints, MPI-Ph/93-66 and MPI-Ph/93-103; P. Langacker and N. Polonsky, Phys. Rev. **D49** (1994) 1454 and UPR-0594T (1994) preprint and references therein.

8 S. Weinberg, in " A Festschrift for I.I. Rabi" [Trans. N.Y. Acad. Sci., Ser. II (1977), v. 38], p. 185; F. Wilczek and A. Zee, Phys. Lett. B **70** (1977) 418.

9 H. Fritsch,Phys. Lett. B **70** (1977) 436; Phys. Lett. B **73** (1978) 317; F.J. Gilman and Y. Nir, Ann. Rev. Nucl. Part. Sci. **40** (1990) 213; P. Kaus and S. Meshkov, Mod. Phys. Lett. **A3**(1988) 1251.

10 P.Ramond, R.G.Roberts and G.G.Ross, Nuclear Physics B406(1993)19.

11 L.Ibanez, G.G.Ross, Oxford/Madrid preprint 1994.

12 For a review of string theories, see M. Green, J. Schwarz and E. Witten, Superstring Theory, Cambridge University Press, 1987.

13 L. Ibanez, Phys. Lett. B 303 (1993) 55.

14 M. Lanzagorta and G. G. Ross, CERN-TH.95-162

15 G.F.Giudice and A.Masiero, Phys. Lett. B206(1988)480.

16 M. Lanzagorta and G. G. Ross, CERN-TH.7522/94 and OUTP 9437P to be published in Phys. Lett. B.

17 G.G.Ross, CERN-TH.95-163

SOFT TERMS FROM DILATON/MODULI SECTORS

C. Muñoz

Departamento de Física Teórica C-XI, Universidad Autónoma de Madrid, Cantoblanco, 28049 Madrid, Spain

We study the structure of the soft SUSY-breaking terms obtained from large classes of 4-D Strings under the assumption of dilaton/moduli dominance in the process of SUSY-breaking. In particular, we first analyze in detail the dilaton-dominated limit because of its finiteness properties and phenomenological predictivity, and second, we consider the new features appearing when several moduli fields contribute to SUSY breaking. Although some qualitative features indeed change in the multimoduli case with respect to the dilaton dominance one, the most natural mass relations at low-energy $m_l < m_q \simeq M_g$ are still similar. We also study the presence of tachyons pointing out that their possible existence may be, in some cases, an interesting advantage in order to break extra gauge symmetries. Finally, we find that the mechanism for generating a "μ-term" by the Kähler potential, as naturally implemented in orbifolds, leads to the prediction $|tg\beta| = 1$ at the String scale, independently of the Goldstino direction. In this connection, it is worth noticing that in the dilaton-dominated case we obtain the remarkable result that the whole SUSY spectrum is completely determined with no free parameters.

1 Introduction

Recently there has been some activity in trying to obtain information about the structure of soft Supersymmetry (SUSY)-breaking terms in effective $N = 1$ theories coming from four-dimensional Strings. The basic idea is to identify some $N = 1$ chiral fields whose auxiliary components could break SUSY by acquiring a vacuum expectation value (vev). No special assumption is made about the possible origin of SUSY-breaking. Natural candidates in four-dimensional Strings are 1) the complex dilaton field $S = \frac{1}{g^2} + ia$ which is present in any four-dimensional String and 2) the moduli fields T^i, U^i which parametrize the size and shape of the compactified variety in models obtained by compactification of a ten-dimensional heterotic String.

The important point in this assumption of locating the seed of SUSY-breaking in the dilaton/moduli sectors, is that it leads to some interesting relationships among different soft terms which could perhaps be experimentally tested.

In ref.[1] a systematic discussion of the structure of soft terms which may be obtained under the assumption of dilaton/moduli dominated SUSY breaking in some classes of four-dimensional Strings was presented, with particular emphasis on the case of Abelian $(0,2)$ orbifold models. It was mostly considered a situation in which only the dilaton S and an "overall modulus T" field contribute to SUSY-breaking. In fact, actual four-dimensional Strings like orbifolds contain several T_i and U_i moduli. Thus it is natural to ask what changes if one relaxes the overall modulus hypothesis and works with the multimoduli case [2]. This is one of the purposes of the present talk. The second one is to analyze in more detail the dilaton-dominated limit, where only the dilaton field contributes to SUSY breaking. This is a very interesting possibility not only due to phenomenological reasons, as universality of the soft terms, but also to theoretical arguments. In this connection it is worth noticing that the boundary conditions $-A = M_{1/2} = \sqrt{3}m$ of dilaton dominance coincide with some boundary conditions considered by Jones, Mezincescu and Yau in 1984 [3]. They found that those same boundary conditions mantain the (two-loop) finiteness properties of certain $N = 1$ SUSY theories. This could perhaps be an indication that at least some of the possible soft terms obtained in the present scheme could have a more general relevance.

In section 2 we present an analysis of the effects of several moduli on the results obtained for soft terms. In the multimoduli case several parameters are needed to specify the Goldstino direction in the dilaton/moduli space, in contrast with the overall modulus case where the relevant information is contained in just one angular parameter θ. The presence of more free parameters leads to some loss of predictivity for the soft terms. This predictivity is recovered and increased in the case of dilaton dominance, where the soft terms eq.(6) are independent of the 4-D String considered and fulfil the low-energy mass relations given by eq.(7). Also we show that, even in the multimoduli case, in some schemes there are certain sum-rules among soft terms eq.(11) which hold independently of the Goldstino direction. The presence of these sum rules cause that, *on average* the *qualitative* results in the dilaton-dominated case still apply. Specifically, if one insists e.g. in obtaining scalar masses heavier than gauginos (something not possible in the dilaton-dominated scenario), this is possible in the multimoduli case, but the sum-rules often force some of the scalars to get negative squared mass. If we want to avoid this, we have to stick to gaugino masses bigger than (or of order) the scalar masses. This would lead us back to the qualitative results obtained in dilaton dominance. In the case of standard model 4-D Strings this tachyonic behaviour may be particularly problematic, since charge and/or colour could be broken. In the case of GUTs constructed from Strings, it may just be the signal of GUT symmetry breaking. However, even in this case one expects the same order of magnitude results for observable scalar and gaugino masses and hence the most natural mass relations *at low-energy* are still similar to the dilaton dominance ones.

Another topic of interest is the B-parameter, the soft mass term which is associated to a SUSY mass term $\mu H_1 H_2$ for the pair of Higgsses $H_{1,2}$ in the Minimal Supersymmetric Standard Model (MSSM). Compared to the other soft terms, the result for the B-parameter is more model-dependent. Indeed, it depends not only on the dilaton/moduli dominance assumption but also on the particular mechanism which could generate the associated "μ-term" [4]. An interesting possibility to generate such a term is the one suggested in ref.[5] in which it was pointed out that in the presence of certain bilinear terms in the Kähler potential an effective μ-term of order the gravitino mass, $m_{3/2}$, is naturally generated. Interestingly enough, such bilinear terms in the Kähler potential do appear in String models and particularly in Abelian orbifolds. In section 3 we compute the μ and B parameters[a] as well as the soft scalar masses of the charged fields which could play the role of Higgs particles in such Abelian orbifold schemes. We find the interesting result that, *independently of the Goldstino direction* in the dilaton/moduli space, one gets the prediction $|tg\beta| = 1$ at the String scale. On the other hand, if we consider the interesting Goldstino direction where only the dilaton breaks SUSY, the whole soft terms and the μ-parameter depend only on the gravitino mass. Imposing the phenomenological requirement of correct electroweak breaking we arrive to the remarkable result that the whole SUSY spectrum is completely determined with no free parameters.

2 Soft terms

2.1 General structure of soft terms: the multimoduli case

We are going to consider $N = 1$ SUSY 4-D Strings with m moduli T_i, $i = 1, .., m$. Such notation refers to both T-type and U-type (Kähler class and complex structure in the Calabi-Yau language) fields. In addition there will be charged matter fields C_α and the complex dilaton field S. In general we will be considering $(0,2)$ compactifications and thus the charged fields do not need to correspond to 27s of E_6.

Before further specifying the class of theories that we are going to consider a comment about the total number of moduli is in order. We are used to think of large numbers of T and U-like moduli due to the fact that in $(2,2)$ (E_6) compactifications there is a one to one correspondence between moduli and charged fields. However, in the case of $(0,2)$ models with arbitrary gauge

[a] The results for B corresponding to other sources for the μ-term can also be found, for the multimoduli case under consideration, in ref.[2]. In particular, the possibility of generating a small μ-term from the superpotential [6] was studied.

group (which is the case of phenomenological interest) the number of moduli is drastically reduced. For example, in the standard $(2,2)$ Z_3 orbifold there are 36 moduli T_i, 9 associated to the untwisted sector and 27 to the fixed points of the orbifold. In the thousands of $(0,2)$ Z_3 orbifolds one can construct by adding different gauge backgrounds or doing different gauge embeddings, only the 9 untwisted moduli remain in the spectrum. The same applies to models with U-fields. This is also the case for compactifications using $(2,2)$ minimal superconformal models. Here all singlets associated to twisted sectors are projected out when proceeding to $(0,2)$. So, as these examples show, in the case of $(0,2)$ compactifications the number of moduli is drastically reduced to a few fields. In the case of generic Abelian orbifolds one is in fact left with only three T-type moduli T_i $(i = 1,2,3)$, the only exceptions being Z_3, Z_4 and Z_6', where such number is 9, 5 and 5 respectively. The number of U-type fields in these $(0,2)$ orbifolds oscillates between 0 and 3, depending on the specific example. Specifically, $(0,2)$ $Z_2 \times Z_2$ orbifolds have 3 U fields, the orbifolds of type $Z_4, Z_6, Z_8, Z_2 \times Z_4, Z_2 \times Z_6$ and Z_{12}' have just one U field and the rest have no untwisted U-fields. Thus, apart from the three exceptions mentioned above, this class of models has at most 6 moduli, three of T-type (always present) and at most three of U-type. In the case of models obtained from Calabi-Yau type of compactifications a similar effect is expected and only one T-field associated to the overall modulus is guaranteed to exist in $(0,2)$ models.

We will consider effective $N = 1$ supergravity (SUGRA) Kähler potentials of the type:

$$\begin{aligned} K(S,S^*,T_i,T_i^*,C_\alpha,C_\alpha^*) &= -\log(S+S^*) + \hat{K}(T_i,T_i^*) \\ &+ \tilde{K}_{\overline{\alpha}\beta}(T_i,T_i^*)C^{*\overline{\alpha}}C^\beta + (Z_{\alpha\beta}(T_i,T_i^*)C^\alpha C^\beta + h.c.) . \end{aligned} \tag{1}$$

The first piece is the usual term corresponding to the complex dilaton S which is present for any compactification whereas the second is the Kähler potential of the moduli fields, where we recall that we are denoting the T- and U-type moduli collectively by T_i. The greek indices label the matter fields and their kinetic term functions are given by $\tilde{K}_{\overline{\alpha}\beta}$ and $Z_{\alpha\beta}$ to lowest order in the matter fields. The last piece is often forbidden by gauge invariance in specific models although it may be relevant in some cases as discussed in section 3. In this section we are going to consider the case of diagonal metric both for the moduli and the matter fields.[b] Then $\hat{K}(T_i,T_i^*)$ will be a sum of contributions (one for each T_i), whereas $\tilde{K}_{\overline{\alpha}\beta}$ will be taken of the diagonal form $\tilde{K}_{\overline{\alpha}\beta} \equiv \delta_{\overline{\alpha}\beta}\tilde{K}_\alpha$. The complete $N = 1$ SUGRA Lagrangian is determined by the

[b] An extensive analysis of the off-diagonal case, including the calculation of the soft terms and their effects on flavour changing neutral currents (FCNC), can be found in ref.[2].

Kähler potential $K(\phi_M, \phi_M^*)$, the superpotential $W(\phi_M)$ and the gauge kinetic functions $f_a(\phi_M)$, where ϕ_M generically denotes the chiral fields S, T_i, C_α. As is well known, K and W appear in the Lagrangian only in the combination $G = K + \log|W|^2$. In particular, the (F-part of the) scalar potential is given by

$$V(\phi_M, \phi_M^*) \ = \ e^G \left(G_M K^{M\bar{N}} G_{\bar{N}} \ - \ 3 \right) \ , \tag{2}$$

where $G_M \equiv \partial_M G \equiv \partial G/\partial\phi_M$ and $K^{M\bar{N}}$ is the inverse of the Kähler metric $K_{\bar{N}M} \equiv \partial_{\bar{N}}\partial_M K$.

The crucial assumption now is to locate the origin of SUSY-breaking in the dilaton/moduli sector. Let us take the following parametrization for the vev's of the dilaton and moduli auxiliary fields $F^S = e^{G/2} G_{\bar{S}S}^{-1} G_S$ and $F^i = e^{G/2} G_{\bar{i}i}^{-1} G_i$:

$$G_{\bar{S}S}^{1/2} F^S \ = \ \sqrt{3} m_{3/2} \sin\theta e^{i\gamma_S} \ \ ; \ \ G_{\bar{i}i}^{1/2} F^i \ = \ \sqrt{3} m_{3/2} \cos\theta \ e^{i\gamma_i} \Theta_i \ , \tag{3}$$

where $\sum_i \Theta_i^2 = 1$ and $e^G = m_{3/2}^2$ is the gravitino mass-squared. The angle θ and the Θ_i just parametrize the direction of the goldstino in the S, T_i field space. We have also allowed for the possibility of some complex phases γ_S, γ_i which could be relevant for the CP structure of the theory. This parametrization has the virtue that when we plug it in the general form of the SUGRA scalar potential eq.(2), its vev (the cosmological constant) vanishes by construction. Notice that such a phenomenological approach allows us to 'reabsorb' (or circumvent) our ignorance about the (nonperturbative) S- and T_i- dependent part of the superpotential, which is responsible for SUSY-breaking. It is now a straightforward exercise to compute the bosonic soft SUSY-breaking terms in this class of theories. Plugging eqs.(3) and (1) into eq.(2) one finds the following results (we recall that we are considering here a diagonal metric for the matter fields):

$$m_\alpha^2 = \ m_{3/2}^2 \ \left[1 \ - \ 3\cos^2\theta \ (\hat{K}_{\bar{i}i})^{-1/2} \Theta_i e^{i\gamma_i} (\log \tilde{K}_\alpha)_{\bar{i}j} (\hat{K}_{\bar{j}j})^{-1/2} \Theta_j e^{-i\gamma_j} \right] \ ,$$

$$A_{\alpha\beta\gamma} = \ -\sqrt{3} m_{3/2} \ \left[e^{-i\gamma_S} \sin\theta \right.$$

$$\left. - \ e^{-i\gamma_i} \cos\theta \ \Theta_i (\hat{K}_{\bar{i}i})^{-1/2} \ \left(\hat{K}_i - \textstyle\sum_{\delta=\alpha,\beta,\gamma} (\log \tilde{K}_\delta)_i + (\log h_{\alpha\beta\gamma})_i \right) \right] \ . \tag{4}$$

The above scalar masses and trilinear scalar couplings correspond to charged fields which have already been canonically normalized. Here $h_{\alpha\beta\gamma}$ is a renormalizable Yukawa coupling involving three charged chiral fields and $A_{\alpha\beta\gamma}$ is its corresponding trilinear soft term.

Physical gaugino masses M_a for the canonically normalized gaugino fields are given by $M_a = \frac{1}{2}(Ref_a)^{-1} e^{G/2} f_{a_M} K^{M\bar{N}} G_{\bar{N}}$. Since the tree-level gauge kinetic function is given for any 4-D String by $f_a = k_a S$, where k_a is the Kac-Moody level of the gauge factor, the result for tree-level gaugino masses is independent of the moduli sector and is simply given by:

$$M \equiv M_a = m_{3/2}\sqrt{3}\sin\theta e^{-i\gamma_S} . \tag{5}$$

The soft term formulae above are in general valid for any compactification as long we are considering diagonal metrics. In addition one is tacitally assuming that the tree-level Kähler potential and f_a-functions constitute a good aproximation. The Kähler potentials for the moduli are in general complicated functions. Before going into specific classes of Superstring models, it is worth studying the interesting limit $\cos\theta = 0$, corresponding to the case where the dilaton sector is the source of all the SUSY-breaking (see eq.(3)).

2.2 The $\cos\theta = 0$ *(dilaton-dominated) limit*

Since the dilaton couples in an universal manner to all particles, *this limit is quite model independent.* Using eqs.(4,5) and neglecting phases one finds the following simple expressions for the soft terms which are independent of the 4-D String considered

$$\begin{aligned} m_\alpha &= m_{3/2} , \\ A_{\alpha\beta\gamma} &= -\sqrt{3} m_{3/2} , \\ M_a &= \sqrt{3} m_{3/2} . \end{aligned} \tag{6}$$

This dilaton-dominated scenario [7,1] is attractive for its simplicity and for the natural explanation that it offers to the universality of the soft terms. Actually, universality is a desirable property not only to reduce the number of independent parameters in the MSSM, but also for phenomenological reasons, particularly to avoid FCNC.

Because of the simplicity of this scenario, the low-energy predictions are quite precise. Since scalars are lighter than gauginos at the String scale, at low-energy ($\sim M_Z$), gluino, slepton and (first and second generation) squark mass relations turn out to be [1,8]

$$M_g : m_Q : m_u : m_d : m_L : M_e \simeq 1 : 0.94 : 0.92 : 0.92 : 0.32 : 0.24 . \tag{7}$$

Although squarks and sleptons have the same soft mass, at low-energy the former are much heavier than the latter because of the gluino contribution to the renormalization of their masses.

In section 3 we will show that even a stronger result than that of eq.(7) is obtained in the context of a natural mechanism for solving the μ problem, namely the whole SUSY spectrum (gluino, squarks, sleptons, Higgses, charginos, neutralinos) is completely determined with no free parameters.

2.3 Orbifold compactifications

To illustrate some general features of the multimoduli case we will concentrate here on the case of generic (0,2) symmetric Abelian orbifolds. As we mentioned above, this class of models contains three T-type moduli and (at most) three U-type moduli. We will denote them collectively by T_i, where e.g. $T_i = U_{i-3}$; $i = 4,5,6$. For this class of models the Kähler potential has the form

$$K(\phi,\phi^*) \;=\; -\log(S+S^*) - \sum_i \log(T_i+T_i^*) + \sum_\alpha |C_\alpha|^2 \Pi_i (T_i+T_i^*)^{n_\alpha^i} \;. \quad (8)$$

Here n_α^i are fractional numbers usually called "modular weights" of the matter fields C_α. For each given Abelian orbifold, independently of the gauge group or particle content, the possible values of the modular weights are very restricted. For a classification of modular weights for all Abelian orbifolds see ref.[2]. Using the particular form (8) of the Kähler potential and eqs.(4,5) we obtain the following results[c] for the scalar masses, gaugino masses and soft trilinear couplings:

$$\begin{aligned} m_\alpha^2 &= m_{3/2}^2 (1 \;+\; 3\cos^2\theta \; \vec{n_\alpha}.\vec{\Theta^2}) \;, \\ M &= \sqrt{3} m_{3/2} \sin\theta e^{-i\gamma_S} \;, \\ A_{\alpha\beta\gamma} &= -\sqrt{3} m_{3/2} \left(\sin\theta e^{-i\gamma_S} \;+\; \cos\theta \textstyle\sum_{i=1}^6 e^{-i\gamma_i} \Theta^i \omega_{\alpha\beta\gamma}^i\right) \;, \end{aligned} \quad (9)$$

where we have defined :

$$\omega_{\alpha\beta\gamma}^i \;=\; (1 + n_\alpha^i + n_\beta^i + n_\gamma^i - Y_{\alpha\beta\gamma}^i) \;\; ; \;\; Y_{\alpha\beta\gamma}^i \;=\; \frac{h_{\alpha\beta\gamma}^i}{h_{\alpha\beta\gamma}} 2ReT_i \;. \quad (10)$$

Notice that neither the scalar nor the gaugino masses have any explicit dependence on S or T_i, they only depend on the gravitino mass and the goldstino angles. This is one of the advantages of a parametrization in terms of such angles. Although in the case of the A-parameter an explicit T_i-dependence may

---

[c]This analysis was also carried out, for the particular case of the three diagonal moduli T_i, in ref.[9] in order to obtain unification of gauge coupling constants. Some particular multimoduli examples were also considered in ref.[10].

appear in the term proportional to $Y^i_{\alpha\beta\gamma}$, it disappears in several interesting cases [2].

With the above information we can now analyze the different structure of soft terms available for each Abelian orbifold.

1) Universality of soft terms

In the dilaton-dominated case ($\cos\theta = 0$) the whole soft terms are universal. However, in general, they show a lack of universality due to the modular weight dependence (see eqs.(9,10)).

2) Soft masses

In the multimoduli case, depending on the goldstino direction, tachyons may appear. For $\cos^2\theta \geq 1/3$, one has to be very careful with the goldstino direction if one is interested in avoiding tachyons. Nevertheless, as we will discuss below, having a tachyonic sector is not necessarily a problem, it may even be an advantage, so one should not disregard this possibility at this point.

Consider now three particles C_α,C_β,C_γ coupling through a Yukawa $h_{\alpha\beta\gamma}$. They may belong both to the untwisted (**U**) sector or to a twisted (**T**) sector, i.e. couplings of the type **UUU**, **UTT**, **TTT**. Then, using the above formulae, one finds [2] that in general for *any choice* of goldstino direction

$$m_\alpha^2 + m_\beta^2 + m_\gamma^2 \leq |M|^2 = 3m_{3/2}^2 \sin^2\theta \tag{11}$$

Notice that if we insist in having a vanishing gaugino mass, the sum-rule (11) forces the scalars to be either all massless or at least one of them tachyonic. Nevertheless we should not forget that tachyons, as we already mentioned above, are not necessarily a problem, but may just show us an instability.

3) Gaugino versus scalar masses

In the multimoduli case on average the scalars are lighter than gauginos but there may be scalars with mass bigger than gauginos. Eq.(11) tells us that this can only be true at the cost of having some of the other three scalars with *negative* squared mass. This may have diverse phenomenological implications depending what is the particle content of the model, as we now explain in some detail:

3-a) Gaugino versus scalar masses in standard model 4-D Strings

Let us suppose we insist in e.g., having tree-level gaugino masses lighter than the scalar masses. If we are dealing with a String model with gauge group $SU(3)_c \times SU(2)_L \times U(1)_Y \times G$ this is potentially a disaster. Some observable particles, like Higgses, squarks or sleptons would be forced to acquire large vev's (of order the String scale). For example, the scalars associated through the Yukawa coupling $H_2 Q_L u_L^c$, which generates the mass of the u-quark, must fulfil the above sum-rule (11). If we allow e.g. the scalars H_2, Q_L to be heavier than gauginos, then u_L^c will become tachyonic breaking charge and color. However, tachyons may be helpful if the particular Yukawa coupling does not involve observable particles. They could break extra gauge symmetries and generate large masses for extra particles. We recall that standard-like models in Strings usually have too many extra particles and many extra U(1) interactions. Although the Fayet-Iliopoulos mechanism helps to cure the problem [11], the existence of tachyons is a complementary solution.

We thus see that, for standard model Strings, if we want to avoid charge and colour-breaking minima (or vev's of order the String scale for the Higgses), we should grosso modo come back to a situation with gauginos heavier than scalars. Thus the low-energy phenomenological predictions of the multimoduli case are similar to those of the dilaton-dominated scenario (see subsect.2.2): due to the sum-rule the tree-level observable scalars are always ligther than gauginos

$$m_\alpha < M \ . \tag{12}$$

Now, at low-energy ($\sim M_Z$), gluino, slepton and (first and second generation) squark mass relations turn out to be

$$m_l < m_q \simeq M_g \ , \tag{13}$$

where gluinos are slightly heavier than scalars. This result is qualitatively similar to the dilaton dominance one, in spite of the different set of (non-universal) soft scalar masses, because the low-energy scalar masses are mainly determined by the gaugino loop contributions. The only exception are the sleptons masses, which do not feel the important gluino contribution, and therefore can get some deviation from the result of eq.(7).

Although String loop corrections, in the particular case that at tree-level $M \to 0$ and $m_\alpha \to 0$, can yield scalars heavier than gauginos [1], it was shown in ref.[2] that this possibility is a sort of fine-tuning. Non-renormalizable Yukawa couplings can also yield scalars heavier than gauginos. However it was shown in ref.[2] that still at low-energy eq.(13) is typically valid, the only difference being that now squarks will be slightly heavier than gluinos.

3-b) Gaugino versus scalar masses in GUT 4-D Strings

What it turned out to be a potential disaster in the case of standard model Strings may be an interesting advantage in the case of String-GUTs. In this case it could well be that the negative squared mass may just induce gauge symmetry breaking by forcing a vev for a particular scalar (GUT-Higgs field) in the model. The latter possibility provides us with interesting phenomenological consequences. Here the breaking of SUSY would directly induce further gauge symmetry breaking. An explicit example of this situation can be found in ref.[2].

Let us finally remark that, in spite of the different possibilities of soft masses in the multimoduli case, the most natural (slepton-squark-gluino) mass relations *at low-energy* will be similar to the ones of the dilaton-dominated case eq.(13) as showed in point *3-a*.

3 The B-parameter and the μ problem

It was pointed out in ref.[5] that terms in a Kähler potential like the one proportional to $Z_{\alpha\beta}$ in eq.(1) can naturally induce a μ-term for the C_α fields of order $m_{3/2}$ after SUSY-breaking, thus providing a rationale for the size of μ. Recently it has been suggested that such type of terms may appear in the Kähler potential of some Calabi-Yau type compactifications [7]. It has also been explicitly shown [12] that the untwisted sector of orbifolds with at least one complex-structure field U possesses the required structure $Z(T_i, T_i^*)C_1C_2 + h.c.$ in their Kähler potentials. Specifically, the Z_N orbifolds based on Z_4, Z_6, Z_8, Z'_{12} and the $Z_N \times Z_M$ orbifolds based on $Z_2 \times Z_4$ and $Z_2 \times Z_6$ do all have a U-type field in (say) the third complex plane. In addition the $Z_2 \times Z_2$ orbifold has U fields in the three complex planes. In all these models the piece of the Kähler potential involving the moduli and the untwisted matter fields $C_{1,2}$ in the third complex plane has the form

$$K = -\log(T_3 + T_3^*) - \log(U_3 + U_3^*) + \frac{(C_1 + C_2^*)(C_1^* + C_2)}{(T_3 + T_3^*)(U_3 + U_3^*)} \tag{14}$$

where one can easily identify the functions $Z, \tilde{K}_1, \tilde{K}_2$ associated to C_1 and C_2:

$$Z = \tilde{K}_1 = \tilde{K}_2 = \frac{1}{(T_3 + T_3^*)(U_3 + U_3^*)} . \tag{15}$$

Plugging back these expressions in eqs.(2,3) one can compute μ and B for this interesting class of models [2]:

$$\mu = m_{3/2} \left(1 + \sqrt{3}\cos\theta(e^{i\gamma_3}\Theta_3 + e^{i\gamma_6}\Theta_6)\right) , \tag{16}$$

$$\begin{aligned} B\mu = 2m_{3/2}^2 \big(1 &+ \sqrt{3}\cos\theta(\cos\gamma_3\Theta_3 + \cos\gamma_6\Theta_6) \\ &+ 3\cos^2\theta\cos(\gamma_3 - \gamma_6)\Theta_3\Theta_6\big) . \end{aligned} \tag{17}$$

In addition, we recall from eq.(9) that the soft masses are

$$m_{C_1}^2 = m_{C_2}^2 = m_{3/2}^2 \left(1 - 3\cos^2\theta(\Theta_3^2 + \Theta_6^2)\right) . \tag{18}$$

In general, the dimension-two scalar potential for $C_{1,2}$ after SUSY-breaking has the form

$$V_2(C_1, C_2) = (m_{C_1}^2 + |\mu|^2)|C_1|^2 + (m_{C_2}^2 + |\mu|^2)|C_2|^2 + (B\mu C_1 C_2 + h.c.) \tag{19}$$

In the specific case under consideration, from eqs.(16,17,18) we find the remarkable result that the three coefficients in $V_2(C_1, C_2)$ are equal, i.e.

$$m_{C_1}^2 + |\mu|^2 = m_{C_2}^2 + |\mu|^2 = B\mu \tag{20}$$

Although the common value of the three coefficients in eq.(20) depends on the Goldstino direction via the parameters $\cos\theta$, Θ_3, Θ_6,... (see expression of $B\mu$ in eq.(17)), we stress that the equality itself holds *independently of the Goldstino direction.*

It is well known that, for a potential of the generic form (19) (+D-terms), the minimization conditions yield

$$\sin 2\beta = \frac{-2B\mu}{m_{C_1}^2 + m_{C_2}^2 + 2|\mu|^2} \; . \tag{21}$$

In particular, this relation embodies the boundedness requirement: if the absolute value of the right-hand side becomes bigger than one, this would indicate that the potential becomes unbounded from below. As we have seen, in the class of models under consideration the particular expressions of the mass parameters lead to the equality (20), which in turns implies $\sin 2\beta = -1$. Thus one finds $\tan\beta = < C_2 > / < C_1 > = -1$ *for any value of* $\cos\theta$,Θ_3,Θ_6 (and of the other Θ_i's of course), i.e. for any Goldstino direction.

As an additional comment, it is worth recalling that in previous analyses of the above mechanism for generating μ and B in the String context [7,1,8] the value of μ was left as a free parameter since one did not have an explicit expression for the function Z. However, if the explicit orbifold formulae for Z are used, one is able to predict both [2] μ and B reaching the above conclusion.[d]

Now that we have computed explicitly the whole soft terms and the μ parameter, it would be interesting to analyze the dilaton-dominated scenario ($\cos\theta = 0$) because of its predictivity. In particular, from eqs.(6,16,17) we obtain

$$\begin{aligned} m_\alpha &= m_{3/2} \; , \\ A_{\alpha\beta\gamma} &= -\sqrt{3} m_{3/2} \; , \\ M_a &= \sqrt{3} m_{3/2} \; , \\ B &= 2m_{3/2} \; , \\ \mu &= m_{3/2} \; . \end{aligned} \tag{22}$$

[d]We should add that situations are conceivable where the above result may be evaded, for example if the physical Higgs doublets are a mixture of the above fields with some other doublets coming from other sectors (e.g. twisted) of the theory.

and therefore the whole SUSY spectrum depends only on one parameter ($m_{3/2}$). If we would know the particular mechanism which breaks SUSY, then we would be able of computing the superpotential and hence $m_{3/2} = e^K|W|$. However, still this parameter can be fixed from the phenomenological requirement of correct electroweak breaking $2M_W/g_2^2 =< H_1 >^2 + < H_2 >^2$. Thus at the end of the day we are left with no free parameters. The whole SUSY spectrum is completely determined in this scenario. This is a remarkable result which deserves further investigation [13]. Of course, if in the next future the mechanism which breaks SUSY is known (i.e. $m_{3/2}$ can be explicitly calculated) and the above scenario is the correct one, the value of $m_{3/2}$ should coincide with the one obtained from the phenomenological constraint.

It is worth noticing here that although the value of μ is compactification dependent even in this dilaton-dominated scenario $\mu = m_{3/2}(\tilde{K}_1\tilde{K}_2)^{-1/2}Z$, the result of eq.(22), $\mu = m_{3/2}$, will be obtained in any compactification scheme with the following property: $\tilde{K}_1 = \tilde{K}_2 = Z$. Of course, this is the case of orbifolds as was shown in eq.(15).

References

1. A. Brignole, L.E. Ibáñez and C. Muñoz, *Nucl. Phys.* **B422** (1994) 125 [Erratum: **B436** (1995) 747].
2. A. Brignole, L.E. Ibáñez and C. Muñoz and C. Scheich, *FTUAM* **95/26**, *hep-ph/* **9508258**.
3. D.R.T. Jones, L. Mezincescu and Y.P. Yao, *Phys. Lett.* **B148** (1984) 317.
4. For a recent review, see: C. Muñoz, *FTUAM* **95/20**, *hep-th/* **9507108**.
5. G.F. Giudice and A. Masiero, *Phys. Lett.* **B206** (1988) 480.
6. J.A. Casas and C. Muñoz, *Phys. Lett.* **B306** (1993) 288.
7. V.S. Kaplunovsky and J. Louis *Phys. Lett.* **B306** (1993) 269.
8. R. Barbieri, J. Louis and M. Moretti, *Phys. Lett.* **B312** (1993) 451.
9. T. Kobayashi, D. Suematsu, K. Yamada and Y. Yamagishi, *Phys. Lett.* **B348** (1995) 402.
10. S. Ferrara, C. Kounnas and F. Zwirner, *Nucl. Phys.* **B429** (1994) 589 [Erratum: **B433** (1995) 255].
11. For a review, see e.g.: J.A. Casas and C. Muñoz *Nucl. Phys. (Proc. Suppl.)* **B16** (1990) 624, and references therein.
12. G. Lopes-Cardoso, D. Lüst and T. Mohaupt, *Nucl. Phys.* **B432** (1994) 68; I. Antoniadis, E. Gava, K.S. Narain and T.R. Taylor, *Nucl. Phys.* **B432** (1994) 187.
13. A. Brignole, L.E. Ibañez and C. Muñoz, to appear.

Low-Energy Supersymmetry: Prospects and Challenges

Howard E. Haber*

CERN, TH Division, CH-1211 Geneva 23, Switzerland

Abstract

An introduction to the minimal supersymmetric Standard Model (MSSM) is given. The motivation for "low-energy" supersymmetry is reviewed, and the structure of the MSSM is outlined. In its most general form, the MSSM can be viewed as a low-energy effective theory parametrized by a set of arbitrary soft-supersymmetry-breaking parameters. A variety of techniques for reducing the parameter freedom of the MSSM are surveyed. The search for supersymmetry below and above the threshold for supersymmetric particle production presents a challenging task for experimentalists at present and future colliders.

1 Motivation for Low-Energy Supersymmetry

The Standard Model of particle physics provides an extremely successful description of all particle physics phenomena accessible to present day accelerators. No unambiguous experimental deviation from the Standard Model have yet been confirmed. However, theorists strongly believe that the success of the Standard Model will not persist to higher energy scales. This belief arises from attempts to embed the Standard Model in a more fundamental theory. We know that the Standard Model cannot be the ultimate theory, valid to arbitrarily high energy scales. Even in the absence of grand unification of strong and electroweak forces at a very high energy scale, it is clear that the Standard Model must be modified to incorporate the effects of gravity at the Planck scale ($M_P \simeq 10^{19}$ GeV). In this context, it is a mystery why the ratio $m_W/M_P \simeq 10^{-17}$ is so small. This is called the hierarchy problem [1–3]. Moreover, in the Standard Model, the scale of the electroweak interactions derives from an elementary scalar field which acquires a vacuum expectation value of $v = 2m_W/g = 246$ GeV. However, if one couples a theory of scalar particles to new physics at some arbitrarily high scale Λ, radiative corrections to the scalar squared-mass are of $\mathcal{O}(\Lambda^2)$, due to

*Permanent address: Santa Cruz Institute for Particle Physics, University of California, Santa Cruz, CA 94064 USA.

the quadratic divergence in the scalar self-energy (which indicates quadratic sensitivity to the largest energy scale, Λ, in the theory). Thus, the "natural" mass for any scalar particle is Λ (which is presumably equal to M_P). Of course, in order to have a successful electroweak theory, the Higgs mass must be of order the electroweak scale. The fact that the Higgs mass must not be equal to its natural value of M_P is called the "naturalness" problem [6]. That is, the bare Higgs squared-mass and the squared-mass shift arising from the radiative corrections are both expected to take on their natural values of $\mathcal{O}(M_P^2)$. To end up with the physical Higgs squared-mass of $\mathcal{O}(m_W^2)$, which is 34 orders of magnitude smaller than M_P^2, requires a miraculous cancelation (or "fine-tuning") among the parameters of the fundamental theory.

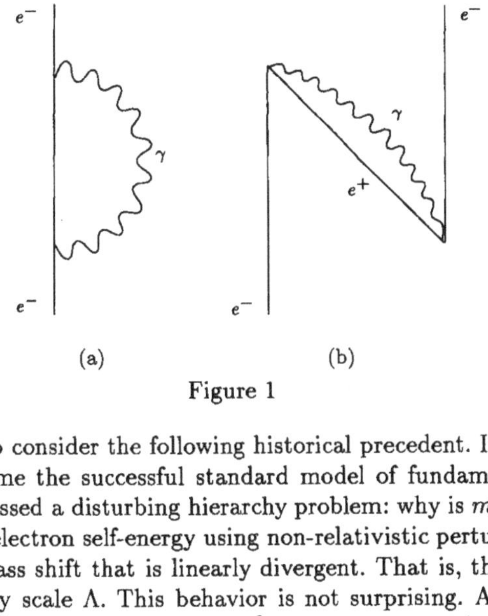

Figure 1

It is instructive to consider the following historical precedent. In the 1920's, quantum mechanics became the successful standard model of fundamental physics. But, this theory also possessed a disturbing hierarchy problem: why is $m_e/M_P \simeq \mathcal{O}(10^{-22})$? A calculation of the electron self-energy using non-relativistic perturbation theory [see Fig. 1(a)] yields a mass shift that is linearly divergent. That is, the natural value for m_e is the high energy scale Λ. This behavior is not surprising. After all, classically, the self-energy of an electron of radius r is e^2/r which diverges linearly as $1/r \to \infty$. The linear divergence persists in the relativistic single-electron quantum theory. How is this naturalness problem solved in quantum electrodynamics? The solution is remarkable. Invent a new symmetry called charge conjugation invariance (C). Now, double the known particle spectrum: for every particle, introduce a partner called an "antiparticle". The C symmetry guarantees that the antiparticle has the same mass and interaction strength as its partner. Now, let us reconsider the perturbation theory computation of the electron self-energy. Now, there is a second diagram to consider [see Fig. 1(b)], in which $e^+e^-\gamma$ is created from the vacuum, the $e^+\gamma$ annihilates the incoming electron, while the e^- just created continues to propagate. In old-fashioned time ordered perturbation theory, both time orderings must be included as shown in Fig. 1.

Due to the C symmetry, the leading linear divergence cancels between the two graphs, leaving a logarithmic divergence. The mass shift of the electron is thus proportional to $e^2 m_e \ln \Lambda$. The naturalness problem is solved, since even for $\Lambda = M_P$, the radiative correction to the electron mass is of the same order as m_e. Of course, antiparticles were not invented to solve the naturalness problem of the single-electron quantum theory. Nevertheless, the cancelation of the linear divergence in Dirac's theory of electrons and positrons, which was discovered by Weisskopf in 1934 [4], was regarded as an important advance in the development of quantum electrodynamics [5].

To solve the naturalness problem of electroweak theory, we mimic the steps just outlined. In this case, we "invent" a new symmetry called supersymmetry, which transforms fermions into bosons and vice versa. Next, we double the particle spectrum: for each particle we introduce a superpartner which differs in spin by half a unit. As in Dirac's theory of electrons and positrons, the quadratic divergence of the scalar squared-mass is exactly cancelled when the virtual exchange of superpartners is added to the contributions of the Standard Model. Thus in a supersymmetric theory, the radiative corrections to the masses of both fermions and bosons are at most logarithmically sensitive to the high energy scale Λ. Of course, the historical precedent does not provide an exact analogy. Because CPT symmetry in quantum field theory must be exactly conserved, antiparticles must be mass-degenerate with their particle partners. In contrast, supersymmetry cannot be an exact symmetry of nature, since experimental data imply that supersymmetric particles are not mass degenerate with their partners. Nevertheless, if the scale of supersymmetry breaking is of order 1 TeV or below, then the naturalness problem of the Standard Model would be resolved. In such theories of "low-energy" supersymmetry, the supersymmetry breaking scale is tied to the scale of electroweak symmetry breaking [1–3].

In addition to providing a potential solution of the naturalness problem of the Standard Model, supersymmetry provides an attractive theoretical framework that may permit the consistent unification of particle physics and gravity [7–9]. It therefore deserves serious consideration as a theory of fundamental particle interactions.

2 The Minimal Supersymmetric Standard Model (MSSM)

The minimal supersymmetric extension of the Standard Model (MSSM) consists of taking the Standard Model and adding the corresponding supersymmetric partners [10]. In addition, the MSSM contains two hypercharge $Y = \pm 1$ Higgs doublets, which is the minimal structure for the Higgs sector of an anomaly-free supersymmetric extension of the Standard Model. The supersymmetric structure of the theory also requires (at least) two Higgs doublets to generate mass for both "up"-type and "down"-type quarks (and charged leptons) [11,12]. All renormalizable supersymmetric interactions consistent with (global) $B-L$ conservation (B =baryon number and L =lepton number) are included. Finally, the most general soft-supersymmetry-breaking terms are added [13]. If supersymmetry is relevant for explaining the scale of electroweak interactions, then the mass parameters associated with the soft-supersymmetry-breaking terms must be of order 1 TeV or below [14]. Some bounds on these parameters exist due to the absence of supersymmetric particle production at current accelerators; see [15] for a complete listing of supersymmetric particle mass limits. Additional constraints arise from limits on the contributions of virtual supersymmetric particle exchange to a variety of Standard Model processes [16]. The impact of precision electroweak measurements at LEP and SLC on the MSSM parameter space is discussed briefly in section 4.

As a consequence of $B-L$ invariance, the MSSM possesses a discrete R-parity invariance, where $R=(-1)^{3(B-L)+2S}$ for a particle of spin S [17]. Note that this formula implies that all the ordinary Standard Model particles have even R-parity, whereas the corresponding supersymmetric partners have odd R-parity. The conservation of R-parity in scattering and decay processes has a crucial impact on supersymmetric phenomenology. For example, starting from an initial state involving ordinary (R-even) particles, it follows that supersymmetric particles must be produced in pairs. In general, these particles are highly unstable and decay quickly into lighter states. However, R-parity invariance also implies that the lightest supersymmetric particle (LSP) is absolutely stable, and must eventually be produced at the end of a decay chain initiated by the decay of a heavy unstable supersymmetric particle. In order to be consistent with cosmological constraints, the LSP is almost certainly electrically and color neutral [18]. Consequently, the LSP is weakly-interacting in ordinary matter, *i.e.* it behaves like a heavy stable neutrino and will escape detectors without being directly observed. Thus, the canonical signature for R-parity conserving supersymmetric theories† is missing (transverse) energy, due to the escape of the LSP.

The parameters of the MSSM are conveniently described by considering separately the supersymmetry-conserving sector and the supersymmetry-breaking sector. Supersymmetry breaking is accomplished by including the most general set of soft-supersymmetry breaking terms; these terms parametrize our ignorance of the fundamental mechanism of supersymmetry breaking. A careful discussion of the conventions used in defining the MSSM parameters can be found in [20]. Among the parameters of the supersymmetry conserving sector are: (i) gauge couplings: g_s, g, and g', corresponding to the Standard Model gauge group SU(3)×SU(2)×U(1) respectively; (ii) Higgs Yukawa couplings: λ_e, λ_u, and λ_d (which are 3×3 matrices in flavor space); and (iii) a supersymmetry-conserving Higgs mass parameter μ. The supersymmetry-breaking sector contains the following set of parameters: (i) gaugino Majorana masses M_3, M_2 and M_1 associated with the SU(3), SU(2), and U(1) subgroups of the Standard Model; (ii) scalar mass matrices for the squarks and sleptons; (iii) Higgs-squark-squark trilinear interaction terms (the so-called "A-parameters") and corresponding terms involving the sleptons; and (iv) three scalar Higgs mass parameters—two diagonal and

†Some model builders attempt to relax the assumption of R-parity conservation. Models of the type must break $B-L$ and are therefore strongly constrained by experiment [19]. In such models, the LSP is unstable and supersymmetric particles can be singly produced and destroyed in association with B or L violation. These features lead to a phenomenology of broken-R-parity models that is very different from that of the MSSM.

one off-diagonal mass terms for the two Higgs doublets. These three mass parameters can be re-expressed in terms of the two Higgs vacuum expectation values, v_1 and v_2, and one physical Higgs mass. Here, v_1 (v_2) is the vacuum expectation value of the Higgs field which couples exclusively to down-type (up-type) quarks and leptons. Note that $v_1^2 + v_2^2 = (246\text{ GeV})^2$ is fixed by the W mass, while the ratio

$$\tan\beta = v_2/v_1 \tag{1}$$

is a free parameter of the model.

The supersymmetric constraints imply that the MSSM Higgs sector is automatically CP-conserving (at tree-level). Thus, $\tan\beta$ is a real parameter (conventionally chosen to be positive), and the physical neutral Higgs scalars are CP-eigenstates. Nevertheless, the MSSM does contain a number of possible new sources of CP violation. For example, gaugino mass parameters, the A-parameters, and μ may be complex. Some combination of these complex phases must be less than of order 10^{-2}–10^{-3} (for a supersymmetry-breaking scale of 100 GeV) to avoid generating electric dipole moments for the neutron, electron, and atoms in conflict with observed data [21]. However, these complex phases have little impact on the direct searches for supersymmetric particles, and are usually ignored in experimental analyses.

Before describing the supersymmetric particle sector, let us consider the Higgs sector of the MSSM [22]. There are five physical Higgs particles in this model: a charged Higgs pair ($H^\pm$), two CP-even neutral Higgs bosons (denoted by h^0 and H^0 where $m_{h^0} \leq m_{H^0}$) and one CP-odd neutral Higgs boson (A^0). The properties of the Higgs sector are determined by the Higgs potential which is made up of quadratic terms [whose squared-mass coefficients were mentioned above eq. (1)] and quartic interaction terms. The strengths of the interaction terms are directly related to the gauge couplings by supersymmetry. As a result, $\tan\beta$ [defined in eq. (1)] and one Higgs mass determine: the Higgs spectrum, an angle α [which indicates the amount of mixing of the original $Y = \pm 1$ Higgs doublet states in the physical CP-even scalars], and the Higgs boson couplings. When one-loop radiative corrections are incorporated, additional parameters of the supersymmetric model enter via virtual loops. The impact of these corrections can be significant [23,24]. For example, at tree-level, the MSSM predicts $m_{h^0} \leq m_Z$ [11,12]. If true, this would imply that experiments to be performed at LEP-2 operating at its maximum energy and luminosity would rule out the MSSM if h^0 were not found. However, this Higgs mass bound can be violated when the radiative corrections are incorporated. For example, in [23], the following approximate upper bound was obtained for m_{h^0} (assuming $m_{A^0} > m_Z$) in the limit of $m_Z \ll m_t \ll M_{\tilde{t}}$ [where top-squark ($\tilde{t}_L$–$\tilde{t}_R$) mixing is neglected]

$$m_{h^0}^2 \lesssim m_Z^2 + \frac{3g^2 m_Z^4}{16\pi^2 m_W^2}\left\{\left[\frac{2m_t^4 - m_t^2 m_Z^2}{m_Z^4}\right]\ln\left(\frac{M_{\tilde{t}}^2}{m_t^2}\right) + \frac{m_t^2}{3m_Z^2}\right\}. \tag{2}$$

More refined computations [25] (which include the effects of top-squark mixing at one-loop, renormalization group improvement, and the leading two-loop contributions) yield $m_{h^0} \lesssim 125$ GeV for $m_t = 175$ GeV and a top-squark mass of $M_{\tilde{t}} = 1$ TeV. Clearly, the radiative corrections to the Higgs masses have a significant impact on the search for the MSSM Higgs bosons at LEP [26].

Consider next the supersymmetric particle sector of the MSSM. The *gluino* is the color octet Majorana fermion partner of the gluon with mass $M_{\tilde{g}} = |M_3|$. The supersymmetric partners of the electroweak gauge and Higgs bosons (the *gauginos* and *higgsinos*) can mix. As a result, the physical mass eigenstates are model-dependent linear combinations of these states, called *charginos* and *neutralinos*, which are obtained by diagonalizing the corresponding mass matrices. The chargino mass matrix depends on M_2, μ, $\tan\beta$ and m_W [27]. The corresponding chargino mass eigenstates are denoted by $\tilde{\chi}_1^+$ and $\tilde{\chi}_2^+$, with masses

$$\begin{aligned} M^2_{\tilde{\chi}_1^+,\tilde{\chi}_2^+} &= \tfrac{1}{2}\Bigg\{|\mu|^2 + |M_2|^2 + 2m_W^2 \mp \Bigg[\left(|\mu|^2 + |M_2|^2 + 2m_W^2\right)^2 \\ &\quad -4|\mu|^2|M_2|^2 - 4m_W^4 \sin^2 2\beta + 8m_W^2 \sin 2\beta \,\mathrm{Re}(\mu M_2)\Bigg]^{1/2}\Bigg\}, \end{aligned} \tag{3}$$

where the states are ordered such that $M_{\tilde{\chi}_1^+} \leq M_{\tilde{\chi}_2^+}$. If CP-violating effects are ignored (in which case, M_2 and μ are real parameters), then one can choose a convention where $\tan\beta$ and M_2 are positive. (Note that the relative sign of M_2 and μ is meaningful. The sign of μ is convention-dependent; the reader is warned that both sign conventions appear in the literature.) The sign convention for μ implicit in eq. (3) is used by the LEP collaborations [28] in their plots of exclusion contours in the M_2 *vs.* μ plane derived from the non-observation of $Z \to \tilde{\chi}_1^+\tilde{\chi}_1^-$. The 4×4 neutralino mass matrix depends on M_1, M_2, μ, $\tan\beta$, m_Z, and the weak mixing angle θ_W [27]. The corresponding neutralino eigenstates are usually denoted by $\tilde{\chi}_i^0$ ($i = 1,\ldots 4$), according to the convention that $M_{\tilde{\chi}_1^0} \leq M_{\tilde{\chi}_2^0} \leq M_{\tilde{\chi}_3^0} \leq M_{\tilde{\chi}_4^0}$. Typically, $\tilde{\chi}_1^0$ is the LSP.

It is common practice in the literature to reduce the supersymmetric parameter freedom by requiring that all three gaugino mass parameters are equal at some grand unification scale. Then, at the electroweak scale, the gaugino mass parameters can be expressed in terms of one of them (say, M_2) and the gauge coupling constants:

$$M_3 = (g_s^2/g^2)M_2\,, \qquad M_1 = (5g'^2/3g^2)M_2\,. \tag{4}$$

Having made this assumption, the chargino and neutralino masses and mixing angles depend only on three unknown parameters: the gluino mass, μ, and $\tan\beta$. However, the assumption of gaugino mass unification could prove false and must eventually be tested experimentally.

The supersymmetric partners of the quarks and leptons are spin-zero bosons: the *squarks*, charged *sleptons*, and *sneutrinos*. For a given fermion f, there are two supersymmetric partners $\tilde{f}_L$ and $\tilde{f}_R$ which are scalar partners of the corresponding left and right-handed fermion. (There is no $\tilde{\nu}_R$.) However, in general, $\tilde{f}_L$ and $\tilde{f}_R$ are not mass-eigenstates since there is $\tilde{f}_L$-$\tilde{f}_R$ mixing which is proportional in strength to the corresponding element of the scalar squared-mass matrix [29]:

$$M_{LR}^2 = \begin{cases} m_d(A_d - \mu\tan\beta), & \text{for "down"-type } f \\ m_u(A_u - \mu\cot\beta), & \text{for "up"-type } f, \end{cases} \tag{5}$$

where m_d (m_u) is the mass of the appropriate "down" ("up") type quark or lepton. Here, A_d and A_u are (unknown) soft-supersymmetry-breaking A–parameters and μ and $\tan\beta$ have been defined earlier. The signs of the A parameters are also convention-dependent; see [20]. Due to the appearance of the *fermion* mass in eq. (5), one expects M_{LR} to be small compared to the diagonal squark and slepton masses, with the possible exception of the top-squark, since m_t is large, and the bottom-squark and tau-slepton if $\tan\beta \gg 1$. The (diagonal) L and R-type squark and slepton masses are given by [8]

$$M_{\tilde{u}_L}^2 = M_{\tilde{Q}}^2 + m_u^2 + m_Z^2\cos 2\beta(\tfrac{1}{2} - \tfrac{2}{3}\sin^2\theta_W) \tag{6}$$

$$M_{\tilde{u}_R}^2 = M_{\tilde{U}}^2 + m_u^2 + \tfrac{2}{3}m_Z^2\cos 2\beta\sin^2\theta_W \tag{7}$$

$$M_{\tilde{d}_L}^2 = M_{\tilde{Q}}^2 + m_d^2 - m_Z^2\cos 2\beta(\tfrac{1}{2} - \tfrac{1}{3}\sin^2\theta_W) \tag{8}$$

$$M_{\tilde{d}_R}^2 = M_{\tilde{D}}^2 + m_d^2 - \tfrac{1}{3}m_Z^2\cos 2\beta\sin^2\theta_W \tag{9}$$

$$M_{\tilde{\nu}}^2 = M_{\tilde{L}}^2 + \tfrac{1}{2}m_Z^2\cos 2\beta \tag{10}$$

$$M_{\tilde{e}_L}^2 = M_{\tilde{L}}^2 + m_e^2 - m_Z^2\cos 2\beta(\tfrac{1}{2} - \sin^2\theta_W) \tag{11}$$

$$M_{\tilde{e}_R}^2 = M_{\tilde{E}}^2 + m_e^2 - m_Z^2\cos 2\beta\sin^2\theta_W\,. \tag{12}$$

The soft-supersymmetry-breaking parameters: $M_{\tilde{Q}}$, $M_{\tilde{U}}$, $M_{\tilde{D}}$, $M_{\tilde{L}}$, and $M_{\tilde{E}}$ are unknown parameters. In the equations above, the notation of first generation fermions has been used and generational indices have been suppressed. Further complications such as intergenerational mixing are possible, although there are some constraints from the nonobservation of flavor-changing neutral currents (FCNC) [30].

3 Reducing the Supersymmetric Parameter Freedom

One way to guarantee the absence of significant FCNC's mediated by virtual supersymmetric particle exchange is to posit that the diagonal soft-supersymmetry-breaking scalar squared-masses are universal in flavor space at some energy scale (normally taken to be at or near the Planck scale) [2,31,32]. Renormalization group evolution is used to determine the low-energy values for the scalar mass parameters listed above. This assumption substantially reduces the MSSM parameter freedom. For example, supersymmetric grand unified models with universal scalar masses at the Planck scale typically give [33] $M_{\tilde{L}} \approx M_{\tilde{E}} < M_{\tilde{Q}} \approx M_{\tilde{U}} \approx M_{\tilde{D}}$ with the squark masses somewhere between a factor of 1–3 larger than the slepton masses (neglecting generational distinctions). More specifically, the first two generations are thought to be nearly degenerate in mass, while $M_{\tilde{Q}_3}$ and $M_{\tilde{U}_3}$ are typically reduced by a factor of 1–3 from the other soft-supersymmetry-breaking masses because of renormalization effects due to the heavy top quark mass. As a result, four flavors of squarks (with two squark eigenstates per flavor) and $\widetilde{b}_R$ will be nearly mass-degenerate and somewhat heavier than six flavors of nearly mass-degenerate sleptons (with two per flavor for the charged sleptons and one per flavor for the sneutrinos). On the other hand, the $\widetilde{b}_L$ mass and the diagonal $\widetilde{t}_L$ and $\widetilde{t}_R$ masses are reduced compared to the common squark mass of the first two generations. In addition, third generation squark masses and tau-slepton masses are sensitive to the strength of the respective $\widetilde{f}_L$–$\widetilde{f}_R$ mixing as discussed below eq. (5).

Two additional theoretical frameworks are often introduced to reduce further the MSSM parameter freedom [7,8,33,34]. The first involves grand unified theories (GUTs) and the desert hypothesis (*i.e.* no new physics between the TeV-scale and the GUT-scale). Perhaps one of the most compelling hints for low-energy supersymmetry is the unification of SU(3)×SU(2)×U(1) gauge couplings predicted by supersymmetric GUT models [2,35] (with the supersymmetry breaking scale of order 1 TeV or below). The unification, which takes place at an energy scale of order 10^{16} GeV, is quite robust (and depends weakly on the details of the GUT-scale theory). For example, a recent analysis [36] finds that supersymmetric GUT unification implies that $\alpha_s(m_Z) = 0.129 \pm 0.010$, not including threshold corrections due to GUT-scale particles (which could diminish the value of $\alpha_s(m_Z)$). This result is compatible with the world average of $\alpha_s(m_Z) = 0.117 \pm 0.005$ [15]. In contrast, gauge coupling unification in the simplest nonsupersymmetric GUT models fails by many standard deviations [37]. Grand unification can impose additional constraints through the unification of Higgs-fermion Yukawa couplings (λ_f). There is some evidence that $\lambda_b = \lambda_\tau$ leads to good low-energy phenomenology [38], and an intriguing possibility that in the MSSM (in the parameter regime where $\tan\beta \simeq m_t/m_b$) $\lambda_b = \lambda_\tau = \lambda_t$ may be phenomenologically viable [39]. However, such unification constraints are GUT-model dependent, and do not address the origin of the first and second generation fermion masses and the CKM mixing matrix. Finally, grand unification imposes constraints on the soft-supersymmetry-breaking parameters. For example, gaugino mass unification leads to the relations given in eq. (4). Diagonal squark and slepton soft-supersymmetry-breaking scalar masses may also be unified at the GUT scale (analogous to the unification of Higgs-fermion Yukawa couplings).

In order to further reduce the number of independent soft-supersymmetry breaking parameters (with or without grand unification), an additional simplifying assumption is required. In the minimal supergravity theory, the soft supersymmetry-breaking parameters are often taken to have the following simple form. Referring to the parameter list given above eq. (1), the Planck-scale values of the soft-supersymmetry-breaking terms depend on the following minimal set of parameters: (i) a universal gaugino mass $m_{1/2}$; (ii) a universal diagonal scalar mass parameter m_0 [whose consequences were described at the beginning of this section]; (iii) a universal A-parameter, A_0; and (iv) three scalar Higgs mass parameters—two common diagonal squared-masses given by $|\mu_0|^2 + m_0^2$ and an off-diagonal squared-mass given by $B_0\mu_0$ (which defines the Planck-scale supersymmetry-breaking parameter B_0), where μ_0 is the Planck-scale value of the μ-parameter. As before, renormalization group evolution is used to compute the low-energy values of the supersymmetry-breaking parameters and determines the supersymmetric particle spectrum. Moreover, in this approach, electroweak symmetry breaking is induced radiatively if one of the Higgs diagonal squared-masses is forced negative by the evolution. This occurs in models with a large Higgs-top quark Yukawa coupling (*i.e.* large m_t). As a result, the two Higgs vacuum expectation values (or equivalently, m_Z and $\tan\beta$) can be expressed as a function of the Planck-scale supergravity parameters. The simplest procedure [33,34] is to remove μ_0 and B_0 in favor of m_Z and $\tan\beta$ (the sign of μ_0 is not fixed in this process). In this case, the MSSM spectrum and its interactions are determined by m_0, A_0, $m_{1/2}$, $\tan\beta$, and the sign of μ_0 (in addition to the parameters of the Standard Model). However, the minimal approach above is probably too restrictive. Theoretical considerations suggest that the universality of Planck-scale soft-supersymmetry breaking parameters is not generic [40]. In the absence of a fundamental theory of supersymmetry breaking, further progress will require a detailed knowledge of the supersymmetric particle spectrum in order to determine the nature of the Planck-scale parameters.

4 Challenges for Supersymmetry Searches

The verification of low-energy supersymmetry requires the discovery of the supersymmetric particles. Once superpartners are discovered, it is necessary to test their detailed properties to verify the supersymmetric nature of their interactions. Furthermore, one can explicitly test many of the additional theoretical assumptions of section 3 that were introduced to reduce the supersymmetric parameter freedom.

The search for supersymmetry at present and future colliders falls into two distinct classes. At colliders whose energies lie below supersymmetric particle production threshold, indirect effects of supersymmetry may be observable. For example, in the Higgs sector, if $m_{A^0} \gg m_{h^0}$, then the properties of h^0 will be nearly indistinguishable from the Higgs boson of the minimal Standard Model [41]. Small deviations from the Standard Model Higgs sector could signal the existence of additional Higgs states, as expected in the MSSM. One can also search for deviations from Standard Model predictions due to the effects of virtual supersymmetric particle exchange. Such effects could be revealed in the measurement of precision electroweak observables. In both cases, one is fighting the decoupling limit. That is, in the limit that soft-supersymmetry-breaking masses (collectively denoted by $M_{\rm SUSY}$) become large, the MSSM below supersymmetric threshold precisely reproduces the predictions of the Standard Model. At colliders whose energies lie above supersymmetric particle production threshold, the direct effects of supersymmetric production and decay are detectable. In this case, once superpartners are discovered, one must elucidate the details of the low-energy supersymmetric theory.

The MSSM (with or without constraints imposed from the theory near the Planck scale) provides a framework that can be tested by precision electroweak data. The level of accuracy of the measured Z decay observables at LEP and SLC is sufficient to test the structure of the one-loop radiative corrections of the electroweak model [42], and is thus potentially sensitive to the virtual effects of undiscovered particles. Combining the most recent LEP and SLC electroweak results [43] with the recent top-quark mass measurement at the Tevatron [44], a weak preference is found [43,45] for a light Higgs boson mass of order m_Z, which is consistent with the MSSM Higgs mass upper bound noted in section 2. Moreover, for Z decay observables, the effects of virtual supersymmetric particle exchange are suppressed by a factor of $m_Z^2/M_{\rm SUSY}^2$, and therefore decouple in the limit of large supersymmetric particle masses. It follows that for $M_{\rm SUSY}^2 \gg m_Z$ (in practice, it is sufficient to have all supersymmetric particle masses above 200 GeV) the MSSM yields an equally good fit to the precision electroweak data as compared to the Standard Model fit. On the other hand, there are a few tantalizing hints in the data for deviations from Standard Model predictions. Indeed, if $R_b \equiv \Gamma(Z \to b\bar{b})/\Gamma(Z \to \text{hadrons})$ is confirmed to lie above its Standard Model prediction due to the presence of new physics, then a plausible candidate for the new physics would be the MSSM with some light supersymmetric particles (*e.g.* a light chargino and top-squark and/or a light CP-odd scalar, A^0) close in mass to their present LEP bounds [46,47]. Such a scenario would be tested by the search for supersymmetric particles at LEP-2 and the Tevatron.

If low-energy supersymmetry exists, it should be discovered at either upgrades of existing colliders or at the LHC [48,49]. Due to its mass reach, the LHC is the definitive machine for discovering or excluding low-energy supersymmetry. Table 1 summarizes the supersymmetry mass discovery potential for hadron colliders. A variety of signatures are considered. Many of the supersymmetry searches rely on the missing energy signature as an indication of new physics beyond the Standard Model. Multi-leptonic signatures also play an important role in supersymmetry searches at hadron colliders. (Such signals can also be exploited in the search for R-parity-violating low-energy supersymmetry.) A comprehensive analysis can be found in [49].

Suppose that a signal is observed in one of the expected channels. This would not be a confirmation of low-energy supersymmetry, unless there is confirming evidence from other expected signatures. This presents a formidable challenge to experimenters at the LHC. Can they prove that a set of signatures of new physics is low-energy supersymmetry? Can they extract parameters of the supersymmetric models with any precision and test the details of the theory? These are questions that have only recently attracted seriously study. It is in this context that a future e^+e^- collider (NLC) can be invaluable. If the lightest supersymmetric particles were produced at LEP-2 or the NLC, precision measurements could begin to map out in detail the parameter space of the supersymmetric model. In particular, beam polarization at the NLC provides an critical tool for studying the relation between chirality and the properties of supersymmetric particles [52]. One can then begin to demonstrate that there is a correlation between the left and right-handed electrons and their slepton partners as expected in supersymmetry. Moreover, the determination of superparticle masses allows one to test theoretical assumptions at various levels. For example, the universality of slepton

Table 1. Discovery reach of various options of future hadron colliders [50]. The numbers are subject to ±15% ambiguity. Also, the clean trilepton signals are sensitive to other model parameters; representative ranges from Ref. [51] are shown where $|\mu|$ is typically much larger than the soft-breaking electroweak gaugino masses. For $\mu > 0$, the leptonic decay of $\widetilde{\chi}_2^0$ may be strongly suppressed so that 3ℓ signals may not be observable even if charginos are just above the LEP bound.

Signal	Tevatron I 0.01 fb^{-1} 1.8 TeV	Tevatron II 0.1 fb^{-1} 1.8 TeV	Main Injector 1 fb^{-1} 2 TeV	Tevatron* 10 fb^{-1} 2 TeV	DiTevatron 1 fb^{-1} 4 TeV	LHC 10 fb^{-1} 14 TeV
$\not\!\!E_T(\tilde{q} \gg \tilde{g})$	$\tilde{g}(150)$	$\tilde{g}(210)$	$\tilde{g}(270)$	$\tilde{g}(340)$	$\tilde{g}(450)$	$\tilde{g}(1300)$
$\ell^\pm\ell^\pm(\tilde{q} \gg \tilde{g})$	—	$\tilde{g}(160)$	$\tilde{g}(210)$	$\tilde{g}(270)$	$\tilde{g}(320)$	$\tilde{g}(1000)$
$all \to 3\ell(\tilde{q} \gg \tilde{g})$	—	$\tilde{g}(150\text{-}180)$	$\tilde{g}(150\text{-}260)$	$\tilde{g}(150\text{-}430)$	$\tilde{g}(150\text{-}320)$	
$\not\!\!E_T(\tilde{q} \sim \tilde{g})$	$\tilde{g}(220)$	$\tilde{g}(300)$	$\tilde{g}(350)$	$\tilde{g}(400)$	$\tilde{g}(580)$	$\tilde{g}(2000)$
$\ell^\pm\ell^\pm(\tilde{q} \sim \tilde{g})$	—	$\tilde{g}(180\text{-}230)$	$\tilde{g}(325)$	$\tilde{g}(385\text{-}405)$	$\tilde{g}(460)$	$\tilde{g}(1000)$
$all \to 3\ell(\tilde{q} \sim \tilde{g})$	—	$\tilde{g}(240\text{-}290)$	$\tilde{g}(425\text{-}440)$	$\tilde{g}(550)$	$\tilde{g}(550)$	$\gtrsim \tilde{g}(1000)$
$\tilde{t}_1 \to c\widetilde{\chi}_1^0$	—	$\tilde{t}_1(80\text{-}100)$	$\tilde{t}_1(120)$			
$\tilde{t}_1 \to b\widetilde{\chi}_1^\pm$	—	$\tilde{t}_1(80\text{-}100)$	$\tilde{t}_1(120)$			
$\Theta(\tilde{t}_1\tilde{t}_1^*) \to \gamma\gamma$	—	—	—	—	—	$\tilde{t}_1(250)$
$\tilde{\ell}\tilde{\ell}^*$	—	$\tilde{\ell}(50)$	$\tilde{\ell}(50)$	$\tilde{\ell}(100)$		$\tilde{\ell}(250\text{-}300)$

masses can be tested at the 1% level. More experimentally challenging is the test of the GUT-relation among gaugino masses [eq. (4)]. However, one can still test eq. (4) at the few percent level by combining slepton and chargino signals, based on the measured masses and polarization dependence of the cross sections. See [49] for further details.

Many theorists believe that the prospects for supersymmetry are excellent. Nevertheless, the search for supersymmetry at future colliders may reveal many surprises and raise new challenges for both theorists and experimentalists. If low-energy supersymmetry is discovered it will have a profound effect on the development of 21st century theories of fundamental particles and their interactions.

Acknowledgments

I would like to thank Hitoshi Murayama for bringing to my attention the naturalness problem of the single-electron quantum theory. I am grateful to Jon Bagger, Vernon Barger, Michael Barnett, Paul Langacker and Michael Peskin for their comments on an earlier version of this manuscript. Finally, I acknowledge the gracious hospitality and support of José Valle and his colleagues during my visit to Valencia. This work was supported in part by the US Department of Energy.

References

1. E. Witten, *Nucl. Phys.* **B188** (1981) 513.
2. S. Dimopoulos and H. Georgi, *Nucl. Phys.* **B193** (1981) 150.
3. L. Susskind, *Phys. Rep.* **104** (1984) 181; N. Sakai, *Z. Phys.* **C11** (1981) 153. See also, R.K. Kaul, *Phys. Lett.* **109B** (1982) 19.
4. V. Weisskopf, *Z. Phys.* **89** (1934) 27 [E: **90** (1934) 817].
5. See A.I. Miller, *Early Quantum Electrodynamics* (Cambridge University Press, Cambridge, England, 1994) for an English translation of [4] and a detailed discussion of the problem of the electron self-energy.
6. S. Weinberg, *Phys. Rev.* **D13** (1976) 974; **D19** (1979) 1277; L. Susskind, *Phys. Rev.* **D20** (1979) 2619; G. 't Hooft, in *Recent Developments in Gauge Theories,* Proceedings of the NATO Advanced Summer Institute, Cargese, 1979, edited by G. 't Hooft *et al.* (Plenum, New York, 1980) pp. 135–157.
7. H.P. Nilles, *Phys. Rep.* **110** (1984) 1.
8. P. Nath, R. Arnowitt, and A. H. Chamseddine, *Applied* $N = 1$ *Supergravity* (World Scientific, Singapore, 1984)
9. M.B. Green, J.S. Schwarz, and E. Witten, *Superstring Theory* (Cambridge University Press, Cambridge, 1987).
10. H.E. Haber and G.L. Kane, *Phys. Rep.* **117** (1985) 75.
11. K. Inoue, A. Kakuto, H. Komatsu, and S. Takeshita, *Prog. Theor. Phys.* **68** (1982) 927 [E: **70** (1983) 330]; **71** (1984) 413; R. Flores and M. Sher, *Ann. Phys. [NY]* **148** (1983) 95.
12. J.F. Gunion and H.E. Haber, *Nucl. Phys.* **B272** (1986) 1 [E: **B402** (1993) 567].
13. L. Girardello and M. Grisaru, *Nucl. Phys.* **B194** (1982) 65.
14. See, *e.g.*, R. Barbieri and G.F. Giudice, *Nucl. Phys.* **B305** (1988) 63; G.W. Anderson and D.J. Castano, *Phys. Lett.* **B347** (1995) 300; *Phys. Rev.* **D52** (1995) 1693.
15. L. Montanet *et al.* [Particle Data Group], *Phys. Rev.* **D50** (1994) 1173.
16. See, *e.g.*, S. Bertolini, F. Borzumati, A. Masiero, and G. Ridolfi, *Nucl. Phys.* **B353** (1991) 591.
17. P. Fayet, *Phys. Lett.* **69B** (1977) 489; G. Farrar and P. Fayet, *Phys. Lett.* **76B** (1978) 575.
18. J. Ellis, J.S. Hagelin, D.V. Nanopoulos, K. Olive, and M. Srednicki, *Nucl. Phys.* **B238** (1984) 453.
19. See, *e.g.*, V. Barger, T. Han and G.F. Giudice, *Phys. Rev.* **D40** (1989) 2987; S. Dimopoulos, R. Esmailzadeh, L.J. Hall, and G.D. Starkman, *Phys. Rev.* **D41** (1990) 2099; H. Dreiner and G.G. Ross, *Nucl. Phys.* **B365** (1991) 597.
20. H.E. Haber, "Introductory Low-Energy Supersymmetry," in *Recent Directions in Particle Theory,* Proceedings of the 1992 Theoretical Advanced Study Institute in Particle Physics, edited by J. Harvey and J. Polchinski (World Scientific, Singapore, 1993) pp. 589–686.
21. W. Fischler, S. Paban, and S. Thomas, *Phys. Lett.* **B289** (1992) 373; S.M. Barr, *Int. J. Mod. Phys.* **A8** (1993) 209.

22. J.F. Gunion, H.E. Haber, G. Kane, and S. Dawson, *The Higgs Hunter's Guide* (Addison-Wesley Publishing Company, Redwood City, CA, 1990).
23. H.E. Haber and R. Hempfling, *Phys. Rev. Lett.* **66** (1991) 1815.
24. Y. Okada, M. Yamaguchi, and T. Yanagida, *Prog. Theor. Phys.* **85** (1991) 1; J. Ellis, G. Ridolfi, and F. Zwirner, *Phys. Lett.* **B257** (1991) 83.
25. M. Carena, J.R. Espinosa, M. Quiros, and C.E.M. Wagner, *Phys. Lett.* **B335** (1995) 209; H.E. Haber, R. Hempfling, and A.H. Hoang, CERN-TH-95-216 (1995).
26. G. Altarelli *et al.*, "Interim Report on the Physics Motivations for an Energy Upgrade of LEP2", CERN-TH/95-151 and CERN-PPE/95-78 (1995).
27. Explicit forms for the chargino and neutralino mass matrices can be found in Appendix A of [12]; see also [20].
28. See, *e.g.*, D. Decamp *et al.* [ALEPH Collaboration], *Phys. Rep.* **216** (1992) 253.
29. J. Ellis and S. Rudaz, *Phys. Lett.* **128B** (1983) 248.
30. For recent works and references to the original literature, see: J. Hagelin, S. Kelley, and T. Tanaka, *Nucl. Phys.* **B415** (1994) 293; D. Choudhury *et al.*, *Phys. Lett.* **B342** (1995) 1980.
31. See *e.g.*, M. Dine, R. Leigh, and A. Kagan, *Phys. Rev.* **D48** (1993) 4269.
32. Other mechanisms for avoiding potentially dangerous FCNC's have been proposed. For example, non-universal scalar masses are not dangerous if squark and slepton mass matrices are diagonal in the same basis as the corresponding fermion mass matrices. See, *e.g.*, Y. Nir and N. Seiberg, *Phys. Lett.* **B309** (1993) 337; S. Dimopoulos, G.F. Giudice, and N. Tetradis, CERN-TH/95-90 (1995).
33. For a recent review, see M. Drees and S.P. Martin, MAD-PH-95-879 [hep-ph/9504324], to appear in *Electroweak Symmetry Breaking and New Physics at the TeV Scale*, edited by T. Barklow, S. Dawson, H.E. Haber, and J. Siegrist (World Scientific, Singapore, to be published).
34. For a recent review, see R. Arnowitt and P. Nath, in *Particles and Fields*, Proceedings of the 7th Summer School Jorge Andre Swieca, Sao Paulo, Brazil, 10–23 January 1993, edited by O.J.P. Eboli and V.O. Rivelles (World Scientific, Singapore, 1994); W. de Boer, *Prog. Part. Nucl. Phys.* **33** (1994) 201.
35. M.B. Einhorn and D.R.T. Jones, *Nucl. Phys.* **B196** (1982) 475; W.J. Marciano and G. Senjanovic, *Phys. Rev.* **D25** (1982) 3092.
36. P. Langacker and N. Polonsky, *Phys. Rev.* **D52** (1995) 3081. Recent related work can be found in: R. Barbieri, P. Ciafaloni, and A. Strumia, *Nucl. Phys.* **B442** (1995) 461; J. Bagger, K. Matchev, and D. Pierce, *Phys. Lett.* **B348** (1995) 443; P.H. Chankowski, Z. Pluciennik and S. Pokorski, *Nucl. Phys.* **B349** (1995) 23.
37. For a recent review comparing unification in supersymmetric and nonsupersymmetric GUTS, see S. Dimopoulos, CERN-TH-7531-94 [hep-ph/9412297].
38. H. Arason *et al.*, *Phys. Rev. Lett.* **67** (1991) 2933; *Phys. Rev.* **D46** (1992) 3945; V. Barger, M.S. Berger, and P. Ohmann, *Phys. Rev.* **D47** (1993) 1093; M. Carena, S. Pokorski, and C.E.M. Wagner, *Nucl. Phys.* **B406** (1993) 59; P. Langacker and N. Polonsky, *Phys. Rev.* **D49** (1994) 1454.

39. M. Olechowski and S. Pokorski, *Phys. Lett.* **B214** (1988) 393; B. Ananthanarayan, G. Lazarides and Q. Shafi, *Phys. Rev.* **D44** (1991) 1613; S. Dimopoulos, L.J. Hall, and S. Raby, *Phys. Rev. Lett.* **68** (1992) 1984; L.J. Hall, R. Rattazzi, and U. Sarid, *Phys. Rev.* **50** (1994) 7048; M. Carena, M. Olechowski, S. Pokorski, and C.E.M. Wagner, *Nucl. Phys.* **B426** (1994) 269.
40. L.E. Ibáñez and D. Lüst, *Nucl. Phys.* **B382** (1992) 305; B. de Carlos, J.A. Casas and C. Munoz, *Phys. Lett.* **B299** (1993) 234; V. Kaplunovsky and J. Louis, *Phys. Lett.* **B306** (1993) 269; A. Brignole, L.E. Ibáñez, and C. Munoz, *Nucl. Phys.* **B422** (1994) 125 [E: **B436** (1995) 747].
41. H.E. Haber, CERN-TH/95-109 and SCIPP-95/15 [hep-ph/9505240]; H.E. Haber and S. Thomas, in preparation.
42. D. Bardin, W. Hollik, and G. Passarino, editors, "Report of the Working Group on Precision Calculations for the Z Resonance", CERN Yellow Report 95-03 (1995).
43. P. Antilogus *et al.* [LEP Electroweak Working Group], LEPEWWG/95-02 (1995), contributions of the LEP Experiments to the 1995 International Europhysics Conference on High Energy Physics, 27 July–2 August, 1995, Brussels, Belgium, and the 17th International Symposium on Lepton-Photon Interactions, 10–15 August, 1995, Beijing, China.
44. F. Abe *et al.* [CDF Collaboration], *Phys. Rev. Lett.* **74** (1995) 2626; S. Abachi *et al.* [D0 Collaboration], *Phys. Rev. Lett.* **74** (1995) 2632.
45. P.H. Chankowski and S. Pokorski, *Phys.Lett.* **B356** (1995) 307; J.Ellis, G.L. Fogli, and E. Lisi, CERN-TH/95-202 [hep-ph/9507424].
46. M. Boulware and D. Finnel, *Phys. Rev.* **D44** (1991) 2054; A. Djouadi *et al.*, *Nucl. Phys.* **B349** (1991) 48; G. Altarelli, R. Barbieri, and F. Caravaglios, *Nucl. Phys.* **B405** (1993) 3; J.D. Wells, C. Kolda, and G.L. Kane, *Phys. Lett.* **B338** (1994) 219.
47. Recent analyses along these lines can be found in: D. Garcia, R.A. Jiménez and J. Solà, *Phys. Lett.* **B347** (1995) 321 [E: **B351** (1995) 602]; D. Garcia and H. Solà, *Phys. Lett.* **B354** (1995) 335; G.L. Kane, R.G. Stuart, and J.D. Wells, *Phys. Lett.* **B354** (1995) 350; A. Dabelstein, W. Hollik and W. Mösle, KA-THEP-5-1995 [hep-ph/9506251]; P.H. Chankowski and S. Pokorski, IFT-95/5 [hep-ph/9505304]; X. Wang, J.L. Lopez, and D.V. Nanopoulos, CTP-TAMU-25/95 [hep-ph/9506217].
48. T.L. Barklow, S.Dawson, H.E. Haber and J.L. Siegrist, SLAC-PUB-95-6893 [hep-ph/9505296], to appear in *Particle Physics–Perspectives and Opportunity*, Report of the DPF Committee on Long Term Planning, edited by R. Peccei *et al.* (World Scientific, Singapore, to be published).
49. H. Baer *et al.*, FSU-HEP-950401 [hep-ph/9503479], to appear in *Electroweak Symmetry Breaking and New Physics at the TeV Scale*, edited by T. Barklow, S. Dawson, H.E. Haber, and J. Siegrist (World Scientific, Singapore, to be published).
50. This table was prepared by H. Baer and X. Tata and appears in [48].
51. H. Baer, C.H. Chen, F. Paige and X. Tata, Phys. Rev. **D50** (1994) 4508; H. Baer, C. H. Chen, C. Kao and X. Tata, Phys. Rev. **D52** (1995) 1565.
52. T. Tsukamoto, K. Fujii, H. Murayama, M. Yamaguchi, and Y. Okada, Phys. Rev. **D51** (1995) 3153.

BOUNDS ON THE HIGGS MASS IN THE STANDARD MODEL AND THE MINIMAL SUPERSYMMETRIC STANDARD MODEL

M. QUIROS

CERN, TH Division, CH-1211 Geneva 23, Switzerland
Instituto de Estructura de la Materia, Serrano 123,
28006-Madrid, Spain

Depending on the Higgs-boson and top-quark masses, M_H and M_t, the effective potential of the **Standard Model** can develop a non-standard minimum for values of the field much larger than the weak scale. In those cases the standard minimum becomes metastable and the possibility of decay to the non-standard one arises. Comparison of the decay rate to the non-standard minimum at finite (and zero) temperature with the corresponding expansion rate of the Universe allows to identify the region, in the (M_H, M_t) plane, where the Higgs field is sitting at the standard electroweak minimum. In the **Minimal Supersymmetric Standard Model**, approximate analytical expressions for the Higgs mass spectrum and couplings are worked out, providing an excellent approximation to the numerical results which include all next-to-leading-log corrections. An appropriate treatment of squark decoupling allows to consider large values of the stop and/or sbottom mixing parameters and thus fix a reliable upper bound on the mass of the lightest CP-even Higgs boson mass. The discovery of the Higgs boson at LEP 2 might put an upper bound (below the Planck scale) on the scale of new physics Λ and eventually disentangle between the Standard Model and the Minimal Supersymmetric Standard Model.

1 Lower bounds on the Standard Model Higgs mass

For particular values of the Higgs boson and top quark masses, M_H and M_t, the effective potential of the Standard Model (SM) develops a deep non-standard minimum for values of the field $\phi \gg G_F^{-1/2}$ [1]. In that case the standard electroweak (EW) minimum becomes metastable and might decay into the non-standard one. This means that the SM might not accomodate certain regions of the plane (M_H,M_t), a fact which can be intrinsically interesting as evidence for new physics. Of course, the mere existence of the non-standard minimum, and also the decay rate of the standard one into it, depends on the scale Λ up to which we believe the SM results. In fact, one can identify Λ with the scale of new physics.

1.1 When the EW minimum becomes metastable?

The preliminary question one should ask is: When the standard EW minimum becomes metastable, due to the appearance of a deep non-standard minimum? This question was addressed in past years [1] taking into account leading-log (LL) and part of next-to-leading-log (NTLL) corrections. More recently, calculations have incorporated all NTLL corrections [2,3] resummed to all-loop by the renormalization group equations (RGE), and considered pole masses for the top-quark and the Higgs-boson. From the requirement of a stable (not metastable) standard EW minimum we obtain a lower bound on the Higgs mass, as a function of the top mass, labelled by the values of the SM cutoff (stability bounds). Our result [3] is lower than previous estimates by $\mathcal{O}(10)$ GeV.

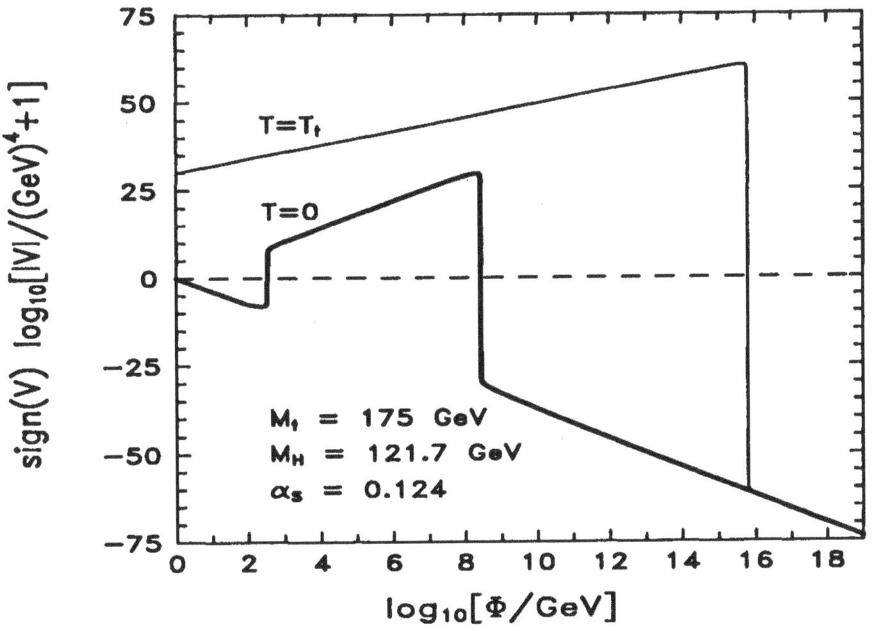

Figure 1: Plot of the effective potential for $M_t = 175$ GeV, $M_H = 121.7$ GeV at $T = 0$ (thick solid line) and $T = T_t = 2.5 \times 10^{15}$ GeV (thin solid line).

The one-loop effective potential of the SM improved by two-loop RGE has been shown to be highly scale independent [4] and, therefore, very reliable for the present study. In Fig. 1 we show (thick solid line) the shape of the effective potential for $M_t = 175$ GeV and $M_H = 121.7$ GeV. We see the appearance of the non-standard maximum, ϕ_M, while the global non-standard minimum has been cutoff at M_P. We can see from Fig. 1 the steep descent from the non-standard maximum. Hence, even if the non-standard minimum is beyond the SM cutoff, the standard minimum becomes metastable and can be destabilized. So for fixed values of M_H and M_t the condition for the standard minimum not to become metastable is

$$\phi_M \gtrsim \Lambda \tag{1}$$

Condition (1) makes the stability condition Λ-dependent. In fact we have plotted in Fig. 2 the stability condition on M_H versus M_t for $\Lambda = 10^{19}$ GeV and 10 TeV. The stability region corresponds to the region above the dashed curves.

1.2 When the EW minimum decays?

In the last subsection we have seen that in the region of Fig. 2 below the dashed line the standard EW minimum is metastable. However we should not draw physical consequences from this fact since we still do not know at which minimum does the Higgs field sit. Thus, the real physical constraint we have to impose is avoiding the Higgs field sitting at its non-standard minimum. In fact the Higgs field can be sitting at its non-standard minimum at zero temperature because:

1. The Higgs field was driven from the origin to the non-standard minimum at finite temperature by thermal fluctuations in a non-standard EW phase transition at high temperature. This minimum evolves naturally to the non-standard minimum at zero temperature. In this case the standard EW phase transition, at $T \sim 10^2$ GeV, will not take place.

2. The Higgs field was driven from the origin to the standard minimum at $T \sim 10^2$ GeV, but decays, at zero temperature, to the non-standard minimum by a quantum fluctuation.

In Fig. 1 we have depicted the effective potential at $T = 2.5 \times 10^{15}$ GeV (thin solid line) which is the corresponding transition temperature. Our finite temperature potential [5] incorporates plasma effects [6] by one-loop resummation of Debye masses [7]. The tunnelling probability per unit time per unit volume was computed long ago for thermal [8] and quantum [9] fluctuations. At finite temperature it is given by $\Gamma/\nu \sim T^4 \exp(-S_3/T)$, where S_3 is the euclidean action evaluated at the bounce solution $\phi_B(0)$. The semiclassical picture is that unstable bubbles are nucleated behind the barrier at $\phi_B(0)$ with a probability given by Γ/ν. Whether or not they fill the Universe depends on the relation between the probability rate and the expansion rate of the Universe. By normalizing the former with respect to the latter we obtain a normalized probability P, and the condition for decay corresponds to $P \sim 1$. Of course our results are trustable, and the decay actually happens, only if $\phi_B(0) < \Lambda$, so that the similar condition to (1) is

$$\Lambda < \phi_B(0) \tag{2}$$

The condition of no-decay (metastability condition) has been plotted in Fig. 2

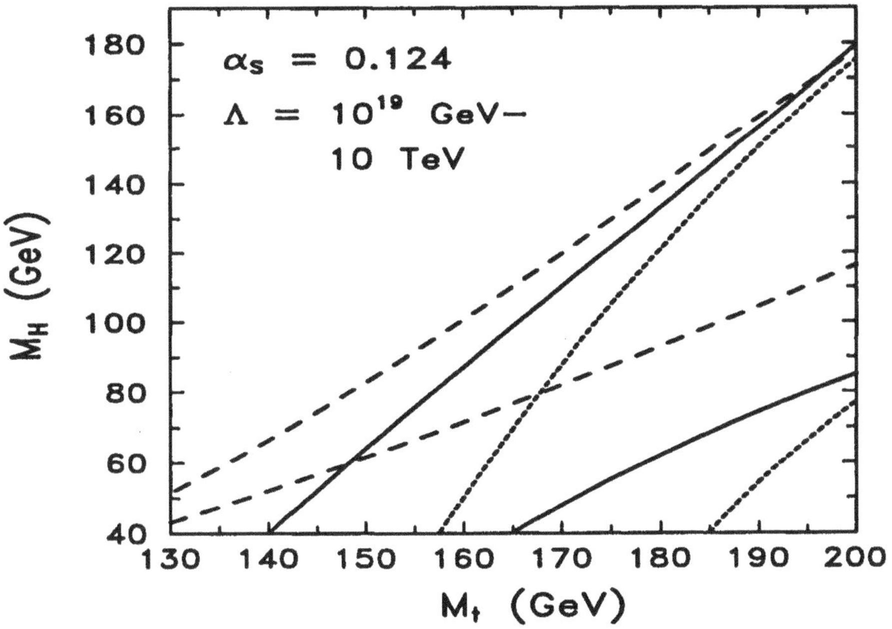

Figure 2: Lower bounds on M_H as a function of M_t, for $\Lambda = 10^{19}$ GeV (upper set) and $\Lambda = 10$ TeV (lower set). The dashed curves correspond to the stability bounds of subsection 1.1 and the solid (dotted) ones to the metastability bounds of subsection 1.2 at finite (zero) temperature.

(solid lines) for $\Lambda = 10^{19}$ GeV and 10 TeV. The region between the dashed and the solid line corresponds to a situation where the non-standard minimum exists but there is no decay to it at finite temperature. In the region below the solid lines the Higgs field is sitting already at the non-standard minimum at $T \sim 10^2$ GeV, and the standard EW phase transition does not happen.

We also have evaluated the tunnelling probability at zero temperature from the standard EW minimum to the non-standard one. The result of the calculation should translate, as in the previous case, in lower bounds on the Higgs mass for differentes values of Λ. The corresponding bounds are shown in Fig. 2 in dotted lines. Since the dotted lines lie always below the solid ones, the possibility of quantum tunnelling at zero temperature does not impose any extra constraint.

As a consequence of all improvements in the calculation, our bounds are lower than previous estimates [10]. To fix ideas, for $M_t = 175$ GeV, the bound reduces by ~ 10 GeV for $\Lambda = 10^4$ GeV, and ~ 30 GeV for $\Lambda = 10^{19}$ GeV.

2 Upper bounds on the Minimal Supersymmetrics Standard Model lightest Higgs boson mass

The effective potential methods to compute the (radiatively corrected) Higgs mass spectrum in the Minimal Supersymmetric Standard Model (MSSM) are useful since they allow to resum (using Renormalization Group (RG) techniques) LL, NTLL,..., corrections to all orders in perturbation theory. These methods [11,12], as well as the diagrammatic methods [13] to compute the Higgs mass spectrum in the MSSM, were first developed in the early nineties.

Effective potential methods are based on the run-and-match procedure by which all dimensionful and dimensionless couplings are running with the RG scale, for scales greater than the masses involved in the theory. When the RG scale equals a particular mass threshold, heavy fields decouple, eventually leaving threshold effects in order to match the effective theory below and above the mass threshold. For instance, assuming a common soft supersymmetry breaking mass for left-handed and right-handed stops and sbottoms, $M_S \sim m_Q \sim m_U \sim m_D$, and assuming for the top-quark mass, m_t, and for the CP-odd Higgs mass, m_A, the range $m_t \leq m_A \leq M_S$, we have: for scales $Q \geq M_S$, the MSSM, for $m_A \leq Q \leq M_S$ the two-Higgs doublet model (2HDM), and for $m_t \leq Q \leq m_A$ the SM. Of course there are thresholds effects at $Q = M_S$ to match the MSSM with the 2HDM, and at $Q = m_A$ to match the 2HDM with the SM.

The neutral Higgs sector of the MSSM contains, on top of the CP-odd Higgs A, two CP-even Higgs mass eigenstates, H_h (the heaviest one) and H (the lightest one). It turns out that the larger m_A the heavier the lightest Higgs H. Therefore the case $m_A \sim M_S$ is, not only a great simplification since the effective theory below M_S is the SM, but also of great interest, since it provides the upper bound on the mass of the lightest Higgs (which is interesting for phenomenological purposes, e.g. at LEP 2). In this case the threshold correction at M_S for the SM quartic coupling λ is:

$$\Delta_{\rm th}\lambda = \frac{3}{16\pi^2} h_t^4 \frac{X_t^2}{M_S^2}\left(2 - \frac{1}{6}\frac{X_t^2}{M_S^2}\right) \tag{3}$$

where h_t is the SM top Yukawa coupling and $X_t = (A_t - \mu/\tan\beta)$ is the mixing in the stop mass matrix, the parameters A_t and μ being the trilinear soft-breaking coupling in the stop sector and the supersymmetric Higgs mixing mass, respectively. The maximum of (3) corresponds to $X_t^2 = 6M_S^2$ which provides the maximum value of the lightest Higgs mass: this case will be referred to as the case of maximal mixing.

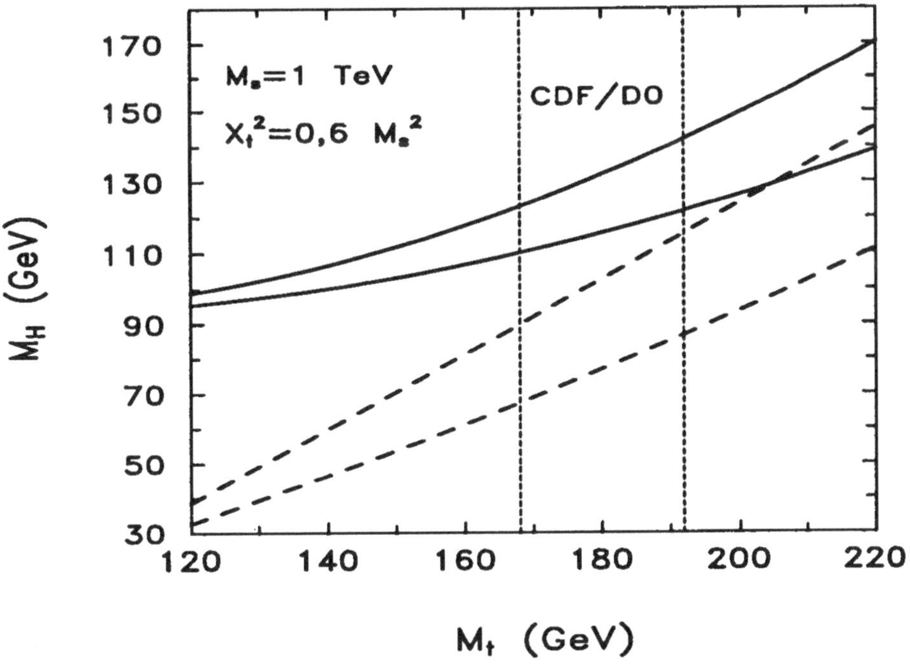

Figure 3: Plot of M_H as a function of M_t for $\tan\beta \gg 1$ (solid lines), $\tan\beta = 1$ (dashed lines), and $X_t^2 = 6M_S^2$ (upper set), $X_t = 0$ (lower set). The experimental band from the CDF/D0 detection is also indicated.

We have plotted in Fig. 3 the lightest Higgs pole mass M_H, where all NTLL corrections are resummed to all-loop by the RG, as a function of M_t [4]. From Fig. 3 we can see that the present experimental band from CDF/D0 for the top-quark mass requires $M_H \lesssim 140$ GeV, while if we fix $M_t = 170$ GeV, the upper bound $M_H \lesssim 125$ GeV follows. It goes without saying that these figures are extremely relevant for MSSM Higgs searches at LEP 2.

2.1 An analytical approximation

We have seen [4] that, since radiative corrections are minimized for scales $Q \sim m_t$, when the LL RG improved Higgs mass expressions are evaluated at the top-quark mass scale, they reproduce the NTLL value with a high level of accuracy, for any value of $\tan\beta$ and the stop mixing parameters [14]

$$m_{H,LL}(Q^2 \sim m_t^2) \sim m_{H,NTLL}. \tag{4}$$

Based on the above observation, we can work out a very accurate analytical approximation to $m_{H,NTLL}$ by just keeping two-loop LL corrections at $Q^2 = m_t^2$, i.e. corrections of order t^2, where $t = \log(M_S^2/m_t^2)$.

Again the case $m_A \sim M_S$ is the simplest, and very illustrative, one. We have found [14,15] that, in the absence of mixing (the case $X_t = 0$) two-loop corrections resum in the one-loop result shifting the energy scale from M_S (the tree-level scale) to $\sqrt{M_S\, m_t}$. More explicitly,

$$m_H^2 = M_Z^2 \cos^2 2\beta \left(1 - \frac{3}{8\pi^2} h_t^2\, t\right) + \frac{3}{2\pi^2 v^2} m_t^4(\sqrt{M_S m_t})t \tag{5}$$

where $v = 246.22$ GeV.

In the presence of mixing ($X_t \neq 0$), the run-and-match procedure yields an extra piece in the SM effective potential $\Delta V_{\rm th}[\phi(M_S)]$ whose second derivative gives an extra contribution to the Higgs mass, as

$$\Delta_{\rm th} m_H^2 = \frac{\partial^2}{\partial\phi^2(t)} \Delta V_{\rm th}[\phi(M_S)] = \frac{1}{\xi^2(t)} \frac{\partial^2}{\partial\phi^2(t)} \Delta V_{\rm th}[\phi(M_S)] \tag{6}$$

which, in our case, reduces to

$$\Delta_{\rm th} m_H^2 = \frac{3}{4\pi^2} \frac{m_t^4(M_S)}{v^2(m_t)} \frac{X_t^2}{M_S^2} \left(2 - \frac{1}{6}\frac{X_t^2}{M_S^2}\right) \tag{7}$$

We have compared our analytical approximation [14] with the numerical NTLL result [4] and found a difference $\lesssim 2$ GeV for all values of supersymmetric parameters.

The case $m_A < M_S$ is a bit more complicated since the effective theory below the supersymmetric scale M_S is the 2HDM. However since radiative corrections in the 2HDM are equally dominated by the top-quark, we can compute analytical expressions based upon the LL approximation at the scale $Q^2 \sim m_t^2$. Our approximation [14] differs from the LL all-loop numerical resummation by $\lesssim$ 3 GeV, which we consider the uncertainty inherent in the theoretical calculation, provided the mixing is moderate and, in particular, bounded by the condition,

$$\left|\frac{m^2_{\tilde{t}_1} - m^2_{\tilde{t}_2}}{m^2_{\tilde{t}_1} + m^2_{\tilde{t}_2}}\right| \lesssim 0.5 \tag{8}$$

where $\tilde{t}_{1,2}$ are the two stop mass eigenstates. In Fig. 4 the Higgs mass spectrum

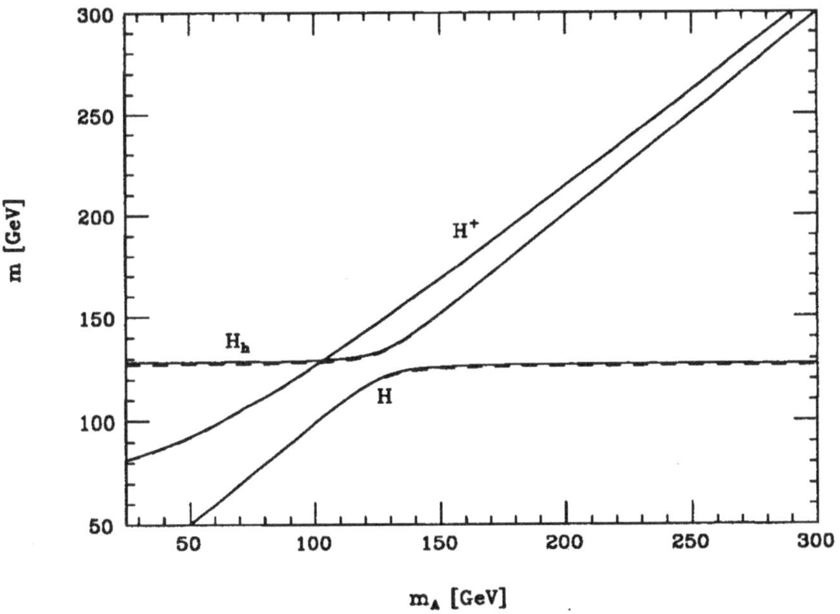

Figure 4: The neutral (H_h, H) and charged (H^+) Higgs mass spectrum as a function of the CP-odd Higgs mass m_A for a physical top-quark mass M_t = 175 GeV and $M_S = 1$ TeV, as obtained from the one-loop improved RG evolution (solid lines) and the analytical formulae (dashed lines). All sets of curves correspond to $\tan\beta = 15$ and large squark mixing, $X_t^2 = 6M_S^2$ $(\mu = 0)$.

is plotted versus m_A.

2.2 *Threshold effects*

There are two possible caveats in the analytical approximation we have just presented: i) Our expansion parameter $\log(M_S^2/m_t^2)$ does not behave properly in the supersymmetric limit $M_S \to 0$, where we should recover the tree-level result. ii) We have expanded the threshold function $\Delta V_{\rm th}[\phi(M_S)]$ to order X_t^4. In fact keeping the whole threshold function $\Delta V_{\rm th}[\phi(M_S)]$ we would be able to go to larger values of X_t and to evaluate the accuracy of the approximation (3) and (7). Only then we will be able to check the reliability of the maximum value of the lightest Higgs mass (which corresponds to the maximal mixing) as provided in the previous sections. This procedure has been properly followed [14,16] for the most general case $m_Q \neq m_U \neq m_D$. We have proved that keeping the exact threshold function $\Delta V_{\rm th}[\phi(M_S)]$, and properly running its value from the high scale to m_t with the corresponding anomalous dimensions as in (6), produces two effects: i) It makes a resummation from M_S^2 to $M_S^2+m_t^2$ and generates as (physical) expansion parameter $\log[(M_S^2+m_t^2)/m_t^2]$. ii) It generates a whole threshold function $X_t^{\rm eff}$ such that (7) becomes

$$\Delta_{\rm th} m_H^2 = \frac{3}{4\pi^2}\frac{m_t^4[M_S^2+m_t^2]}{v^2(m_t)}X_t^{\rm eff} \tag{9}$$

and

$$X_t^{\rm eff} = \frac{X_t^2}{M_S^2+m_t^2}\left(2-\frac{1}{6}\frac{X_t^2}{M_S^2+m_t^2}\right)+\cdots \tag{10}$$

The numerical calculation shows [16] that $X_t^{\rm eff}$ has the maximum very close to $X_t^2 = 6(M_S^2+m_t^2)$, what justifies the reliability of previous upper bounds on the lightest Higgs mass.

3 Would a light Higgs detection imply new physics?

From the previous sections it should be clear by now that the Higgs and top mass measurements could serve to discriminate between the SM and its extensions, and to provide information about the scale of new physics Λ. In Fig. 5 we give the SM lower bounds on M_H for $\Lambda \gtrsim 10^{15}$ (thick lines) and the upper bound on the mass of the lightest Higgs boson in the MSSM (thin lines) for $M_S \sim 1$ TeV. Taking, for instance, $M_t = 180$ GeV, which coincides with the central value recently reported by CDF+D0 [17], and $M_H \gtrsim 130$ GeV, the SM is allowed and the MSSM is excluded. On the other hand, if $M_H \lesssim 130$ GeV, then the MSSM is allowed while the SM is excluded. Likewise there are regions where the SM is excluded, others where the MSSM is excluded and others where both are permitted or both are excluded.

Finally from the bounds $M_H(\Lambda)$ (see Fig. 6) one can easily deduce that a measurement of M_H might provide an upper bound (below the Planck scale) on the scale of new physics provided that

$$M_t > \frac{M_H}{2.25\ \text{GeV}} + 123\ \text{GeV} \tag{11}$$

Thus, the present experimental bound from LEP, $M_H > 64$ GeV, would imply,

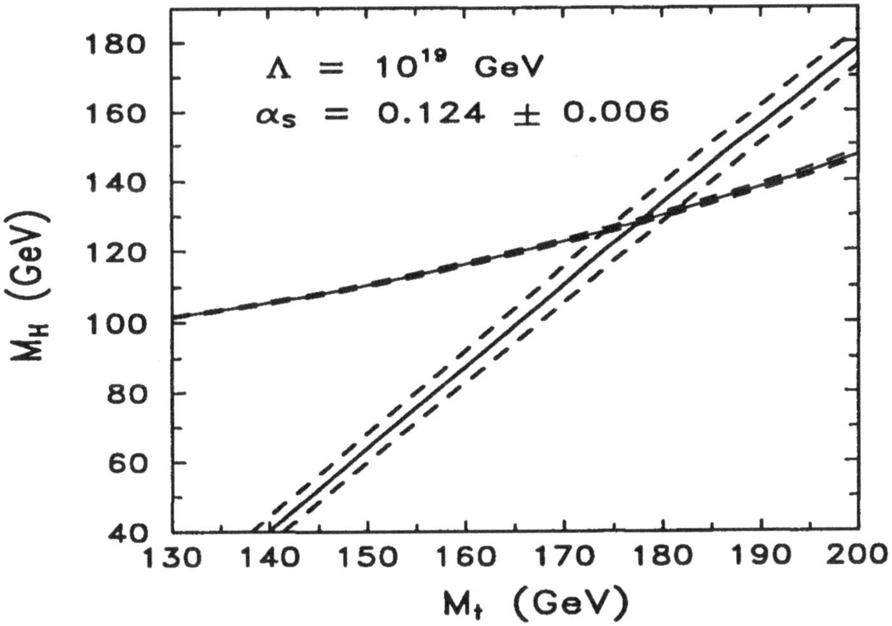

Figure 5: SM lower bounds on M_H (thick lines) as a function of M_t, for $\Lambda = 10^{19}$ GeV, from metastability requirements, and upper bound on the lightest Higgs boson mass in the MSSM (thin lines) for $M_S = 1$ TeV.

from (11), $M_t > 152$ GeV, which is fulfilled by experimental detection of the top [17]. Even non-observation of the Higgs at LEP 2 (i.e. $M_H \gtrsim 95$ GeV), would leave an open window ($M_t \gtrsim 163$ GeV) to the possibility that a future Higgs detection at LHC could lead to an upper bound on Λ. Moreover, Higgs detection at LEP 2 would put an upper bound on the scale of new physics. Taking, for instance, $M_H \lesssim 95$ GeV and 170 GeV $< M_t <$ 180 GeV, then

$\Lambda \lesssim 10^7$ GeV, while for 180 GeV $< M_t <$ 190 GeV, then $\Lambda \lesssim 10^4$ GeV, as can be deduced from Fig. 6.

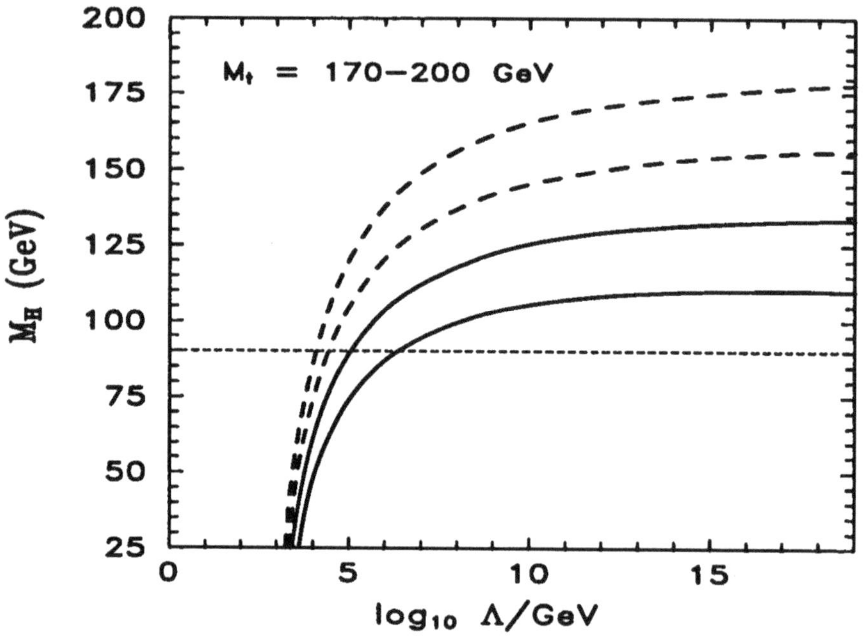

Figure 6: SM lower bounds on M_H from metastability requirements as a function of Λ for different values of M_t.

Acknowledgments

Work supported in part by the European Union (contract CHRX-CT92-0004) and CICYT of Spain (contract AEN94-0928).

References

1. N. Cabibbo, L. Maiani, G. Parisi and R. Petronzio, *Nucl. Phys.* **B158** (1979) 295; M. Lindner, *Z. Phys.* **C31** (1986) 295; M. Sher, *Phys. Rep.* **179** (1989) 273; M. Lindner, M. Sher and H.W. Zaglauer, *Phys. Lett.* **B228** (1989) 139; M. Sher, *Phys. Lett.* **B317** (1993) 159; Addendum: *Phys. Lett.* **B331** (1994) 448; C. Ford, D.R.T. Jones, P.W. Stephenson and M.B. Einhorn, *Nucl. Phys.* **B395** (1993) 17

2. G. Altarelli and I. Isidori, *Phys. Lett.* **B337** (1994) 141
3. J.A. Casas, J.R. Espinosa and M. Quirós, *Phys. Lett.* **B342** (1995) 171
4. J.A. Casas, J.R. Espinosa, M. Quirós and A. Riotto, *Nucl. Phys.* **B436** (1995) 3; (E) **B439** (1995) 466
5. J.R. Espinosa and M. Quirós, *Phys. Lett.* **B355** (1995) 257
6. For a recent review, see, e.g.: M. Quirós, *Helv. Phys. Acta* **67** (1994) 451
7. L. Dolan and R. Jackiw, *Phys. Rev.* **D9** (1974) 3320; S. Weinberg, *Phys. Rev.* **D9** (1974) 3357; D.J. Gross, R.D. Pisarski and L.G. Yaffe, *Rev. Mod. Phys.* **53** (1981) 43; M.E. Carrington, *Phys. Rev.* **D45** (1992) 2933; M.E. Shaposhnikov, *Phys. Lett.* **B277** (1992) 324 and (E) *Phys. Lett.* **B282** (1992) 483; M. Dine, R.G. Leigh, P. Huet, A. Linde and D. Linde, *Phys. Lett.* **B283** (1992) 319 and *Phys. Rev.* **D46** (1992) 550; J.R. Espinosa and M. Quirós, *Phys. Lett.* **B305** (1993) 98; J.R. Espinosa, M. Quirós and F. Zwirner, *Phys. Lett.* **B314** (1993) 206; C.G. Boyd, D.E. Brahm and S.D. Hsu, *Phys. Rev.* **D48** (1993) 4963; P. Arnold and O. Espinosa, *Phys. Rev.* **D47** (1993) 3546; W. Buchmüller, T. Helbig and D. Walliser, *Nucl. Phys.* **B407** (1993) 387
8. A.D. Linde, *Phys. Lett.* **B70** (1977) 306; *Phys. Lett.* **B100** (1981) 37; *Nucl. Phys.* **B216** (1983) 421
9. S. Coleman, *Phys. Rev.* **D15** (1977) 2929
10. P. Arnold and S. Vokos, *Phys. Rev.* **D44** (1991) 3620
11. Y. Okada, M. Yamaguchi and T. Yanagida, *Prog. Theor. Phys.* **85** (1991) 1; *Phys. Lett.* **B262** (1991) 54; J. Ellis, G. Ridolfi and F. Zwirner, *Phys. Lett.* **B257** (1991) 83; *Phys. Lett.* **B262** (1991) 477; R. Barbieri and M. Frigeni, *Phys. Lett.* **B258** (1991) 395; R. Barbieri, M. Frigeni and F. Caravaglios, *Phys. Lett.* **B258** (1991) 167
12. J.R. Espinosa and M. Quirós, *Phys. Lett.* **B266** (1991) 389
13. H.E. Haber and R. Hempfling, *Phys. Rev. Lett.* **66** (1991) 1815; A. Yamada, *Phys. Lett.* **B263** (1991) 233
14. M. Carena, J.R. Espinosa, M. Quirós and C.E.M. Wagner, *Phys. Lett.* **B355** (1995) 209
15. H.E. Haber, R. Hempfling and A.H. Hoang, CERN preprint, in preparation
16. M. Carena, M. Quirós and C.E.M. Wagner, preprint CERN-TH/95-157
17. F. Abe et al., CDF Collaboration, *Phys. Rev.* **D50** (1994) 2966; *Phys. Rev. Lett.* **73** (1994) 225; preprint FERMILAB-PUB-95/022-E (2 March 1995); S. Abachi et al., D0 Collaboration, *Phys. Rev. Lett.* **72** (1994) 2138; *Phys. Rev. Lett.* **74** (1995) 2422; preprint FERMILAB-PUB-95/028-E (3 March 1995)

IMPLICATIONS OF SPONTANEOUS R-PARITY BREAKING

J. C. Romão

Departamento de Física, Instituto Superior Técnico,
Av. Rovisco Pais, 1, 1096 Lisboa Codex, Portugal

If supersymmetry is realized with spontaneous breaking of R-parity, there will be important consequences in several different areas which can be tested through different types of experiments.These range from single chargino production at LEP, to new decay modes of the tau and muon leptons as well as the possibility of explaining, at the same time, the observed deficit in the solar neutrino flux. Also the Higgs sector in spontaneously broken R-parity supersymmetry shows interesting features that require new search techniques. For some parameter choices, the dominant Higgs boson decay can be invisible, leading to events with large missing transverse momentum carried by superweakly interacting majorons. This weakens the Higgs boson mass limits from LEP I. The conclusions for LEP II and LHC are similar and should be taken in account in the planning of Higgs boson search strategies. For the chargino and neutralino sectors, the effects of R-parity breaking at LEP II and LHC, will be mainly in the decays that will differ from those of the MSSM.

1 Introduction

Most studies of supersymmetric phenomenology have been made in the framework of the *Minimal Supersymmetric Standard Model (MSSM)* which assumes the conservation of a discrete symmetry called R-parity (R_p)[1]. Under this symmetry all the standard model particles are R-even while their superpartners are R-odd. R_p is related to the spin (S), total lepton (L), and baryon (B) number according to $R_p = (-1)^{(3B+L+2S)}$. Therefore the requirement of baryon and lepton number conservation implies the conservation of R_p. Under this assumption the supersymmetric (SUSY) particles must be pair-produced, every SUSY particle decays into another SUSY particle and the lightest of them is absolutely stable. These three features underlie all the experimental searches for new supersymmetric states.

However, neither gauge invariance nor SUSY require R_p conservation. The most general supersymmetric extension of the standard model contains explicit R_p violating interactions that are consistent with both gauge invariance and supersymmetry. Detailed analysis of the constraints on these models and their possible signals have been made[2]. In general, there are too many independent couplings and some of these couplings have to be set to zero to avoid the proton to decay too fast.

For these reasons we restrict our attention to the possibility that R_p can be an exact symmetry of the Lagrangean, broken spontaneously through the Higgs mechanism[3,4,5]. This may occur via nonzero vacuum expectation values for scalar neutrinos, such as

$$v_R = \langle \tilde{\nu}_{R\tau} \rangle \qquad ; \qquad v_L = \langle \tilde{\nu}_{L\tau} \rangle \ . \tag{1}$$

If spontaneous R_p violation occurs in absence of any additional gauge symmetry, it leads to the existence of a physical massless Nambu-Goldstone boson, called Majoron (J)[3]. In these models there is a new decay mode for the Z^0 boson, $Z^0 \to \rho + J$, where ρ is a light scalar. This decay mode would increase the invisible Z^0 width by an amount equivalent to 1/2 of a light neutrino family[6]. The LEP measurement on the number of such neutrinos is enough to exclude any model where the Majoron is not mainly an isosinglet[6]. The simplest way to avoid this limit is to extend the *MSSM*, so that the R_p breaking is driven by isosinglet VEVs, so that the Majoron is mainly a singlet[4].

Here we will review the phenomenological implications of a model with Spontaneous Broken R-Parity (SBRP).

2 A Viable Model for SBRP

In order to set up our notation we recall the basic ingredients of the model for spontaneous violation of R parity and lepton number proposed by Masiero and Valle[4] The superpotential is given by

$$\begin{aligned} W &= h_u Q H_u u^c + h_d H_d Q d^c + h_e \ell H_d e^c \\ &+ (h_0 H_u H_d - \varepsilon^2)\Phi + \hat{\mu} H_u H_d + h_\nu \ell H_u \nu^c \\ &+ h \Phi S \nu^c + M \nu^c S + m_\Phi \Phi\Phi \end{aligned} \tag{2}$$

This superpotential conserves *total* lepton number and R_p. The superfields $\Phi, {\nu^c}_i$ and S_i are singlets under $SU_2 \otimes U(1)$ and carry a conserved lepton number assigned as $(0,-1,1)$ respectively. All couplings $h_u, h_d, h_e, h_\nu, h, M$ are described by arbitrary matrices in generation space which explicitly break flavor conservation.

It has been shown[4,7] that these singlets may drive the spontaneous violation of R parity leading to the existence of a Majoron given by the imaginary part of

$$\frac{v_L^2}{V v^2}(v_u H_u - v_d H_d) + \frac{v_L}{V}\tilde{\nu_\tau} - \frac{v_R}{V}\tilde{\nu^c}_\tau + \frac{v_S}{V}\tilde{S}_\tau \tag{3}$$

where the isosinglet VEVs

$$v_R = \langle \tilde{\nu}_{R\tau} \rangle \quad , \quad v_S = \left\langle \tilde{S}_\tau \right\rangle \tag{4}$$

with $V = \sqrt{v_R^2 + v_S^2}$ characterize R_p or lepton number breaking and the isodoublet VEVS

$$v_u = \langle H_u \rangle \quad , \quad v_d = \langle H_d \rangle \quad , \quad v_L = \langle \tilde{\nu}_{L\tau} \rangle \tag{5}$$

drive electroweak breaking and the fermion masses.

3 Main Features of the Model

3.1 Charged Fermion Sector

The most important features of this model in the fermionic sector are the mixings of the *normal* leptons with the supersymmetric ones. In the charged lepton sector the electron, muon and tau leptons are mixed with the charginos in a 5×5 mass matrix[5]. This in turn gives place to new couplings like, for instance, the coupling of the Z^0 to the chargino and tau lepton that can be important at LEP I, as we will see.

3.2 Neutral Fermion Sector

In the neutral lepton sector the neutrinos are mixed with the neutralinos. The full matrix is now 14×14, but with an appropriate choice of parameters the heavy fermion singlets decouple giving raise to an effective 7×7 mass matrix[5] where the 7 fermions (Majorana) are the neutrinos and the usual neutralinos of MSSM. An important consequence of this mixing is that although no tree level mass term is given for the neutrinos, two of them will get mass after diagonalization of the mass matrix. For typical values of the parameters we get masses in the following range[8] ($m_1 = 0$)

$$\begin{aligned} 10^{-4}\ eV \leq m_2 \leq 10^{-2}\ eV \\ 10\ keV \leq m_\tau \leq 31\ MeV \end{aligned} \tag{6}$$

where we have identified the tau neutrino with the third eigenstate. This is nice because we get m_2 in an interesting range for the MSW mechanism but we have to be careful with m_{ν_τ}. In fact if the τ_ν were stable (or long-lived) we would have problems with the limit on stable neutrinos from cosmology

$$\sum m_i \leq 100\ eV \tag{7}$$

where the sum runs over stable neutrinos. In our model this problem is avoided through mixing among the generations. This is achieved by having a non-diagonal coupling matrix $h_{\nu ij}$. Then the ν_i will mix and the ν_τ will decay sufficiently fast to avoid any conflict with cosmology. We see then that lepton flavor violation is needed for consistency of the model. In doing this analysis we included all the known experimental constraints coming from astrophysics, cosmology and laboratory. For instance, the constraints on lepton flavor violation and universality but also the LEP constraints on the chargino masses and invisible Z^0 width. A complete discussion of the experimental constraints as well as of the results is given in[5,8,9,10].

3.3 The Higgs Sector

Here we will review briefly the main features of the Higgs bosons in our model. First of all the charged Higgs are exactly as in the *MSSM*. This is because we also have just two Higgs doublets. On the contrary, the neutral Higgs sector is far more complicated. Here we are going to make the simplification of considering just a 1-generation model. As we have said in the previous section, a phenomenologically consistent model requires the presence of flavor nondiagonal couplings in order to assure that the ν_τ decays fast enough. In this approximation the M_R^2 and M_I^2 mass matrices are 6×6. This gives 6 *CP* even states and 5 *CP* odd, including the massless majoron. Radiative corrections to the mass matrices have been also studied[11]. The consequences of their inclusion is similar to *MSSM* and we will not discuss them further here.

The two main points regarding the Higgs sector have to do with different production and decay rates. The production mechanisms are the Bjorken process

$$Z \to H_i f\bar{f} \tag{8}$$

and

$$Z \to H_i A_j \tag{9}$$

In the *MSSM* where $H_1 = h$, $H_2 = H$ and $A_2 = A$ (there is no majoron in this limit) there is a complementarity between the two types of decays and this has been used as the basis of the experimental analysis used to place limits on the SUSY Higgs spectrum. In the present model there is a breakdown in the sum rule involving the coupling strengths characterizing the processes in eq. (8) and eq. (9) with respect to *MSSM* expectations. As a result the overall Higgs production rates can be weaker than in *MSSM*.

Regarding the decays of the Higgs, there is in the models with spontaneously broken R_p, a new Higgs decay channel with majoron emission

$$H_i \to JJ \tag{10}$$

In order to evaluate the rates we need to evaluate the corresponding trilinear couplings. This was done in[11] where the decay channel eq. (10) was compared with the other channels. The conclusion was that normally the dominant decay mode is the invisible channel given in eq. (10) over most of the kinematical range available. Since the majoron is weakly coupled, it will escape detection. Thus the decay eq. (10) will lead to events with missing energy carried by the majorons. This has to taken in account when establishing limits on the Higgs mass.

3.4 An Effective Model: EBRP

For most practical applications at LEP and LHC our model is equivalent to a SUSY model with an *explicit* R-parity breaking term. More precisely most of the phenomenology can be studied for a model defined by the superpotential

$$\begin{aligned} W &= h_u Q H_u u^c + h_d H_d Q d^c + h_e \ell H_d e^c \\ &+ (h_0 H_u H_d - \varepsilon^2)\Phi + \epsilon \ell H_u \nu^c \end{aligned}$$

where the last term explicitly breaks R-parity. The connection between this model of explicitly broken R-parity (EBRP) and the SBRP is given in simple terms by the relation

$$\epsilon = h_\nu v_R \tag{11}$$

The advantage of this model is to reproduce in a simpler way the physics of the fermion sector at LEP and LHC. However we know that the model is not consistent. This is due to the lack of lepton flavor violation which has as consequence that the ν_τ is stable leading to problems with cosmology. Therefore it should be considered as an effective model, good to study the fermion sector at LEP and LHC. In the following some results will be presented for this effective model.

4 Laboratory Experiments

Here we will describe the implications of SBRP in non accelerator experiments.

4.1 Solar Neutrinos

To a good approximation we can write

$$\begin{aligned}\nu_1 &= \cos\theta\nu_e - \sin\theta\nu_\mu \\ \nu_2 &= \sin\theta\nu_e + \sin\theta\nu_\mu \\ \nu_3 &= \nu_\tau\end{aligned} \tag{12}$$

where ν_i, $i = 1,2,3$ and ν_e, ν_μ and ν_τ are, respectively, the mass and weak interaction eigenstates. The mixing angle θ is given in terms of the model parameters by

$$\tan\theta = \frac{h_{\nu_{13}}}{h_{\nu_{23}}} \tag{13}$$

The constraints on $h_{\nu_{13}}$ and $h_{\nu_{23}}$ do not restrict much their ratio. Therefore a large range of mixing angles is allowed. For the masses we get the values given in Eq 6, just in the right range for the MSW mechanism.

4.2 SUSY Signals in μ and τ Decays

The existence of a massless scalar particle, the majoron, can affect the decay spectra of the μ and τ leptons through the emission of the Majoron in processes such as

$$\mu \to e + J \quad ; \quad \tau \to e + J \quad ; \quad \tau \to \mu + J \tag{14}$$

These are flavor violating decays that are present in our model because the matrix $h_{\nu ij}$ is not flavor diagonal. As we saw before, this is an essential ingredient for the consistency of the model[8] and not something that we are adding now.

After a careful sampling of the parameter space, we found out[9] that the rates can be close to the present experimental limits. In Figure 1 we present the branching ratios for $\mu \to eJ$ as a function of m_{ν_τ}. We can see that for some values of the ν_τ mass we can get up to the present experimental limit[12], $BR(\mu \to e + J) < 2.6 \times 10^{-6}$.

5 LEP I Physics

In this section we review some implications of SBRP for LEP physics.

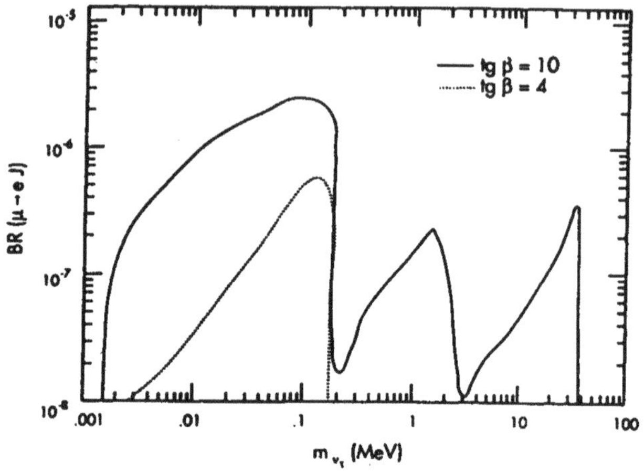

Figure 1: $BR(\mu \to eJ)$ **as a function of the mass of the** ν_τ **for various values of** $\tan\beta = v_u/v_d$.

5.1 Single Chargino Production

The first is the possibility of the decay

$$Z^0 \to \chi^{\pm}\tau^{\mp} \tag{15}$$

This decay is possible because R_p is broken. We have shown[5] that this branching ratio can be as high as 5×10^{-5}. This range is of interest for LEP I in the near future. The attainable branching ratios for single chargino production are shown in Figure 2, as a function of the chargino and ν_τ masses.

Another important point is that the chargino has different decay modes with respect to the MSSM. This is due to in addition to the 3-body decay $\chi \to \chi^0 + f\overline{f'}$ we also have know a 2-body decay $\chi \to \tau + J$. Therefore the branching ratios in the several final states will differ from those of the MSSM.

5.2 New Higgs Physics

The structure of the neutral Higgs sector is more complicated than in the MSSM. However the main points are simple to understand:

- Like in the MSSM the coupling of the Higgs to the Z^0 is reduced by a factor ε defined by

$$\varepsilon = \left| \frac{g_{ZZH}}{g_{ZZH}^{SM}} \right| < 1 \tag{16}$$

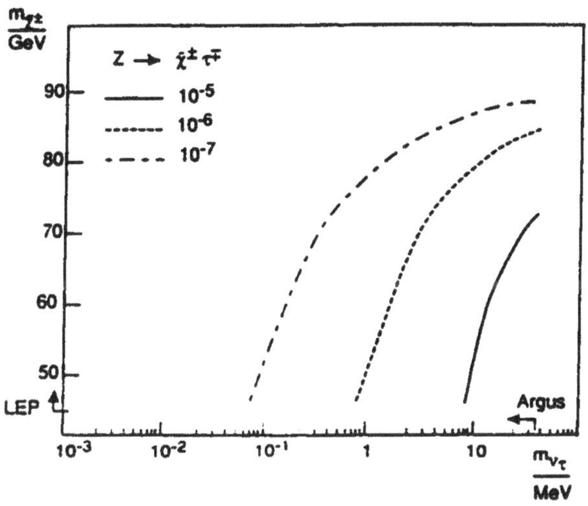

Figure 2: Attainable $BR(Z^0 \to \chi^\pm \tau^\mp)$ as a function of the chargino and ν_τ masses.

- Unlike the SM and the MSSM where the Higgs decays mostly in $b\bar{b}$, here it can have *invisible* decay modes like

$$H \to J + J \tag{17}$$

Depending on the parameters, the $BR(H \to \text{invisible})$ can be large. This will relax the mass limits obtained from LEP. We performed a model independent analysis of the LEP data[13] taking m_H, ε and $BR(H \to \text{invisible})$ as independent parameters. We show this in Figure 4 where the extension for LEP II is also included.

5.3 Single Photon Decays of the Z^0

In the SM the decay $Z^0 \to H + \gamma$ only wins over the usual $Z^0 \to H + f + \bar{f}$ for large m_H[14]. But when this happens the branching ratio is too small ($\leq 10^{-6}$). In the MSSM there are new contributions coming from the loop of charginos and the loop of charged Higgs. But in the end the final conclusion is that the resulting branching ratio is not much different from that of the SM. In the SBRP case there are three new aspects to this problem:

- If the ratio ε given in Eq. 16 is less than 1 we can have a low m_H. Then $Z^0 \to H + \gamma$ is not reduced by phase space.

- In the SBRP model we can also have the decay

$$Z^0 \to \gamma + J \tag{18}$$

This process needs R-parity to be broken. The branching ratios can be[15] as large as 10^{-5}.

- We can have also the decay

$$Z^0 \to \gamma + J + J \tag{19}$$

This would lead to a continuous photon spectrum in contrast with the monoenergetic spectrum of the previous example. A rough estimate indicates that this process can be sizeable. A detailed study is under way[16].

5.4 Lepton Monojets

Here we consider processes of the type

$$Z^0 \to \ell + \overline{\ell} + \not{p}_T \tag{20}$$

These processes are not unique to the SBRP model. In fact, they can occur in the SM

$$Z^0 \to \ell + \overline{\ell} + \nu + \overline{\nu} \tag{21}$$

where the missing energy is carried by the neutrinos, and also in the MSSM in processes like

$$Z^0 \to \chi_0' + \chi_0 \quad ; \quad \chi_0' \to \chi_0 + f + \overline{f} \tag{22}$$

where the missing energy is carried by the LSP, the χ_0. However there is not much phase space for this process to occur, because we must have

$$m_Z > m_{\chi_0} + m_{\chi_0'} \tag{23}$$

In the SBRP the main point is that the χ_0 is *not* the LSP. It can decay through *invisible* channels[5],

$$\chi_0 \to \nu_\tau + J \quad \text{or} \quad \chi_0 \to \nu_\tau + \nu + \overline{\nu} \tag{24}$$

or *visible* ones like,

$$\chi_0 \to \nu_\tau + \mu^+ + \mu^- \quad \text{or} \quad \chi_0 \to \tau^\pm + \nu + \mu^\mp \tag{25}$$

Then decays like

$$Z^0 \to \chi_0 + \chi_0 \tag{26}$$

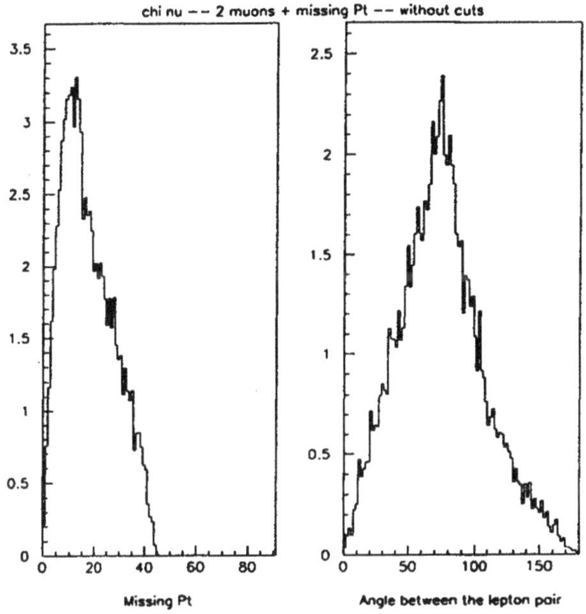

Figure 3: Distributions of missing p_T and of the angle between the 2 leptons.

where one of the χ_0 decays visibly and the other is invisible can take place even for

$$m_{\chi_0} + m_{\chi_0'} > m_Z \tag{27}$$

The signal will be 2 leptons in one side of the detector and nothing in the other side. It is important to notice that there is the possibility of lepton flavor violating decays

$$Z^0 \to \mu^{\pm} + \tau^{\mp} + \not{p}_T \tag{28}$$

We have studied these channels in some detail and the conclusion[17] is that the branching ratios can be as large as 10^{-5}, even for the lepton violating decays. These should be looked at LEP I. In order to study the possibility of detection of these processes, we constructed an event generator[17]. In Figure 3 is shown the missing p_T distribution as well the distribution of the angle between the two leptons. They clearly show that a cut of $\not{p}_T > 5GeV$ will not affect the signal. Also we can ask for an angle bigger than 10 degrees without losing much of the signal.

5.5 *Comment on the ALEPH search*

The ALEPH collaboration at LEP has studied[18] violating decays of the type

$$Z^0 \to \chi_0 + \chi_0 \to 4 \text{ leptons} + \not{p}_T \tag{29}$$

where the leptonic decays of the χ_0 are coming from *explicit* R-parity breaking. Although we are here considering *spontaneous* R-parity breaking their result does have some implication in our results. The first thing to notice is that the origin of R-parity is irrelevant in their analysis, provided that the χ_0 decays inside the detector. We have explicitly verified that this is true in our case. The second point is that they assume a 100% branching ratio into leptons. That is a consequence of the type of couplings that they consider. In our case the χ_0 can have other decay modes. Therefore the limits that they get for $Z^0 \to \chi_0 + \chi_0$ should in our case be divided by $BR(\chi_0 \to \text{leptons})^2$. This puts some constraints on the parameters of SBRP, but we have verified that they are not too strong. We still get large branching ratios for our processes.

6 Implications for LEP II and LHC

6.1 *Higgs Physics*

This is an extension of the results for LEP I. The main points are again that the production can be reduced with respect to the SM and that the decay can be invisible.

For LEP II we have extended[13,19] our analysis to map out the accessible region in the plane (ε^2, m_H). This is shown in Figure 4.

6.2 *Chargino and Neutralino Physics*

At LEP II and LHC the production rates will not be very large, compared with those at LEP I. Therefore we expect that the production rates will be like in the MSSM, via non R-parity breaking processes. However the decays will be modified much in the same way as in the LEP I case. This is specially important for the χ_0 because it is invisible in the MSSM but visible here. Also the R-parity violating decays of the charginos

$$\chi^- \to \tau^- + J \tag{30}$$

can have a substantial decay fraction compared with the MSSM decays

$$\chi^- \to \chi^0 + f\overline{f'} \tag{31}$$

Therefore care is needed in interpreting SUSY signals at those colliders.

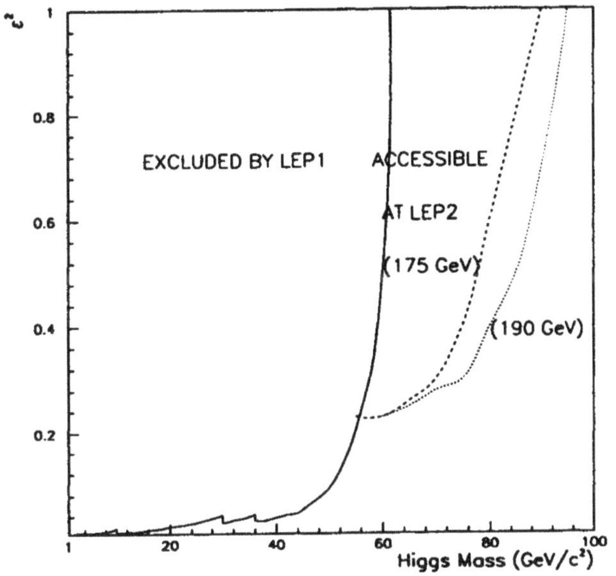

Figure 4: Excluded and accessible regions for LEP I and II in the ϵ^2 vs. m_H plane.

7 Conclusions

We have reviewed a viable model for spontaneous broken R-parity. This model leads to very rich phenomenology, both in laboratory experiments as well as at present and future accelerators. We also showed that some of these processes will have features that will allow to distinguish them from MSSM. At LEP I we are reaching the point where limits can be placed in some of the processes. Particularly interesting channels are

$$Z^0 \to \chi + \tau \tag{32}$$

$$Z^0 \to 2\text{leptons} + \not{p}_T \tag{33}$$

Acknowledgments

This work was partly supported by EU under contract number CHRX-CT92-004.

References

1. H Haber and G Kane, *Phys. Rep.* **117**, 75 (1985); H P Nilles *Phys. Rep.* **110**, 1 (1984).
2. V Barger, G F Giudice, T Han, *Phys. Rev.* D **40**, 2987 (1989); H Dreiner and G G Ross, Oxford preprint OUTP-91-15P.
3. C Aulakh, R Mohapatra, *Phys. Lett.* B **119**, 136 (1983); A Santamaria, J W F Valle, *Phys. Lett.* B **195**, 423 (1987); PRL **60**, 397 (1988); *Phys. Rev.* D **39**, 1780 (1989)
4. A Masiero, J W F Valle, *Phys. Lett.* B **251**, 273 (1990)
5. P Nogueira, J C Romão, J W F Valle, *Phys. Lett.* B **251**, 142 (1990)
6. M C Gonzalez-Garcia, Y Nir, *Phys. Lett.* B **232**, 383 (1990); P Nogueira, J C Romão, *Phys. Lett.* B **234**, 371 (1990).
7. J C Romão, C A Santos, J W F Valle, *Phys. Lett.* B **228**, 311 (1992).
8. J. C. Romão and J. W. F. Valle, *Phys. Lett.* B **272**, 436 (1991); *Nucl. Phys.* B **381**, 87 (1992).
9. J C Romão, N Rius and J W F Valle, *Nucl. Phys.* B **363**, 369 (1991).
10. M C Gonzalez-Garcia, J C Romão, J W F Valle, *Nucl. Phys.* B **391**, 100 (1993).
11. J.C.Romão, F. de Campos and J.W.F.Valle, *Phys. Lett.* B **292**, 329 (1992).
12. A Jodidio *et al.*, , *Phys. Rev.* D **34**, 1967 (1986); MARK II Collaboration, *Phys. Rev. Lett.* **55**, 1842 (1985); Argus Collaboration, *Phys. Lett.* B **246**, 278 (1990).
13. A Lopez-Fernandez, J. Romão, F. de Campos and J. W. F. Valle, *Phys. Lett.* B **312**, 240 (1993).
14. See F. Richard's talk at this Workshop.
15. J. C. Romão, J. Rosiek and J. W. F. Valle, *Phys. Lett.* B , (1995).
16. M. Magro, P. Nogueira, S. Novaes, J. C. Romão and J. W. F. Valle, in preparation.
17. F. de Campos, M. Garcia-Jareno, M. Magro, J. C. Romão and J. W. F. Valle, in preparation.
18. ALEPH Collaboration, CERN PPE 95/15. See also B. Michel's talk at this Workshop.
19. A Lopez-Fernandez, J. Romão, F. de Campos and J. W. F. Valle, to appear in the Proceedings of Moriond '94.

EFFECTS OF R PARITY BREAKING IN THE HIGGS SECTOR

J. ROSIEK

Institut für Theoretische Physik, Universität Karlsruhe

We discuss possible manifestations of R parity breaking in the Higgs sector. We illustrate this with three examples: high-energy mono-photon production at LEP, calculation of limits on associated production of invisibly decaying Higgs bosons and discussion of non-standard scalar decays in the MSSM with broken R parity.

1 Introduction

In analysing the supersymmetric theories one usually assumes the conservation of a discrete symmetry, called R parity[1], distinguishing matter fields from their superpartners. The R parity violation could lead to phenomenologically interesting effects, such as mixing of gauginos with leptons[2] and Higgs bosons with sleptons[3,4] (and hence breaking of the lepton number), non-zero neutrino masses[5,4] and their decays[6], Z^0 decays to single charginos or neutralinos[7], existence of a massless Goldstone boson called majoron[8] and many others. In contrast to the R parity breaking in the fermion sector, its consequences in the scalar sector are relatively less explored. We present three examples of possible manifestations of R-parity breaking in the Higgs sector: high-energy mono-photon production at LEP[9], associated production of invisibly decaying Higgs bosons[10] and novel scalar decays in the MSSM with broken R parity[11].

2 Single Photon Decays of the Z^0 and SUSY with Broken R-Parity

Recently the OPAL collaboration has published a high statistics single photon spectrum that shows some excess of high energy photons above the expectations from the initial state radiation (ISR)[12]. This excess could be explained in a class of SUSY models with spontaneous violation of R parity, which pre-

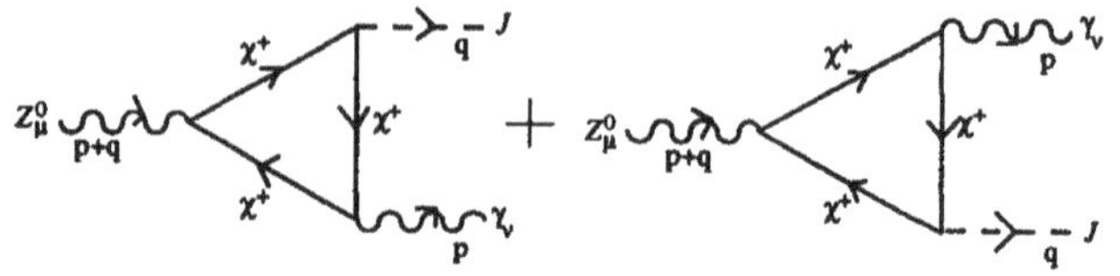

Figure 1: Diagrams contributing to the monochromatic single photon emission $Z^0 \to \gamma J$

dict the existence of new Z^0 decay modes involving single ($Z^0 \to J\gamma$) or double ($Z^0 \to JJ\gamma$) invisible massless majoron emission.

Single majoron emission would produce monochromatic photons emitted with energy $\frac{M_Z}{2}$. In order to estimate the attainable values of Br($Z^0 \to J\gamma$) we adopt as a model for the spontaneous R parity violation the one proposed in[13]. Diagrams contributing at the 1-loop level to the process $Z^0 \to J\gamma$ are shown in Fig. 1. We obtained the explicit expression for Br($Z^0 \to \gamma J$) by calculation of the 3-point Green functions of Fig. 1. We have varied model parameters, rejecting points violating constraints imposed by existing laboratory observations, cosmology and astrophysics. We have found that the branching ratio for the 2-body Z^0 decay to a single photon + missing momentum in the considered model can reach 10^{-5}, being within the sensitivity of LEP[9].

Double emission process $Z^0 \to JJ\gamma$ would give rise to a continuous photon spectrum[a] Imposing gauge and CP invariance and Bose symmetry, one can express the on-shell amplitude in terms of a single form-factor V_0:

$$A(Z^0 \to JJ\gamma) = V_0[p^\mu(q_1+q_2)^\nu - p(q_1+q_2)g^{\mu\nu}]\epsilon_\mu(p+q_1+q_2)\epsilon_\nu(p) \quad (1)$$

As the simplest illustration we derive the γ-spectrum following from eq. (1) in the approximation of constant V_0. We obtain

$$\frac{d\Gamma}{dE_\gamma} = \frac{|V_0|^2 M_Z}{96\pi^3} E_\gamma^3 \qquad 0 \le E_\gamma \le \frac{M_Z}{2} \quad (2)$$

The details of the shape of the γ spectrum depend upon the specific model and SUSY parameters choice, but its general characteristic is determined by kinematics and is different from the SM process $Z^0 \to \nu\bar{\nu}\gamma$. Superimposed

[a]We do not consider here other possible mechanisms, like LSP pair production followed by the radiative decay of one of them $\chi^0 \to \gamma\nu$ and the invisible decay of the other $\chi^0 \to J\nu$.

upon the continuous spectrum we have, in addition, a spike at its endpoint, corresponding to the emission of the monochromatic γ in the $Z^0 \to \gamma J$ decay.

3 Limits on Associated Production of Invisibly Decaying Higgs Bosons from Z^0 Decays

In a class of models with a spontaneously broken global symmetry the CP-even Higgs boson(s) are expected to have sizeable invisible decay modes to the majorons[14]. This decay could contribute to the signal looked for at LEP – two acoplanar jets + missing momentum. The existing analyses of the invisible Higgs search have concentrated on a minimally extended SM Higgs sector with the addition of a Higgs singlet and hence on the Bjorken process[15]. We extend this analysis to include the model containing two Higgs doublets and a singlet (carrying lepton number)[13] and the associated production of the CP-even and CP-odd Higgs bosons[10]. For the illustration, we display the constraints obtained for the published ALEPH data sample[16] based on a statistics of 1.23×10^6 hadronic Z^0 events.

After spontaneous $SU(2) \times U(1) \times U(1)_L$ breaking, the Higgs sector of the considered model contains 3 massive CP-even scalars H_i, a massive CP-odd scalar A and the massless majoron J. We assume that at LEP only the lightest CP-even scalar $H_1 \equiv H$ can be produced. The relevant couplings are:

$$\mathcal{L}_{HZZ} = (\sqrt{2}G_F)^{1/2} M_Z^2 \, \epsilon_B \, Z_\mu Z^\mu H \tag{3}$$

$$\mathcal{L}_{HAZ} = -\frac{e}{\sin\theta_W \cos\theta_W} \, \epsilon_A \, Z_\mu H \overleftrightarrow{\partial^\mu} A. \tag{4}$$

where $\epsilon_A^2, \epsilon_B^2 \leq 1$. The main H decay modes are $b\bar{b}$, JJ and AA (if $m_H > 2m_A$). HJJ coupling strength is unconstrained and can be effectively parameterized by $B = \mathrm{Br}(H \to JJ)$, $0 \leq B \leq 1$.[b] The decay $A \to JJJ$ does not exist since the (CP-allowed) couplings AJ^3 or AHJ vanish at the tree level in our model. Since A can decay only visibly ($B_A = \mathrm{Br}(A \to b\bar{b}) \approx 90\%$ for $m_A \geq 20$ GeV),

[b]The parameterization of the expected number of the invisible Higgs boson events in terms of m_H, m_A, ϵ_A, ϵ_B and B is quite general and not limited only to the model[13].

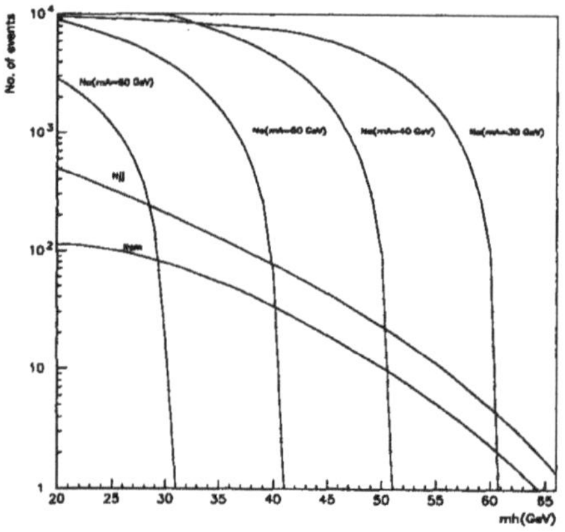

Figure 2: Numbers of expected dijet + missing momentum events after imposing ALEPH cuts.

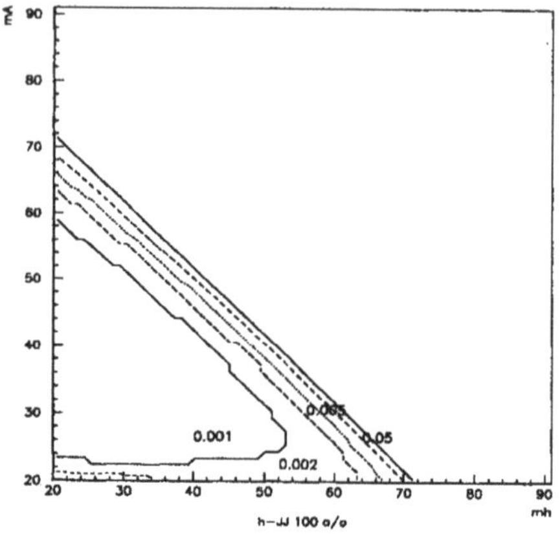

Figure 3: Limits on ϵ_A^2 in the m_A-m_H plane based on $e^+e^- \to HA \to JJb\bar{b}$ channel. We have assumed $\mathrm{Br}(H \to JJ) = 100\%$.

one expects dijets + missing momentum as a signal of the Higgs production. This signal arises from three processes: 1) $Z^{0\star} \to \nu\bar{\nu}$, $H \to b\bar{b}$, 2) $Z^{0\star} \to q\bar{q}$, $H \to JJ$ and 3) $H \to JJ$, $A \to b\bar{b}$. For each process one has a sizeable missing momentum which is aligned neither along the beam nor along the jets. For the SM background missing momentum arises from i) jet fluctuation (including escaping ν from b, c and τ jets) in which case it is aligned along the jets, and ii) ISR or $e^+e^- \to (e^+e^-)\gamma\gamma$ process in which case it is aligned along the beam direction. This enables one to eliminate the SM background by a combination of kinematic cuts[17]. We denote the number of expected signal events for the processes 1)-3), after the cuts, by N_{SM}, N_{JJ} and N_A respectively, assuming no suppression due to the mixing angles or branching fractions in each case. The expected number of signal events, after incorporating these effects, is given by

$$N_{2j} = \epsilon_B^2 \left[B N_J + (1 - B) N_{SM} \right] + \epsilon_A^2 B_A B N_A \tag{5}$$

As can be seen in Fig. 2, $N_A \gg N_{SM}, N_{JJ}$, implying that, if not suppressed kinematically or by mixing angles, associated production gives the strong limits on ϵ_A^2 (see Fig. 3, where we have assumed fully invisible H decay). The limit on ϵ_B^2 is given by the Bjorken process and is the same as obtained in[15]. A decay mode independent limits on ϵ_A^2, ϵ_B^2 can be obtained by varying B from 0 to 1 and combining the data on dijet + missing momentum with those from 4 and 6 b-jet searches. The overall limit is dominated by visible channels, which give weaker constraints on ϵ_A^2, ϵ_B^2 then those obtained for the invisible H decay.

4 Novel Scalar Boson Decays in SUSY with Broken R-Parity

The standard MSSM R parity conserving superpotential can be generalized by adding the following R parity (and lepton number) violating terms:

$$W_R = \epsilon_{ab}\left[\lambda_{ijk}\hat{L}_i^a\hat{L}_j^b\hat{E}_k^C + \lambda'_{ijk}\hat{L}_i^a\hat{Q}_j^b\hat{D}_k^C + \epsilon_i\hat{L}_i^a\hat{H}_2^b\right] \quad (6)$$

We focus on the effect of the last term in eq. (6), assuming in addition $\epsilon_{1,2} \approx 0$. The remaining $\epsilon_3\hat{L}_3\hat{H}_2$ term induces a non-zero VEV for the tau sneutrino: $\langle\tilde{\nu_\tau}\rangle \equiv \frac{v_3}{\sqrt{2}}$ and leads to mixing of gauginos with leptons, Higgs scalars with the tau sneutrino and hence to the R parity violating Higgs bosons decay modes.

All the elements of the various mixing matrices can be expressed in terms of six independent parameters which we choose as $\tan\beta = \frac{v_2}{\sqrt{v_1^2+v_3^2}}$, μ, ϵ_3, m_A^2, the gaugino and soft sneutrino mass parameters M_2 and m_{L_3}. We have taken into account the following constraints on the model parameters:

1) The major constraint on ϵ_3 and hence on R parity violating mixings comes from bound on ν_τ mass[18], induced by the non-zero ϵ_3: $m_{\nu_\tau} \leq 30$ MeV[c]

2) Decay widths $\Gamma(Z^0 \to \chi^0\chi^0, \chi^+\chi^-)$ should obey the LEP restrictions.

The decay $h \to \chi^0\nu_\tau$ occurs either through the sneutrino component of h,

$$h = a_{31}(\tilde{\nu}_\tau)_R + a_{21}(\phi_2)_R + a_{11}(\phi_1)_R \quad (7)$$

or through ν_τ admixture in the LSP in the $h\chi\chi$ vertex. The mixing a_{31} appearing in eq. (7) is of the order of $\mathcal{O}\left(\frac{\mu\epsilon_3}{m_h^2 - m_{\tilde{\nu}_\tau}^2}\right)$ and become large if the *sneutrino* mass is close to the mass of the relevant *Higgs* boson. The relative importance of the SUSY decay mode $h \to \chi^0\nu_\tau$ follows from the ratio:

---

[c]see comments in[11] on possible cosmological constraints.

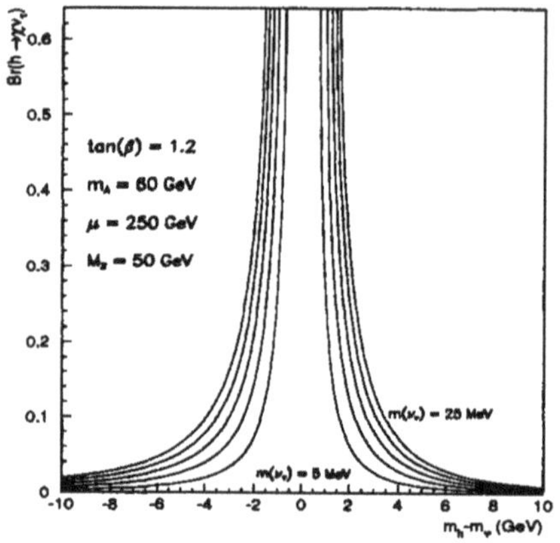

Figure 4: $\mathrm{Br}(h \to \chi^0 \nu_\tau)$ as a function of the difference of the physical h and $\tilde{\nu}_\tau$ masses. The ν_τ mass labels the different curves.

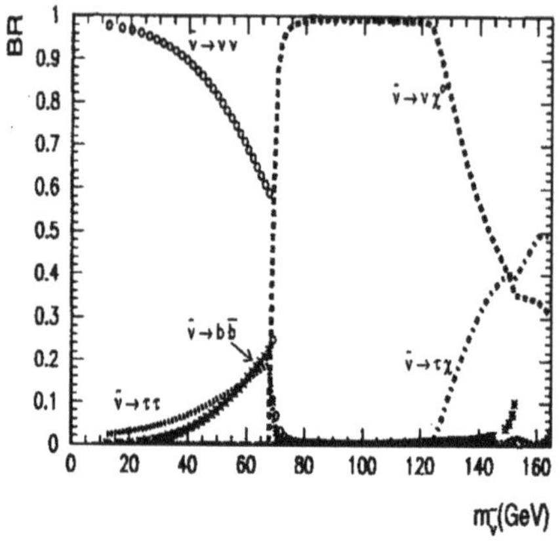

Figure 5: Branching ratios for $\tilde{\nu}_\tau$ decays as a function of its mass for $\tan\beta$=10,M_2=70 GeV,μ=200 GeV,ϵ_3=1 GeV,m_A=250 GeV.

$$R_0 = \frac{\Gamma(h \to \chi^0 \nu_\tau)}{\Gamma(h \to b\bar{b})} \approx \frac{\tan^2\theta_W}{2} \frac{M_W^2}{m_b^2} \frac{(1 - m_\chi^2/M_H^2)^2}{(1 - 4m_b^2/M_H^2)^{3/2}} \frac{a_{31}^2}{a_{11}^2} \cos^2\beta |\xi|^2 \qquad (8)$$

where ξ denotes the appropriate gaugino-LSP mixing element. R_0 and also $R_+ = \frac{\Gamma(h \to \chi^\pm \tau^\mp)}{\Gamma(h \to b\bar{b})}$ can be estimated to be of the order of $\mathcal{O}\left(\frac{\epsilon_3^2}{m_b^2}\right) \times \frac{\mu^2}{(m_h - m_{\tilde{\nu}_\tau})^2}$ In Fig. 4 we display the branching ratios for h decays to LSP + ν_τ as a function of the h-$\tilde{\nu}_\tau$ mass difference, for a suitable choice of SUSY parameters. Clearly, for relatively small h-$\tilde{\nu}_\tau$ mass differences of a few GeV, the supersymmetric channel can dominate over the SM ones (similarly for $A \to \chi^0 \nu_\tau, \chi^\pm \tau^\mp$ decays). Conversely, $\tilde{\nu}_\tau$ may decay into R parity violating SM channels such as $b\bar{b}$, $\tau^+\tau^-$ or the invisible mode $\nu\bar{\nu}$. These decays are dominant when the phase space for the R parity conserving channels such as $\chi\nu$ is closed (see Fig. 5 where $m_{LSP} \approx 65$ GeV) and may be non-negligible even above the LSP threshold if $m_h \approx m_{\tilde{\nu}_\tau}$, leading to a resonant enhancement of $b\bar{b}, \tau^+\tau^-$ modes (see small rise of $\mathrm{Br}(\tilde{\nu}_\tau \to b\bar{b})$ in Fig. 5 for $m_{\tilde{\nu}_\tau} \approx m_h = 155$ GeV). Finally, the R parity breaking terms lead to the unstable LSP, which would decay inside the detector if $\epsilon_3 \sim$ few GeV. Folding the R parity violating Higgs boson decays to the LSP with the standard decays of Z^0 one gets the signatures in e^+e^- collisions which do not occur in the SM, such as τe or $\tau\mu$ pairs + missing momentum[11].

Acknowledgements

This work was supported in part by postdoctoral fellowship of the Spanish MEC and by Alexander von Humboldt Stiftung.

References

1. S.Weinberg, *Phys. Rev.* **D26** (1982) 533; N.Sakai, T.Yanagida, *Nucl. Phys.* **B197** (1982) 533; for a review see J.W.F.Valle, *Nucl. Phys. B (Proc. Suppl.)* **13** (1990) 520 and *ibid.* **13** (1990) 195
2. G.G.Ross, J.W.F. Valle, *Phys. Lett.* **B151** (1985) 375; J.Ellis et al, *Phys. Lett.* **B150** (1985) 142
3. D.Comelli et al, *Phys. Lett.* **B224** (1994) 397
4. A.S.Joshipura, M.Nowakowski, FTUV/94-42, hep-ph 9408224
5. L.Hall, M.Suzuki, *Nucl. Phys.* **B231** (1984) 419; I.Lee *Nucl. Phys.* **B246** (1984) 120; K.Enqvist et al, *Nucl. Phys.* **B373** (1992) 95
6. J.C.Romão, J.W.F.Valle *Nucl. Phys.* **B381** (1992) 87
7. P.Nogueira et al, *Phys. Lett.* **B251** (1990) 142; R.Barbieri,L.Hall, *Phys. Lett.* **B238** (1990) 86; M.C.Gonzalez-Garcia, J.W.F.Valle, *Nucl. Phys.* **B355** (1991) 330; D.E.Brahm et al, *Phys. Rev.* **D42** (1992) 1860
8. Y.Chikashige, R.Mohapatra, R.Peccei, *Phys. Rev. Lett.* **45** (1980) 1926; for a review see J.W.F.Valle, *Prog. Part. Nucl. Phys.* **26** (1991) 91
9. J.C.Romão, J.Rosiek, J.W.F.Valle, *Phys. Lett.* **B351** (1995) 497
10. F.de Campos et al, *Phys. Lett.* **B336** (1994) 446
11. F.de Campos et al, hep-ph 9502237, in print in *Nucl. Phys.* **B** (1995)
12. OPAL collaboration, *Zeit. für Physik* **C65** (1995) 47
13. A.Masiero, J.W.F.Valle, *Phys. Lett.* **B251** (1990) 273; J.C.Romão, C.A.Santos, J.W.F.Valle, *Phys. Lett.* **B288** (1992) 311
14. J.D.Bjorken, report SLAC-PUB-5673 (1991); R.E.Schrock, M.Suzuki, *Phys. Lett.* **10B** (1982) 250; L.F.Li, Y.Liu, L.Wolfenstein, *Phys. Lett.* **B159** (1985) 45; J.C.Romão et al, *Phys. Lett.* **B292** (1992) 329
15. A.Lopez-Fernandez et al, *Phys. Lett.* **B312** (1993) 240; O.Eboli et al, *Nucl. Phys.* **B421** (1994) 65; D.P.Roy et al, *Phys.Rev.* **D48** (1993) 4224
16. ALEPH Collaboration, *Phys. Lett.* **B313** (1993) 299; **B313** (1993) 312
17. ALEPH collaboration, *Phys. Rep.* **216** (1992) 253
18. Particle Data Group, *Phys. Rev.* **D50** (1994) 1173

NATURAL FLAVOUR MIXING IN THE MSSM AND $\mu \to e, \gamma$

B. DE CARLOS.

Department of Physics, Theoretical Physics, 1 Keble Road, Oxford OX1 3NP, England

J.A. CASAS and J.M. MORENO

Instituto de Estructura de la Materia, CSIC, Serrano 123, 28006 Madrid, Spain

In the absence of any additional assumption it is natural to conjecture that sizeable flavour-mixing mass entries, Δm^2, may appear in the mass matrices of the scalars of the MSSM, i.e. $\Delta m^2 \sim O(m^2)$. This flavour violation can still be reconciled with the experiment if the gaugino mass, $M_{1/2}$, is large enough to yield (through the renormalization group running) a sufficiently small $\Delta m^2/m^2$ at low energy. This leads to a *gaugino dominance* framework (i.e. $M_{1/2}^2 \gg m^2$), which permits a remarkably model-independent analysis. We study this possibility focussing our attention on the $\mu \to e, \gamma$ decay. In this way we obtain very strong and general constraints, in particular $\frac{M_{1/2}^2}{\Delta m} \gtrsim 34$ TeV.

It is well-known that FCNC processes are very sensitive tests to physics beyond the standard model (SM) and, in particular, to supersymmetric extensions of the SM (SSM) [1]. Furthermore supersymmetry provides new direct sources of flavour violation, namely the possible (and even natural as we will see) presence of *off-diagonal* terms (say generically Δm^2) in the squark and slepton mass matrices[2,3,4,5,6,7]. In the present talk we will focus all our attention on the constraints on Δm^2 from the $\mu \to e, \gamma$ process because they are very strong and, as we will see, their evaluation is remarkably model-independent.

The minimal supersymmetric standard model (MSSM) is defined by the superpotential, W (from which the supersymmetric part of the Lagrangian is readily obtained), and the soft supersymmetry breaking terms coming from the (unknown) supersymmetry breaking mechanism

$$\begin{aligned} -\mathcal{L}_{\rm soft} &= \frac{1}{2} M_a \lambda_a \lambda_a + \left(m_L^2\right)_{ij} \bar{L}_i L_j + \left(m_{e_R}^2\right)_{ij} \bar{e}_{Ri} e_{Rj} \\ &+ \left(m_Q^2\right)_{ij} \bar{Q}_i Q_j + \left(m_{u_R}^2\right)_{ij} \bar{u}_{Ri} u_{Rj} + \left(m_{d_R}^2\right)_{ij} \bar{d}_{Ri} d_{Rj} \quad (1) \end{aligned}$$

$$+ \left[A^u_{ij} h^u_{ij} Q_i H_2 u_{Rj} + A^d_{ij} h^d_{ij} Q_i H_1 d_{Rj} + A^e_{ij} h^e_{ij} L_i H_1 e_{Rj} + B\mu H_1 H_2 + \text{h.c.}\right] ,$$

where i, j (a) are generation (gauge group) indices, λ_a are the gauginos, and the remaining fields in the formula denote just their corresponding scalar components in a standard notation. In the simplest version of the MSSM the soft breaking parameters are taken as *universal* (at the unification scale M_X). Then, the independent parameters of the theory are

$$\mu, m, M_{1/2}, A, B \tag{2}$$

(the rest of the parameters can be worked out demanding a correct unification of the gauge coupling constants and correct masses for all the observed particles). However this simplification is not at all a general principle. In particular there is no theoretical argument against non-vanishing off-diagonal $(m^2_{ij})_{i\neq j} \equiv \Delta m^2_{ij}$ terms. From the previous arguments, it is natural to assume that these off-diagonal entries can be *sizeable*, or even of the same order as the diagonal terms

$$\Delta m^2 \sim O(m^2) \ . \tag{3}$$

Certainly, there are proposed mechanisms to avoid this, for example the above-mentioned assumption of universality [8,9]. However one should wonder whether, in the absence of any additional assumption, the perfectly possible and even natural situation of eq. (3) could still be compatible with the experimental data and, more precisely, with the present experimental bound [10] on $\mu \to e, \gamma$

$$BR(\mu \to e, \gamma) \leq 5 \times 10^{-11} \ , \tag{4}$$

The expression of $BR(\mu \to e, \gamma)$ in the MSSM depends on several low-energy quantities, namely $\tan\beta \equiv \langle H_2\rangle/\langle H_1\rangle$, μ, A, and the spectrum of masses of sleptons and gauginos. These can be obtained from the initial parameters of the theory (see eq. (2)) through the corresponding RGEs (see ref. 11). On the other hand the ratio $\Delta m^2/m^2$ will in general be small at low energies (even if it is $O(1)$ at M_X), provided that gaugino masses are bigger than scalar masses, $M^2_{1/2} \gg m^2$, because of the contribution of the former in the RGEs of the diagonal parts of the latter, which is not the case for the off-diagonal entries. This is the reason why the RGEs have the potential to "cure" initial sizeable values of Δm. Therefore, the assumption of naturally large flavour mixing at M_X leads us necessarily to a $M^2_{1/2} \gg m^2$ ("gaugino dominance") [12] scenario, where *all* the soft breaking parameters are essentially determined *at low energies* by the value of $M_{1/2}$ at M_X independently of their initial values. Note that all this does *not* apply to the μ parameter, as it renormalises proportional to itself. However, the further requirement of a correct electroweak breaking fixes the value of μ, giving us the *whole* spectrum and other relevant low-energy quantities (such as A and $\tan\beta$) in terms of a *unique* parameter $M_{1/2}$. In particular, the values of $\tan\beta$ obtained in this framework tend to be rather large (ranging from 11 to 26 as $M_{1/2}$ increases from 150 GeV to 10 TeV. The gaugino dominance is a very interesting fact that makes the subsequent analysis rather accurate and model–independent.

At lowest order, the $\mu \to e,\gamma$ process is induced by one–loop diagrams that involve a flip of the leptonic flavour triggered by the slepton mixing, besides the propagation of a neutralino or chargino (see ref. 11 for details). Since the electron and muon Yukawa couplings are very suppressed, only the gauge part of the couplings of the charginos and neutralinos will play a role in the diagrams. In fact, one important consequence of the gaugino dominance framework is that for large enough values of $M_{1/2}$ the neutralinos (charginos) are almost pure neutral (charged) gaugino and higgsino. Then, the relevant diagrams correspond to bino ($\tilde{B}$) and wino ($\tilde{W}^0$, $\tilde{W}^-$) exchange. We have evaluated *all* of them. The expressions given in the previous literature are either incomplete or not directly applicable to our case.

Although in principle all the diagrams can have a similar magnitude (e.g. if we assume $\Delta m^2_{\tilde{\nu}_e\tilde{\nu}_\mu} \sim \Delta m^2_{\tilde{e}_L\tilde{\mu}_L} \sim \Delta m^2_{\tilde{e}_R\tilde{\mu}_R}$), in practice the bino diagram is the dominant one. This comes from the coefficient of proportionality ($A+\mu\tan\beta$), that appears in its evaluation and turns out to be very important in the gaugino dominance framework due to the large $\tan\beta$ value.

The theoretical $BR(\mu \to e,\gamma)$ depends on two different sets of parameters. First, the different masses involved in the game ($m^2_{\tilde{\nu}}$, $m^2_{\tilde{l}_L}$, $m^2_{\tilde{l}_R}$, $M_{\tilde{B}}$, $M_{\tilde{W}}$) and certain relevant low-energy quantities ($A, \mu, \tan\beta$). Second, the three independent flavour-mixing mass entries: $\Delta m^2_{\tilde{\nu}_e\tilde{\nu}_\mu}$, $\Delta m^2_{\tilde{e}_L\tilde{\mu}_L}$, $\Delta m^2_{\tilde{e}_R\tilde{\mu}_R}$. As explained in before, once we are working in the framework of gaugino dominance, $M^2_{1/2} \gg m^2$, the first set is completely determined in terms of the initial gaugino mass, $M_{1/2}$. Recall that we were led to this framework by the mere assumption of naturally large flavour mixing at M_X (see eq. (3)). The three flavour-mixing mass parameters, however, remain independent.

The constraints on the MSSM from $BR(\mu \to e,\gamma)$ arise by evaluating the previous diagrams and comparing them with the present experimental bound, eq. (4). We have illustrated this in Fig. 1, where an overall mass-mixing parameter $\Delta m^2_{\tilde{\nu}_e\tilde{\nu}_\mu} = \Delta m^2_{\tilde{e}_L\tilde{\mu}_L} = \Delta m^2_{\tilde{e}_R\tilde{\mu}_R} \equiv \Delta m^2$ has been taken for simplicity. Then we have plotted $BR(\mu \to e,\gamma)$ *vs* $M_{1/2}$ for different values of Δm. From this figure we can derive the maximum allowed value of Δm (or, equivalently, the minimum allowed value of $M_{1/2}/\Delta m$) for each value of $M_{1/2}$. This is represented in Fig. 2 for four different cases: *a)* $\Delta m^2_{\tilde{\nu}_e\tilde{\nu}_\mu} = \Delta m^2_{\tilde{e}_L\tilde{\mu}_L} =$

$\Delta m^2_{\tilde{e}_R\tilde{\mu}_R} \equiv \Delta m^2$; *b)* only $\Delta m^2_{\tilde{e}_R\tilde{\mu}_R} \neq 0$; *c)* only $\Delta m^2_{\tilde{e}_L\tilde{\mu}_L} \neq 0$ and *d)* only $\Delta m^2_{\tilde{\nu}_e\tilde{\nu}_\mu} \neq 0$, which gives a complete picture of the results. Notice that the *(d)* case is the less restrictive one.

The constraints are in general extremely strong. For case *(a)*, which is the most representative one, the corresponding curve can be approximately fitted by the simple constraint

$$\frac{M^2_{1/2}}{\Delta m} \gtrsim 34 \text{ TeV} \tag{5}$$

(similar equations can be written for the other curves). Under the assumption of eq. (3), i.e. $\Delta m = O(m)$, the results of Fig. 2 or eq. (5) imply that, indeed, a very large hierarchy between the scalar and gaugino masses is needed in order to reconcile the theoretical and experimental results. This gives full justification to our assumption of a gaugino dominance framework once eq. (3) has been conjectured. For example, for $M_{1/2} \sim 500$ GeV the assumption $\Delta m \sim m$ demands $M_{1/2}/\Delta m > 65$. Actually, it is hard to think of a scenario where such a dramatical hierarchy can naturally arise. Consequently, we can conclude at this point that a naturally large flavour mixing, as that conjectured in eq. (3), can hardly be reconciled with the experiment in a natural way.

We can now summarize our work. In the absence of any additional assumption it is natural to conjecture that sizeable flavour-mixing mass entries, Δm^2, may appear in the mass matrices of the scalars of the MSSM, i.e. $\Delta m^2 \sim O(m^2)$. This flavour violation can still be reconciled with the experiment if the gaugino mass, $M_{1/2}$, is large enough to yield (through the renormalization group running) a sufficiently small $\Delta m^2/m^2$ at low energy. We have analyzed this possibility, focussing our attention on the leptonic sector, particularly on the $\mu \to e, \gamma$ decay, which is by far the FCNC process with higher potential to restrict the value of the off-diagonal terms, Δm^2. The results are the following:

1. The $\Delta m^2 \sim O(m^2)$ conjecture automatically leads to a *gaugino dominance* framework (i.e. $M^2_{1/2} \gg m^2$), where, apart from Δm^2 itself, *all* the relevant low-energy quantities (mass spectrum, A, μ, $\tan\beta$) are determined in terms of a unique parameter, $M_{1/2}$. This makes the subsequent analysis and results remarkably model–independent.

2. The resulting constraints in the MSSM, obtained by comparing the calculated $BR(\mu \to e, \gamma)$ with the experimental bound, are very strong (see Figs. 1, 2 and eq. (5)). This makes, in our opinion, the natural flavour mixing conjecture $\Delta m^2 \sim O(m^2)$ extremely hard to be reconciled with the experiment in a natural way. Hence, $\Delta m/m$ should be small already at the unification scale.

Finally, let us comment that the need of starting with small $\Delta m/m$ can be satisfied in some theoretically well-founded scenarios, which become favoured from this point of view. In particular, we would like to stress that many string constructions can be consistent with that requirement [13,14,15]. Other scenarios, however, can produce a larger non-universality of the scalar masses, with potentially dangerous contributions to FCNC processes [5]. In any case, these non-universality effects are to produce non-vanishing off-diagonal terms in the scalar mass matrices once the usual rotation of fields to get diagonal fermionic mass matrices is carried out. The phenomenological viability of these physically relevant scenarios undoubtedly deserves further investigation.

References

1. S. Dimopoulos and H. Georgi, Nucl. Phys. B193 (1981) 150;
   J. Ellis and D.V. Nanopoulos, Phys. Lett. B110 (1982) 44;
   R. Barbieri and R. Gatto, Phys. Lett. B110 (1982) 211.
2. F. Gabbiani and A. Masiero, Nucl. Phys. B322 (1989) 235.
3. J.S. Hagelin, S. Kelley and T. Tanaka, Nucl. Phys. B415 (1994) 293.
4. D. Choudhury, F. Eberlein, A. König, J. Louis and S. Pokorski, Phys. Lett. B342 (1995) 180.
5. P. Brax and M. Chemtob, Phys. Rev. D51 (1995) 6550;
   P. Brax and C.A. Savoy, Nucl. Phys. B447 (1995) 227.
6. R. Barbieri and L.J. Hall, Phys. Lett. B338 (1994) 212.
7. S. Dimopoulos and D. Sutter, preprint CERN-TH-95-101 (1995).
8. Y. Nir and N. Seiberg, Phys. Lett. B309 (1993) 337.
9. M. Dine, R. Leigh and A. Kagan, Phys. Rev. D48 (1993) 4269.
10. R.D. Bolton et al., Phys. Rev. D38 (1988) 2077.
11. B. de Carlos, J.A. Casas, J.M. Moreno, preprint IEM-FT-108/95, hep-ph/9507337.
12. M. Dine, A. Kagan and S. Samuel, Phys. Lett. B243 (1990) 250.
13. V. Kaplunovsky and J. Louis, Phys. Lett. B306 (1993) 269.
14. A. Brignole, L.E. Ibáñez and C. Muñoz, Nucl. Phys. B422 (1994) 125; erratum-ibid. B436 (1995) 747.
15. B. de Carlos, J.A. Casas and C. Muñoz, Phys. Lett. B299 (1993) 234.

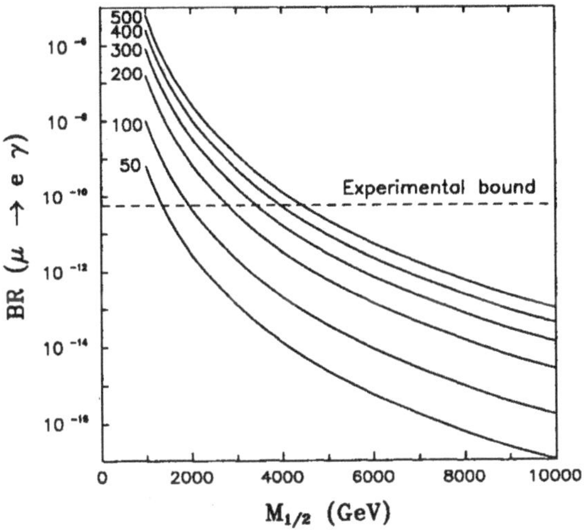

Figure 1: Plot of $BR(\mu \to e, \gamma)$ vs. $M_{1/2}$, taking for simplicity $\Delta m^2_{\tilde{\nu}_e \tilde{\nu}_\mu} = \Delta m^2_{\tilde{e}_L \tilde{\mu}_L} = \Delta m^2_{\tilde{e}_R \tilde{\mu}_R} \equiv \Delta m^2$. The different curves correspond to $\Delta m = 50$, 100, 200, 300, 400, 500 GeV respectively.

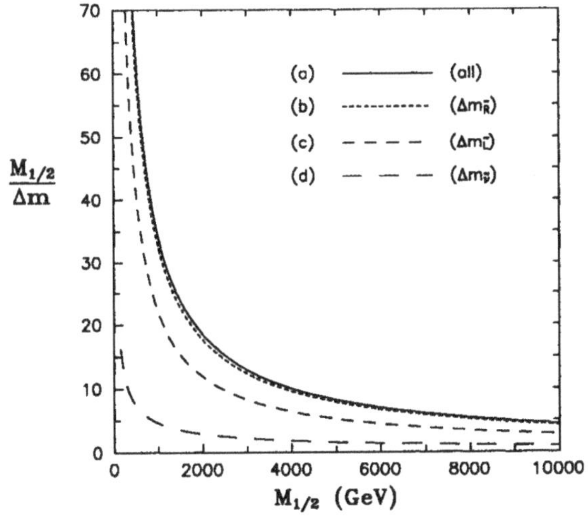

Figure 2: Plot of the *minimum* allowed value of $M_{1/2}/\Delta m$ vs. $M_{1/2}$ in four different cases: *a)* $\Delta m^2_{\tilde{\nu}_e \tilde{\nu}_\mu} = \Delta m^2_{\tilde{e}_L \tilde{\mu}_L} = \Delta m^2_{\tilde{e}_R \tilde{\mu}_R} \equiv \Delta m^2$; *b)* only $\Delta m^2_{\tilde{e}_R \tilde{\mu}_R} \neq 0$; *c)* only $\Delta m^2_{\tilde{e}_L \tilde{\mu}_L} \neq 0$ and *d)* only $\Delta m^2_{\tilde{\nu}_e \tilde{\nu}_\mu} \neq 0$.

FERMION MASSES IN SUSY GUTS

B. C. Allanach[a] and S. F. King[b]

Department of Physics, University of Southampton,
Southampton SO17 1BJ, England

We discuss the fermion mass problem in SUSY GUTs, including such ideas as texture zeroes, and Georgi-Jarlskog textures. We focus on a specific supersymmetric model based on the gauge group $SU(4) \otimes SU(2)_L \otimes SU(2)_R$. In this model the gauge group is broken to that of the standard model at 10^{16} GeV, and supersymmetry is broken at low-energy. The model may be regarded as a "surrogate SUSY GUT", and has several advantages over the $SU(5)$ and $SO(10)$ SUSY GUTs, such as the absence of the doublet-triplet splitting problem, and a simpler Higgs sector which does not involve any large Higgs representations. The model predicts full quadruple Yukawa unification (top-bottom-tau-tau neutrino), leading to the prediction of large top mass and $\tan\beta$. An operator analysis leads to schemes where the strange and down mass are predicted, and $|V_{ub}| > 0.004$.

1 Introduction

The Standard Model involves three gauge couplings g_i, $(i = 1, 2, 3)$, a Higgs mass parameter μ and quartic coupling λ, and the fermion Yukawa couplings. In the Standard Model Lagrangian one must specify the three Yukawa matrices λ_{ij}^E, λ_{ij}^D, λ_{ij}^U, corresponding to up to 54 real parameters, which after diagonalisation lead to 9 physical masses (6 quark masses and 3 charged lepton masses) and 4 physical quark mixing parameters. Thus the fermion sector of the Standard Model involves either 54 or 13 unknown free parameters, depending on how you choose to count them. Either way, from a fundamental point of view the situation is unacceptable and the fermion mass problem, as it has been called, is one motivation for going beyond the Standard Model. The big question of course is what lies beyond it?

We have not yet experimentally studied the mechanism of electroweak symmetry breaking, so one might argue that it is premature to study the fermion mass problem. Unless we can answer this, we have no hope of understanding anything about fermion masses since we do not have a starting point from which to analyse the problem. However LEP has taught us that whatever breaks electroweak symmetry must do so in a way which very closely

---

[a] Address after 1 October 1995: Rutherford Appleton Laboratory, Chilton, Didcot OX11 0QX.

[b] Talk presented by S. F. King.

resembles the standard model. This observation by itself is enough to disfavour many dynamical models involving large numbers of new fermions. By contrast the minimal supersymmetric standard model (MSSM) mimics the standard model very closely. Furthermore, by accurately measuring the strong coupling constant, LEP has shown that the gauge couplings of the MSSM merge very accurately at a scale just above 10^{16} GeV, thus providing a hint for possible unification at this scale. On the theoretical side, supersymmetry (SUSY) and grand unified theories (GUTs) fit together very nicely in several ways, providing a solution to the technical hierarchy problem for example. When SUSY GUTs are extended to supergravity (SUGRA) the beautiful picture of universal soft SUSY breaking parameters and radiative electroweak symmetry breaking via a large top quark yukawa coupling emerges. Finally, there is an on-going effort to embed all of this structure in superstring models, thereby allowing a complete unification, including gravity.

Given the promising scenario mentioned above, it is hardly surprising that many authors have turned to the SUSY GUT framework as a springboard from which to attack the problem of fermion masses [1]. Indeed in recent years there has been a flood of papers on fermion masses in SUSY GUTs. Although the approaches differ in detail, there are some common successful themes which have been known for some time. For example the idea of bottom-tau Yukawa unification in SUSY GUTs [2] works well with current data [3]. A more ambitious extension of this idea is the Georgi-Jarlskog (GJ) ansatz (see later) which provides a successful description of all down-type quark and charged lepton masses [4], and which also works well with current data [5]. In order to understand the origin of the texture zeroes, one must consider the details of the model at the scale $M_X \sim 10^{16}$ GeV. The alternative is to simply make a list of assumptions about the nature of the Yukawa matrices at M_X [6]. For example Ramond, Roberts and Ross (RRR) [7] assumed symmetric Yukawa matrices at M_X, together with the GJ ansatz for the lepton sector. It is difficult to proceed beyond this without specifying a particular model. Indeed, this model dependence may be a good thing since it may mean that the fermion mass spectrum at low energies is sensitive to the theory at M_X, so it can be used as a window into the high-energy theory. Therefore in what follows we shall restrict ourselves to a very specific gauge group at M_X.

Twenty-one years ago Pati and Salam proposed a model in which the standard model was embedded in the gauge group $SU(4)\otimes SU(2)_L \otimes SU(2)_R$ [8]. More recently a superpersymmetric (SUSY) version of this model was proposed in which the gauge group is broken at $M_X \sim 10^{16}$ GeV [9]. The model [9] does not involve adjoint representations and later some attempt was made to derive it from four-dimensional strings, although there are some difficulties with the current formulation [10]. The absence of adjoint representations is not an essential prerequisite for the model to descend from the superstring, but it leads to some technical simplifications. Also in the present model, the colour triplets which are in separate representations from the Higgs doublets, become heavy in a very simple way so the Higgs doublet-triplet splitting problem does not arise. These two features (absence of adjoint representations and absence of the doublet-triplet splitting problem,) are shared by flipped SU(5)⊗U(1) [11], which also has a superstring formulation. Although the present model and flipped SU(5)⊗U(1) are similar in many ways, there are some important differences. Whereas the Yukawa matrices of flipped SU(5)⊗U(1) are completely unrelated at the level of the effective field theory at M_X (although they may have relations coming from the string model) in the present model there is a constraint that the top, bottom and tau Yukawa couplings must all unify at that scale. In addition there will be Clebsch relations between the other elements of the Yukawa matrices, assuming they are described by non-renormalisable operators, which would not be present in flipped SU(5)⊗U(1). In these respects the model resembles the SO(10) model recently analysed by Anderson et al [12]. However it differs from the SO(10) model in that the present model does not have an SU(5) subgroup which is central to the analysis of the SO(10) model. In addition the operator structure of the present model is totally different. Thus the model under consideration is in some sense similar to flipped SU(5)⊗U(1), but has third family Yukawa unification and precise Clebsch relationships as in SO(10). We find this combination of features quite remarkable, and it seems to us that this provides a rather strong motivation to study the problem of fermion masses in this model [13].

2 The 422 Model

Here we briefly summarise the parts of the model which are relevant for our analysis. The gauge group is,

$$\mathrm{SU}(4) \otimes \mathrm{SU}(2)_L \otimes \mathrm{SU}(2)_R. \tag{1}$$

The left-handed quarks and leptons are accommodated in the following representations,

$$F^{i\alpha a} = (4,2,1) = \begin{pmatrix} u^R & u^B & u^G & \nu \\ d^R & d^B & d^G & e^- \end{pmatrix}^i \tag{2}$$

$$\bar{F}^i_{x\alpha} = (\bar{4},1,\bar{2}) = \begin{pmatrix} \bar{d}^R & \bar{d}^B & \bar{d}^G & e^+ \\ \bar{u}^R & \bar{u}^B & \bar{u}^G & \bar{\nu} \end{pmatrix}^i \tag{3}$$

where $\alpha = 1 \ldots 4$ is an SU(4) index, $a, x = 1, 2$ are $\mathrm{SU}(2)_{L,R}$ indices, and $i = 1 \ldots 3$ is a family index. The Higgs fields are contained in the following representations,

$$h_a^x = (1, \bar{2}, 2) = \begin{pmatrix} {h_2}^+ & {h_1}^0 \\ {h_2}^0 & {h_1}^- \end{pmatrix} \tag{4}$$

(where h_1 and h_2 are the low energy Higgs superfields associated with the MSSM.) The two heavy Higgs representations are

$$H^{\alpha b} = (4, 1, 2) = \begin{pmatrix} u_H^R & u_H^B & u_H^G & \nu_H \\ d_H^R & d_H^B & d_H^G & e_H^- \end{pmatrix} \tag{5}$$

and

$$\bar{H}_{\alpha x} = (\bar{4}, 1, \bar{2}) = \begin{pmatrix} \bar{d}_H^R & \bar{d}_H^B & \bar{d}_H^G & e_H^+ \\ \bar{u}_H^R & \bar{u}_H^B & \bar{u}_H^G & \bar{\nu}_H \end{pmatrix}. \tag{6}$$

The Higgs fields are assumed to develop VEVs,

$$< H >=< \nu_H >\sim M_X, \quad < \bar{H} >=< \bar{\nu}_H >\sim M_X \tag{7}$$

leading to the symmetry breaking at M_X

$$\mathrm{SU}(4) \otimes \mathrm{SU}(2)_L \otimes \mathrm{SU}(2)_R \longrightarrow \mathrm{SU}(3)_C \otimes \mathrm{SU}(2)_L \otimes \mathrm{U}(1)_Y \tag{8}$$

in the usual notation. Under the symmetry breaking in Eq.8, the Higgs field h in Eq.4 splits into two Higgs doublets h_1, h_2 whose neutral components subsequently develop weak scale VEVs,

$$< h_1^0 >= v_1, \quad < h_2^0 >= v_2 \tag{9}$$

with $\tan\beta \equiv v_2/v_1$.

Below M_X the part of the superpotential involving quark and charged lepton fields is just

$$W = \lambda_U^{ij} Q_i \bar{U}_j h_2 + \lambda_D^{ij} Q_i \bar{D}_j h_1 + \lambda_E^{ij} L_i \bar{E}_j h_1 + \ldots \tag{10}$$

with the boundary conditions at M_X,

$$\lambda_U^{ij} = \lambda_D^{ij} = \lambda_E^{ij}. \tag{11}$$

The same Yukawa relations also occur in minimal $SO(10)$.

3 The Basic Strategy

Such Yukawa relations as in Eq.11 taken at face value are a phenomenological disaster. For example consider the minimal $SU(5)$ prediction $\lambda_D^{ij} = \lambda_E^{ij}$. After diagonalisation this leads to $\lambda_e = \lambda_d$, $\lambda_\mu = \lambda_s$, $\lambda_\tau = \lambda_b$, (at the scale M_X) and hence

$$\frac{\lambda_s}{\lambda_d} = \frac{\lambda_\mu}{\lambda_e} \tag{12}$$

which is RG invariant and fails badly at low-energy. On the other hand the third family relation leads to the low-energy prediction (assuming the SUSY RG equations) $\lambda_b/\lambda_\tau \approx 2.4$ which works well.

A possible fix is provided by the GJ texture,

$$\lambda^E = \begin{pmatrix} 0 & \lambda_{12} & 0 \\ \lambda_{21} & 3\lambda_{22} & 0 \\ 0 & 0 & \lambda_{33} \end{pmatrix} \tag{13}$$

$$\lambda^D = \begin{pmatrix} 0 & \lambda_{12} & 0 \\ \lambda_{21} & \lambda_{22} & 0 \\ 0 & 0 & \lambda_{33} \end{pmatrix}, \tag{14}$$

With predictions (at M_X)

$$\lambda_s = \lambda_\mu/3, \quad \lambda_d = 3\lambda_e, \tag{15}$$

which is viable.

As it turns out, the idea of full top-bottom-tau Yukawa unification works rather well for the third family [14], leading to the prediction of a large top quark mass $m_t > 165$ GeV, and $\tan\beta \sim m_t/m_b$ where m_b is the bottom quark mass. However Yukawa unification for the first two families is not successful, since it would lead to unacceptable mass relations amongst the lighter fermions, and zero mixing angles at M_X. In order to cure these problems, we require something akin to the GJ texture, in which the Yukawa relations are altered by group theoretical Clebsch coefficients, leading to enhanced predictivity.

One interesting proposal has recently been put forward to account for the fermion masses in an SO(10) SUSY GUT with a single Higgs in the 10 representation [12]. According this approach, only the third family is allowed to receive mass from the renormalisable operators in the superpotential. The remaining masses and mixings are generated from a minimal set of just three specially chosen non-renormalisable operators whose coefficients are suppressed by some large scale. Furthermore these operators are only allowed to contain adjoint 45 Higgs representations, chosen From a set of fields denoted 45_Y, 45_{B-L}, $45_{T_{3R}}$,

45_X whose VEVs point in the direction of the generators specified by the subscripts, in the notation of [12]. This is precisely the strategy we wish to follow. We shall assume that only the third family receives its mass from a renormalisable Yukawa coupling. All the other renormalisable Yukawa couplings are set to zero. Then non-renormalisable operators are written down which will play the role of small effective Yukawa couplings. The effective Yukawa couplings are small because they originate from non-renormalisable operators which are suppressed by powers of the heavy scale M. We shall restrict ourselves to all possible non-renormalisable operators which can be constructed from different group theoretical contractions of the fields:

$$O_{ij} \sim (F_i \bar{F}_j) h \left(\frac{H\bar{H}}{M^2} \right)^n + \text{h.c.} \tag{16}$$

where we have used the fields $H, \bar{H}$ in Eqs.5,6 and M is the large scale $M > M_X$. The idea is that when $H, \bar{H}$ develop their VEVs such operators will become effective Yukawa couplings of the form $hF\bar{F}$ with a small coefficient of order M_X^2/M^2. Although we assume no intermediate symmetry breaking scale (i.e. SU(4)⊗SU(2)$_L$⊗SU(2)$_R$ is broken directly to the standard model at the scale M_X) we shall allow the possibility that there are different higher scales M which are relevant in determining the operators. For example one particular contraction of the indices of the fields may be associated with one scale M, and a different contraction may be associated with a different scale M'. We shall either appeal to this kind of idea in order to account for the various hierarchies present in the Yukawa matrices, or to higher dimensional operators which are suppressed by a further factor of M.

In the present model, although there are no adjoint representations, there will in general be non-renormalisable operators which closely resemble those in SO(10) involving adjoint fields. The simplest such operators correspond to $n = 1$ in Eq.16, with the $(H\bar{H})$ group indices contracted together.

These operators are similar to those of ref. [12] but with $H\bar{H}$ playing the rôle of the adjoint Higgs representations. It is useful to define the following combinations of fields, corresponding to the different $H\bar{H}$ transformation properties under the gauge group in Eq.1,

$$\begin{aligned} (H\bar{H})_A &= (1,1,1) \\ (H\bar{H})_B &= (1,1,3) \\ (H\bar{H})_C &= (15,1,1) \\ (H\bar{H})_D &= (15,1,3) \end{aligned} \tag{17}$$

It is straightforward to construct the operators of the form of Eq.16 explicitly, and hence deduce the effect of each operator. For example for $n = 1$

the four operators are, respectively,

$$O_{ij}^{A,B,C,D} \sim F_i \bar{F}_j h \frac{(H\bar{H})_{A,B,C,D}}{M^2} + H.c. \tag{18}$$

where we have suppressed gauge group indices.

These operators lead to quark-lepton and isospin splittings, as shown explicitly below:

$$\begin{aligned} O_{ij}^A &= a_{ij}(Q_i\bar{U}_j h_2 + Q_i\bar{D}_j h_1 + L_i\bar{E}_j h_1 + H.c.) \\ O_{ij}^B &= b_{ij}(Q_i\bar{U}_j h_2 - Q_i\bar{D}_j h_1 - L_i\bar{E}_j h_1 + H.c.) \\ O_{ij}^C &= c_{ij}(Q_i\bar{U}_j h_2 + Q_i\bar{D}_j h_1 - 3L_i\bar{E}_j h_1 + H.c.) \\ O_{ij}^D &= d_{ij}(Q_i\bar{U}_j h_2 - Q_i\bar{D}_j h_1 + 3L_i\bar{E}_j h_1 + H.c.) \end{aligned} \tag{19}$$

where the coefficients of the operators $a_{ij}, b_{ij}, c_{ij}, d_{ij}$ are all of order $\frac{M_X^2}{M^2}$.

4 Results and Conclusions

Using operators such as those above, together with more complicated $n = 2$ operators, it is possible to account for the entire fermion mass spectrum. The successful ansatze [13] involve 8 real parameters plus an unremovable phase. With these 8 parameters we can describe the 13 physical masses and mixing angles. Third family Yukawa unification leads to a prediction for m_t(pole) $=$ $130 - 200$ GeV and $\tan\beta = 35 - 65$, depending on $\alpha_S(M_Z)$ and $\bar{m}_b$. More accurate predictions could be obtained if the error on $\alpha_S(M_Z)$ and $\bar{m}_b$ were reduced. The analysis of the lower 2 by 2 block of the Yukawa matrices leads to 2 possible predictions for $\lambda_\mu/\lambda_s = 3, 4$ at the scale M_X (3 is the GJ prediction). In the upper 2 by 2 block analysis we are led to 5 possible predictions for $\lambda_d/\lambda_e = 2, 8/3, 3, 4, 16/3$. (again 3 is the GJ prediction.) Finally, we have a prediction that $|V_{ub}| > 0.004$[13].

The high values of $\tan\beta$ required by our model (also predicted in SO(10)) can be arranged by a suitable choice of soft SUSY breaking parameters as discussed in ref.[15], although this leads to a moderate fine tuning problem [14]. The high value of $\tan\beta$ is not stable under radiative corrections unless some other mechanism such as extra approximate symmetries are invoked. m_t may have been overestimated, since for high $\tan\beta$, the equations for the running of the Yukawa couplings in the MSSM can get corrections of a significant size from Higgsino–stop and gluino–sbottom loops. The size of this effect depends upon the mass spectrum and may be as much as 30 GeV. For our results to be quantitatively correct, the sparticle corrections to m_b must be small. This could happen in a scenario with non-universal soft parameters, for example. Not included in our analysis are threshold effects, at low or high energies. These could alter our results by several per cent and so it should be borne in mind that all of the mass predictions have a significant uncertainty in them. It is also unclear how reliable 3 loop perturbative QCD at 1 GeV is.

Despite a slight lack of predictivity of the model compared to SO(10), the SU(4) $\otimes SU(2)_L \otimes SU(2)_R$ model has the twin advantages of having no doublet-triplet splitting problem, and containing no adjoint representations, making the model technically simpler to embed into a realistic string theory. Although both these problems can be addressed in the SO(10) model [16], we find it encouraging that such problems do not arise in the first place in the $SU(4) \otimes SU(2)_L \otimes SU(2)_R$ model. Of course there are other models which also share these advantages such as flipped SU(5) or even the standard model. However, at the field theory level, such models do not lead to Yukawa unification, or have precise Clebsch relations between the operators describing the light fermion masses. It is the combination of all of the attractive features mentioned above which singles out the present model for serious consideration.

Acknowledgments

I would like to thank the organisers for inviting me to a very stimulating and enjoyable workshop.

References

1. S. Raby, Preprint No. OHSTPY-HEP-T-95-024; S. Raby, Preprint No. OHSTPY-HEP-T-95-001.
2. M. B. Einhorn and D. R. T. Jones, Nucl. Phys. **B196** (1982) 475; J. Ellis, D. V. Nanopoulos and S. Rudaz, Nucl. Phys. **B202** (1982) 43.
3. B. C. Allanach and S. F. King, Phys. Lett. **B328**, (1994) 360; H. Arason *et al*, Phys. Rev. Lett. **67** (1991) 2933; *ibid* Phys. Rev. **D47** (1993) 232; P. Langacker and N. Polonsky, UPR-0556T. V.Barger, M.S.Berger and P.Ohmann, Phys. Rev. **D47**, 1093 (1993).
4. H. Georgi and C. Jarlskog, Phys. Lett. **B86** (1979) 297.
5. S. Dimopoulos, L. Hall and S. Raby, Phys. Rev. **D45** (1992) 4192.
6. C. D. Froggatt and H. B. Nielson, Nucl. Phys. **B147** (1979) 277; Z. G. Berezhiani, Phys. Lett. **B129** (1983) 99; ibid., **B150** (1985) 177; S. Dimopoulos, Phys. Lett. **B129** (1983) 417.
7. P. Ramond, R. Roberts and G. Ross, Nucl. Phys. **B406** (1993) 19.
8. J. Pati and A. Salam, Phys. Rev. **D10** (1974) 275.

9. I. Antoniadis and G. K. Leontaris, Phys. Lett. **B216** (1989) 333.
10. I. Antoniadis, G. K. Leontaris and J. Rizos, Phys. Lett. **B245** (1990) 161.
11. I. Antoniadis, J. Ellis, J. Hagelin and D. Nanopoulos, Phys. Lett. **B194** (1987) 231.
12. G. Anderson, S. Raby, S. Dimopoulos, L. Hall and G. Starkman, Phys. Rev **D49** (1994) 3660
13. S. F. King Phys. Lett. **B325** (1994) 129;
B. C. Allanach and S. F. King, hep-ph/9502219, submitted to Nucl. Phys. **B.**
14. B. Ananthanarayan, G. Lazarides and Q. Shafi, Phys. Rev. **D44** (1991) 1613; H. Arason *et al*, Phys. Rev. Lett. **67** (1991) 2933; L. Hall, R. Rattazzi and U. Sarid, LBL preprint 33997.
15. M. Carena, M. Olechowski, S. Pokorski and C. E. M. Wagner, Nucl. Phys. **B426** (1994) 269.
16. S. Chaudhuri, S. Chung and J. Lykken, FERMILAB PUB-94/137-T; L. Hall and S. Raby, OHSTPY-HEP-T-94-023, LBL-36357, UCB-PTH-94/27.

FERMION MASSES AND MIXINGS FROM HORIZONTAL ABELIAN SYMMETRIES

C. A. SAVOY

Service de Physique Théorique, C.E. de SACLAY
F-91191 Gif-sur-Yvette, FRANCE

Recent work on the Froggatt-Nielsen mechanism to understand the hierarchies in the quark and leptons masses and mixings is reviewed. Collateral issues such as anomaly cancellation, squark and slepton mixings and FCNC effects are also discussed.

The fermion unit in the Standard Model is a family of fifteen quarks and leptons that provide a non-trivial representation of the $SU(3) \times SU(2) \times U(1)$ gauge group consistent with the observed maximal violation of P and C in weak interactions. The anomaly cancellation requirement gives a partial justification for the set of quantum numbers in a fermion family. A more complete explanation is provided by the assumption of a GUT, in the usual sense of **vertical** unification of the elements of the family within a representation of a larger gauge theory at very high energies. Instead, the observed quarks and leptons form a **society** of three such families, three replicas of each set of Standard Model quantum numbers. However these fermion replicas do not look as clones since they quite differ in their masses. The natural explanation of this situation is to hypothetize that quarks/leptons of the same charge have different quantum numbers of some new symmetries at high energies (symmetries that commute with the Standard Model symmetries have been called **horizontal**). The otherwise appealing idea of spontaneously broken horizontal unification to explain the three families seems rather unappropriate to understand the pattern of fermion masses and mixings, and the related FCNC experimental constraints.

As in many particle physics issues, hints come from superstrings models, where one can find examples of compactifications with three (or more) families of quarks and leptons and neither vertical nor horizontal unification. Instead, there are in general additional abelian $U(1)$ symmetries that differentiate between fermions with identical Standard Model quantum numbers. As another hint, the superstring theory particle masses are field dependent dynamically determined quantities. Interestingly enough, the paradoxical fact that horizontal $U(1)$ symmetries could in principle explain the fermion mass pattern has been suggested some time ago by **Froggatt and Nielsen** [1] by assuming that the dependence of the fermion masses on new fields is prescribed by the new ad hoc symmetries.

In the Standard Model the quark/lepton masses are given by the Yukawa couplings of the fermions to the Higgs scalars,

$$\sum_{ij} \left[Y^U_{ij} q^i u^j H_2 + Y^D_{ij} q^i d^j H_1 + Y^E_{ij} l^i e^j H_1 \right] \tag{1}$$

where we have allowed for two Higgs doublets to anticipate the use of supersymmetry herebelow as in most of the recent work [3,4,5,6,7,8,9] on this subject. The physical content of the three Yukawa coupling 3×3 matrices is given by their eigenvalues $Y_i (i = u, c, t; d, s, b; e, \mu, \tau)$ as well as the Cabibbo-Kobayashi-Maskawa matrix V_{CKM}. The observed quark masses and mixings and lepton masses reveal a strong hierarchy conveniently displayed in terms of a small parameter which we choose to be the Cabibbo angle, $\lambda = .22$: $Y_t : Y_c : Y_u = \lambda^8 : \lambda^4 : 1, Y_b : Y_s : Y_d = \lambda^4 : \lambda^2 : 1, Y_\tau : Y_\mu : Y_e = \lambda^4 : \lambda^2 : 1, V_{us} = \lambda, V_{cb} \sim \lambda^2, V_{ub} \sim \lambda^3$.

The smalness of these mass ratios and mixing angles faces us with a problem of **naturalness**. A parameter is naturally small when the symmetry of the system increases as the parameter vanishes. The direction initiated by Froggatt and Nielsen [1] to understand such a hierarchical pattern goes as follows:

(i) The key assumption is a **gauged horizontal** $U(1)_X$ **symmetry** which would be broken by the small quark masses so that small Yukawa couplings are protected by this symmetry. The effective $U(1)_X$ symmetric theory below some scale M is supposed to be natural to the extent that all parameters are of $O(1)$. The scale M is the limit of validity of the effective theory and presumably it is of $O(M_{Planck})$ if one adopts a superstring point of view [2]. The X-charges of quarks, leptons and Higgses are free parameters to be fixed **a posteriori** and simply denoted q_i,u_i, d_i,l_i,e_i,h_1,h_2, for the different flavours, where $i = 1, 2, 3$ is the family index.

(ii) One introduces (at least) one Froggatt-Nielsen (FN) field Φ, a Standard Model **gauge singlet** with X-charge ϕ and we normalize the $U(1)_X$ so that $\phi = -1$. The effective (non-renormalizable) $U(1)_X$ allowed couplings are then of the form $g^U_{ij}(\Phi/M)^{q_i+u_j+h_2} q_i u_j H_2$, with analogous expressions for the H_1 couplings to down quarks and leptons. The coefficients g^U_{ij}, etc, are taken to be **natural**, *i.e.*, of $O)1)$, unless they are required to vanish by the $U(1)_X$ symmetry.

(iii) The **small parameter** λ is identified with the ratio $(\frac{<\Phi>}{M})$ as the $U(1)_X$ symmetry is broken by the Φ v.e.v.. Below the scale $< \Phi >= \lambda M$, one recovers the Standard Model with the effective Yukawa coupling matrices given by

$$\begin{aligned} Y_{ij}^U &= \theta\,(q_i + u_j + h_2)\,\lambda^{q_i+u_j+h_2} \\ Y_{ij}^D &= \theta\,(q_i + d_j + h_1)\,\lambda^{q_i+d_j+h_1} \\ Y_{ij}^E &= \theta\,(l_i + e_j + h_1)\,\lambda^{l_i+e_j+h_1} \end{aligned} \tag{2}$$

The X-charges are now chosen to fit the hierarchy in the eigenvalues and mixing angles. In particular, one has from the experimental masses (at $O(M_{Planck})$) of the third families the constraints: $h_2 + q_3 + u_3 = 0$ and $x = h_1 + q_3 + d_3 = h_1 + l_3 + e_3$, where the parameter x depends on the assumed value for $\tan\beta =< H_2 > / < H_1 >$. Notice that with this restriction the Yukawa couplings depend only on the charge differences $q_{i3} = q_i - q_3$, $u_{i3} = u_i - u_3$, ..., $e_{i3} = e_i - e_3$ and x.

The Yukawa matrix entries corresponding to negative charge differences would vanish because of the θ function. However, it has been pointed out [8] that the same symmetry and naturalness principles have to be applied to the effective field dependent metrics of the effective theory. The metrics for q_i will then take the form

$$Z_{ij}\left[\theta\,(q_i - q_j)\left(\frac{\Phi}{M}\right)^{q_i-q_j} + \theta\,(q_j - q_i)\left(\frac{\Phi^+}{M}\right)^{q_j-q_i}\right] \tag{3}$$

where Z_{ij} are $O(1)$ numbers, with analogous exprewssions for the metrics of the other fermions. Once one redefines the fermion fields to obtain canonical metrics in the corresponding low energy Standard Model the renormalized Yukawa coupling matrices will have their zero entries removed, so that $\theta(q_{i3} + u_{j3})\lambda^{q_{i3}+u_{j3})} \rightarrow \lambda^{|q_{i3}+u_{j3}|}$. Therefore the recent analysis [4] of phenomenologically allowed zero entries in Yukawa matrices, or **textures**, is not quite relevant for the present approach. Indeed, the physical content of the Yukawa matrices may change after the zero filling due to the non-diagonal metrics [8].

To search for all possible acceptable Yukawa or mass matrices, we have scanned over all possible charge assignments [8]. This analysis has been done by allowing at most a few deviations of $O(\lambda)$ in the mass ratios and CKM matrix elements. On the other hand, we have considered two cases: only one FN field or two FN fields of opposite charges. Only a few solutions were obtained, but their patterns are quite different.

The new abelian gauge group $U(1)_X$ is potentially anomalous. It generates, through triangle diagrams, mixed anomalies with the standard model gauge group. The mixed anomalies to be considered are the following: $[SU(3)]^2U(1)_X, [SU(2)]^2U(1)_X, [U(1)_Y]^2U(1)_X$ and $U(1)_Y[U(1)_X]^2$, whose algebraic coefficients are denoted $A_3, /A_2, /A_1$ and A_1', respectively. It is useful to use the variable x as well the constraints on the charges of the third family discussed hereabove to write the following linear combinations of the anomalies

$$
\begin{aligned}
A_1 + A_2 - 2A_3 &= 2x + \sum_a \left[\frac{2}{3}(q_{a3} + u_{a3}) - \frac{4}{3}(q_{a3} + d_{a3}) + 2(l_{a3} + e_{a3}) \right] \\
A_3 + 3(h_1 + h_2) &= 3x + \sum_a \left[(q_{a3} + u_{a3}) + (q_{a3} + d_{a3}) \right]
\end{aligned}
\tag{4}
$$

where $q_{a3} = q_a - q_3$, etc. For a given set of mass matrices, the charge differences $(q_{a3} + u_{a3})$, $(q_{a3} + d_{a3})$, $(l_{a3} + e_{a3})$ are fixed for $a = 1, 2.$ and so is the r.h.s. of eq.(4). The remaining freedom in the charge assignment can then be utilized to cancel the anomalies. It is appropriate to approach this issue in terms of the familiar invariances of the Standard Model fermion mass matrices: $SU(3) \times SU(2) \times U(1)_Y$, the baryon number B, the lepton number L and the Peccei-Quinn symmetry P, with a charge 1 for all the matter fields and -2 for the two Higgs doublets. Since the charge X of the $U(1)_X$ has to commute with the Standard Model gauge group the allowed shifts in X are given by a linear combination of 4 abelian charges: $X \to X + \beta Y + \gamma(B - L) + \delta(3B + L) + \zeta P$ without changing the mass matrices.

This redefinition can be used to achieve an anomaly free theory by implementing either the Green-Schwarz condition [10] (with $sin^2\theta_W = \frac{3}{8}$) through the relations, $\frac{3}{5}A_1 = A_2 = A_3$, or the usual cancellation, $A_1 = A_2 = A_3 = 0$. Of course, the eqs.(4) are to be satisfied, what amounts to fixing the values of x and $h_1 + h_2$. In both cases one must also impose $A_1' = 0$.

We now turn to the implementation of the anomaly cancellation in the specific models. Let us first assume only one FN field. Then we can readily write the relations between the Yukawa (mass) determinants and anomaly combinations as follows [5,6,7,8,9]

$$
\begin{aligned}
det\,(Y_U Y_D^{-2} Y_L^3) &\sim \lambda^{3/2(A_2 + A_1 - 2A_3)} \\
det\,(Y_U Y_D) &\sim \lambda^{A_3 + 3(h_1 + h_2)}
\end{aligned}
\tag{5}
$$

This result is valid as well if there are negative charge differences [8] since the determinant of the Kähler metrics is $O(1)$ and the determinants of the mass matrices are not changed by this renormalization at the leading order in λ. Clearly, the anomaly combinations appearing in these mass determinants are precisely those that are invariant under the symmetries of the mass matrices.

Replacing the powers of λ corresponding to the mass determinants, one obtains the constraints on the coefficients A_i. The vanishing of the anomalies would require the physically excluded value $x = -6$. However, it is possible to implement the Green-Schwarz mechanism [5,6] with $sin^2\theta_W = \frac{3}{8}$ by shifting X so that, $A_3 = A_2 = \frac{3}{5}A_1 = 3(6+x)$ and $h_1 + h_2 = 0$, as required by eqs.(5).

In the case of two FN fields if there are negative charge differences, eqs.(5) will change. In general one can only implement the Green-Schwarz cancellation, but now $h_1 + h_2 \neq 0$. Interestingly enough, it is easy to show that with FN fields with opposite charges all the anomalies can be put to zero if and only if $h_1 + h_2 = 8$ and $x = 2$. From the ratio m_b/m_t one then get $\tan\beta \sim 2$ as a necessary condition for the vanishing of the anomaly coefficients, $A_i = 0$ $(i = 1, 2, 3)$.

These anomaly cancellation conditions can now be imposed on the phenomenologically allowed Yukawa coupling matrices to determine the X-charge solutions. They are displayed in ref.[[8]]. It goes without saying that the main task remains to find models that can reproduce the Frogatt-Nielsen scenario with the allowed charge assignments.

Because the solution of the hierarchy in the Yukawa coupling matrices appears to require some new physics at much higher energies, supersymmetry would be mostly welcome to solve the hierarchy problem between the new scale and the weak (Fermi) scale. It has been invoked indeed in the recent studies of horizontal symmetries. Then one cannot underestimate the low energy effects of the supersymmetric partners, squarks and sleptons. They will also be submitted to the horizontal symmetry constraints as they share the $U(1)_X$ charges of their corresponding fermions. In the spirit of our approach here, one should assume that the mass matrices of squarks and leptons is field dependent, with powers of the FN fields needed to implement the symmetries, so that below the scale (λM) the mass matrix for the doublet $Q^i = (U^i_L, D^i_L)$ of "left-handed" squarks will be of the form (for simplicity, we neglect here LR analytic mass terms, as well as D-terms, in the squark mass matrices.),

$$(M^2_{Q_L})_{ij} = f_{ij}\lambda^{|q_i - q_j|} \tag{6}$$

with similar expressions for the "right handed" squarks U_i and D_i and for the sleptons, L_i and E_i. In the absence of any other symmetry principle, the coefficients f_ij are all of the order of the effective symmetry breaking parameter, which we characterize by an average squark mass, $\bar{m}^2$.

In general, even in the flavour basis that diagonalizes quark mass matrices, the squark mass matrices will still be of the non-diagonal form (6). Therefore one could induce large FCNC effects from loop diagrams with the exchange of neutral sfermions (gluino, photino,...) in possible disagreement with experiments. This issue has been recently addressed [11] by using the results of the general analysis of Yukawa matrices in ref.[[8]]. It turns out that at least one of the general predictions from eq.(6) could imply much too large FCNC effects in K-physics, namely,

$$(M^2_{D_L})_{12}(M^2_{D_R})_{12} \sim (\frac{m_d}{m_s})^n \bar{m}^4 \tag{7}$$

with $n = 1$ with one FN field and $n = 2$ with two of them.

Superstring theory puts forward a new symmetry of effective supergravity models, **modular invariance.** It has been shown [11] that by assigning modular weights n_i to quarks, leptons, Higgses and FN fields, modular invariance together with $U(1)_X$ invariance of the theory implies that the **modular weights are proportional to their X-charges,** $n_i = kX_i$.

As a consequence, the squark masses in the diagonal quark mass flavour basis will have non-diagonal matrix elements of the form,

$$M^2_{ij} \propto (X_i - X_j)\lambda^{|X_i - X_j|} \tag{8}$$

Therefore, by selecting the models with X-charges such that $d_1 = d_2$ and $e_1 = e_2$, the most constraining experimental bounds on FCNC effects in both the hadron and the lepton physics are fulfilled.

Acknowledgements

I thank José Wagner Valle and the organizers for their kind invitation and for providing an enjoyable workshop. This work was partially supported by the EU through a grant of the Flavourdynamics Network.

References

1. C.D. Froggatt and H.B. Nielsen, *Nucl. Phys.* B **147**, 277 (1979).
2. M. Dine, N. Seiberg and E. Witten, *Nucl. Phys.* B **289**, 585 (1987).
3. Y. Nir and N. Seiberg, *Phys. Lett.* B **309**, 337 (1993).
4. P. Ramond, R.G. Roberts and G.G. Ross,*Nucl. Phys.* B **406**, 19 (1993).
5. L.E. Ibàñez and G.G. Ross,*Phys. Lett.* B **332**, 100 (1994).
6. P. Binétruy and P. Ramond, *Phys. Lett.* B , (19)
7. V. Jain and R. Shrock, *Phys. Lett.* B , (19).
8. E. Dudas, S. Pokorski and C.A. Savoy, *Phys. Lett.* B , (19).
9. Y. Nir, *Phys. Lett.* B , (1995).
10. L.E. Ibàñez, *Phys. Lett.* B **303**, 55 (1993).
11. E. Dudas, S. Pokorski and C.A. Savoy, in preparation.

SUPERGRAVITY AND THE $b \to s, \gamma$ DECAY

B. DE CARLOS

Department of Theoretical Physics, 1 Keble Road, Oxford OX1 3NP, UK

J.A. CASAS

Instituto de Estructura de la Materia (CSIC), Serrano 123, 28006 Madrid, Spain

We evaluate the branching ratio BR($b \to s, \gamma$) in the minimal supersymmetric standard model (MSSM), determining the corresponding phenomenological restrictions on two attractive supergravity scenarios, namely minimal supergravity and a class of models with a natural solution to the μ problem. We have included in the calculation some one–loop refinements that have a substantial impact on the results. It is stressed the fact that an eventual improvement of the experimental bounds of order 10^{-4} would strengthen the restrictions on the MSSM dramatically. This would be enough to discard these supergravity scenarios with $\mu < 0$ if no discrepancy is found with the standard model prediction, while for $\mu > 0$ there will remain low-energy windows.

1 The $b \to s, \gamma$ decay in SUGRA models

The new experimental data available[1] place stringent upper and lower bounds on the branching ratio (BR) for the process $b \to s, \gamma$: $1 \times 10^{-4} <$ BR($b \to s, \gamma$) $< 4 \times 10^{-4}$. On the other hand it is well known that this process has the potential to put relevant constraints to physics beyond the Standard Model (SM). In particular, we want to study the prediction that for this BR give two different supergravity (SUGRA) models whose particle content at low energies is that of the Minimal Supersymmetric Standard Model (MSSM).[a] That is, we will evaluate the expression (in units of the BR for the semileptonic b decay):

$$\frac{BR(b \to s\gamma)}{BR(b \to ce\bar{\nu})} = \frac{|K_{ts}^* K_{tb}|^2}{|K_{cb}|^2} \frac{6\alpha_{QED}}{\pi} \times \frac{\left[\eta^{16/23} A_\gamma + \frac{8}{3}(\eta^{14/23} - \eta^{16/23}) A_g + C\right]^2}{I(z)} F \qquad (1)$$

[a] A detailed description of the calculation is given elsewhere [2].

where $z = \frac{m_c}{m_b}$, $\eta = \frac{\alpha_s(M_W)}{\alpha_s(m_b)}$, $I(z)$ is the phase space factor, C stands for the leading logarithmic QCD corrections, and F contains NLO effects. Finally, $A_{\gamma,g} = A^{SM}_{\gamma,g} + A^{H^-}_{\gamma,g} + A^{\chi^-}_{\gamma,g}$ are the coefficients of the effective operators for the $bs\gamma$ and bsg interactions[3]; in our case we consider as relevant the contributions coming from the SM diagram (top quark and W^-) plus those with top quark and charged Higgs, and stops/scharms and charginos running in the loop. It is interesting to note that both $A^{SM}_{\gamma,g}$ and $A^{H^-}_{\gamma,g}$ have always the same sign, giving therefore a total amplitude which is bigger than the SM one. However the presence of the chargino contribution which, for a wide range of the parameter space has opposite sign to that of the other two amplitudes, softens this effect. This is not enough, in general, to lower the total BR below the SM prediction, although in some cases it might be possible to have such big values for $A^{\chi^-}_{\gamma,g}$ as to drive the total BR even below the CLEO lower bound.

The MSSM is defined by the superpotential, W, and the form of the soft supersymmetry breaking terms. W is given by

$$W = \sum_i \{h_{u_i} Q_i H_2 u_i^c + h_{d_i} Q_i H_1 d_i^c + h_{e_i} L_i H_1 e_i^c\} + \mu H_1 H_2 \ , \qquad (2)$$

where i is a generation index, Q_i (L_i) are the scalar partners of the quark (lepton) SU(2) doublets, u_i^c, d_i^c (e_i^c) are the quark (lepton) singlets and $H_{1,2}$ are the two SUSY Higgs doublets; the h–factors are the Yukawa couplings and μ is the usual Higgs mixing parameter. The soft breaking terms have the form

$$\begin{aligned} \mathcal{L}_{\rm soft} &= \frac{1}{2} M \lambda_a \lambda_a - \sum_j m^2 |\phi_j|^2 - \sum_i A\,[h_{u_i} Q_i H_2 u_i^c + h_{d_i} Q_i H_1 d_i^c \\ &+ h_{e_i} L_i H_1 e_i^c + \text{h.c.}] - [B\mu H_1 H_2 + \text{h.c.}] \ , \end{aligned} \qquad (3)$$

where a is gauge group index and λ_a are the gauginos. m, M, A, B are the scalar and gaugino masses and the coefficients of the trilinear and bilinear scalar terms, respectively. They are assumed to be universal at the unification scale M_X.

The aim of our calculation is to obtain the restrictions on the MSSM from the $b \to s, \gamma$ decay in the two SUGRA scenarios defined by:

SUGRA I

This is just the minimal SUGRA theory. It is defined by a Kähler potential $K = \sum_j |\phi_j|^2$ and a gauge kinetic function $f_{ab} = \delta_{ab}$, so that all the kinetic terms are canonical, whereas the superpotential W is assumed to be as in eq. (2), μ being an inicial parameter. Then all the soft terms are automatically universal and the coefficients A and B are related to each other by the well known relation $B = A - m$. This scenario has a serious drawback, namely the unnaturally small (electroweak) size of the initial μ parameter in the superpotential (often known as μ problem); that leads us to the next scenario.

SUGRA II

Recently, there have appeared several attractive mechanisms to solve the μ problem [4] that, quite remarkably, lead, in the presence of a Kähler potential as in minimal SUGRA, to a similar prediction for the value of B, namely $B = 2m$. This can be achieved also if, as suggested by many superstring constructions, the kinetic terms are non-universal [5]. From now on we will refer (any of) these scenarios as SUGRA II. The corresponding MSSM emerging from them is as in minimal SUGRA except for the value of B.

So our starting point is the MSSM with initial parameters $\alpha_X, M_X, h_t, h_b, h_\tau, \mu, m, M, A, B$. By consideration of one of the two previous SUGRA theories we eliminate the B parameter, and then we demand consistency with all the experimental data, that is:

i) Correct unification of the gauge coupling constants

This can only be achieved when the renormalization group equations (RGE) of the gauge couplings are taken at two–loop order. For consistency, all the supersymmetric thresholds (and the top quark one) have to be taken into account in the running in a separate way. Therefore, given a choice of the initial parameters, an iterative process is necessary to achieve an agreement between the resulting prediction for $\alpha_1(M_Z), \alpha_2(M_Z), \alpha_3(M_Z)$ and their experimental values [6].

ii) Correct masses for all the observed particles.

Concerning the masses of the fermions, the boundary conditions for Yukawa couplings of the top and bottom quarks and the tau lepton, h_t, h_b, h_τ, have to be chosen so that the experimental masses are properly fitted (note that the running masses $m_{t,b,\tau}(Q)$ do not coincide with the physical (pole) masses $M_{t,b,\tau}$). Notice also that, as mentioned above, the masses in the BR($b \to s, \gamma$) are the running masses at the electroweak scale.

iii) Masses for the unobserved particles compatible with the experimental bounds [7].

iv) Correct electroweak breaking, i.e. $M_Z = M_Z^{exp}$.

The vacuum expectation values (VEVs) of the two Higgses, $v_1 = \langle H_1 \rangle$, $v_2 = \langle H_2 \rangle$ (upon which M_Z depends) are to be obtained from the minimization of the Higgs potential. The tree level part of this in the MSSM has the form $V_o = m_1^2|H_1|^2 + m_2^2|H_2|^2 + 2\mu B H_1 H_2 + \frac{1}{8}(g^2 + g'^2)(|H_2|^2 - |H_1|^2)^2$, where all the parameters are understood to be running parameters evaluated at the renormalization scale Q. By a suitable redefinition of the phases of the fields it is always possible to impose $v_1, v_2 > 0$. As was clarified some time ago [8], V_o and the corresponding tree level VEVs $v_1^o(Q), v_2^o(Q)$ are strongly Q–dependent. In

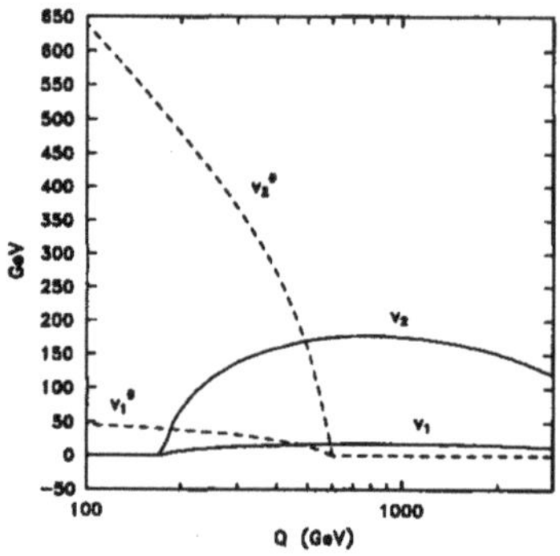

Figure 1: Plot of $v_1 \equiv \langle H_1 \rangle$, $v_2 \equiv \langle H_2 \rangle$ vs the Q scale between 100 GeV and 3 TeV for the model defined by $m = M = 300$ GeV, $A = 317$ GeV, $B = A - m$, $\mu = 403.76$ GeV, $h_t = 0.568$, $h_b = 0.063$, $h_\tau = 0.072$, $\alpha_X = 0.0404$ (all quantities defined at $M_X = 1.6 \times 10^{16}$ GeV). Solid lines: complete one–loop results, dashed lines: (renormalization improved) tree level results.

order to get a much more scale independent potential the one–loop correction ΔV_1 is needed. This is given by

$$\Delta V_1 = \sum_j \frac{n_j}{64\pi^2} M_j^4 \left[\log \frac{M_j^2}{Q^2} - \frac{3}{2} \right] \quad , \tag{4}$$

where $M_j^2(\phi, t)$ are the tree-level (field–dependent) mass eigenstates and n_j are spin factors. In this way, the minimization of $V = V_o + \Delta V_1$ gives one-loop VEVs $v_1(Q), v_2(Q)$ much more stable against variations of the Q scale. In general, there is a region of Q where $v_1(Q), v_2(Q)$ are remarkably Q–stable and a particular scale, $\hat{Q}$, always belonging to that region, at which $v_1(\hat{Q}), v_2(\hat{Q})$ essentially coincide with $v_1^o(\hat{Q}), v_2^o(\hat{Q})$. This is illustrated in Fig. 1.

Here we have evaluated v_1, v_2 at $\hat{Q}$ and then obtained $v_1(Q)$, $v_2(Q)$ at any scale by using the RG running of the H_1, H_2 wave functions. Furthermore, we have included in eq. (4) *all* the supersymmetric spectrum since some approximations can lead to wrong results [9].

Conditions *(i)–(iii)* allow to eliminate $\alpha_X, M_X, h_t, h_b, h_\tau$, while *(iv)* allows to eliminate one of the remaining parameters, which we choose to be μ. This finally leaves three independent parameters, namely m, M and A. Actually, for many choices of these parameters there is no value of μ capable of producing the correct electroweak breaking, a feature which restricts the parameter space substantially. On the other hand, there can be two branches of solutions depending on the sign of μ.

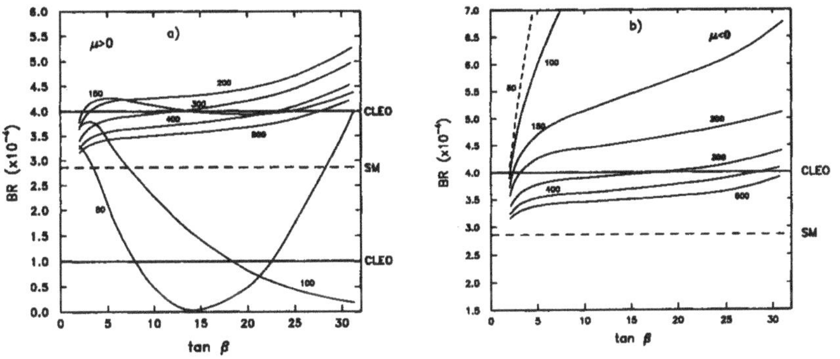

Figure 2: Plot of the branching ratio $BR(b \to s,\gamma)$ vs $\tan\beta$ for the SUGRA I scenario (minimal SUGRA) with $m = M$ for different values of m (80, 100, 150, 200, 300, 400, 500 GeV) : a) branch $\mu > 0$; b) branch $\mu < 0$. The dashed curves indicate that the model becomes incompatible with the experimental lower bounds on supersymmetric particles. The Standard Model prediction (SM) and the CLEO bounds are also shown in the figure.

2 Results

SUGRA I

As discussed above, our parameter space is now restricted to m, M, A. In order to present the results in a comprehensible way, let us make the assumption $m = M$, and trade the high energy parameter A by the low energy one $\tan\beta = v_2/v_1$. The corresponding plots of the branching ratio $BR(b \to s,\gamma)$ versus the remaining parameter, $\tan\beta$ for different values of m are given in Fig. 2a (branch $\mu > 0$) and Fig. 2b (branch $\mu < 0$). We can see that for $m \geq 200$ GeV and both $\mu > 0$, $\mu < 0$, for each value of m there is a maximum acceptable value for $\tan\beta$. In general for $m \geq 200$ GeV the restrictions are stronger for positive values of μ and small values of m. For $m \leq 200$ GeV, however, the situation is different: whereas for $\mu < 0$ the restrictions become very strong (only a narrow range of $\tan\beta$ is allowed), for $\mu > 0$ there appear large windows of allowed values of $\tan\beta$. Here the CLEO lower bound plays a relevant role.

It is clear that the trend to the SM result is so slow that an improvement of the CLEO bounds (particularly the upper one) on BR of order 10^{-4} would push the lower limits on m to the TeV region (except for the above-mentioned windows in the $\mu > 0$ case).

SUGRA II

We note here that the sign of μ is always negative since (adopting the convention $v_1, v_2 > 0$) for $\mu > 0$ and $B = 2m$ the necessary electroweak breaking cannot be achieved. The SUGRA II results present a similar pattern to those

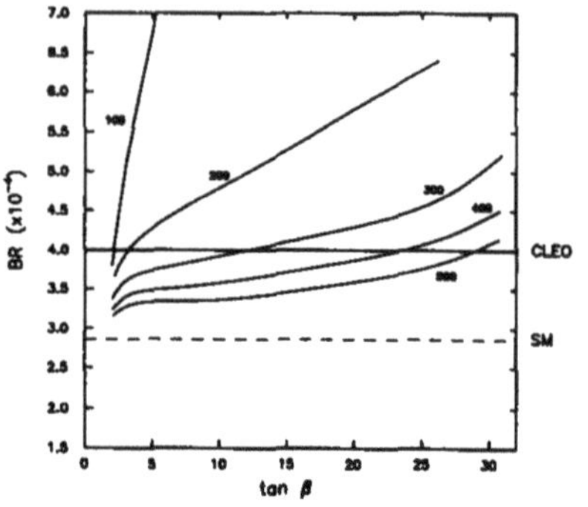

Figure 3: The same as in Fig. 2, but for the SUGRA II scenario.

of SUGRA I with $\mu < 0$. Again, we find from Fig. 3 that for a given value of m there is a maximum acceptable value for $\tan\beta$. The bound becomes less stringent as m increases; therefore, once again, an improvement of the CLEO bounds (especially the upper one) of order 10^{-4} would amount to a dramatical improvement of the MSSM constraints.

References

1. B. Barish et al. (CLEO collaboration) in *Proceedings of the XXVII ICHEP, Glasgow*, (Institute of Physics Publishing, 1994).
2. B. de Carlos and J.A. Casas, *Phys. Lett.* B **349**, 1995 (300); erratum ibid. **351**, 1995 (604).
3. S. Bertolini, F. Borzumati, A. Masiero and G. Ridolfi, *Nucl. Phys.* B **353**, 591 (1991).
4. J.A. Casas and C. Muñoz, *Phys. Lett.* B **306**, 1993 (288);
   G.F. Giudice and E. Roulet, *Phys. Lett.* B **315**, 1993 (107).
5. V. Kaplunovsky and J. Louis, *Phys. Lett.* B **306**, 1993 (269);
   A. Brignole, L.E. Ibáñez and C. Muñoz, *Nucl. Phys.* B **422**, 1994 (125); erratum ibid. **436**, 1995 (747).
6. U. Amaldi, W. de Boer and H. Fürsternau, *Phys. Lett.* B **260**, 1991 (447);
   P. Langacker, invited talk at the *22nd INS International Symposium on Physics with High Energy Collisions*, Tokyo (Japan) 8-10 March 1994.
7. Particle Data Group, *Phys. Rev.* D **50**, 1994 (1173).
8. G. Gamberini, G. Ridolfi and F. Zwirner, *Nucl. Phys.* B **331**, 1990 (331).
9. B. de Carlos and J.A. Casas, *Phys. Lett.* B **309**, 1993 (320).

DYNAMICAL BREAKING OF LEFT–RIGHT SYMMETRY

E.Kh. AKHMEDOV

International Centre for Theoretical Physics,
Strada Costiera 11, I-34100, Trieste, Italy
and
National Research Center "Kurchatov Institute",
123182 Moscow, Russia

Left–right symmetry breaking in a model with composite Higgs scalars is discussed. It is assumed that the Higgs bosons are generated dynamically through a set of gauge– and parity–invariant 4-fermion operators. It is shown that in a model with composite bi-doublet and two triplet scalars there is no parity breaking at low energies, whereas in the model with two doublets instead of two triplets parity is broken automatically regardless of the choice of the parameters of the model. For phenomenologically allowed values of the right–handed scale the tumbling symmetry breaking mechanism is realized in which parity breaking at a high scale μ_R propagates down and eventually causes the electroweak symmetry breaking at the scale $\mu_{EW} \sim 100$ GeV. The model yields a viable top quark mass and exhibits a number of low and intermediate mass Higgs bosons which may lead to observable lepton flavour violation effects. In particular, $\mu \to e\gamma$ decay rate may be close to its present experimental upper bound.

Several years ago, a novel approach to electroweak symmetry breaking was suggested, so called "top condensate" model [1,2,3,4]. In this model no fundamental Higgs boson is present; instead, it is assumed that there is a strong attractive interaction between the top quarks, which can lead to the formation of the $t\bar{t}$ bound state playing a role of the Higgs scalar. This interaction is assumed to result from a new physics at some high–energy scale Λ, the origin and precise nature of which are not specified. At low energies this new physics would manifest itself through non-renormalizable interactions between usual fermions and gauge bosons. At the energies $E \ll \Lambda$ the lowest dimensional operators are most important, which are just the four–fermion (4-f) operators. These 4-f interactions can be studied analytically in the large N_c (number of colours) limit in the fermion bubble approximation. For large enough 4-f coupling G ($G > G_{cr} = 8\pi^2/N_c\Lambda^2$) the electroweak symmetry is spontaneously broken, $W^{\pm}$ and Z^0 bosons and top quark acquire masses, and a composite Higgs scalar doublet $H \sim \bar{t}_R Q_L$ is formed.

The top condensate approach reproduced correctly the structure of the low–energy effective Lagrangian of the standard model and demonstrated how the electroweak symmetry breaking can result from a high–energy dynamics.

The question naturally arises as to whether a similar approach can work in more complicated cases, i.e. whether the Higgs sectors and symmetry breaking patterns of more complicated models can always be successfully described starting from a set of gauge–invariant 4-f operators. In this talk I will report the main results of the analysis of dynamical symmetry breaking in the left–right symmetric models done in collaboration with M. Lindner, E. Schnapka and J.W.F. Valle [5].

We studied dynamical symmetry breaking in the left–right symmetric (LR) models based on the gauge group $SU(2)_L \times SU(2)_R \times U(1)_{B-L}$ following the BHL approach to the standard model [3]. The Higgs sector of the most popular LR model consists of a bi-doublet $\phi \sim (2,2,0)$ and two triplets, $\Delta_L \sim (3,1,2)$ and $\Delta_R \sim (1,3,2)$. Assuming that these scalars are composite, their fermionic content is

$$\phi_{ij} \sim \alpha(\bar{Q}_{Rj}Q_{Li}) + \beta(\tau_2\bar{Q}_L Q_R \tau_2)_{ij} + \text{leptonic terms} \ ,$$

$$\vec{\Delta}_L \sim (\Psi_L^T C\tau_2\vec{\tau}\Psi_L), \quad \vec{\Delta}_R \sim (\Psi_R^T C\tau_2\vec{\tau}\Psi_R) \ . \qquad (1)$$

Here Q_L, Ψ_L (Q_R, Ψ_R) are left–handed (right–handed) doublets of quarks and leptons, respectively; i and j are isospin indices.

In models with Higgs bosons generated by 4-f operators the composite scalars are, roughly speaking, "square roots" of these 4-f operators. One can therefore obtain the above composite Higgs bosons starting from the 4-f operators which are "squares" of the expressions in eq. (1). A convenient way to study models with composite Higgs bosons is the auxiliary field technique, in which one introduces the static auxiliary scalar fields (with appropriate quantum numbers) with Yukawa couplings and mass terms but no kinetic terms and no quartic couplings. One can use the equations of motion for these fields to express them in terms of the fermionic degrees of freedom and recover the initial 4-f structures.

The static auxiliary fields can acquire gauge invariant kinetic terms and quartic self–interactions through the radiative corrections and become physical propagating scalar fields at low energies provided that the corresponding gap equations are satisfied [3]. The kinetic terms and mass corrections can be derived from the 2–point Green function, whereas the quartic couplings are given by the 4–point functions.

It is usually assumed that the Higgs potential of the LR model is exactly symmetric with respect to the discrete parity symmetry. However, parity can be spontaneously broken by $\langle\Delta_R\rangle > \langle\Delta_L\rangle$ provided $\lambda_2 > \lambda_1$ where λ_1 and λ_2 are the coefficients of the $[(\Delta_L^\dagger\Delta_L)^2 + (\Delta_R^\dagger\Delta_R)^2]$ and $2(\Delta_L^\dagger\Delta_L)(\Delta_R^\dagger\Delta_R)$ quartic couplings in the Higgs potential. In the conventional approach, λ_1 and λ_2 are free parameters and one can always choose $\lambda_2 > \lambda_1$. On the contrary, in the composite Higgs approach based on a certain set of the effective 4-f couplings, the parameters of the effective Higgs potential are not arbitrary: they are all calculable in terms of the 4-f couplings G_a and the scale of new physics Λ [3]. In particular, in the fermion bubble approximation at one loop level the quartic couplings λ_1 and λ_2 are induced through the Majorana–like Yukawa couplings $f(\Psi_L^T C\tau_2\vec{\tau}\vec{\Delta}_L\Psi_L + \Psi_R^T C\tau_2\vec{\tau}\vec{\Delta}_R\Psi_R) + h.c.$.

It is easy to see that to induce the λ_2 term one needs the Ψ_L–Ψ_R mixing in the fermion line in the loop, i.e. the lepton Dirac mass term insertions. However, the Dirac mass terms are generated by the VEVs of the bi-doublet ϕ; they are absent at the parity breaking scale which is supposed to be higher than the electroweak scale.

One is therefore led to consider a model with a different composite Higgs content. The simplest LR model includes two doublets, $\chi_L \sim (2, 1, -1)$ and $\chi_R \sim (1, 2, -1)$, instead of the triplets Δ_L and Δ_R. As we shall see, the model with composite doublets will automatically lead to the correct pattern of the dynamical breaking of parity. In conventional LR models, one can have the doublet Higgs bosons without introducing any other new particles. In our model all scalars are composite, thus we must introduce additional singlet fermions – otherwise it would not be possible to build up the composite χ_L and χ_R fields. We assume that in addition to the usual quark and lepton doublets there is a gauge–singlet fermion $S_L \sim (1,\ 1,\ 0)$. To maintain the discrete parity symmetry one needs a right–handed counterpart of S_L. This can be either another particle, S_R, or the right–handed antiparticle of S_L, $(S_L)^c \equiv C\bar{S}_L^T = S_R^c$. The latter choice is more economical and, as we shall see, leads to the desired symmetry breaking pattern. We therefore assume that under parity operation $S_L \leftrightarrow S_R^c$. With this new singlet and usual quark and lepton doublets one can introduce the following 4-f interactions [5]:

$$\begin{aligned}\mathcal{L}'_{4f} \;=\;& G_1(\bar{Q}_{Li}Q_{Rj})(\bar{Q}_{Rj}Q_{Li}) + [G_2(\bar{Q}_{Li}Q_{Rj})(\tau_2\bar{Q}_L Q_R\tau_2)_{ij} + h.c.]\\ &+G_3(\bar{\Psi}_{Li}\Psi_{Rj})(\bar{\Psi}_{Rj}\Psi_{Li}) + [G_4(\bar{\Psi}_{Li}\Psi_{Rj})(\tau_2\bar{\Psi}_L\Psi_R\tau_2)_{ij} + h.c.]\\ &+[G_5(\bar{Q}_{Li}Q_{Rj})(\bar{\Psi}_{Rj}\Psi_{Li}) + h.c.] + [G_6(\bar{Q}_{Li}Q_{Rj})(\tau_2\bar{\Psi}_L\Psi_R\tau_2)_{ij} + h.c.]\\ &+G_7[(S_L^T C\Psi_L)(\bar{\Psi}_L C\bar{S}_L^T) + (L \to R)] + G_8(S_L^T C S_L)(\bar{S}_L C\bar{S}_L^T)\;. \end{aligned} \qquad (2)$$

These interactions are not only gauge invariant, but also (for hermitean G_2, G_4, G_5 and G_6) symmetric with respect to the discrete parity symmetry.

The composite Higgs scalars which can be induced by these 4-f couplings include, in addition to the bi-doublet ϕ of the structure given in eq. (1), two doublets χ_L and χ_R and also a singlet σ:

$$\chi_L \sim S_L^T C\Psi_L, \quad \chi_R \sim \bar{S}_L\Psi_R = (S_R^c)^T C\Psi_R, \quad \sigma \sim \bar{S}_L C\bar{S}_L^T. \qquad (3)$$

Under parity $\chi_L \leftrightarrow \chi_R$, $\sigma \leftrightarrow \sigma^\dagger$.

In the auxiliary field formalism the scalars χ_L, χ_R, ϕ and σ have the following bare mass terms and Yukawa couplings:

$$\begin{aligned}\mathcal{L}_{aux} &= -M_0^2(\chi_L^\dagger\chi_L + \chi_R^\dagger\chi_R) - M_1^2\,\mathrm{tr}\,(\phi^\dagger\phi) - \frac{M_2^2}{2}\,\mathrm{tr}\,(\phi^\dagger\tilde{\phi} + h.c.) - M_3^2\sigma^\dagger\sigma \\ &\quad - \left[\bar{Q}_L(Y_1\phi + Y_2\tilde{\phi})Q_R + \bar{\Psi}_L(Y_3\phi + Y_4\tilde{\phi})\Psi_R + h.c.\right] \\ &\quad - \left[Y_5(\bar{\Psi}_L\chi_L S_R^c + \bar{\Psi}_R\chi_R S_L) + Y_6(S_L^T C S_L)\sigma + h.c.\right] \end{aligned} \quad (4)$$

where the field $\tilde{\phi} \equiv \tau_2\phi^*\tau_2$ has the same quantum numbers as ϕ: $\tilde{\phi} \sim (2,\ 2,\ 0)$. By integrating out the auxiliary scalar fields one can reproduce the 4-f structures of eq. (2).

Consider now parity breaking in the present LR model. Using the Yukawa couplings of the doublets χ_L and χ_R, one can calculate the fermion-loop contributions to the quartic couplings $\lambda_1[(\chi_L^\dagger\chi_L)^2+(\chi_R^\dagger\chi_R)^2]$ and $2\lambda_2(\chi_L^\dagger\chi_L)(\chi_R^\dagger\chi_R)$ in the effective Higgs potential. One can easily make sure that the fermion–loop diagrams yield $\lambda_1 = \lambda_2$. Recall that one needs $\lambda_2 > \lambda_1$ to have spontaneous parity breaking in the LR models. As we shall see, taking into account the gauge-boson loop contributions to λ_1 and λ_2 will automatically secure this relation.

Both λ_1 and λ_2 obtain corrections from $U(1)_{B-L}$ gauge boson loops, whereas only λ_1 is corrected by diagrams with W_L^i or W_R^i loops. Since all these contributions have a relative minus sign compared to the fermion loop ones, one finds $\lambda_2 > \lambda_1$ irrespective of the values of the Yukawa or gauge couplings or any other parameter of the model, provided that the $SU(2)$ gauge coupling $g_2 \neq 0$. Thus the condition for spontaneous parity breaking is automatically satisfied in our model.

We have a very interesting situation here. In a model with composite triplets Δ_L and Δ_R parity is never broken, i.e. the model is not phenomenologically viable. At the same time, in the model with two composite doublets χ_L and χ_R instead of two triplets (which requires introduction of an additional singlet fermion S_L) parity is broken automatically. This means that, unlike in the conventional LR models, in the composite Higgs approach *whether or not parity is spontaneously broken depends on the particle content of the model rather than on the choice of the parameters of the Higgs potential.*

Although there are no triplet Higgs bosons in the present version of the model, a modified seesaw mechanism is operative which ensures the smallness of the masses of neutrinos taking part in the usual electroweak interactions. This is because the neutrinos mix with the gauge–singlet fermion S_L. Minimization of the effective Higgs potential shows that for the choice of the 4-f couplings resulting in the dynamical breaking of the LR gauge symmetry, the χ_L and σ fields do not develop VEVs; for this reason the lightest neutrinos remain massless in our model.

Analysis of the vacuum structure of the model shows that for the right-handed scale $v_R = \langle\chi_R^0\rangle$ to lie in the phenomenologically allowed domain, the effective $(\text{mass})^2$ term of the composite bi-doublet ϕ must always be positive; the electroweak symmetry is broken because of the mixing of ϕ with χ_R. Thus we have a tumbling scenario where the breakdown of parity and $SU(2)_R$ occurring at the scale μ_R causes the electroweak symmetry breaking at a lower scale μ_{EW}.

The physical Higgs boson sector of the model includes 4 charged scalars, 4 neutral CP–even and 2 CP–odd scalars. In conventional LR models only one scalar, which is the analog of the standard model Higgs boson, is light (at the electroweak scale), all the others have their masses of the order of the right-handed scale M. In our case, the masses of those scalars are also proportional to M, but all of them except the mass of H_1 have some suppression factors. The masses of χ_L are suppressed because of their pseudo–Goldstone nature. It was already mentioned that at the fermion–bubble level $\lambda_1 = \lambda_2$ which means that the (χ_L, χ_R) sector of the Higgs potential depends on χ_L and χ_R only through the combination $(\chi_L^\dagger\chi_L + \chi_R^\dagger\chi_R)$. This, in turn, implies that the Higgs potential has a global $SU(4)$ symmetry which is bigger than the original gauge symmetry. In fact, the origin of this $SU(4)$ symmetry can be traced back to the 4-f operators of eq. (2). It is an accidental symmetry resulting from the gauge invariance and parity symmetry of the G_7 term. Note that no such symmetry occurs in conventional LR models. After χ_R^0 acquires a VEV, $SU(4)$ symmetry is broken down to $SU(3)$, and the components of χ_L are the corresponding Goldstone bosons. The $SU(4)$ symmetry of the Higgs potential is not exact: it is broken by the $SU(2)$ gauge–boson loop contributions which make $\lambda_2 > \lambda_1$ (and also by ϕ–dependent terms). As a result, χ_L are pseudo-Goldstone bosons with their mass vanishing in the limit $g_2 \to 0$, $m_\tau \to 0$. In fact, though the $SU(2)$ gauge coupling constant g_2 is smaller than the typical Yukawa constants in our model, it is not too small; estimates of the χ_L mass give $M_{\chi_L} \sim 10^{-1}M$.

The bi-doublet ϕ can be viewed as consisting of two doublets, ϕ_1 which develops a VEV and is similar to the standard model Higgs doublet, and the orthogonal field ϕ_2 which is VEVless. In the conventional LR models ϕ_2 is heavy, $M_{\phi_2} \sim M$. In our case the mass of its charged components $\phi_2^\pm$ is suppressed by the factor m_τ/m_t and is therefore of the order $10^{-2}M$. The masses of the neutral CP–even and CP–odd components are even smaller; they are related to the masses of $\phi_2^\pm$ and the standard model Higgs boson H_2

by

$$M^2_{\phi^0_{2r}} = M^2_{\phi^0_{2i}} = M^2_{\phi_2^\pm} - \frac{M^2_{H_2}}{2} = \frac{2}{3}M^2\frac{m_\tau^2}{m_t^2} - \frac{M^2_{H_2}}{2}. \quad (5)$$

This equation imposes an upper limit on the standard model Higgs boson mass M_{H_2} (for a given M) or a lower limit on the right–handed mass M (for a given M_{H_2}). These limits follow from the requirement that $M^2_{\phi_2^\pm}$ be positive, i.e. from the vacuum stability condition. For example, for $M_{H_2} \approx 200$ GeV we find $M \gtrsim 17$ TeV.

Assuming that only one of the neutral components of the bi-doublet develops a non-vanishing VEV, one can readily relate the top quark mass to the scale of new physics Λ (on which it depends logarithmically) and the electroweak VEV. For example, for $\Lambda \simeq 10^{15}$ GeV, one finds $m_t \simeq 165$ GeV. This result holds in the bubble approximation; the renormalization group improved result depends substantially on the ratio $\tan\beta \equiv \kappa/\kappa'$ of the VEVs of the bi-doublet. Phenomenologically acceptable values of m_t can be obtained for $\tan\beta \simeq 1.3 - 4$ [5].

To summarize, we have a successful dynamical LR model in which a tumbling symmetry breaking mechanism is operative. The model exhibits a number of low and intermediate scale Higgs bosons and predicts the relations between masses of various scalars and between fermion and Higgs boson masses which are in principle testable. If the right–handed scale μ_R is of the order of a few tens of TeV, the neutral CP–even and CP–odd scalars ϕ^0_{2r} and ϕ^0_{2i} can be even lighter than the electroweak Higgs boson. In fact, they can be as light as ~ 50 GeV and so might be observable at LEP2. Such light ϕ^0_{2r} and ϕ^0_{2i} can also provide a positive contribution to $R_b = \Gamma(Z \to b\bar{b})/\Gamma(Z \to hadrons)$ which is necessary to account for the discrepancy between the LEP observations and the standard model predictions.

References

1. Y. Nambu, in *New Theories in Physics, Proc. XI Int. Symposium on Elementary Particle Physics*, eds. Z. Ajduk, S. Pokorski and A. Trautman (World Scientific, Singapore, 1989) and EFI report No. 89-08 (1989), unpublished.
2. A. Miransky, M. Tanabashi and K. Yamawaki, *Mod. Phys. Lett.* A **4**, 1043 (1989); PLB **221**, 177 (1989).
3. W.A. Bardeen, C.T. Hill and M. Lindner, PRD **41**, 1647 (1990).
4. W.J. Marciano, PR **62**, 2793 (1989).
5. E.Kh. Akhmedov, M. Lindner, E. Schnapka and J.W.F. Valle, hep-ph/9509255.

MIRROR WORLD AND PRESENT NEUTRINO PUZZLES

Z.G. BEREZHIANI

INFN Sezione di Ferrara, I-44100 Ferrara, Italy, and
Institute of Physics, Georgian Academy of Sciences, Tbilisi, Georgia

In reconciling the present neutrino puzzles that are the solar and atmospheric neutrino deficits, the recent LSND oscillation and the need in the ~eV mass neutrino as the hot dark matter, one unavoidably has to introduce the extra light sterile neutrino(s). We suggest that this role can be naturally played by neutrinos of the mirror (shadow) world which is completely analogous to the visible one except that its 'electroweak' scale of the $[SU(2)\times U(1)]'$ gauge symmetry breaking is larger by about factor of 30 than the scale of the standard $SU(2)\times U(1)$ model. In this case the sterile neutrinos should be light by the same gauge symmetry reasons as the usual ones in the standard model. The needed pattern of the neutrino mass matrix can be obtained by assuming an approximate ZKM-type global lepton number symmetry $\bar{L} = L_e + L_\mu - L_\tau$. The Planck scale terms mix ν_e with its mirror partner ν_e' and induce oscillation with parameters naturally in the MSW solution range to the solar neutrino problem. Some cosmological aspects and the idea of asymmetric postinflationary reheating of two worlds are also discussed. [Talk is based on refs.[1,2] done in collaboration with R.N. Mohapatra and A.D. Dolgov].

Present neutrino puzzles. These are:

A. *Solar neutrino problem* (SNP). The solar ν_e deficit indicated by the solar neutrino experiments[3] cannot be explained by astrophysical reasons. The most popular and natural solution is provided by the MSW oscillation[4] of ν_e into another neutrino ν_x in solar medium. The required parameter range corresponds to $\delta m^2_{ex} \sim 10^{-5}$ eV^2 and, in the preferable "small mixing angle" scenario, $\sin^2 2\theta_{ex} \sim 10^{-3} - 10^{-2}$.

B. *Atmospheric neutrino problem* (ANP). There is an evidence[5] for a significant depletion of the atmospheric ν_μ flux by almost a factor of 2. This points to $\nu_\mu - \nu_x$ oscillation, with $\delta m^2_{\mu x} \sim 10^{-2}$ eV^2 and $\sin^2 2\theta_{\mu x} \sim 1$.

C. *Dark matter problem* (DMP). The COBE measurements of the cosmic microwave background anisotropy suggests that cosmological dark matter consists of two components, cold dark matter (CDM) being a dominant component and hot dark matter (HDM) being a smaller admixture[6,7]. The latter role can be naturally played by neutrinos with mass of few eV's. As for the CDM, it can be of interest to think of it as also consisting of neutrinos, this time heavier (keV range), so called warm dark matter (WDM)[8].

D. *LSND result:* Direct evidence of $\bar{\nu}_\mu - \bar{\nu}_e$ oscillation from the recent Los Alamos experiment[9], with $\delta m^2_{e\mu} \geq 0.3$ eV^2 and $\sin^2 2\theta_{e\mu} = 10^{-3} - 10^{-2}$.

An optimistic approach to the situation would be to assume that all these hints (or at least first three of them) will indeed be confirmed in future and one should therefore study their implications for the texture of neutrino masses. It was shown[10] that only one possibility is compatible with all the above mentioned data. It requires an extra light, sterile neutrino ν_s beyond the three known neutrinos ν_e, ν_μ and ν_τ. Then, assuming that $m_{\nu_{e,s}} \ll m_{\nu_{\mu,\tau}}$, the SNP can be explained by the $\nu_e - \nu_s$ oscillation and the ANP by the $\nu_\mu - \nu_\tau$ oscillation. In addition, the ν_μ and ν_τ with mass $\simeq 2.4$ eV provide the cosmological HDM and can also explain the LSND result.

Indeed, the small mixing angle MSW oscillation $\nu_e - \nu_s$ is the only place where the extra sterile neutrino could show up. Obviously, ν_s cannot constitute HDM, neither it can be applied to ANP, due to primordial nucleosynthesis bound on the effective number of neutrino species $N_\nu < 3.1$.[11] In particular, this bound implies that the oscillation of active neutrinos into the sterile one should obey the limit[12] $\delta m^2 \sin^2 2\theta \leq 1.6 \cdot 10^{-6}\,\mathrm{eV}^2$, which excludes the $\nu_\mu - \nu_s$ oscillation as a candidate for the ANP solution. The large mixing angle MSW oscillation $\nu_e - \nu_s$ is also excluded as a SNP solution by the same reason.

Concerning the sterile neutrinos, an immediate question that arises is "where do they all come from, and where do they all belong?"[13]. In particular, how they can be so light when their masses are allowed by the gauge symmetry?[a] Here we suggest that the sterile neutrinos are in fact the neutrinos of a shadow world which is the mirror duplicate of our visible world but its 'electroweak' scale v' is by some factor ζ larger than the standard electroweak scale v. Thus, the mirror neutrinos $\nu'_{e,\mu,\tau}$ should be light by the same reason as the usual ones $\nu_{e,\mu,\tau}$: their mass terms are suppressed by the accidental B-L symmetry resulting from the gauge structure of the theory.

We show that this framework can provide a plausible solution to all present neutrino puzzles. We suggest that the dominant entries ($\sim$ eV) in the neutrino mass matrix have origin in some intermediate scale physics which respects an approximate global ZKM-type lepton number $L_e + L_\mu - L_\tau$ in both sectors. This fixes the neutrino mass texture needed for reconciling the HDM requirement, the ANP and the LSND oscillation, while the ν_e and ν'_e states are left massless. The masses and mixing of the latter and thus $\nu_e - \nu'_e$ oscillation then can be induced by the Planck scale effects, which for $\zeta \sim 30$ or so properly reproduce the parameter range required by the MSW solution to the SNP. In addition, as far as $\nu_{\mu,\tau}$ with $\sim$ eV mass play a HDM role, then $\nu'_{\mu,\tau}$ being $\zeta^2 \sim 1000$ times heavier fall in the keV range and thus can constitute WDM of the universe.

We also address some other cosmological implications of the mirror world.

[a]For some proposals for light ν_s invoking extra global symmetries see [14,15].

Without the mirror world. Apparently, three known neutrinos are not enough to reconcile all these hints[10]. In fact, the key difficulty is related to SNP. The other puzzles can be easily reconciled by assuming that in flavour basis $\nu_{e,\mu,\tau}$ the neutrino mass matrix has a following texture obeying the approximate $\bar{L} = L_e + L_\mu - L_\tau$ conservation, with say $m \simeq 2.4$ eV and $\varepsilon \simeq 10^{-3}$:

$$\hat{m}_\nu = \begin{pmatrix} 0 & 0 & m\sin\theta_{e\mu} \\ 0 & 0 & m\cos\theta_{e\mu} \\ m\sin\theta_{e\mu} & m\cos\theta_{e\mu} & \varepsilon m \end{pmatrix} \tag{1}$$

This matrix has one massless eigenstate $\nu_1 \simeq \nu_e$ and two almost degenerate eigenstates $\nu_{2,3} \simeq (\nu_\mu \pm \nu_\tau)/\sqrt{2}$ with masses $m_{2,3} \approx m(1 \pm \varepsilon/2)$. Then the latter will play a role of the HDM[7], the $\nu_\mu - \nu_\tau$ oscillation with $\sin 2\theta_{\mu\tau} \approx 1$ and $\delta m^2_{\mu\tau} \approx 2\varepsilon m^2 \simeq 10^{-2}$ eV^2 will explain the ANP, and the $\nu_e - \nu_\mu$ oscillation with $\delta m^2_{e\mu} = m^2 \simeq 6\,\text{eV}^2$ could explain the LSND result if $\sin\theta_{e\mu} \simeq 2 \cdot 10^{-2}$.

To be less vague, let us suggest a simple seesaw model that could naturally implement the desired structure. We introduce two "right-handed" neutrinos N and S, and prescribe the global lepton number $\bar{L} = 1$ to $l_{e,\mu}$ (lepton doublets) and S, and $\bar{L} = -1$ to l_τ and N. Let us also introduce a gauge singlet scalar Φ with $\bar{L} = 2$ and nonzero VEV $\langle\Phi\rangle = V$. Then the relevant piece of Lagrangian takes a form

$$f_1\phi l_e N + f_2\phi l_\mu N + f_3\phi l_\tau S + MNS + G\Phi NN + \bar{G}\Phi^* SS + h.c. \tag{2}$$

where we assume that the Yukawa constants obey the 'horizontal' hierarchy $f_1 \ll f_2 \ll f_3$, and $GV \ll M$. Clearly, this model provides a texture resembling (1). One neutrino eigenstate ν_1 (an admixture of ν_e and ν_μ with angle $\theta_{e\mu} \simeq f_1/f_2$) is left massless while two other eigenstates $\nu_{2,3}$ get masses $m \simeq f_2 f_3 v^2/M$ with small splitting: $\varepsilon \simeq (f_3/f_2)(GV/M)$.

Thus, only the SNP remains unresolved.

Introducing the mirror world.[16] Having in mind the $E_8 \times E_8'$ string, one can imagine a theory given by the product of two identical gauge groups $G \times G'$, where $G = SU(3) \times SU(2) \times U(1)$ stands for the standard model describing particles of the visible world: quarks and leptons q_i, u_i^c, d_i^c; l_i, e_i^c and the Higgs doublet ϕ, and $G' = [SU(3) \times SU(2) \times U(1)]'$ is its mirror counterpart with analogous particle content: q_i', $u_i'^c$, $d_i'^c$; l_i', $e_i'^c$ and ϕ' ($i = 1,2,3$ is a family index)[b] Let us suppose also that two worlds communicate only through

[b]For the naturalness it is essential that at higher energies that the $SU(3) \times SU(2) \times U(1)$ factors in both sectors should be embedded into some simple gauge groups. E.g. one can consider (SUSY) GUT like $SU(5) \times SU(5)'$ without mixed representations, which then breaks down to $G \times G'$. Otherwise, kinetic terms of the two $U(1)$ gauge fields can mix which would induce arbitrary electric charges to the particles[16].

gravity and possibly also via some heavy gauge singlet matter, and there exists a discrete symmetry $P(G \leftrightarrow G')$ interchanging all particles in corresponding representations of G and G'. It implies that all coupling constants (gauge, Yukawa, Higgs) have the same pattern in both of these worlds. Let us also assume that there is some mechanism that spontaneously breaks P parity at lower energies and thus allows the weak interaction scales $\langle\phi\rangle = v$ and $\langle\phi'\rangle = v'$ to be different, say $v' \gg v$. Thus the fermion mass and mixing pattern (except the neutrinos) in the mirror world is completely analogous to that of the visible one, with corresponding fermion masses just scaled up by the factor $\zeta = v'/v$. The W', Z' and Higgs$'$ boson masses are also scaled up by ζ while photons and gluons remain massless in both systems.

As for neutrinos, they can acquire nonzero masses only via operators bilinear in the Higgs fields and cutoff by some large scale M. We assume that P parity breaking is a lower energy phenomenon and it is respected by these operators. For example, one can directly extend the model of previous section, by introducing along with the heavy neutral states N, S also their mirror partners N', S', with the same mass M. Then the neutrino mass operators are

$$\frac{h_{ij}}{M}(l_i\phi)(l_j\phi) + \frac{h_{ij}}{M}(l'_i\phi')(l'_j\phi') + h.c. \tag{3}$$

with constants h_{ij} obeying the approximate $L_e + L_\mu - L_\tau$ symmetry. In doing so, the $\nu_{2,3}$ states get almost degenerate masses $m \sim$ few eV and thus can constitute HDM, and the ANP and LSND problems are also solved. Then for $\zeta \sim 30$, mass $m' = \zeta^2 m$ of their mirror partners $\nu'_{2,3}$ is in the keV range, and thus they can consitute WDM of the universe.

The ν_e and ν'_e are left massless. However, they can get masses from the Planck scale operators[17,18] which also can induce thir mixing[18]. These are

$$\frac{\alpha_{ij}}{M_{Pl}}(l_i\phi)(l_j\phi) + \frac{\alpha_{ij}}{M_{Pl}}(l'_i\phi')(l'_j\phi') + \frac{\beta_{ij}}{M_{Pl}}(l_i\phi)(l'_j\phi') + h.c. \tag{4}$$

with constants $\alpha, \beta \sim 1$. Then ν_e and ν'_e acquire masses respectively $\sim \overset{o}{m}$ and $\sim \zeta^2 \overset{o}{m}$, and their mixing term is $\sim \zeta \overset{o}{m}$, where $\overset{o}{m} = v^2/M_{Pl} = 3 \cdot 10^{-6}$ eV. Hence, the $\nu_e - \nu'_e$ oscillation emerges with parameters in the range

$$\delta m^2 \sim (\zeta/30)^4 \times 8 \cdot 10^{-6}\,\text{eV}^2, \qquad \sin^2 2\theta \sim (30/\zeta)^2 \times 5 \cdot 10^{-3} \tag{5}$$

which for $\zeta \sim 30$ perfectly fit the small mixing angle MSW solution to the SNP. More in general, by taking into account the solar model uncertainties[4], as well as possible order of magnitude spread in the constants α, β, the possible range for ζ can be extended to $\zeta \sim 10 - 100$.[1]

Spontaneous P parity breaking and mirror cosmology. The simplest possibility to break the discrete $P(G \leftrightarrow G')$ symmetry is to introduce a P-odd real scalar η with VEV $\langle\eta\rangle = \mu$ at some intermediate scale[19]. Its interaction with the Higgs doublets ϕ and ϕ' then has a form

$$g\mu\eta(|\phi|^2 - |\phi'|^2) + h\eta^2(|\phi|^2 + |\phi'|^2) \tag{6}$$

Hence, after substituting the VEV of η the effective mass terms of ϕ and ϕ' become different, and their VEVs $v \neq v'$ can be nonzero. As for the coupling constants, they will not be affected and thus will maintain the same pattern in both worlds. Hence, the particle spectrum in the mirror world will have the same shape as that of the visible one but scaled up by the factor $\zeta = v'/v$.

Concerning the strong interactions, it is clear that a big difference between the electroweak scales v' and v will not cause as big difference between the confinement scales in two worlds. Indeed, as far as P parity is valid at higher scales, the strong coupling constant will evolve down in energies with same value in both sectors until it reaches the value of the mirror-top (t') mass. Below it α'_s will have a different slope than α_s. It is then very easy to calculate the value of the scale Λ' at which α'_s becomes large. This value will of course depend on ζ. By taking $\Lambda_{QCD} = 200$ MeV for the ordinary QCD, then say for $\zeta \sim 30$ we find $\Lambda'_{QCD} \simeq 280$ MeV. On the other hand, for $\zeta \sim 30$ the masses of 'light' mirror quarks $m'_{u,d} = \zeta m_{u,d}$ do not exceed Λ'_{QCD} and thus they should develope condensates $\langle\bar{q}'q'\rangle$ as their visible partners do. Then the mirror π-mesons should have masses rather close to that of our K-mesons. Also the mirror nucleons are not much heavier than the usual proton and neutron, and the typical nuclear binding energies are not altered substantially. However, the mirror light quarks have ~ 200 MeV masses and thus we expect the mass difference between the mirror neutron and proton of 100 MeV or so, while the mirror electron mass is 15 MeV. Then unlike in our world, in the mirror world all bound neutrons will be unstable against β decay and the mirror hydrogen will be the only stable nucleus.

An immediate challenge to our approach is to reconcile the big bang nucleosynthesis (BBN) constraints. In standard cosmology effective number of the light degrees of freedom at the BBN era is $g_* = 10.75$, as it is contributed by photons, electrons and three neutrino species. It is known that $N_\nu = 3$ is in remarkable agreement with the observed abundances of light elements[11]. In our case, the mirror photons γ' and neutrinos $\nu'_{e,\mu,\tau}$ would also contribute as an effective number of extra neutrino species ΔN_ν:

$$\Delta g_* = (2 + 5.25x^4)\left(\frac{T'}{T}\right)^4 = 1.75\Delta N_\nu \tag{7}$$

Here T and T' respectively are the usual and mirror photon temperatures at the BBN era and $x = T'_\nu/T'$, where T'_ν is a temperature of the mirror neutrinos. Its value is determined by the temperature T'_D at which ν' decouple from the mirror photons, which in turn depends on the value of ζ: $T'_D/T_D \sim \zeta^{4/3}$ where $T_D = 3 - 5$ MeV is the decoupling temperature of the usual neutrinos. For $\zeta \sim 30$, T'_D appears to be larger than $m'_{u,d} = \zeta m_{u,d}$ and $\Lambda'_{QCD}(\simeq 280$ MeV), so that the mirror quarks u', d' will also contribute in heating of γ' after the ν' freezing out. Then $x = (4/53)^{1/3}$ and by taking $\Delta N_\nu < 0.1$ (or < 0.01) we obtain from (7) $T'/T = 0.95(\Delta N_\nu)^{1/4} < 0.53$ (or < 0.30).

The similar bound can be derived from the universe overclosure constraint. Indeed, if the $\nu_{\mu,\tau}$ mass is in the eV range required by the HDM: $2m = \Omega_{HDM} \cdot 92h^2$ eV, then their mirror partners $\nu'_{\mu,\tau}$ being ζ^2 times heavier would constitute WDM: $2rm\zeta^2 = \Omega_{WDM} \cdot 92h^2$eV, where $r = n'_\nu/n_\nu$ is a present abundance of mirror neutrinos relative to ordinary ones: $r = (xT'/T)^3$. By taking rather conservatively $\Omega_{HDM} < 0.8$ (since CDM component could exist also in other forms), then we obtain a bound $r < 4/\zeta^2$ which in turn e.g. for $\zeta \simeq 30$ implies $T'/T < 1.6/x\zeta^{2/3} < 0.4$.

Thus, at the BBN epoch the mirror photons should be at least 2-3 times 'cooler' than the usual ones. How one could achieve this situation? For solving this issue, one can suggest the idea of asymmetric postinflationary "reheating" between two worlds[1,2]. Indeed, as far as P-parity is broken, then it should not be respected also by the inflaton couplings to visible and mirror matter. (In fact, η can itself play a role of inflaton[2]). Then the inflaton decays into the visible and mirror particles occur with different rates, so that the ordinary and mirror thermal bathes can be established having different temperatures T_R and T'_R.[2] One has also to assume that already at the reheating stage two worlds are decoupled from each other. In particular, this means that couplings like $a|\phi|^2|\phi'|^2$, in principle allowed by symmetry, should be strongly suppressed: $a < 10^{-7}$. In this way, the initial cosmological abundance of the mirror particles can be smaller than that of their visible partners.

As far as the two worlds are decoupled, during the universe expansion they evolve with separately conserved entropy[c]. Then the T_R, T'_R are related to the temperatures T, T' at the BBN epoch as

$$\frac{T'_R}{T_R} = \left(\frac{2 + 5.25x^3}{10.75}\right)^{\frac{1}{3}} \frac{T'}{T} \simeq 0.6\,\frac{T'}{T} \tag{8}$$

Thus, postinflationary reheating temperature in the mirror sector has to be

[c] In fact, this applies if the expansion goes adiabatically in both sectors and only the second order phase transitions occur. At the presence of the first order phase transitions in both sectors this relation would change due to additional entropy production[2].

few times smaller than that of the visible world. For more detailed discussion of asymmetric reheating and possible (supersymmetric) models see [2].

As we remarked earlier, the concept of the mirror universe suggests a natural possibility for solving the DMP with dark matter consisting completely of neutrinos. In particular, the bulk of the dark matter can be due the mirror neutrinos with few keV mass (the WDM component), while usual neutrinos with few eV mass form the HDM component. Implications of the WDM for the shape of the large scale inhomogeneities are rather similar[8] to that of the standard CDM made upon the heavy ($m \sim 100$ GeV) particles. However, the WDM consisting of sterile neutrinos implies negative predictions for the direct search of the dark matter particles. One can also hope that in future the observational data on the cosmological large scale structure will allow to discriminate some peculiar features of the WDM from the standard CDM picture (in fact, in the supersymmetric version of our scheme the CDM could also exist in the form of lightest neutralino).

It is of specific interest also the structure of the mirror baryonic matter. Most likely, baryogenesis in the mirror world should proceed by the same mechanism as in our visible universe, so that the baryon asymmetry in both worlds should be nearly the same. On the other hand, mirror nucleons are not substantially heavier than the usual ones, so that fraction of mirror baryons in the present cosmological density can be few percents: $\Omega_{B'} \sim \Omega_B$. On the other hand, recombination of the mirror matter occurs earlier so that it will evolve into inhomogeneities more efficiently than the visible one, and being a dissipative "dark" matter its big fraction can be accumulated into galaxy cores and mostly collapse into black holes (since the mirror hydrogen is the only stable nucleus, the big stars in the mirror universe cannot be formed due to absence of nuclear burning). The remaining fraction could form cold "Jupiters" which would be observed as machos in microlensing experiments.

In conclusion, we have shown that if the existence of the light sterile neutrinos is to be taken seriously for solving the neutrino puzzles, then their lightness is most easily understood if they are identified with the neutrinos of the mirror universe. We have also discussed how the basic texture could emerge in this picture which would reconcile all present neutrino puzzles.

Acknowledgments

I thank Sasha Dolgov and Rabi Mohapatra for collaboration on these issues, and also wish to express my ♡ gratitude to Jose Valle and other organizers of the wonderful meeting in Valencia.

References

1. Z.G. Berezhiani and R.N. Mohapatra, hep-ph/9505385; see also rather different model by R. Foot and R. Volkas, hep-ph/9505359.
2. Z.G. Berezhiani, A.D. Dolgov and R.N. Mohapatra, in preparation.
3. Homestake Collaboration, *Nucl. Phys.* **B38** (Proc. Suppl.), 47 (1995); Kamiokande Collaboration, *ibid.*, 55; SAGE Collaboration, *ibid.*, 60; GALLEX Collaboration, *ibid.*, 68.
4. S.P. Mikheyev and A.Yu. Smirnov, *Yad. Fiz.* **42**, 1441 (1985); L. Wolfenstein, *Phys. Rev.* D **17**, 2369 (1978). for the recent status see P. Krastev and A. Smirnov, *Phys. Lett.* B **338**, 282 (1994); V. Berezinsky, G. Fiorentini and M. Lissia, *ibid.* **341**, 38 (1994).
5. Y. Fukuda et al., *Phys. Lett.* B **335**, 237 (1994).
6. R. Shaefer and Q. Shafi, *Nature* **359**, 199 (1992); E.L. Wright et al, *Astrop. J.* **396**, L13 (1992); A. Klypin et al., *ibid.* **416**, 1 (1993).
7. J.R. Primack, J. Holtzman, A. Klypin and D.O. Caldwell, *Phys. Rev. Lett.* **74**, 2160 (1995).
8. S. Dodelson and L. Widrow, *Phys. Rev. Lett.* **72**, 17 (1994); R. Malaney, G. Starkman and L. Widrow, preprint CITA-95-9 (1995).
9. C. Athanassopoulos et al., nucl-ex/9504002.
10. D.O. Caldwell and R.N. Mohapatra, *Phys. Rev.* D **48**, 3259 (1993); J. Peltoniemi and J.W.F. Valle, *Nucl. Phys.* B **406**, 409 (1993).
11. T. Walker et al, *Ap. J.* **376**, 51 (1991); P. Kernan and L. Krauss, *Phys. Rev. Lett.* **72**, 3309 (1994); K. Olive and G. Steigman, OSU-TA-2/95.
12. K. Enqvist, J. Maalampi and V. Semikoz, hep-ph/9505210.
13. J. Lennon and P. McCartney, "Eleanor Rigby", *Revolver,* EMI, 1966.
14. A. Smirnov and J. Valle, *Nucl. Phys.* B **375**, 649 (1992); J. Peltoniemi, D. Tommasini and J. Valle, *Phys. Lett.* B **298**, 383 (1993); E. Akhmedov, Z. Berezhiani, G. Senjanović+ Z. Tao, *Phys. Rev.* D **47**, 3245 (1993).
15. E.J. Chun, A. Joshipura and A. Smirnov, hep-ph/9505275, hep-ph/9507371; E. Ma, hep-ph/9507348.
16. The concept of a hidden mirror world of particles has been considered in several earlier papers, e.g. T.D. Li and C.N. Yang, *Phys. Rev.* **104** (1956) 254; Y. Kobzarev, L. Okun and I. Pomeranchuk, *Yad. Fiz.* **3** (1966) 1154; B. Holdom, *Phys. Lett.* B **166**, 196 (1985); S.L. Glashow, *ibid.* **167**, 35 (1986); R. Foot, H. Lew, R. Volkas, *ibid.* **272**, 67 (1991).
17. R. Barbieri, J. Ellis and M.K. Gaillard, *Phys. Lett.* **90B**, 249 (1980).
18. E. Akhmedov, Z. Berezhiani and G. Senjanović, *Phys. Rev. Lett.* **69**, 3013 (1992).
19. D. Chang, R.N. Mohapatra, M. Parida, *Phys. Rev.* D **30**, 1052 (1984).

LIGHT SINGLET FERMIONS AND NEUTRINO PHYSICS

EUNG JIN CHUN

International Center for Theoretical Physics
P. O. Box 586, 34100 Trieste, Italy

The existence of a light singlet fermion mixed with the electron neutrino is hinted by the simultaneous explanation of various neutrino anomalies. We show that supersymmetry can provide a natural framework for the existence and the desired properties of such a fermion. Quasi Goldstone fermions (QGF) of spontaneously broken global symmetries like the Peccei-Quinn symmetry or lepton number can mix properly with the neutrinos provided the presence of the R-parity breaking term ϵLH_2. The lightness of QGF can be a consequence of non-minimal Kähler potentials like that of no-scale supergravity. In order to keep R-parity, such a sterile component has to be placed in a new singlet superfield with no vacuum expectation value. In the context of the standard seesaw mechanism the lightness of such a singlet can be understood by imposing a R-symmetry.

1 Introduction

All the experimentally known fermions transform non-trivially under the gauge group $SU(3) \times SU(2) \times U(1)$ of the standard model (SM). However there are experimental hints in the neutrino sector which suggest the existence of $SU(3) \times SU(2) \times U(1)$ - singlet fermions mixing appreciably with the known neutrinos. These hints come from (a) the deficits in the solar [1] and atmospheric [2] neutrino fluxes (b) possible need of significant hot component [3] in the dark matter of the universe and (c) some indication of $\bar{\nu}_e - \bar{\nu}_\mu$ oscillations in the laboratory [4]. These hints can be reconciled with each other if there exists a fourth very light ($< \mathcal{O}(\text{eV})$) neutrino mixed with some of the known neutrinos preferably with the electron one. The fourth neutrino is required to be sterile in view of the strong bounds on number of neutrino flavours coming both from the LEP experiment and from the primordial nucleosynthesis [5].

The existence of a very light sterile neutrino demands theoretical justification since unlike the active neutrinos, the mass of a sterile state is not protected by the gauge symmetry of the SM and hence could be very large. Usually a sterile neutrino is considered on the same footing as the active neutrinos and some ad hoc symmetry is introduced to keep this neutrino light. Recently there are several attempts to construct models for sterile neutrinos which have their origin beyond the usual lepton structure [6,7,8,9,10].

In this report, we discuss the role of supersymmetry (SUSY) in explaining both the existence and the lightness of a singlet fermion S which can mix with the neutrinos. As a case of special interest we will concentrate on the mass of S and its mixing with the electron neutrino in the range:

$$\begin{aligned} m_S &\simeq (2-3)\cdot 10^{-3}\,\text{eV} \\ \sin\theta_{es} &\simeq \tan\theta_{es} \simeq (2-6)\cdot 10^{-2}\,. \end{aligned} \quad (1)$$

These values of parameters allow one to solve the solar neutrino problem through the resonance conversion $\nu_e \to S$ [11]. More discussions on simultaneous reconciliations of the diverse neutrino problems can be found in refs. [6,7] on which this report is based.

2 Quasi Goldstone Fermion

The existence of SM-singlet fields is a common property in physics beyond the standard model. The most interesting examples are the Goldstone bosons of spontaneously broken global symmetries required to solve the strong CP problem (the Peccei-Quinn symmetry) [12] and to explain the origin of neutrino masses (the lepton number symmetry) [13]. In the SUSY limit, a spontaneously broken global symmetry automatically generates a massless singlet (Goldstone) fermion being a superpartner of a Goldstone boson. However, SUSY breakdown results in generation of mass of a Goldstone fermion. While the existence of these quasi Goldstone fermions (QGF) is logically independent of neutrino physics, there are good reasons to expect that these fermions will couple to neutrinos. Indeed, in the case of lepton number symmetry the superfield which is mainly responsible for the breakdown of the lepton number symmetry carries nontrivial lepton number and therefore it can directly couple to leptons if the charge is appropriate. In the case of the PQ symmetry, this superfield could couple to the Higgs supermultiplet. If theory contains small violation of R-parity then this mixing with the Higgs gets communicated to the neutrino sector. Thus the occurrence of a QGF can have implications for neutrino physics. In the following subsections we elaborate upon the expected properties of the QGF: their masses arising after SUSY breaking and the mixing of these fermions with the electron neutrino.

2.1 masses of QGF

The supersymmetric standard model with some global symmetry $U(1)_G$ can be characterized by the following superpotential:

$$W = W_{MSSM} + W_S + W_{mixing}\,, \quad (2)$$

where W is assumed to be invariant under $U(1)_G$. As we outlined in the above, this symmetry may be identified with the PQ symmetry, lepton number symmetry or combination thereof. The first term in eq. (2) refers to the superpotential of the minimal supersymmetric standard model (MSSM). The second term contains $SU(3) \times SU(2) \times U(1)$ singlet superfields which are responsible for the breakdown of $U(1)_G$. The minimal choice for W_S is

$$W_S = \lambda(\sigma\sigma' - f_G^2)y \, , \tag{3}$$

where σ, σ' carry non trivial G-charges and f_G sets the scale of $U(1)_G$ breaking. The last term of eq. (2) describes mixing of the singlet fields with the superfields of the MSSM.

In the case (3) the Goldstone fermion is contained in $S \sim \sigma - \sigma'$ and is massless in the SUSY limit. Broken SUSY itself cannot automatically protect the mass of a QGF. It depends on the structure of the superpotential W_S [14] and on the pattern of soft-terms [15]. It also depends on the way this breaking is communicated to the singlet S and the scale f_G [7]. The most natural framework for light QGF is no-scale supergravity [16]. No-scale models contain only one kind of soft-terms, namely, gaugino masses. Therefore, th soft SUSY-breaking terms corresponding to W_S in eq. (2) are absent at tree-level and thus QGF remains massless. However, the radiative mass can be triggered by the $SU(3) \times SU(2) \times U(1)$ gaugino masses through a set of interactions. A realistic example can be found in the context of the seesaw mechanism. The vacuum expectation value (VEV) of the field σ (or σ') may give rise to large masses of right-handed (RH) neutrinos N as in the following superpotential invariant under $U(1)_G$:

$$W = \frac{m^D}{\langle H_2 \rangle} LNH_2 + \frac{M}{f_G} NN\sigma \, , \tag{4}$$

where we have omitted the generation indices. The generation structure of the superpotential (4) will depend on the $U(1)_G$-charge assignment to the fields [7]. This $U(1)_G$ symmetry is not necessarily the lepton number symmetry as we will discuss in subsection 2.2. The first term in eq. (4) gives rise to the Dirac masses of the neutrinos, whereas the second one gives the Majorana masses of RH neutrino components. The scale $f_G \sim 10^{10} - 10^{12}$ GeV generates $M \sim 10^{10} - 10^{11}$ GeV required by the hot dark matter and atmospheric neutrinos. If the soft-term $A_N NN\sigma$ with $A_N \sim m_{3/2}$ is present, there appears one-loop mass of the QGF proportional to A_N [17]. But in no-scale models $A_N = 0$ at tree-level and the QGF mass is indeed generated in three loops as shown in Figure 1. This three-loop mass can be estimated as

$$m_S \simeq \frac{\alpha_2}{(4\pi)^5} \frac{m_\nu M^3}{v_2^2 f_G^2} m_{1/2} \, . \tag{5}$$

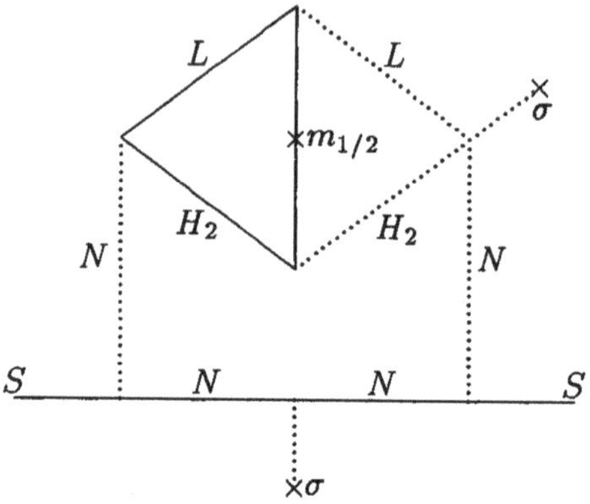

Figure 1: Three-loop diagram for the QGF mass. The cross with $m_{1/2}$ denotes gaugino mass insertion.

Here α_2 and $m_{1/2}$ are the $SU(2)$ fine structure constant and gaugino mass respectively. For $m_\nu \simeq 3$ eV, $m_{1/2} \simeq v_2 \simeq 100$ GeV, and $f_G \simeq 10^{12}$ GeV, one gets $m_S \simeq 3 \cdot 10^{-3}$ eV with a value of $M \simeq 10^{10}$ GeV.

A contribution to the mass of the QGF can follow also from interactions, W_{mixing}, which mix S with usual neutrinos (subsection 2.2).

2.2 Neutrino-QGF mixing

We now discuss how the QGF can mix with neutrinos. Such a mixing implies the violation of R-parity conventionally imposed in the MSSM [18]. This is simply because that the leptons being ordinary matter fields are R-even and the QGF being a fermionic partner of a Goldstone boson is R-odd. The violation of R-parity may destabilize the lightest supersymmetric particle (LSP) which is usually considered as the cold dark matter (CDM) of the Universe. For this reason, we consider the PQ symmetry as a good candidate for $U(1)_G$ since the coherent oscillation of the axion can provide the CDM for $f_{PQ} \sim 10^{12}$ GeV [19]. Therefore, the PQ mechanism required for a resolution of the strong CP problem can supply both the CDM and the sterile neutrino.

The best way to implement the PQ symmetry in the MSSM is to extend the Higgs mass term in such a way that the smallness of the Higgs mass parameter μ can be naturally obtained. For instance, let us consider the non-renormalizable term [20]

$$\lambda H_1 H_2 \frac{\sigma^2}{M_P}\,, \tag{6}$$

where M_P is the Planck mass.[a] Here the VEV of σ, $\langle\sigma\rangle \sim f_{PQ}$, spontaneously breaks the PQ symmetry. In this case, $\mu = \lambda\frac{\langle\sigma\rangle^2}{M_P}$ can be about the weak scale. When the axion superfield S is predominantly consists of σ, the PQ symmetry breaking yields the Higgs mass term and the coupling of S to the Higgs superfields

$$W_{mixing} = c_\mu \frac{\mu}{f_{PQ}} H_1 H_2 S + \mu H_1 H_2 \tag{7}$$

with c_μ being $\mathcal{O}(1)$. In order to have the mixing of S with neutrinos, one needs the lepton number violating term $\epsilon L H_2$. It is remarkable to notice that the PQ scale is in the right range for the RH neutrino masses. The PQ symmetry can indeed play a role of the lepton number symmetry if both the Higgs and leptons transform non-trivially under the PQ symmetry as in ref. [22]. In this case one can correlate the origin of ϵ and μ to the same symmetry breaking scale f_{PQ}. The neutrino and Higgs coupling to QGF is then given by

$$\begin{aligned} W_{mixing} &= \mu H_1 H_2 + \epsilon L_e H_2 + \\ & c_\mu \frac{\mu}{f_{PQ}} H_1 H_2 S + c_\epsilon \frac{\epsilon}{f_{PQ}} L_e H_2 S \ , \end{aligned} \tag{8}$$

where L_e is the electron doublet. If the PQ symmetry is the standard one unrelated to the lepton sector, the parameter ϵ vanishes. On the other hand, the global $U(1)$ symmetry becomes the usual lepton number symmetry when $c_\mu = 0$ and the bare μ-term is introduced.

An example of models which leads to the mixing terms of eq. (8) can be obtained by the PQ-charge prescription $(-1,-1,1,-1,-2)$ for $(H_1, H_2, \sigma, \sigma', L_e)$. It permits the following $U(1)_{PQ}$ invariant superpotential:

$$W = \lambda(\sigma\sigma' - f_{PQ}^2)y + \frac{\delta_\mu}{M_P} H_1 H_2 \sigma^2 + \frac{\delta_\epsilon}{M_P^2} L_e H_2 \sigma^3 \ , \tag{9}$$

which gives the terms displayed in eq. (8) with $c_\epsilon = \frac{3}{\sqrt{2}}, c_\mu = \sqrt{2}$.

The W_{mixing} in eq. (8) generates the following effective mass matrix for ν_e and S

$$\begin{pmatrix} 0 & (c_\epsilon - c_\mu)\epsilon v \sin\beta / f_{PQ} \\ (c_\epsilon - c_\mu)\epsilon v \sin\beta / f_{PQ} & m_S^0 - c_\mu^2 \mu v^2 \sin 2\beta / f_{PQ}^2 \end{pmatrix} , \tag{10}$$

[a] One can also introduce the renormalizable term to generate $\mu \simeq m_{3/2}$ [21].

where we added the direct mass m_S^0 which can be generated by the mechanism of subsection 2.1. According to eq. (10) the $\nu_e - S$ mixing angle θ_{es} is determined by

$$\tan\theta_{es} \sim \frac{(c_\mu - c_\epsilon)\epsilon v \sin\beta}{m_S^0 f_{PQ} - c_\mu^2 \mu v^2 \sin 2\beta / f_{PQ}} . \tag{11}$$

For $f_{PQ} \simeq 10^{12}$ GeV, $m_S^0 \simeq 3 \cdot 10^{-3}$ eV is the dominant contribution to the mass of S. In this case one obtains from eq. (11) for the $\nu_e - S$ mixing

$$\tan\theta_{es} \sim \frac{\epsilon v \sin\beta}{m_S^0 f_{PQ}} . \tag{12}$$

Then the desired value, $\tan\theta_{es} \sim (2-6)\cdot 10^{-2}\,\mathrm{eV}$ (1), can be obtained if the R-parity breaking parameter ϵ equals

$$\epsilon \sim \frac{m_S^0 f_{PQ} \tan\theta_{es}}{v \sin\beta} \approx (2-6)\cdot 10^{-16} \frac{f_{PQ}}{\sin\beta} . \tag{13}$$

For $f_{PQ} \sim 10^{12}$ GeV one has $\epsilon \sim 0.1$ MeV.

Let us remark the other possibilities for the QGF mass. If $m_S^0 = 0$ in eq. (10), the QGF mass, $m_S = (2-3)\cdot 10^{-3}\,\mathrm{eV}$ can be obtained for the marginally allowed value of the PQ scale:

$$f_{PQ} \approx v\sqrt{\frac{\mu \sin 2\beta}{m_S}} \lesssim 4\cdot 10^{9}\,\mathrm{GeV} . \tag{14}$$

For $f_{PQ} > 10^{10}$ GeV the QGF mass generated via μ-term is too small for the MSW solution. For $f_{PQ} \sim 10^{11}$ GeV, $m_S \approx 10^{-5}\,\mathrm{eV}$ is in the region of "just-so" solution of the solar neutrino problem. In these cases, however, axions cannot provide the CDM as we noted before.

3 A light singlet in the standard seesaw structure

In the previous case, the QGF mixes with the electron neutrino directly ($\epsilon c_\epsilon \neq 0$) or via its coupling to the Higgses ($c_\mu \neq 0$). The small mass of the QGF was related to the multi-loop effect or the suppression by $1/f_{PQ}^2$ due to the Goldstone property. An important consequence was the R-parity violation leading to destabilization of the LSP.

In this section, we will suggest another scheme in which R-parity is preserved. For this, one should place the singlet S in the superfield with zero VEV. This implies that the singlet has to be introduced from outside. Being a singlet S can mix with neutrinos via its coupling to the right-handed neutrinos. In this case, the existence of S cannot be explained but the smallness of its mass can be understood in terms of the seesaw mechanism. In order to implement a light singlet fermion in the standard seesaw structure, we will suggest to use R-symmetry which occurs in many SUSY theories. The (unbroken) R-parity is then embedded in the R-symmetry.

Let us first determine the parameters appearing in the phenomenological superpotential

$$W = \frac{m_e}{\langle H_2 \rangle} L_e N_e H_2 + \frac{M_e}{2} N_e N_e + m_{es} N_e S \ , \tag{15}$$

where N_e is the right-handed neutrino component. The Dirac mass m_e and the mixing mass m_{es} are much smaller than the Majorana mass M_e: $m_e, m_{es} \ll M_e$. The superpotential (15) leads to the mass matrix in the basis (S, ν_e, N_e):

$$\mathcal{M} = \begin{pmatrix} 0 & 0 & m_{es} \\ 0 & 0 & m_e \\ m_{es} & m_e & M_e \end{pmatrix} . \tag{16}$$

The diagonalization of (16) is straightforward. One combination of ν_e and S is massless and the orthogonal combination acquires a mass via the see-saw mechanism:

$$m_1 \simeq -\frac{m_e^2 + m_{es}^2}{M_e} . \tag{17}$$

The mass of the heavy neutrino is $\simeq M_e$. The ν_e–S mixing angle is determined by

$$\tan \theta_{es} = \frac{m_e}{m_{es}} . \tag{18}$$

Taking for m_e the typical Dirac mass of the first generation: $m_e \sim (1-5)$ MeV, and suggesting that $\nu_e \to S$ conversion explains the solar neutrino problem with $m_1 = m_S$ as in (1), we find

$$m_{es} = \frac{m_e}{\tan \theta_{es}} \simeq (0.02 - 0.3)\,\mathrm{GeV} . \tag{19}$$

According to (17) the RH mass scale is

$$M_e \simeq m_{es}^2 / m_1 = \frac{m_e^2}{m_1 \tan^2 \theta_{es}} \simeq (10^8 - 3 \cdot 10^{10})\,\mathrm{GeV} . \tag{20}$$

One has now to understand how the mixing mass (19) arises without introducing new mass scales. One also has to ensure that there is no direct coupling of S with L_e, and the mass term SS is absent or negligibly small.

Our prescription is quite simple. Consider the superpotential

$$W = \frac{m_e}{\langle H_2 \rangle} L_e N_e H_2 + f N_e N_e \sigma + f' N_e S y - \frac{\lambda}{2}(\sigma^2 - M^2) y \,. \tag{21}$$

whose structure is determined by the R-symmetry under which the fields (L_e, N_e, S, y, σ, H_2) carry the R-charges (1, 1,−1, 2, 0, 0). Note that the R-symmetry forbids the bare mass terms SS as well as the coupling $SS\sigma$. The last term in eq. (21) can be replaced by $(\sigma\sigma' - M^2)y$ to implement the lepton number symmetry. In the global SUSY limit, σ gets non-zero VEV $\langle\sigma\rangle \simeq M \sim 10^{11}$ GeV which generates the Majorana mass of N_e: $M_e = f\langle\sigma\rangle$. The point is that y develops a VEV as a consequence of SUSY breaking. Broken SUSY produces the following soft-breaking terms in the scalar potential:

$$\begin{aligned} V_{soft} &= \{A_L \frac{m_e}{\langle H_2 \rangle} L_e N_e H_2 + f A_\nu N_e N_e \sigma + f' A_S N_e S y - \\ &\quad \frac{\lambda}{2}(A_y \sigma^2 - B_y M^2) y + \text{h.c.}\} + \sum_i m_i^2 |z_i|^2 \,, \end{aligned} \tag{22}$$

where z_i denotes the fields appearing in the superpotential (21) and A_L, etc., are the soft-breaking parameters. Minimization of the potential shows the following: (1) The fields L_e, N_e, S do not develop VEV and therefore R-parity is unbroken. (2) The field y acquires non-zero VEV due to the soft-breaking terms. Consequently, the mixing mass for S and N_e appears:

$$m_{es} = \frac{f'}{2\lambda}(A_y - B_y) \tag{23}$$

Since $m_{es} \gg m_1$, no strong tunning of $A_y - B_y$ is needed. For $A_y - B_y \sim \mathcal{O}(m_{3/2})$, the desired value of m_{es} (19) can be obtained by choosing $f'/\lambda \sim 10^{-3} - 10^{-2}$. However, more elegant possibility is that $A_y = B_y$ at tree level but a non-zero value for $A_y - B_y$ is generated due to radiative corrections through the differences in interactions of σ and y. In this case one expects

$$m_{es} \sim \frac{\bar{\lambda}^2}{16\pi^2} m_{3/2} \,, \tag{24}$$

where $\bar{\lambda}$ represents a combination of the constants λ, f and f'. As a consequence, the value $m_{es} \sim 0.1$ GeV does not require smallness of $\bar{\lambda}$ or f'.

The equality $A_y = B_y$ at tree level can be achieved by the introduction of non-minimal Kähler potential allowing mixings between the observable and hidden sectors. Let us introduce the following Kähler potential:

$$K = C\overline{C} + C\overline{C}(a\frac{Z}{M_{Pl}} + \overline{a}\frac{\overline{Z}}{M_{Pl}}) + Z\overline{Z} \,, \tag{25}$$

where C and Z represent an observable and hidden sector field, respectively. Then usual assumption that the observable sector has no direct coupling to the hidden sector in superpotential, $W = W(C) + W(Z)$, leads to the universal soft-terms:

$$V_{soft} \sim m_{3/2} W(C) + \text{h.c.}\,, \tag{26}$$

provided $\overline{a} = \langle W(Z)\rangle / \langle M_{Pl}\partial W/\partial Z + W(Z)\overline{Z}/M_{Pl}\rangle$. Note also that the field C does not acquire a soft-breaking mass. This mechanism can be generalized to arbitrary number of observable sector fileds. For our purpose $C \equiv \sigma, y$, i.e., we couple σ and y to the hidden sector field Z with the above-mentioned choice for a.

4 Conclusions

Simultaneous presence of different neutrino anomalies points to the existence of a sterile neutrino. In particular, the resonance conversion of the electron neutrino into such a singlet fermion S can explain the solar neutrino problem provided its mass and mixing are appropriate (1). Supersymmetry is shown to provide a framework within which the existence and the desired properties of such a light fermion follow naturally.

We have considered first a possibility that the sterile neutrino is a quasi Goldstone fermion appearing in supersymmetric theories as a result of spontaneous breaking of a global $U(1)_G$ symmetry. This global $U(1)_G$ symmetry can be identified with the PQ symmetry, the lepton number symmetry. The smallness of m_S can be attributed in supergravity theory to no-scale kinetic terms for certain superfields. The mixing of QGF with the neutrinos implies spontaneous or explicit violation of R-parity. QGF can mix with neutrino via interaction with Higgs multiplets (in the case of PQ symmetry) or directly via coupling with the combination LH_2 (in the case of lepton number symmetry). In the case of the PQ symmetry, the PQ-scale $f_{PQ} \sim 10^{10} - 10^{12}$ GeV determines several features of the model presented here. It provides simultaneous explanation of the parameters ϵ and μ and thus leads to small R-parity violation (ϵLH_2 with $\epsilon \sim 0.1$ MeV) required in order to solve the solar neutrino problem in our approach. It also provides the intermediate scale for the right-handed neutrino masses which is required in order to solve the dark matter and the atmospheric neutrino problem. Furthermore, it controls the magnitude of the radiatively generated mass of the QGF and allows it to be in the range needed for the MSW solution of the solar neutrino problem. Finally, the CDM can consist of the axion if $f_{PQ} \sim 10^{12}$ GeV. Thus the basic scenario presented here is able to correlate variety of phenomena.

The conservation of R-parity requires for the fermion S to be a component of singlet superfield which has no VEV. This allows to construct simple model (21) in which the properties (mass and mixing) of S follow from the conservation of R-symmetry. The singlet field is mixed with RH neutrinos by the interaction with the field y which can acquire VEV radiatively after soft SUSY breaking.

Let us finally comment on the other phenomenological consequences of the existence of such a sterile state S. An $U(1)_G$ symmetry being generation-dependent [6,7] can provide simultaneous explanations for the predominant coupling of S to the first generation (thus satisfying the nucleosynthesis bound) and for the pseudo-Dirac structure of ν_μ–ν_τ needed in solving the atmospheric neutrino and the hot dark matter problem. In this case, it appears nontrivial to accommodate the parameters of $\bar{\nu}_\mu \to \bar{\nu}_e$ oscillations in the region of sensitivity of LSND and KARMEN experiments. The simplest way is to introduce a slight violation of the $U(1)_G$ symmetry through which such parameters can be incorporated.

References

1. GALLEX Collaboration, *Phys. Lett.* B **327**, 377 (1994); SAGE Collaboration, *Phys. Lett.* B **328**, 234 (1994); Homestake Collaboration, *Nucl. Phys.* B **38** (Proc. Suppl.), 47 (1995); Kamiokande Collaboration, *Nucl. Phys.* B **38** (Proc. Suppl.), 55 (1995).
2. Kamiokande Collaboration, H. S. Hirata *et al.*, *Phys. Lett.* B **205**, 416 (1988) and *Phys. Lett.* B **280**, 146 (1992); Kamiokande Collaboration, Y. Fukuda *et al.*, *Phys. Lett.* B **335**, 237 (1994).
3. J. R. Primack, J. Holtzman, A. Klypin, and D. O. Caldwell, *Phys. Rev. Lett.* **74**, 2160 (1995); and references therein.
4. LSND collaboration, LA-UR-95-1238, nucl-ex/9504002.
5. K. Enqvist, J. Maalampi and V. B. Semikoz, HU-TFT-95/28, hep-ph/9505210; X. Shi, D. N. Schramm and B. D. Fields, *Phys. Rev.* D **48**, 2563 (1993); and references therein.
6. E. J. Chun, A. S. Joshipura and A. Yu. Smirnov, IC/95/76, PRL-TH/95-7, hep-ph/9505275, (to appear in *Phys. Lett.* B) .
7. E. J. Chun, A. S. Joshipura and A. Yu. Smirnov, IC/95/164, PRL-TH/95-11, hep-ph/9507371.
8. R. Foot and R. R. Volkas, UM-P-95/49, hep-ph/9505359.
9. Z.G. Berezhiani and R.N. Mohapatra, INFN-FE-06-95, hep-ph/9505385.
10. E. Ma, UCRHEP-T149, hep-ph/9507348.
11. S. P. Mikheyev and A. Yu. Smirnov, *Sov. Journal of Nucl. Phys.* **42**,

913 (1985); *Sov. Phys. JETP* **64**, 4 (1986); L. Wolfenstein, *Phys. Rev.* D **17**, 2369 (1978).
12. R. D. Peccei and H. R. Quinn, *Phys. Rev. Lett.* **38**, 1440 (1977).
13. Y. Chikashige, R. N. Mohapatra and R. D. Peccei, *Phys. Lett.* B **98**, 265 (1981).
14. T. Goto and M. Yamaguchi, *Phys. Lett.* B **276**, 103 (1992); E. J. Chun, J. E. Kim and H. P. Nilles, *Phys. Lett.* B **287**, 123 (1992).
15. E. J. Chun and A. Lukas, IC/95/29, TUM-HEP 215/95, hep-ph/9503233 (to appear in *Phys. Lett.* B).
16. For a review, see A. B. Lahanas and D. V. Nanopoulos, *Phys. Rep.* **145**, 1 (1987).
17. E. J. Chun, H. B. Kim and A. Lukas, *Phys. Lett.* B **328**, 346 (1994).
18. For a recent review, see L. Hall, *Mod. Phys. Lett.* **A7**, 467 (1990).
19. For a review, see E. Kolb and M. Turner, *The Early Universe* (Addison–Wesley, 1990).
20. J. E. Kim and H. P. Nilles, *Phys. Lett.* B **138**, 150 (1984); E. J. Chun, J. E. Kim and H. P. Nilles, *Nucl. Phys.* B **370**, 105 (1992); E. J. Chun and A. Lukas, *Phys. Lett.* B **297**, 298 (1992).
21. E. J. Chun, *Phys. Lett.* B **348**, 111 (1995).
22. P. Langacker, R. D. Peccei and T. Yanagida, *Mod. Phys. Lett.* **A1**, 541 (1986); A. S. Joshipura, *Zeit. für Phys.* C **38**, 479 (1988).

NEW ISSUES IN QUARK-LEPTON SYMMETRY

R.R. VOLKAS

Research Centre for High Energy Physics, School of Physics, University of Melbourne, Parkville 3052, Australia

Two models incorporating different forms of spontaneously broken quark-lepton symmetry are discussed. Both models are constructed so that quark-lepton symmetry can be broken at as low an energy scale as phenomenology allows, thus maximising their testability. The first model uses the Pati-Salam gauge group SU(4)$\otimes$SU(2)$_L\otimes$SU(2)$_R$. The main analysis presented here concerns large threshold corrections to the $m_b = m_\tau$ tree-level mass relation. The second model is a new and simplified theory using discrete quark-lepton symmetry which has a very economical Higgs sector and elegantly avoids a potential quark-lepton mass relation problem.

1 Introduction

The idea that quarks and leptons are related in a pairwise fashion has a long history. In the late 1950's Marshak discussed an evident "quark-hadron correspondence" whereby the leptons e, ν_e and μ were associated with the hadrons p, n and Λ. This turned into a "quark-lepton correspondence" between e, ν_e, μ and d, u, s once quark substructure was revealed. With the discovery of ν_μ, the quark-lepton correspondence idea was used to postulate the existence of a fourth quark c. This remarkable prediction bore fruit with the discovery of charm in the mid 1970's. The correspondence between quarks and leptons was subsequently found to also hold for third generation fermions.

The pairwise association between quarks and leptons is so striking that it makes sense to suppose that the fundamental Lagrangian of the world actually treats quarks and leptons on an equal footing, with their disparate properties ascribed to a spontaneous violation of the exact quark-lepton symmetry. Such a symmetry can either be continuous or discrete. Candidates for a continuous symmetry of this nature are Pati-Salam SU(4) [1] and the various grand unification groups. Alternatively, a discrete symmetry between quarks and leptons can be introduced by postulating a spontaneously broken colour SU(3) group for leptons [2].

In this talk I will describe two pieces of recent work in this area. The first piece of work analyses some of the implications of having Pati-Salam SU(4) broken at the lowest energy scale that phenomenology allows [3] (1000 TeV). I will discuss how best to treat the neutrino mass problem, followed by the core of the analysis which is a detailed computation of radiative corrections to the tree-level mass relation $m_b = m_\tau$. I will show here that threshold corrections due to charged Higgs boson graphs can be very large. The second piece of work is a reconstruction of the quark-lepton discrete symmetry idea in such a way that the Higgs sector of the model is kept as simple as possible. I will also demonstrate that the resulting theory deals with the quark-lepton mass relation "problem" in an elegant way.

2 Low-scale Pati-Salam SU(4)

The following discussion is based on Ref.[4]. Under the Pati-Salam gauge group $G_{\rm PS}$ =SU(4)⊗SU(2)$_L$⊗SU(2)$_R$, a fermion generation appears as follows:

$$f_L \sim (4,2,1), \qquad f_R \sim (4,1,2). \tag{1}$$

The SU(4) factor transforms quarks into leptons in a continuous way. It contains the subgroup SU(3)$_c$⊗U(1)$_{B-L}$. The linear combination $2I_R + (B-L)$, where I_R is the diagonal generator of SU(2)$_R$, is identified as weak hypercharge Y. I do not impose a discrete left-right symmetry on the model.

The model has a two stage symmetry breaking chain. At a relatively high scale $M_{\rm PS}$ SU(4) and SU(2)$_R$ are broken simultaneously, which is followed by electroweak symmetry breaking at its usual scale. Let us discuss electroweak symmetry breaking first. This is achieved by introducing a complex bidoublet Higgs field $\Phi \sim (1,2,2)$. The Yukawa Lagrangian

$$\mathcal{L}_{\rm Yuk} = \lambda_1 {\rm Tr}[\overline{f}_L \Phi f_R] + \lambda_2 {\rm Tr}[\overline{f}_L \Phi^c f_R] + {\rm H.c.} \tag{2}$$

where $\Phi^c \equiv \tau_2 \Phi^* \tau_2$ yields the tree-level mass relations

$$m_b = m_\tau \quad \text{and} \quad m_t = m_{\nu_3}^{\rm Dirac} \tag{3}$$

given a nonzero vacuum expectation value for Φ (the fermion multiplets are written as 2×4 matrices in the above). Similar relations hold for the first two generations, but the present analysis will confine itself to the third generation. The deep problem of generation structure will not be tackled here.

What is one to do with these mass relations? The equality between m_t and the neutrino Dirac mass is usually taken care of through the see-saw mechanism [5] by using a $\Delta \sim (10,1,3)$ Higgs multiplet to generate a large Majorana mass for the right-handed tau neutrino. This also breaks SU(4)⊗SU(2)$_R$ down to SU(3)$_c$⊗U(1)$_Y$. However, the cosmological closure bound on the light mass eigenstate $m_t^2/\langle\Delta\rangle < 30$ eV leads to $\langle\Delta\rangle \sim 10^{12}$ TeV. This is much higher than the phenomenological lower bound of 1000 TeV, and thus use of the standard see-saw mechanism would undermine the motivation for the analysis.

An interesting alternative is to use what I will call the "3×3 see-saw mechanism" [6]. One introduces a singlet fermion $S_L \sim (1,1,1)$ and the simple Higgs multiplet $\chi \sim (4,1,2)$. The non-electroweak Yukawa Lagrangian

$$\mathcal{L}_{\text{Yuk}} = n\overline{S}_L \text{Tr}[\chi^\dagger f_R] + \text{H.c.} \tag{4}$$

when combined with the electroweak Yukawa terms above yield a neutrino mass matrix of the form

$$\begin{pmatrix} 0 & m_t & 0 \\ m_t & 0 & n\langle\chi\rangle \\ 0 & n\langle\chi\rangle & 0 \end{pmatrix} \tag{5}$$

in the $[\nu_L, (\nu_R)^c, S_L]$ basis. This produces one massless eigenstate which we identify with the standard neutrino, and a massive Dirac neutrino. For $n\langle\chi\rangle \gg m_t$, the massless state has approximately standard electroweak interactions. Because the light eigenstate is massless for all values of the nonzero entries of the mass matrix, $\langle\chi\rangle$ can be reduced to about 1000 TeV.

The $m_b = m_\tau$ relation receives radiative corrections of two types: those associated with large logarithms which are best calculated through renormalisation group equations, and threshold corrections which depend on either heavy/heavy or light/light mass ratios. It is interesting that renormalisation group evolution using standard particles in the loops merges m_b with m_τ at about 1000 TeV [7]! Therefore if Pati-Salam SU(4) is important for the fermion mass problem, an indirect indication through $K^0 \to e^\pm \mu^\mp$ may be just around the corner. But this renormalisation group result is only relevant if threshold corrections are not too large. It turns out that there is a generically large high-scale threshold correction due to loops containing the physical charged Higgs boson. See Ref.[4] for details. These graphs produce a threshold correction given by

$$m_\tau - m_b \simeq -\frac{1}{16\pi^2} \frac{m_s^2 - m_t^2}{m_H^2 - m_s^2} \frac{m_t(m_t - m \sin 2\omega)(m_t \sin 2\omega - m)}{(u_1^2 + u_2^2)\cos^2 2\omega} \ln \frac{m_s^2}{m_H^2}, \tag{6}$$

where m_s is the heavy Dirac neutrino mass, m_H is the charged Higgs boson mass, m is the "common" tree-level mass for b and τ, $u_{1,2}$ are the electroweak breaking VEVs and $\tan\omega = u_2/u_1$.

This threshold correction can clearly produce a mass difference between m_τ and m_b of the order of a GeV, provided an accidental cancellation between m and $m_t \sin 2\omega$ does not occur. The "common" mass m of τ and b at M_{PS}

must be about the same as the measured m_τ, namely 1.8 GeV. The above threshold correction can therefore alter the initial ratio m_b/m_τ by up to 50%. This correction is thus as numerically significant as those incorporated through the renormalisation group. The sign of the correction depends on the unknown parameter ω and therefore cannot be predicted. It can either raise or lower the mass ratio by up to 50%.

This calculation demonstrates that generally speaking one must take care in the use of renormalisation group evolution to predict low-energy masses.

3 New and improved quark-lepton symmetric model

The idea of discrete quark-lepton symmetry was introduced in Ref.[2]. It is based on the gauge group

$$G_{q\ell} = \mathrm{SU}(3)_\ell \otimes \mathrm{SU}(3)_q \otimes \mathrm{SU}(2)_L \otimes \mathrm{U}(1)_X, \tag{7}$$

where $\mathrm{SU}(3)_\ell$ is leptonic colour, $\mathrm{SU}(3)_q$ is ordinary colour and X is an Abelian charge different from Y. A generation of fermions is placed into the following multiplet pattern:

$$\begin{aligned} &Q_L \sim (1,3,2)(1/3), \quad && u_R \sim (1,3,1)(4/3), \quad && d_R \sim (1,3,1)(-2/3), \\ &F_L \sim (3,1,2)(-1/3), \quad && E_R \sim (3,1,1)(-4/3), \quad && N_R \sim (3,1,1)(2/3). \end{aligned} \tag{8}$$

The standard leptons lie within F_L, E_R and N_R in a manner specified by the way $G_{q\ell}$ is spontaneously broken. The multiplet assignment above allows us to define a discrete quark-lepton symmetry by

$$\begin{aligned} &Q_L \leftrightarrow F_L, \quad u_R \leftrightarrow E_R, \quad d_R \leftrightarrow N_R, \\ &G_q^\mu \leftrightarrow G_\ell^\mu \quad \text{and} \quad C^\mu \leftrightarrow -C^\mu, \end{aligned} \tag{9}$$

where $G^\mu_{q,\ell}$ are the gauge bosons of $\mathrm{SU}(3)_{q,\ell}$ and C^μ is the gauge boson of $\mathrm{U}(1)_X$.

The original quark-lepton symmetric models were based on $G_{q\ell}$ breaking achieved through the Higgs multiplet $\chi \sim (3,1,1)(2/3)$ that appears in the Yukawa Lagrangian

$$\mathcal{L}_{\text{Yuk}} = h_1 \overline{(F_L)^c} F_L \chi + h_2 \overline{(E_R)^c} N_R \chi + \text{H.c.} \tag{10}$$

(A quark-lepton symmetric partner for χ is also introduced with a corresponding Yukawa Lagrangian.) A nonzero VEV for χ breaks $G_{q\ell}$ down to $\mathrm{SU}(2)' \otimes G_{SM}$, where $\mathrm{SU}(2)'$ is a subgroup of leptonic colour that is necessarily unbroken. The weak hypercharge is identified as $Y = X + T_8/3$, where

$T_8 = \mathrm{diag}(-2, 1, 1)$ is a generator of leptonic colour. The standard leptons are then the $T_8 = -2$ components, while the $T_8 = 1$ states are exotic fermions.

The minimal electroweak Higgs sector consists of a single doublet $\phi \sim (1, 1, 2)(1)$ as in the Standard Model. Under discrete quark-lepton symmetry it transforms into its charge conjugate field $\phi^c = i\tau_2\phi^*$. The electroweak Yukawa Lagrangian

$$\mathcal{L}_{\mathrm{Yuk}} = \lambda_1(\bar{Q}_L u_R \phi^c + \bar{F}_L E_R \phi) + \lambda_2(\bar{Q}_L d_R \phi + \bar{F}_L N_R \phi^c) + \mathrm{H.c.} \qquad (11)$$

then yields the tree-level mass relations $m_u = m_e$ and $m_d = m_\nu^{\mathrm{Dirac}}$.

The second of these relations can be made phenomenologically acceptable by using the standard see-saw mechanism through the analogue of the Higgs field Δ of the Pati-Salam model discussed previously. In this case, $\Delta \sim (\bar{6}, 1, 1)(-4/3)$. The $m_u = m_e$ relation can only be dealt with by adding a second electroweak Higgs doublet which allows one to completely evade mass relations between quarks and leptons.

A Higgs sector consisting of two ϕ's, a χ and a Δ is rather cumbersome. It turns out that there is a much simpler way of constructing the model, which I now describe. This work is more fully explained in Ref.[8] which in turn is based in part on Ref.[9]. The new Higgs sector consists of only one ϕ and two χ-like fields I will call χ and ξ (plus their discrete symmetry partners). There is no Δ field because I am again going to use singlet fermions S_L together with an analogue of the 3×3 see-saw mechanism employed in the previous section. The VEV pattern for χ and ξ is the most general one possible, namely $\langle\chi\rangle = (v, 0, 0)^T$ and $\langle\xi\rangle = (w_1, w_2, 0)^T$ which completely breaks leptonic colour $SU(3)_\ell$. There is no unbroken SU(2)$'$ subgroup, and all components of the leptonic triplets become integrally charged [$Y = X + T_8/3 - T_3$ here where $T_3 = \mathrm{diag}(0, 1, -1)$]. The standard leptons are then a linear combination of more than one leptonic colour.

The Yukawa Lagrangian of the model contains the electroweak Yukawa terms of Eq. 11, the non-electroweak terms of Eq. 10 extended to include ξ in an obvious way, together with mixing terms between the S_L's and N_R's given by

$$\mathcal{L}_{\mathrm{Yuk}} = n_1 \overline{S}_L \chi^\dagger N_R + n_2 \overline{S}_L \xi^\dagger N_R + \mathrm{H.c.} \qquad (12)$$

The neutrino mass matrix consists of a 7×7 version of the 3×3 see-saw mechanism used in my Pati-Salam model. I will not display it here, except to say that the electroweak breaking entries in it are specified in terms of the quark masses and that the lightest mass eigenstate has close to standard electroweak interactions.

The charged lepton mass matrix is given by

$$\begin{pmatrix} m_u & 0 & M_3 \\ 0 & m_u & M_1 \\ M_4 & M_2 & m_d \end{pmatrix} \tag{13}$$

in the basis (e_1, e_2, e_3) where the subscript labels the three leptonic colours. The M_i are large masses proportional the the VEVs of χ and ξ. The smallest mass eigenvalue is given by $m_e = m_u \cos(\beta_1 - \beta_2) \leq m_u$ where $\tan\beta_1 \equiv M_2/M_4$ and $\tan\beta_2 \equiv M_1/M_3$. Note that the $m_u = m_e$ mass relation does not arise here, and furthermore that m_e is necessarily less than m_u assuming large M_i and sufficiently small intergenerational mixing (this type of result was first found in Ref.[9]). It is straightforward to check that the fields corresponding to m_e have the correct electroweak interactions to a good approximation.

This reconstruction of the idea of discrete quark-lepton symmetry produces a simpler model with regard to the Higgs sector than previous versions, and it has no quark-lepton mass relation problems. In this sense the above theory can be regarded as an improvement on earlier models.

Acknowledgments

This work was supported by the Australian Research Council.

References

1. J. C. Pati and A. Salam, Phys. Rev. **D8**. 1240 (1973); *ibid.* **D10**, 275 (1974).
2. R. Foot and H. Lew, Phys. Rev. **D41**, 3502 (1990).
3. G. Valencia and S. Willenbrock, Phys. Rev. **D50**, 6845 (1994).
4. R. R. Volkas, Cambridge preprint DAMTP-94-106, hep-ph/9507215.
5. T. Yanagida, in *Proceedings of the Workshop on Unified Theory and baryon Number of the Universe*, edited by O. Sawada and A. Sugamoto (KEK, Tsukuba, Japan, 1979); M. Gell-Mann, P. Ramond and R. Slansky, in *Supergravity*, edited by P. van Nieuwenhuizen and D. Freedman (North-Holland, Amsterdam, 1980); R. N. Mohapatra and G. Senjanovic, Phys. Rev. Lett. **44**, 912 (1980).
6. D. Wyler and L. Wolfenstein, Nucl. Phys. **B218**, 205 (1983).
7. H. Arason et al., Phys. Rev. **D46**, 3945 (1992).
8. R. Foot and R. R. Volkas, Melbourne preprint UM-P-95/48, hep-ph/9505331.
9. R. R. Volkas, Phys. Rev. **D50**, 4625 (1994).

LOW ENERGY YUKAWA INPUT PARAMETERS FOR YUKAWA COUPLING UNIFICATION

G. RODRIGO

Dept. de Física Teòrica, Univ. de València, E-46100 Burjassot (València), Spain

I make a short review of the most recent determinations of the quark masses and run them to the M_Z energy scale

GUT and SUSY theories predict some relations among the fermion masses (or more properly among Yukawa couplings) at the unification scale, $M_{GUT} \sim 10^{16}(GeV)$. For instance, in $SU(5)$ we have the usual lepton-bottom quarks unification, $h_b = h_\tau$, $h_s = h_\mu$, $h_d = h_e$, or the modified Georgi-Jarlskog relation, $h_b = h_\tau$, $h_s = h_\mu/3$, $h_d = 3h_e$, while for $SO(10)$ typically we get unification of the third family, $h_t = h_b = h_\tau = h_{\nu_\tau}$. This together with the RGE's provide us a powerful tool for predicting quark masses at low energies.

In the following I will not talk about Unification but rather try to make a short review of the most recent determinations of the quark masses and to do the running to M_Z, the mass of the Z boson. Why M_Z?, because for model building purposes it is a good idea to have a reference scale, because extensions of the SM appear above M_Z and because below M_Z the strong coupling constant α_s is really strong and then special care has to be taken on the running and the matching in passing a heavy quark threshold.

Because of confinement to define what is the mass of a quark is not an easy task. For leptons it is clear that the physical mass is the pole of the propagator but for quarks we need a precise theoretical framework and the masses appear more like coupling constants. Dealing with quark masses we can find the Euclidean mass, $M_E(p^2 = -M^2)$, defined as the mass renormalized at the Euclidean point $p^2 = -M^2$, it is gauge dependent but softly dependent on Λ_{QCD}. However, it does not appear in the most recent works. We can talk about the "perturbative" pole mass, $M(p^2 = M^2)$, perturbative because only order by order in perturbation theory the pole of the propagator is well defined. It is gauge and scheme independent but appears to be ambiguous because of non-perturbative renormalons. Finally $\bar{m}(\mu)$, the mass renormalized in the Modified Minimal Substraction scheme or its corresponding Yukawa coupling related to it through the vev of the Higgs, $\bar{m}(\mu) = v(\mu)\bar{h}(\mu)$.

For the light quarks, up, down and strange, chiral perturbation theory [3,4]

Table 1: Recent determinations of the light quark masses from second order χPT, QCD Sum Rules and lattice.

J. Gasser H. Leutwyler	χPT	$O(p^4)$	$\frac{m_d - m_u}{m_s - \hat{m}} \frac{2\hat{m}}{m_s + \hat{m}} = 2.35 \times 10^{-3}$ $m_s/\hat{m} = 25.7 \pm 2.6$
Donoghue et al.	χPT	$O(p^4)$	$\frac{m_d - m_u}{m_s - \hat{m}} \frac{2\hat{m}}{m_s + \hat{m}} = 2.11 \times 10^{-3}$ $m_s/\hat{m} = 31.$

Bijnens Prades de Rafael	FESR Laplace SR (pseudo)	NNLO	$(\bar{m}_u + \bar{m}_d)(1GeV)$ $= 12.0 \pm 2.5$	$\alpha_s(M_Z) = 0.117(5)$
Ioffe et al.	Isospin viol. in QCD SR	NNLO	$(\bar{m}_d - \bar{m}_u)(1GeV)$ $= 3.0 \pm 1.0$	$\Lambda = 150.$

S. Narison	τ-like SR	NNLO NLO	$\bar{m}_s(1GeV) = 197(29)$ $\bar{m}_s(1GeV) = 222(22)$	$\alpha_s(M_Z) = 0.118(6)$
M. Jamin M. Münz	QSSR (scalar)	NNLO	$\bar{m}_s(1GeV) = 189(32)$	$\alpha_s(M_Z) = 0.118(6)$
Chetyrkin et al.	QCD SR	NNLO	$\bar{m}_s(1GeV) = 171(15)$	$\alpha_s(M_Z) = 0.117(5)$
Allton et al.	quenched Lattice	NLO	$\bar{m}_s(2GeV) = 128(18)$	$\Lambda^5 = 240. \pm 90.$

provide us a powerful tool for determining renormalization group invariant quark mass ratios. The absolute values, usually the running mass at $1(GeV)$, can be extracted from different QCD Sum Rules [5,6,7,8,9] or for the strange quark mass from lattice [10]. For the heavy quarks, bottom and charm, we can deal either with QCD Sum Rules [11,12,13,14], lattice calculations [15,16,17,18] or may be soon for the bottom quark with jet physics at LEP [20,21]. For the top quark we have the recent measurements from CDF an DØ at FERMILAB [24] I will identify it with the pole mass.

I have summarized in tables 1 and 2 all of these recent determinations of the quark masses. Of course the final result depends on the strong gauge coupling constant used in the analysis, for this reason I quote it too. In the running I will take $\alpha_s^{(5)}(M_Z) = 0.118 \pm 0.006$ for masses obtained from QCD Sum Rules but for lattice masses I will run with the lattice [17] result $\alpha_s^{(5)}(M_Z) = 0.115 \pm 0.002$. This values are consistent with almost all the references. For those that differ an update is needed but this is beyond the goals of this paper. For instance,

Table 2: Recent determinations of the heavy quark masses from QCD Sum Rules, lattice and FERMILAB.

S. Narison	QSSR Ψ, Υ	NLO	$\bar{m}_b(M_b) = 4.23(4)$ $\bar{m}_c(M_c) = 1.23^{+(4)}_{-(5)}$ $M_b = 4.62(2)$ $M_c = 1.42(3)$	$\alpha_s(M_Z) = 0.118(6)$
S. Narison	non-rel Laplc.SR	NLO	$M_b^{NR} = 4.69^{+(3)}_{-(2)}$ $M_c^{NR} = 1.45^{+(5)}_{-(4)}$	$\alpha_s(M_Z) = 0.118(6)$
Dominguez et al.	rel,non-rel Laplc.SR J/Ψ, Υ	LO $1/m_q^2$	$M_b = 4.70(7)$ $M_c = 1.46(7)$	$\Lambda^4 = 200 - 300$ $\Lambda^5 = 100 - 200$
Titard Ynduráin	$q\bar{q}$ potential		$\bar{m}_b(\bar{m}_b) = 4.397^{+(18)}_{-(33)}$ $\bar{m}_c(\bar{m}_c) = 1.306^{+(22)}_{-(35)}$	$\alpha_s(M_Z) = 0.117(5)$
M. Neubert	QCD SR	NLO $1/m_q$	$M_b = 4.71(7)$ $M_c = 1.30(12)$	

Crisafulli et al. [15]	Lattice in B-meson	$\bar{m}_b(\bar{m}_b) = 4.17(6)$	
Crisafulli et al. [16]	Lattice in B-meson	$\bar{m}_b(\bar{m}_b) = 4.15(7)$	
Davies et al.	NRQCD + leading rel and Lattice spacing, $b\bar{b}$	$M_b = 5.0(2)$ $\bar{m}_b(M_b) = 4.0(1)$	$\alpha^{(5)}_{\overline{MS}} = 0.115(2)$
El-Khadra Mertens	Fermilab action in quenched Lat	$M_c = 1.5(2)$	

CDF	$M_t = 176. \pm 8.(stat) \pm 10.(sys)$	*mean*
DØ	$M_t = 199.^{+19.}_{-21.}(stat) \pm 22.(sys)$	$M_t = 180. \pm 12.$

S. Narison [11] makes two different determinations for the bottom and the charm quark masses. In the first, and for the first time, he gets directly the running mass avoiding then the renormalon ambiguities associated with the pole mass. The second one, from non-relativistic Laplace Sum Rules, is in fact an update of the work of Dominguez et al. [12].

The $O(\alpha_s^2)$ strong correction to the relation between the perturbative pole mass and the running mass was calculated in [1]

$$\frac{M}{\bar{m}(M)} = 1 + \frac{4}{3}\frac{\alpha_s(M)}{\pi} + K\left(\frac{\alpha_s(M)}{\pi}\right)^2 + O(\alpha_s^3(M)), \tag{1}$$

where $K_t \simeq 10.95$ for the top quark, $K_b \simeq 12.4$ for the bottom and $K_c \simeq 13.3$ for the charm. As pointed out by S. Narison [11] equation 1 is consistent with three loops running but for two loop running we can drop the $O(\alpha_s^2)$ term. Recently, the electroweak correction to the relation between the perturbative pole mass and the Yukawa coupling has been calculated [2]. However, this correction is small, for instance for the top quark it is less than 0.5% in the SM for a mass of the Higgs lower than $600(GeV)$ and at most 3.% for $M_H \simeq 1(TeV)$, and for consistency one has to include it only if two loop electroweak running is done.

Instead of expressing the solution of the QCD renormalization group equations for the strong gauge coupling constant and the quark masses in terms of Λ_{QCD} we can solve the running as an expansion in the strong coupling constant at one loop [19]. At three loops

$$\alpha_s(\mu) = \alpha_s^{(1)}(\mu)\left(1 + c_1(\mu)\alpha_s^{(1)}(\mu) + c_2(\mu)(\alpha_s^{(1)}(\mu))^2\right), \tag{2}$$

$$\bar{m}(\mu) = \bar{m}^{(1)}(\mu)\left(1 + d_1(\mu)\alpha_s^{(1)}(\mu) + d_2(\mu)(\alpha_s^{(1)}(\mu))^2\right), \tag{3}$$

where $\alpha_s^{(1)}(\mu)$ and $\bar{m}^{(1)}(\mu)$ are the one loop solutions

$$\alpha_s^{(1)}(\mu) = \frac{\alpha_s(\mu_0)}{1 + \alpha_s(\mu_0)\beta_0 t}, \qquad \bar{m}^{(1)}(\mu) = \bar{m}(\mu_0)K(\mu)^{-2\gamma_0/\beta_0}, \tag{4}$$

with $t = 1/(4\pi)\log \mu^2/\mu_0^2$, $K(\mu)$ the ratio $K(\mu) = \alpha_s(\mu_0)/\alpha_s^{(1)}(\mu)$, and

$$\begin{aligned} c_1(\mu) &= -b_1 \log K(\mu), \\ c_2(\mu) &= b_1^2 \log K(\mu)\left[\log K(\mu) - 1\right] - (b_1^2 - b_2)\left[1 - K(\mu)\right], \\ d_1(\mu) &= -\frac{2\gamma_0}{\beta_0}\left[(b_1 - g_1)\left[1 - K(\mu)\right] + b_1 \log K(\mu)\right], \end{aligned} \tag{5}$$

$$d_2(\mu) = \frac{\gamma_0}{\beta_0^2}\Big\{[\beta_0(b_2 - b_1^2) + 2\gamma_0(b_1 - g_1)^2][1 - K(\mu)]^2$$
$$+ \quad \beta_0(g_2 - b_1 g_1)[1 - K^2(\mu)]$$
$$+ \quad \Big[4\gamma_0 b_1(b_1 - g_1)[1 - K(\mu)] - 2\beta_0 b_1 g_1 + b_1^2(\beta_0 + 2\gamma_0)\log K(\mu)\Big]\log K(\mu)\Big\},$$

where

$$b_1 = \frac{\beta_1}{4\pi\beta_0}, \qquad b_2 = \frac{\beta_2}{(4\pi)^2\beta_0}, \qquad g_1 = \frac{\gamma_1}{\pi\gamma_0}, \qquad g_2 = \frac{\gamma_2}{\pi^2\gamma_0}, \qquad (6)$$

are the ratios of the well known beta and gamma functions in the $\overline{MS}$ scheme

$$\beta_0 = 11 - \frac{2}{3}N_F, \qquad \gamma_0 = 2,$$
$$\beta_1 = 102 - \frac{38}{3}N_F, \qquad \beta_2 = \frac{1}{2}\left(2857 - \frac{5033}{9}N_F + \frac{325}{27}N_F^2\right),$$
$$\gamma_1 = \frac{101}{12} - \frac{5}{18}N_F, \qquad \gamma_2 = \frac{1249}{32} - \frac{277 + 180\zeta(3)}{108}N_F - \frac{35}{648}N_F^2, \quad (7)$$

and $\zeta(3) = 1.2020569\ldots$ is the Riemann zeta-function. Our initial condition for the strong coupling constant will be $\alpha_s(M_Z)$. Then we will run α_s from M_Z to lower scales, i.e. for instance $\mu_0 = M_Z$ or the upper threshold. On the other side, for the masses we will run from low to higher scales then we need the inverted version of equation 3

$$\bar{m}(\mu_0) = \bar{m}(\mu)K(\mu)^{2\gamma_0/\beta_0}\left(1 - d_1(\mu)\alpha_s^{(1)}(\mu) + (d_1^2(\mu) - d_2(\mu))(\alpha_s^{(1)}(\mu))^2\right). \quad (8)$$

The beta and gamma functions depend on the number of flavours N_F therefore we have to decide where we have five where we have four flavours. The trick is as was done by [22] and recently corrected by [23] to built below the heavy quark threshold an effective theory where the heavy quark has been integrated out. Imposing agreement of both theories, the full and the effective, at low energies they wrote μ dependent matching conditions that express the parameters of the effective theory, with $N-1$ quark flavours, as a perturbative expansion in terms of the parameters of the full theory with N flavours

$$\alpha_s^{N-1}(\mu) = \alpha_s^N(\mu)\left[1 + \frac{x}{6}\frac{\alpha_s^N(\mu)}{\pi} + \frac{1}{12}\left(\frac{x^2}{3} + \frac{11x}{2} + \frac{11}{6}\right)\left(\frac{\alpha_s^N(\mu)}{\pi}\right)^2\right],$$

$$\bar{m}_l^N(\mu) = \bar{m}_l^{N-1}(\mu)\left[1-\frac{1}{12}\left(x^2+\frac{5x}{3}+\frac{89}{36}\right)\left(\frac{\alpha_s^{N-1}(\mu)}{\pi}\right)^2\right], \qquad (9)$$

with $x = \log \bar{m}^2(\mu)/\mu^2$, where $\bar{m}(\mu)$ is the mass of the heavy quark we decouple at the energy scale μ and $\bar{m}_l(\mu)$ are the masses of the light quarks. This matching conditions make the strong coupling constant and the mass of the light quarks discontinuous at the thresholds. However, taking the matching in this way we ensure, as pointed out explicitly in [19], that the final result is independent of the particular matching point we choose for passing the threshold. As it is independent, the easiest way to implement a heavy quark decoupling is to take the threshold as the running mass at the running mass scale, i.e. $\mu_{th} = \bar{m}(\bar{m})$ or equivalently $x = 0$, then the discontinuity appears only at two loops matching.

Table 3: Running at the NLO and NNLO of the top quark mass to M_Z, $\alpha_s^{(5)}(M_Z) = 0.118 \pm 0.006$, $\alpha_s^{(6)}(M_Z) = 0.117 \pm 0.006$.

	$\bar{m}_t(M_t)$	$\bar{m}_t(\bar{m}_t)$	$\bar{m}_t(M_Z)$
NLO	$172. \pm 12.$	$173. \pm 12.$	$182. \pm 13.$
NNLO	$170. \pm 12.$	$171. \pm 12.$	$180. \pm 13.$

I have summarized in tables 3, 4 and 5 the result for the running of the quark masses to M_Z. I mean by NLO connection between the perturbative pole mass and the running mass dropping the $O(\alpha_s^2)$ term, running to two loops and matching at one loop, i.e. strong gauge coupling and masses continuous at $\mu_{th} = \bar{m}(\bar{m})$. Three loops running and matching as expressed in equation 9 with $x = 0$ correspond to NNLO. For consistency with the original works we can only do the running for the bottom and the charm quarks just to NLO. For the light quarks the running is consistent to NNLO using the threshold masses, $\bar{m}(\bar{m})$, of the bottom and charm quarks determined at NLO. I propagate the errors in the running in such a way we maximize them.

It is informative to notice that the running mass of the top quark is shifted about $7(GeV)$ down from its perturbative pole mass that is of the order of its error. Therefore it is important to clarify which mass CDF and DØ are talking about. I have decoupled the top quark at M_Z otherwise it makes no sense to run the top down. This fact shifts down slightly the strong coupling constant in M_Z, from $\alpha_s^{(5)}(M_Z) = 0.118 \pm 0.006$ we get $\alpha_s^{(6)}(M_Z) = 0.117 \pm 0.006$ but has no effect on the masses because the errors screen the difference between the theory with 5 and 6 flavours. Curiously the running of the top to M_Z

Table 4: Running at the NLO of the bottom and charm quarks masses to M_Z and running masses at the running mass scale needed for thresholds. For masses extracted from QCD SR $\alpha_s^{(5)}(M_Z) = 0.118 \pm 0.006$, for lattice $\alpha_s^{(5)}(M_Z) = 0.115 \pm 0.002$

	$\bar{m}(\bar{m})$	$\bar{m}(M_Z)$
S. Narison	$\bar{m}_b(\bar{m}_b) = 4.29 \pm 0.04$	$\bar{m}_b(M_Z) = 2.97 \pm 0.13$
	$\bar{m}_c(\bar{m}_c) = 1.28 \pm 0.04$	$\bar{m}_c(M_Z) = 0.52 \pm 0.09$
S. Narison	$\bar{m}_b(\bar{m}_b) = 4.35 \pm 0.05$	$\bar{m}_b(M_Z) = 3.03 \pm 0.13$
	$\bar{m}_c(\bar{m}_c) = 1.31 \pm 0.06$	$\bar{m}_c(M_Z) = 0.54 \pm 0.10$
S. Titard	$\bar{m}_b(\bar{m}_b) = 4.397^{+0.018}_{-0.033}$	$\bar{m}_b(M_Z) = 3.07 \pm 0.11$
F.J. Ynduráin	$\bar{m}_c(\bar{m}_c) = 1.306^{+0.022}_{-0.035}$	$\bar{m}_c(M_Z) = 0.52 \pm 0.08$
M. Neubert	$\bar{m}_b(\bar{m}_b) = 4.37 \pm 0.09$	$\bar{m}_b(M_Z) = 3.04 \pm 0.17$
	$\bar{m}_c(\bar{m}_c) = 1.17 \pm 0.12$	$\bar{m}_c(M_Z) = 0.45 \pm 0.14$
Crisafulli et al. [15]	$\bar{m}_b(\bar{m}_b) = 4.17 \pm 0.06$	$\bar{m}_b(M_Z) = 2.93 \pm 0.08$
Crisafulli et al. [16]	$\bar{m}_b(\bar{m}_b) = 4.15 \pm 0.07$	$\bar{m}_b(M_Z) = 2.91 \pm 0.09$
El-Khadra et al.	$\bar{m}_c(\bar{m}_c) = 1.36 \pm 0.19$	$\bar{m}_c(M_Z) = 0.61 \pm 0.15$
Davies et al.	$\bar{m}_b(\bar{m}_b) = 4.13 \pm 0.11$	$\bar{m}_b(M_Z) = 2.89 \pm 0.12$
mean	$\bar{m}_b(\bar{m}_b) = 4.33 \pm 0.06$	$\bar{m}_b(M_Z) = 3.00 \pm 0.12$
	$\bar{m}_c(\bar{m}_c) = 1.30 \pm 0.08$	$\bar{m}_c(M_Z) = 0.52 \pm 0.10$

cancels the difference between the running and the pole mass.

One has to be very careful in comparing the running of the masses obtained from QCD Sum Rules and those obtained from lattice because I took different values for the strong coupling constant at M_Z. In addition, we have to remember that the error in the running is dominated by the error in the strong coupling constant. However it is impressive to notice the good agreement of the results obtained in lattice [17,15] with the running of the masses from QCD Sum Rules. Even, the APE-Collaboration [15], has improved recently its result by increasing the statistics on the lattice [16].

We can now play a game combining the light quark masses of table 5 with the ratios obtained from χPT. The mean value of the strange quark mass together with the Bijnens et at. [5] result gives

$$\frac{2\bar{m}_s}{\bar{m}_u + \bar{m}_d} = 33. \pm 12., \tag{10}$$

in agreement with the χPT result. Being conservative we can also get for the up and down quarks $\bar{m}_u(1GeV) = (3. \pm 2.)MeV$ and $\bar{m}_d(1GeV) = (9. \pm 2.)MeV$ that translate into $\bar{m}_u(M_Z) = (1.5 \pm 1.2)MeV$ and $\bar{m}_d(M_Z) = (4.1 \pm 1.7)MeV$.

Table 5: Running of the light quark masses to M_Z. For masses extracted from QCD SR $\alpha_s^{(5)}(M_Z) = 0.118 \pm 0.006$, for lattice $\alpha_s^{(5)}(M_Z) = 0.115 \pm 0.002$. First box is NLO running, second and third boxes are NNLO running.

	$\bar{m}(1GeV)$	$\bar{m}(M_Z)$
S. Narison	$\bar{m}_s(1GeV) = 222. \pm 22.$	$\bar{m}_s(M_Z) = 105. \pm 28.$
Allton et al.	$\bar{m}_s(1GeV) = 156. \pm 17.$	$\bar{m}_s(M_Z) = 78. \pm 15.$

	$\bar{m}(1GeV)$	$\bar{m}(M_Z)$
S. Narison	$\bar{m}_s(1GeV) = 197. \pm 29.$	$\bar{m}_s(M_Z) = 88. \pm 31.$
Jamin / Münz	$\bar{m}_s(1GeV) = 189. \pm 32.$	$\bar{m}_s(M_Z) = 85. \pm 32.$
Chetyrkin et al.	$\bar{m}_s(1GeV) = 171. \pm 15.$	$\bar{m}_s(M_Z) = 75. \pm 23.$
mean	$\bar{m}_s(1GeV) = 186. \pm 30.$	$\bar{m}_s(M_Z) = 83. \pm 30.$

Bijnens et al.	$(\bar{m}_u + \bar{m}_d)(1GeV) = 12.0. \pm 2.5$	$(\bar{m}_u + \bar{m}_d)(M_Z) = 5.4 \pm 2.2$
Ioffe et al.	$(\bar{m}_d - \bar{m}_u)(1GeV) = 3. \pm 1.$	$(\bar{m}_d - \bar{m}_u)(M_Z) = 1.4 \pm 0.7$

To summarize, the running of the quark masses to the M_Z energy scale gives us a running top mass that is around its perturbative pole mass, $\bar{m}_t(M_Z) = (180. \pm 13.)GeV$, for the bottom and the charm quark we get $\bar{m}_b(M_Z) = (3.00 \pm 0.12)GeV$ and $\bar{m}_c(M_Z) = (0.52 \pm 0.10)GeV$ respectively, while for the strange quark we have a result affected by a big error $\bar{m}_s(M_Z) = (83. \pm 30.)MeV$. The same happens for the up and down quarks, we get $\bar{m}_u(M_Z) = (1.5 \pm 1.2)MeV$ and $\bar{m}_d(M_Z) = (4.1 \pm 1.7)MeV$.

Acknowledgements

I would like to acknowledge M. Neubert, A. Pich, K. Hornbostel, V. Giménez and F. Botella for very useful discussions and A. Santamaria for carefully reading the manuscript. This work was supported by the Conselleria d'Educació i Ciència de la Generalitat Valenciana, CICYT (Spain) under grant AEN93-0234 and I.V.E.I. under grant 03-007.

References

1. N. Gray, D.J. Broadhurst, W. Grafe and K. Schilcher, Z. Phys. C **48** (90) 673.
2. R. Hempfling and B. A. Kniehl, Phys. Rev. **D51** (95) 1386. A. I. Bochkarev and R. S. Willey, Phys. Rev. **D51** (95) 2049.

3. J. Gasser and H. Leutwyler, Nucl. Phys. **B250** (85) 465, 517, 539.
4. J. Donoghue, B. R. Holstein and D. Wyler, Phys. Rev. **D47** (93) 2089, Phys. Rev. Lett. **69** (92) 3444.
5. J. Bijnens, J. Prades and E. de Rafael, Phys. Lett **B348** (95) 226.
6. V. L. Eletsky and B. L. Ioffe, Phys. Rev. **D48** (93) 1441. C. Adami,E. G. Drukarev and B. L. Ioffe, Phys. Rev. **D48** (93) 2304.
7. S. Narison, PM 95/06, hep-ph/9504333.
8. M. Jamin and M. Münz, CERN-TH.7435/94, hep-ph/9409335.
9. K. G. Chetyrkin, C. A. Dominguez, D. Pirjol and K. Schilcher, Phys. Rev. **D51** (95) 5090.
10. C. R. Allton, M. Ciuchini, M. Crisafulli, E. Franco, V. Lubicz and G. Martinelli, Nucl. Phys. **B431** (94) 667.
11. S. Narison, Phys. Lett **B341** (94) 73.
12. C. A. Dominguez, G. R. Gluckman and N. Paver, Phys. Lett **B333** (94) 184. C. A. Dominguez, and N. Paver, Phys. Lett **B293** (92) 197. C. A. Dominguez, UCT-TP-219/94, hep-ph/9410362.
13. S. Titard and F. J. Ynduráin, Phys. Rev. **D49** (94) 6007.
14. M. Neubert, Phys. Rep. **245** (94) 259, Boulder 1993, Proceedings, The building blocks of creation, p. 125, and CERN-TH.7225/94, hep-ph/9404296.
15. M. Crisafulli, V. Giménez, G. Martinelli and C. T. Sachrajda, CERN-TH.7521/94, hep-ph/9506210.
16. M. Crisafulli, V. Giménez, G. Martinelli and C. T. Sachrajda, talk presented at the '95 Lattice, Melbourne, July 11-15, 1995, to appear in the Proceedings. V. Giménez, APE-Collaboration, Private communication.
17. C. T. H. Davies et al., Phys. Lett **B345** (95) 42, Phys. Rev. Lett **73** (94) 2654.
18. A. X. El-Khadra and B. P. Mertens, hep-lat/9501022.
19. G. Rodrigo and A. Santamaria, Phys. Lett **B313** (93) 441.
20. M. Bilenky, G. Rodrigo and A. Santamaria, Nucl. Phys. **B439** (95) 505.
21. J. Fuster, *Recent results on QCD at LEP*. Proc. XXII International Meeting on Fundamental Physics, Jaca, 1994, Spain.
22. W. Bernreuther, Ann. of Phys. **151** (83) 127. W. Bernreuther and W. Wetzel, Nucl. Phys. **B197** (82) 228.
23. S. A. Larin, T. van Ritbergen and J. A. M. Vermaseren, Nucl. Phys. **B438** (95) 278.
24. F. Abe et al., CDF Collaboration, FERMILAB Pub-95/022-E CDF, hep-ex/9503002. S. Abachi et al., DØ Collaboration, FERMILAB Pub-95/028-E DØ hep-ex/9503003.

Chapter 4

FROM THE BIG BANG TO NOW, AND THE JOURNEY BACK

EDWARD W. KOLB

NASA/Fermilab Theoretical Astrophysics Group
Fermi National Accelerator Laboratory, Batavia, IL 60510, USA
and
Department of Astronomy and Astrophysics, Enrico Fermi Institute
The University of Chicago, 5640 South Ellis Ave., Chicago, IL 60637, USA

By observing the Universe today, we may be able to understand something of the events that occurred in the early Universe, and in so doing learn something about the fundamental physics behind the processes leading to the formation of structure in the Universe.

1 Introduction

Observations of the present Universe can provide information about conditions in the early Universe. Since the evolution of the early Universe depended upon the fundamental laws of physics, if we can infer something about the early Universe, we might be able to discover something about the physics that determined the evolution at early times. In this way, observations of the present-day Universe may provide information about physics at energy scales inaccessible any other way. In particular, if the existence of structure in the Universe, both in the distribution of matter on large scales and in the fluctuations of the cosmic background radiation (CBR), can be traced to events in the first 10^{-35}seconds of the Universe, we may learn something about energy scales of 10^{16}GeV.

But before we can hope to execute such an ambitious program, it is necessary to obtain detailed information about 1) the large-scale structure of the Universe, 2) fluctuations in the temperature of the cosmic microwave background radiation, 3) the matter content of the Universe, and 4) the cosmological parameters that describe our world model.

In this talk I will outline the basic features of the observation program necessary to learn something about fundamental physics, and the potential information that might be obtained.

2 Fluctuations

One might say that there are two basic views of the Universe. The observer's view is of a Universe full of bumps and lumps, with matter gathered together into astronomical bodies. The observer looks through the eyepiece of her telescope and sees stars, galaxies, clusters, superclusters, etc. On the other hand, the theorist's view is of a Universe that is homogeneous and isotropic, with matter uniformly spread throughout space with a density exactly equal to the critical density, $\rho_C = 1.88h^2 \times 10^{-29}\text{g cm}^{-3}$.[a]

In fact, both of these seemingly antithetical views are reasonable. Of course there are objects in the Universe: we do not live in a Universe that is exactly homogeneous and isotropic because the density field has a spatial dependence. But on sufficiently large scales, the Universe *is* approximately homogeneous and isotropic. Even on scales where structure is discernible, the view of an average density for the Universe is still a useful concept, because the density field is typically described in terms of fluctuations about the mean density $\langle\rho\rangle$. The fluctuations in the density field, $\delta\rho(\mathbf{x}) \equiv \rho(\mathbf{x}) - \langle\rho\rangle$ can be expanded in terms of Fourier components δ_k:

$$\frac{\delta\rho(\mathbf{x})}{\rho} \propto \int d^3k\, e^{-i\mathbf{k}\cdot\mathbf{x}}\delta_k. \tag{1}$$

So long as the power spectrum is isotropic, the *rms* density fluctuation can be expressed as

$$\left\langle \left(\frac{\delta\rho(\mathbf{x})}{\langle\rho\rangle}\right)^2 \right\rangle \propto \int \frac{dk}{k}\, k^3 \left|\delta_k\right|^2. \tag{2}$$

The power spectrum, either expressed as $P(k)$ or $\Delta(k)$, where

$$P(k) \equiv |\delta_k|^2, \qquad \Delta(k) \equiv k^3|\delta_k|^2, \tag{3}$$

is directly related to the Fourier transform of the two-point autocorrelation function of the mass distribution. $\Delta(k)$ is the power per logarithmic interval in k space, and is what is usually meant by the amplitude of the mass fluctuation on scale λ, where $k\lambda \sim 2\pi$

Determination of the power spectrum, over a range of say a few h^{-1}Mpc to a few $100h^{-1}$Mpc, is one of the primary goals of observational cosmology.

---

[a]The ubiquitous factor of h represents our uncertainty in the value of the Hubble constant: $H_0 = 100h$ km s^{-1}Mpc^{-1}, with $0.4 < h < 1$.

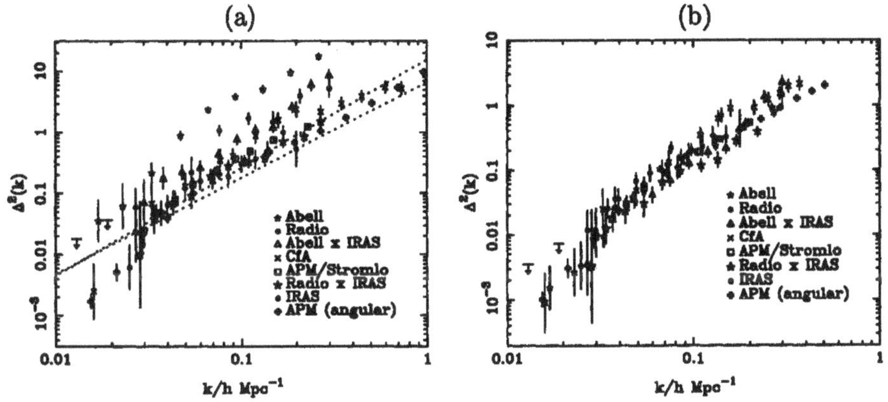

Figure 1: On the left is the power spectrum inferred from the indicated surveys of large-scale structure. On the right is the same data, corrected for red-shift space distortions, including individual bias factors, and properly linearized. Recall that small k corresponds to large scales. The figure and analysis is from Peacock and Dodds; references to the different surveys can be found there.

2.1 Large-Scale Structure.

The traditional method to obtain the power spectrum is through large-scale structure surveys. The procedure is straightforward in concept, but exceedingly complex in execution. One looks at a sample of objects in a survey, determines a mean density, then calculates the *rms* fluctuations in the number counts of objects on some scale λ, performs the Fourier transform, and obtains the power spectrum.

A survey of surveys of power spectra is displayed in figure 1 (left), taken from the work of Peacock and Dodds.[3] On the left-hand side of the figure is the "raw" data from various surveys. The goal is to extract from the data a consistent value for the linear power spectrum of the mass fluctuations, which is what is predicted by early-Universe cosmology. There are three major corrections that must be applied to the data: (i) For large k (small scales) the spectrum has undergone some degree of non-linear evolution and the observed spectrum deviates from a linear extrapolation of the initial spectrum. (ii) Three-dimensional data sets, which use redshift as the radial coordinate, must be corrected for the existence of peculiar velocities. (iii) Actually measured are the fluctuations in the *number density* of the objects surveyed, while what is desired are the fluctuations in the *mass density*. There are astrophysical effects that suggest that the relationship between the number density and the mass density might be different for different objects. This is known as the bias factor.

Peacock and Dodds have attempted to make these corrections; their result is shown in figure 1 (right). While the procedure is involved, and while some corrections such as the relative bias factors are phenomenological and yet not well understood on a physical basis, the resulting power spectrum is fairly well determined. Of course one of the goals in large-scale structure is to understand better the corrections necessary to convert the observed power spectra into a single power spectrum. However we have every expectation that eventually we will be able to extract an accurate power spectrum from large-scale structure surveys.

2.2 Microwave Background Fluctuations.

The Cosmic Background Radiation (CBR) is a fundamental probe of the early Universe. The photons in the CBR last interacted around the time of recombination (redshift $1 + z = 1100$; age about 300,000 years) when it became cool enough for the electrons and nuclei to form atoms. Although the CBR provides direct evidence about conditions in the Universe at recombination, small fluctuations in the temperature of the Universe at recombination may be the result of process that occurred much earlier in the history of the Universe.

Since a map of temperature variations forms a two-sphere, the temperature pattern on the sky on the sky can be expressed in terms of the temperature as a function of angle. One can form the temperature–temperature autocorrelation function at an angle θ (similar to, say, the galaxy-galaxy autocorrelation function) by measuring the temperature at some angle α, comparing it to the temperature at angle $\alpha + \theta$, and averaging over positions α In equation form rather than word calculus, the temperature-temperature aotocorrelation function is $C(\theta) = \langle \delta T(\alpha)\, \delta T(\alpha + \theta)\rangle_\alpha$. Just as the two-point autocorrelation function in large-scale structure is expanded in a Fourier series, it is useful to expand $C(\theta)$ in Legendre polynomials: $C(\theta) = (4\pi)^{-1} \sum_l (2l+1) C_l P_l(\cos\theta)$. The C_l are the multipole moments, and C_l as a function of l is known as the spectrum of the temperature fluctuations.

There are now over a dozen experiments that have detected microwave anisotropies, and the spectrum is beginning to fill in (see Fig. 2). Of course we are still in the early game, and over the next few years the experimental situation should steadily improve. What will eventually emerge is detailed knowledge of the angular spectrum of microwave fluctuation.

Temperature fluctuations in the CBR arise due to five distinct physical effects: (i) our peculiar velocity with respect to the cosmic rest frame; (ii) fluctuations in the gravitational potential on the last scattering surface; (iii)

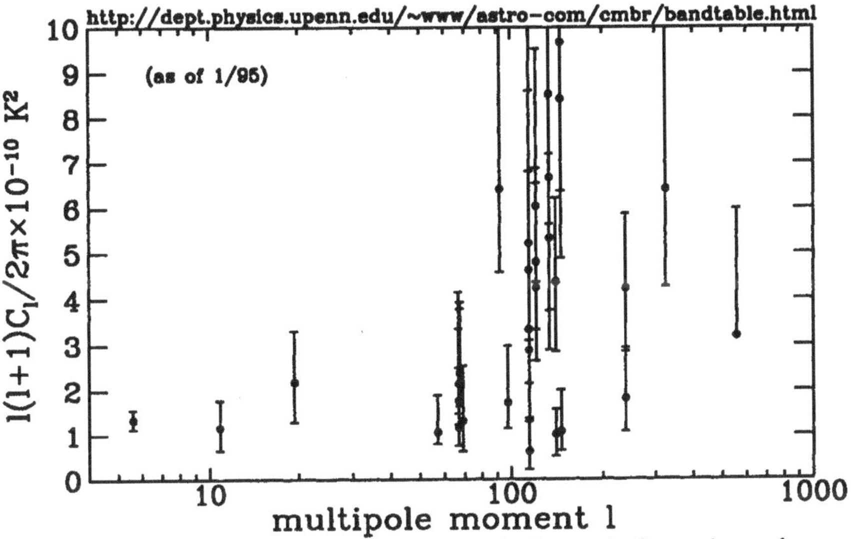

Figure 2: Multipole moments of microwave background fluctuations along with 1σ errors. A key to connect experiments with the data may be found at `http://dept.phys.upenn.edu/~www/astro-cosmo/cmbr/bandtable.html`.

fluctuations intrinsic to the radiation field itself on the last-scattering surface; (iv) the peculiar velocity of the last-scattering surface; and (v) damping of anisotropies if the Universe should be re-ionized after decoupling.

Of particular interest are the temperature fluctuations on large-angular scales that arise from (ii), know as the "Sachs-Wolfe" effect. These anisotropies are simple to understand because they probe length scales that were larger than the Hubble radius at decoupling, and therefore insensitive to microphysical processes. Furthermore, because of their insensitivity to microphysical processing, they provide a probe of the "primordial" spectrum of primeval fluctuations. In the synchronous gauge and for a flat cosmology, the large-angle fluctuations in the CBR temperature are given in terms of fluctuations in the gravitational potential on the last scatter surface, ϕ_1:

$$\frac{\delta T(\hat{\mathbf{x}})}{T} \propto -\int k^{-2}\delta_k \exp(-i\vec{\mathbf{k}}\cdot\vec{\mathbf{x}})d^3k = \frac{1}{3}\phi_1(\vec{\mathbf{x}}) \sim \Delta(k), \tag{4}$$

where the vector $\vec{\mathbf{x}}$ points to the last-scattering surface and has length $|\vec{\mathbf{x}}| = 2H_0^{-1}$ (the distance to the horizon). From intuition one would have guessed that $\delta T/T$ should be given by $\phi_1(\vec{\mathbf{x}})$; when all the general relativistic effects are taken into account, the result is one third of that.

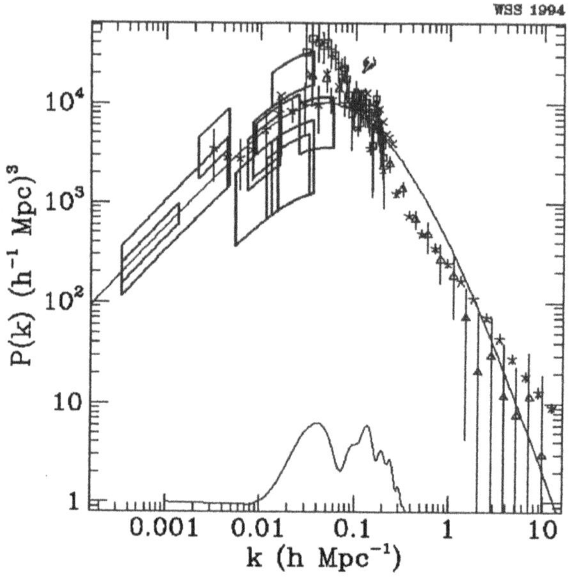

Figure 3: The power spectrum as constructed by White, Scott, and Silk. The solid line is the prediction of a simple CDM model. Clearly what is needed is more information around the peak in the area indicated by the finger.

Thus microwave background fluctuations probe a part of the power spectrum, just as large-scale structure surveys. In fact the microwave fluctuations probe a region (in k space) of the spectrum on larger scales (smaller k) than accessible to large-scale structure surveys. On the microwave sky, an angle of θ (for small enough θ) samples linear scales of $100h^{-1}(\theta/1^\circ)$Mpc.[b] The l-th multipole is often taken as corresponding roughly to a scale $k_l \simeq lH_0/2\text{Mpc}^{-1} = lh/6000\text{Mpc}^{-1}$.

2.3 *The Big Picture*

One may envisage in the near future the emergence of a picture of the spectrum of perturbations, expressed as the power spectrum, pieced together from large-scale structure surveys and microwave background fluctuation experiments. In fact even today the picture is beginning to emerge. An illustration of what is emerging is shown in figure 3. Boxes are from microwave fluctuations, while the points are from a particular large-scale structure survey.

[b] The surface of last scattering is located some $200h^{-1}$ Mpc inside the horizon distance.

The solid line in the figure is from the CDM model, the Rasputin of cosmological models. While it has been "killed" by many people, clearly the general features seem to be a good approximation to reality.

3 Primordial Origin of Fluctuations

There are two well developed scenarios for the generation of primordial fluctuations: inflation and defects. Both are capable of producing fluctuations on large angular scales as seen by microwave background fluctuations, and both are capable of producing large-scale structure (whether the large-scale structure produced by either model resembles the structure observed in our Universe is not completely clear).

3.1 Inflation:

Just as fluctuations in the density field may be expanded in a Fourier series, fluctuations in the scalar field driving inflation, known as the ***inflaton*** field, may be expanded in terms of its Fourier coefficients $\delta\phi_{\mathbf{k}}$: $\delta\phi(\mathbf{x}) \propto \int \delta\phi_{\mathbf{k}} \exp(-i\mathbf{k}\cdot\mathbf{x})d^3k$. During inflation there is an event horizon, and quantum-mechanical fluctuations in the Fourier components of the inflaton field are given by[4] $k^3 \left|\delta\phi_{\mathbf{k}}\right|^2 / 2\pi^2 = (H/2\pi)^2$, where $H/2\pi$ plays a role similar to the Hawking temperature of black holes. Thus, when a given mode of the inflaton field leaves the Hubble radius during inflation, it has impressed upon it quantum mechanical fluctuations. Fluctuations in the inflaton field on scale k are proportional to $k^{3/2}|\delta\phi_{\mathbf{k}}|$, which in turn is proportional to $H/2\pi$. Fluctuations in ϕ lead to perturbations in the energy density: $\delta\rho_\phi = \delta\phi(\partial V/\partial\phi)$.

Fluctuations as a particular mode leaves the Hubble radius during inflation leads to fluctuations in the intrinsic curvature of the co-moving spatial hypersurfaces. These fluctuations are constant for Fourier modes well outside the Hubble radius, and will be the same when the re-enter the Hubble radius in the radiation or matter-dominated era as when they went outside the Hubble radius during inflation. The fluctuation in the intrinsic curvature is given by[10] (prime denotes $d/d\phi$)

$$\left(\frac{\delta\rho}{\rho}\right)_\lambda^{\mathrm{HOR}} \equiv \frac{m}{\sqrt{2}}A_S(\phi) \sim \frac{m\kappa^2}{8\pi^{3/2}}\frac{H^2(\phi)}{|H'(\phi)|} \propto \frac{V^{3/2}(\phi)}{m_{Pl}^3 V'(\phi)}, \tag{5}$$

where $H(\phi)$ and $H'(\phi)$ are to be evaluated when the scale λ crossed the Hubble radius *during* inflation. The constant m equals $2/5$ or 4 if the perturbation re-enters during the matter or radiation dominated eras respectively, and $\kappa^2 = 8\pi G$.

In addition to the scalar density perturbations caused by de Sitter fluctuations in the inflaton field, there are gravitational mode perturbations, $g_{\mu\nu} \to g_{\mu\nu}^{\rm FRW} + h_{\mu\nu}$, caused by de Sitter fluctuations in the metric tensor[5]. Here, $g_{\mu\nu}^{\rm FRW}$ is the Friedmann–Robertson–Walker metric and $h_{\mu\nu}$ are the metric perturbations. That de Sitter space fluctuations should lead to fluctuations in the metric tensor is not surprising, since after all, gravitons are the propagating modes associated with transverse, traceless metric perturbations, and they too behave as minimally coupled scalar fields. The dimensionless tensor metric perturbations can be expressed in terms of two graviton modes we will denote as h. Performing a Fourier decomposition of h, $h(\vec{x}) \propto \int \delta h_k \exp(-i\vec{\mathbf{k}} \cdot \vec{x}) d^3k$, we can use the formalism for scalar field perturbations simply by the identification $\delta\phi_{\mathbf{k}} \to h_{\mathbf{k}}/\kappa\sqrt{2}$, with resulting quantum fluctuations $k^3|h_{\mathbf{k}}|^2/2\pi^2 = 2\kappa^2(H/2\pi)^2$. While outside the Hubble radius, the amplitude of a given mode remains constant, so the amplitude of the dimensionless strain on scale λ when it crosses the Hubble radius after inflation is

$$\left|k^{3/2} h_{\mathbf{k}}\right|_{\lambda}^{\rm HOR} \equiv A_T(\phi) = \frac{\kappa}{4\pi^{3/2}} H(\phi) \sim \frac{V^{1/2}(\phi)}{m_{Pl}^2}, \tag{6}$$

where once again $H(\phi)$ is to be evaluated when the scale λ crossed the Hubble radius *during* inflation.

The basic picture of fluctuations in the evolving inflaton field giving rise to scalar and tensor fluctuations is illustrated in figure 4.

All inflationary calculations rely on the use of a *slow-roll* approximation, which assumes that quantities like A_S and A_T may be expressed as an expansion in terms of derivatives of the Hubble parameter H. If one defines $\epsilon_0(\phi) = 2\left[H'(\phi)/H(\phi)\right]^2/\kappa^2$, then the expansion variables are $\epsilon_n \equiv \sqrt[n]{2}\,\kappa^{-n}\epsilon_0^{-n/2} d^n\epsilon_0/d\phi^n$. In general there are an infinite hierarchy of these derivatives which can in principle all enter at the same order in an expansion. In practice one assumes that since the potential is flat, only the first few terms are important. The slow-roll approximation applies when these slow-roll parameters are small in comparison to unity. The lowest-order expressions for the scalar (A_S) and tensor (A_T) amplitudes above assume only ϵ_0 and ϵ_1 are important, and that they are much less than one. Improved expressions for the scalar and tensor amplitudes for finite but small $\{\epsilon_0,\ \epsilon_1\}$ were found by Stewart and Lyth.[6]

Since the potential must be flat in order for inflation to occur, H and H' will not vary much over accessible length scales. Therefore it is often convenient to assume that the scalar and tensor fluctuations are given by a power law: $d\ln A_S^2(\lambda)/d\ln\lambda \equiv 1 - n_S$ and $d\ln A_T^2(\lambda)/d\ln\lambda \equiv -n_T$, where n_S and n_T are called the spectral indices. The CDM model assumes $n_S = n_T = 0$.

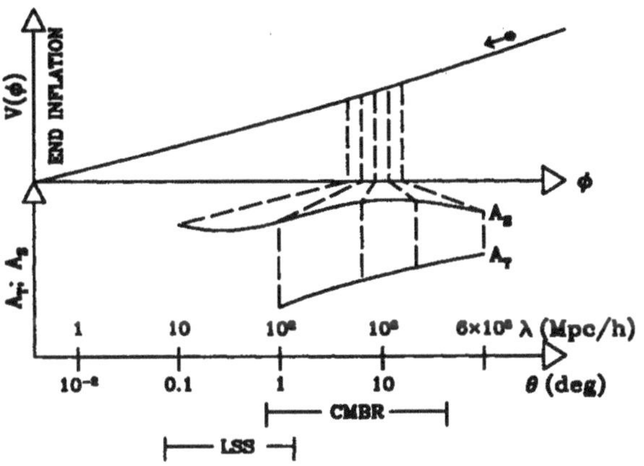

Figure 4: The basic idea of inflation is that as the field evolves in the potential, quantum fluctuations in the inflaton field produce scalar density perturbations A_S, while fluctuations in the transverse, traceless metric components produce tensor gravitational wave perturbations, A_T. Also indicated is the main observational information from the cosmic microwave background, roughly over the range indicated by CBR, and structure surveys, roughly over the range noted LSS.

3.2 Defects:

Perhaps the most important concept in modern particle theory is that of spontaneous symmetry breaking (SSB). The idea that there are underlying symmetries of nature that are not manifest in the structure of the vacuum appears to play a crucial role in the unification of the forces. In all unified gauge theories—including the standard electroweak model—the underlying gauge symmetry is larger than the unbroken $SU(3)_C \otimes U(1)_{EM}$. Of particular interest for cosmology is the theoretical expectation that at high temperatures, symmetries that are spontaneously broken today were restored, and that during the evolution of the Universe there were phase transitions associated with spontaneous breakdown of gauge (and perhaps global) symmetries. For example, we can be reasonably confident that there was such a phase transition at a temperature of order 100 GeV and a time of order 10^{-11} sec, associated with the breakdown of $SU(2)_L \otimes U(1)_Y \rightarrow U(1)_{EM}$. Moreover, the vacuum structure in many spontaneously broken gauge theories is very rich: topologically stable configurations of gauge and Higgs fields exist as domain walls, cosmic strings, and monopoles. In addition, classical configurations that are not topologically stable, so-called non-topological solitons, may exist and be stable for dynamical reasons. Interesting examples include soliton stars, Q-balls, non-topological cosmic strings, sphalerons, and so on.

How can we tell if the Universe underwent a series of SSB phase transitions? One possibility is that symmetry-breaking transitions were not "perfect," and that false vacuum remnants were left behind, frozen in the form of topological defects.

Domain walls arise from the spontaneous breaking of a discrete symmetry. The general mathematical criterion for the existence of topologically stable domain walls for the symmetry-breaking pattern $\mathcal{G} \rightarrow \mathcal{H}$ is that $\mathbf{\Pi_0}(\mathcal{M}) \neq \mathcal{I}$, where $\mathcal{M}$ is the manifold of equivalent vacuum states, $\mathcal{M} \equiv \mathcal{G}/\mathcal{H}$, and $\mathbf{\Pi_0}$ is the homotopy group that counts disconnected components.

In general, there will be string solutions associated with the symmetry breaking of a gauge group $\mathcal{G} \rightarrow \mathcal{H}$, if the manifold of degenerate vacuum states, $\mathcal{M} = \mathcal{G}/\mathcal{H}$, contains unshrinkable loops, i.e., if the mapping of $\mathcal{M}$ onto the circle is non-trivial. This is formally expressed by the statement that topologically stable string solutions exist if $\mathbf{\Pi_1}(\mathcal{M}) \neq \mathcal{I}$. Such a mapping is typically characterized by the winding number of the mapping, i.e., $\theta \rightarrow N\theta$ ($N = 0,\ 1,\ \cdots$), so that $\Pi_1(\mathcal{M}) = \mathcal{Z}$, the set of integers.

Gauge and Higgs field configurations corresponding to a magnetic monopole exist if the vacuum manifold ($\mathcal{M} = \mathcal{G}/\mathcal{H}$) associated with the symmetry-breaking pattern $\mathcal{G} \rightarrow \mathcal{H}$ contains non-shrinkable surfaces, i.e., if the mapping of $\mathcal{M}$ onto the two-sphere is non-trivial. Mathematically, this is expressed by the statement that monopole solutions arise in the theory if $\mathbf{\Pi_2}(\mathcal{M}) \neq \mathcal{I}$. If $\mathcal{G}$ is simply connected, then $\mathbf{\Pi_2}(\mathcal{G}/\mathcal{H}) = \mathbf{\Pi_1}(\mathcal{H})$. If $\mathcal{G}$ is not simply connected, then the generalization of the above expression is $\mathbf{\Pi_2}(\mathcal{G}/\mathcal{H}) = \mathbf{\Pi_1}(\mathcal{H})/\mathbf{\Pi_1}(\mathcal{G})$.

Defects can also arise in the spontaneous breaking of *global* symmetries. The analogies of the defects discussed above are global strings and global monopoles. The global field configurations look like their local counterparts for the scalar field, but of course there is no vector field. This means that formally the string and monopole solutions have infinite energy (recall for the local defects the energy in the gauge fields cancels the energy in the Higgs field far from the defect.) This is really not a problem, because there the divergence in the energy is only logarithmic, and there are many physical effects to cut it off (such as the inter-defect separation). There are just two main differences in the behavior of gauge and global defects: (1) the energy of the global defects are slightly more spread out, (2) the global strings can radiate energy by the emission of Nambu–Goldstone bosons.

However there are new types of defects in global symmetry breaking that do not appear in the breaking of gauge symmetries. For example, in the spontaneous breaking of a global O(N) model to O($N-1$), for $N = 1$ walls appear, for $N = 2$ global strings result, for $N = 3$ global monopoles are produced. These all have counterparts in local theories. However for $N > 3$ global defects also exist: for $N = 4$ the defect is called global texture (for gauge theories the texture configuration resembles instantons), and for $N > 4$ they are called Kibble gradients. Texture corresponds to knots in the Higgs field that arise when the field winds around the three sphere. These knots are generally formed by misalignment of the field on scales larger than the horizon at the symmetry-breaking phase transition because of the Kibble mechanism.[7] As the knots enter the horizon, they collapse at roughly the speed of light, giving rise to nearly spherical energy density perturbations. New knots are constantly coming into the horizon and collapsing, leading to a scale invariant spectrum of density perturbations. The magnitude of the perturbations is set by the scale of the symmetry breaking, and for scenarios of structure formation involving texture, the scale of symmetry breaking must be about 10^{16}GeV.

In the past few years defects produced in the electroweak transition has attracted a lot of interest. Although the breaking scheme $SU(2)_L \otimes U(1)_Y \rightarrow U(1)_{EM}$ does not produce stable topological defects, unstable defects are produced: strings that end on monopoles, called electroweak strings. These non-perturbative objects are quite interesting, although their rôle in cosmology (and particle physics for that matter) is still unclear (for a review, see [8]).

4 What can be learned

Let's suppose that future determinations of the power spectrum favor either inflation or defects. Then what can we say about physics at high energy scales?

If defects are correct then we would know that there is spontaneous symmetry breaking at energy scales of about 10^{16}GeV, because if the scale is much different than that, the mass scale of the defects could not account for large-scale structure. For instance if cosmic strings lead to large-scale structure, then the mass per unit length of the string must correspond to a symmetry breaking scale of 10^{16}GeV. Furthermore, the pattern of microwave fluctuations are different if for defects of different spatial dimension. The pattern from strings look different than the pattern for monopoles, which look different than the pattern for texture, etc. So perhaps we could learn something of the size of the symmetry that is broken. Finally, since local defects evolve differently than defects from global theories, we could learn whether the symmetry breaking responsible for the defects was a global or a local symmetry.

If fluctuations from inflation are the source of fluctuations, then we would know that all structure in the Universe has a quantum-mechanical origin. We would know that gravity must be quantized. If a tensor component is detected, then we would know that gravitons exist, and finally, we may be able to reconstruct a small part of the potential of the inflaton.

One of the most remarkable features of the Inner Space/Outer Space connection is the possibility that the largest structures in the Universe may have an origin that can be traced to scalar field dynamics at energy scales of 10^{16}GeV.

Acknowledgements

This work was supported by DOE and NASA grant NAG 5-2788 at Fermilab

References

1. A. R. Liddle and D. H. Lyth, Phys. Rep. **X**, X (1994).
2. R. K. Sachs and A. M. Wolfe, Ap. J. **147**, 73 (1967); P. J. E. Peebles, Ap. J. **263**, L1 (1982).
3. J. A. Peacock and S. J. Dodds, Mon. Not. R. astr. Soc. **267**, 994 (1020).
4. T. Bunch and P. C. W. Davies, Proc. Roy. Soc. London **A360**, 117 (1978).
5. V. A. Rubakov, M. Sazhin, and A. Veryaskin, Phys. Lett. **115B**, 189 (1982); R. Fabbri and M. Pollock, Phys. Lett. **125B**, 445 (1983); B. Allen, Phys. Rev. D **37**, 2078 (1988); and L. Abbott and M. Wise, Nucl. Phys. **B244**, 541 (1984).
6. E. D. Stewart and D. H. Lyth, Phys. Lett. **302B**, 171 (1993).
7. See e.g., T. W. B. Kibble, Phys. Rep. **67**, 183 (1980).
8. T. Vachaspati and A. Achucarro, Phys. Rev. D **44**, 3067 (1991).
9. M. White, D. Scott, and J. Silk, Ann. Rev. Astron. and Astrophys. **32**, 319 (1994) .
10. A. H. Guth and S.-Y. Pi, Phys. Rev. Lett. **49**, 1110 (1982); A. A. Starobinskii, Phys. Lett. **117B**, 175 (1982); and S. W. Hawking, Phys. Lett. **115B**, 295 (1982); J. M. Bardeen, P. J. Steinhardt, and M. S. Turner, Phys. Rev. D **28**, 679 (1983).

SOLAR NEUTRINOS: EXPECTING 1996

A. Yu. SMIRNOV

Yukawa Institute for Theoretical Physics, Kyoto University, Kyoto 606-01, Japan, and
International Centre for Theoretical Physics, I-34100, Trieste, Italy

These are remarks (mainly on the solar neutrinos) written in anticipation of 1996 - the year which can be crucial for the neutrino physics. Recent results on solar neutrinos are discussed. The topics include: (solar) model independent approach to the solar neutrino problem, status of different solutions of the problem, standard and non-standard scenarios of the lepton mixing, light singlet fermions and the neutrino phenomenology.

1 Introduction

According to time schedule [1] in January - March '96 stainless tank placed in the Kamioka mine will be filled by 50.000 tons of water and in April 96 the SuperKamiokande – the detector of new generation of the underground experiments – will start to take data. Later Sudbury detector [2] will to be "on the line". CHORUS [3] is expected to announce first results. CHOOZ [4] - the first long base line reactor experiment will begin to operate soon.

The results from these experiments may resolve neutrino puzzles we are discussing now. In any case they will have strong impact not only on the neutrino physics and astrophysics but also on particle physics as whole.

Taking this into account it may be wise to abstain from further theoretical speculations and just wait (laying a bet) for new experimental results. On the other hand it is a good time to summarize what we have learned and to understand what is our starting point before new results will arrive. It is a good time to formulate *a priori* criteria according to which we will make the conclusions.

2 Experiment and Theory

2.1 Experiment

1. Results from all solar neutrino experiments [5,6,7,8] are stable. There are small changes of the average signals within 1σ: Homestake result has increased by

1σ [5], Kamiokande flux decreased by statistical 1σ [6], the change of GALLEX result is even smaller [8].

2. The error bars slowly decrease indicating that even Gallium experiments become "old". When experiments become old it is a proper moment to speak about time variations. Search for time variations is one of justifications to continue the experiments. Recent analysis of Gallium data shows very good agreement with constant original flux and even admit-ion of different time variations practically does not improve the goodness of the fit [8].

Another justification is accuracy. In fact, one can imagine a situation when, even 50% decrease of error bars in Gallium experiment could be decisive for the problem.

Kamiokande does not see any time variations (day-night, seasonal, anti-correlation with solar activity etc..) at least with more than 30% amplitude. Of course, an experienced eye can find about 4 years period "wave" (especially if one removes systematical errors).

During last 5 years Homestake data did not show neither correlations nor anticorrelations with sunspot numbers (and in this sense there is a good agreement with Kamiokande negative result). The confidence level of the anticorrelation during all the period of observation is approximately the same as in 1980 [9]. That is the effects is essentially due to very low signal during one year: 1979 - 1980. This was the year of change of the magnetic field polarity in the Sun. (Accidental coincidence? This can be check in the year 2002). Runs with high counting rate in the period of low activity 1986 - 1987 have rather peculiar statistical distribution [10].

Anyway, SuperKamiokande will continue the Kamiokande job. Iodine experiment can probably substitute Chlorine monitoring, and it will be certainly worthwhile to have Gallium experiment working during the operation of SuperKamiokande and SNO (just in case).

3. It is difficult to overestimate the importance of the result of GALLEX experiment with ^{51}Cr source [11]. It gives not only the overall check of the solar experiment and especially, the efficiency of detection in the important energy region of the ^{7}Be neutrinos. It gives the credit to radiochemical method, and therefore the additional credit to the Homestake result.

4. One more result on the solar neutrinos: recently new bound on the antineutrino flux from the Sun $\Phi(\bar{\nu}_e)$ has been obtained by LVD collaboration [12]

$$\Phi(\bar{\nu}_e) = (2-4)\cdot 10^{-3} cm^{-2} s^{-1}, \tag{1}$$

i.e. $\Phi(\bar{\nu}_e)/\Phi_B^{SSM} = 0.3 - 0.6$ % , where Φ_B^{SSM} is the boron neutrino flux according to the SSM. The bound has an important implications to the spin-flavor precession of the solar neutrinos.

2.2 Theory

Standard Solar Model (SSM) predictions where updated recently in a number of aspects. New values of input parameters like nuclear cross sections, solar age, abundance of the elements, radiative opacities etc.. are used. The uncertainties related to the pre-Main sequence evolution, to the depth of the convection zone etc., were studied (see [15] for review). The most important changes are related to taking in to account the diffusion of heavy elements (C, N, O, ... Fe, Ni ...). The diffusion leads to increase of the opacity in the radiative zone, and consequently, in increase of the central temperature of the Sun by about 1.1 %. As a result the boron and the beryllium neutrino fluxes increase by 17 % and 6 % correspondingly.

The solution of the Li-problem - one of the origin of doubt in the reliability of the SSM itself and its predictions of the neutrino fluxes - probably has been found. The reason of the Li- deficit (whose surface concentration is about 200 times smaller than expected) could be in plasma physics [16]. It has been pointed out that in the effective screening potential of the electron cloud the nuclei become more transparent to each other. This increases the effective energy of collisions by 600 - 700 eV which leads to decrease of temperature of the ^7Li burning. Also diffusion becomes more efficient.

3 Experiment without Theory

3.1 Solar neutrino problem without solar model

In spite of serious progress in the solar modelling and very good agreement of SSM and helioseismological data, some predicted solar neutrino fluxes still have rather large uncertainties. Mainly, they are related to the nuclear cross-sections (first of all, for the reaction $p +^7 Be \rightarrow^8 B + \gamma$) and probably to some plasma effects which have not yet been properly taken into account [17].

These uncertainties will hardly be fixed before new experiments on solar neutrinos start to operate. In this connection the approach to the problem has been elaborated which does not use the absolute values of neutrino fluxes from the SSM. Main points of the approach which can be called "solar neutrino problem without solar model" are the following [18 - 30]:

1. Only general notion is used about the solar neutrinos: the composition and the energy spectra of components, but not the absolute values of fluxes. These absolute values are considered as *free parameters to be found from the solar neutrino experiments.* In particular, the boron neutrino flux can be

represented as

$$\Phi_B = f_B \cdot \Phi_B^{SSM}, \tag{2}$$

where f_B is free parameter, and Φ_B^{SSM} is the flux in the reference SSM e.g. [31]. Similarly, the parameters f_i ($i = Be, pp, NO$) for other important fluxes can be introduced.

2. The data from different experiments are confronted immediately.
3. The solar neutrino fluxes are normalized on the solar luminosity (the normalization in based on the condition of thermal equilibrium of the Sun).

Already existing data allow one (i) to formulate the problem in practically model independent way, (ii) to restrict not only neutrino parameters but also original neutrino fluxes. In future precision solar neutrino data will be used to get the information on solar model [28]. Thus we will turn to the original proposal to study the interior of the Sun by neutrinos.

There are two key points in the analysis of present experimental situation.

Kamiokande versus Homestake [18–26]. Suppose first that neutrino flux consists of ν_e only. Then boron neutrino flux measured by Kamiokande gives the contribution to Ar-production rate $Q_{Ar,B} = 3.00 \pm 0.45$ SNU which exceeds the total signal observed by Homestake: $Q_{Ar}^{obs} = 2.55 \pm 0.25$ SNU. This means that the contributions of all other fluxes to Q_{Ar}, and in particular, of Beryllium neutrinos should be strongly suppressed.

Gallium Experiment Results versus Solar Luminosity [20,21,32,23]. The luminosity of the Sun allows one to estimate the pp- neutrino flux, and consequently its contribution to Ge-production rate: $Q_{Ge,pp} \approx 71$ SNU. This value plus small (~ 5 SNU) contribution of boron neutrinos coincides with total signal observed by GALLEX. Consequently, gallium results can be reproduced if the beryllium neutrino flux as well as all other fluxes of the intermediate energies are strongly suppressed.

Thus both these points indicate on strong suppression of the ^{7}Be- neutrino flux (if there is no neutrino conversion). Statistical analysis gives $f_{Be} < 0.3$ (2σ) (for more detail see [25,26]).

The solar neutrino problem becomes more detailed. It can be formulated as

1. Deficit of boron neutrinos
2. Deficit of the beryllium neutrinos.

The first one is strongly model dependent. In fact, it may have the astrophysical or/and nuclear physics explanation: say, 25% decrease of the S_{17} and 1% decrease of the central temperature (due to the plasma effects) are enough to accommodate the Kamiokande result. The second deficit is essentially (solar) model independent and it is almost impossible to explain it by reasonable variations of the parameters.

3.2 Best fit of the data

If nothing happens with neutrinos and the flux at the Earth consists of the electron neutrinos, then the data fix uniquely values of fluxes which give the *best fit* [29]:
1. Boron neutrino flux should be $\approx 0.35 - 0.40\Phi_B^{SSM}$.
2. Beryllium neutrino flux as well as other fluxes of the intermediate energies (pep, N, O) give negligible contributions to the signals.
3. There is little or no suppression of the pp-flux.

Thus the energy dependence of the suppression factor $P(E)$ can be represented as

$$P(E < 0.5\ MeV) \equiv P_{pp} = 0.9 - 1, \quad P(E = 0.7 - 1.5 MeV) \equiv P_{Be} \sim 0,$$

$$P(E7MeV) \equiv P_B = 0.4 - 1. \tag{3}$$

Large uncertainty of suppression in high energy region is related to the uncertainty in the original boron neutrino flux. Kamiokande admits a mild distortion of the recoil electron spectrum.

Evidently the astrophysics can not reproduce such a picture [18,19, 21 – 26]. Typically one gets more strong suppression of the boron neutrino flux than the beryllium neutrino flux.

(To reproduce central values of signals one should suggest that there is an additional flux which contributes to the Kamiokande signal, $\Delta\Phi_B \approx 0.09\Phi_B^{SSM}$, but does not contribute to the Ar-production rate. This however implies the conversion of the electron neutrinos to muon or tau neutrinos).

The suppression profile can be strongly changed if one admits the existence of muon or/and tau neutrino components in the solar neutrino flux (which already implies some kind of neutrino transformations). In this case, especially if the original boron neutrino flux is higher than in SSM, ν_μ and ν_τ scattering on electrons can give big (main) contribution to the Kamiokande result and the statement that beryllium neutrino flux should be strongly suppressed is not true. Denoting by P_{Be} the suppression factor in the region of Be-neutrinos (at the intermediate energies) we find the suppression factor for the boron neutrinos needed to reproduce the Homestake result:

$$P_B = \frac{Q_{Ar}^{obs} - P_{Be} Q_{Ar,int}^{SSM}}{f_B Q_{Ar,B}^{SSM}}, \tag{4}$$

where Q_{Ar}^{obs} is the measured Ar-production rate, $Q_{Ar,int}^{SSM}$ and $Q_{Ar,B}^{SSM}$ are respectively the contributions of fluxes of intermediate energies (Be, pep, N, O) and the boron neutrino flux to the Ar-production rate according to the reference SSM [31]. The original boron neutrino flux which is needed to reproduce the Kamiokande signal can be found from condition $R_{\nu e} = f_B[P_B + (1 - P_B)/6]$, where $R_{\nu e} \equiv \Phi_B^{obs}/\Phi_B^{SSM}$ is the suppression factor observed by Kamiokande. Substituting P_B from (4) in this condition one gets

$$f_B \approx 6R_{\nu e} - 5\frac{Q_{Ar}^{obs}}{Q_{Ar,B}^{SSM}} + P_{Be}\frac{Q_{Ar,int}^{SSM}}{Q_{Ar,B}^{SSM}} \tag{5}$$

Then predicted values of Q [31] and the central values of experimental signals give according to (5) $f_B \approx 1.4$ for $P_{Be} = 0.5$ and $f_B \approx 2.1$ for $P_{Be} = 1$. That is to avoid any suppression of the Be-neutrino flux one needs two times larger original boron neutrino flux. For 2σ smaller value of $R_{\nu e}$ and $f_B = 1$ the suppression as weak as $P_{Be} \sim 0.7$ becomes allowed. Thus if the neutrinos undergo conversion the Be-neutrino flux may not be suppressed.

In the case of weak suppression of the beryllium line the pp-neutrino flux should be suppressed according to the Gallium result:

$$P_{pp} = \frac{1}{Q_{Ge,pp}^{SSM}}\left[Q_{Ge}^{obs} - P_{Be} Q_{Ge,int}^{SSM} - P_B f_B Q_{Ge,B}^{SSM}\right], \tag{6}$$

where $Q_{Ge,pp}^{SSM}$ and $Q_{Ge,int}^{SSM}$ are predicted contributions to the Ge-production rate from pp - flux and the intermediate energy fluxes. For $P_{Be} \sim 0.7$ one gets from (6) $P_{pp} \sim 0.6$.

Such a situation (weak suppression of the beryllium flux and appreciable suppression of the pp-flux) is realized e.g. in the case of vacuum oscillations [36].

4 Neutrino parameters and neutrino fluxes

Although the solar neutrino problem can be formulated in practically model independent way the implications to the neutrino physics strongly depend on the original fluxes. There are several recent studies of the particle physics solutions of the solar neutrino problem according to the (solar) model independent approach [27,29,30,33,34,35].

4.1 Long length vacuum oscillations

These oscillations can reasonably well reproduce the desired suppression. For $\Delta m^2 7 \cdot 10^{-11}$ eV^2 the pp-neutrino flux is in the region of averaged oscillations, where $P = 1 - 0.5\sin^2 2\theta$, the Beryllium neutrinos are in the fastly oscillating part of the $P(E)$ (so that one expects an appreciable time variations of the Be-neutrino flux due to annual change of distance between the Sun and the Earth). Boron neutrinos are in the first (high energy) minimum of $P(E)$. This allows one to reach the inequality $P_{pp} P_B B_{Be}$ implied by (3). However, there is an obvious relation between maximal suppression of the Be-line and suppression of pp-neutrinos: $P_{Be,min} = 2P_{pp} - 1$, and due to this the best fit configuration (3) is not realized. Good fit can be obtained for moderate suppression of the Be-line and ~ 0.6 suppression of the pp-neutrinos. The fit becomes better for increased values of f_B [33].

With diminishing f_B the needed suppression of B-neutrino flux due to the oscillations becomes weaker. Therefore for fixed values of Δm^2 the allowed regions of parameters shift to smaller $\sin^2 2\theta$ [33,34]. In particular, for $f_B = 0.7$, the region is at $\sin^2 2\theta < 0.7$ thus satisfying the bound from SN87A [37]. For $f_B \sim 0.4$ mixing can be as small as $\sin^2 2\theta < 0.5 - 0.6$. Moreover, for $f_B = 0.5$ the allowed region appears at $\Delta m^2 \sim 5 \cdot 10^{-12}$ eV^2 which corresponds to the Be-neutrino line in the first high energy minimum of P, pp-neutrinos in the first maximum of the P and high energy part of the boron neutrino spectrum out of suppression pit. No appreciable time variations are expected. Such a configuration is quit similar to that from very small mixing MSW solution which further increases the ambiguity of situation. Distortion of pp-neutrino spectrum is the signature of the solution [35].

Depending on neutrino parameters and f_B, f_{Be} ... one can get variety of distortions of the boron neutrino energy spectrum.

Being excluded at $f_B = 1$, the oscillations into sterile neutrino are allowed for $f_B < 0.7$ [35].

4.2 Resonance flavor conversion

It can precisely reproduce the desired energy dependence of the suppression factor (3). In the region of small mixing angles one has

$$P_{pp} \sim 1, \quad P_{Be} \sim 0, \quad P_B \sim exp(-E_{na}/E), \tag{7}$$

where $E_{na} \equiv \Delta m^2 l_n \sin^2 2\theta$. Additional contribution to Kamiokande $\Delta f_B \approx 0.09$, follows from scattering of the converted ν_μ (ν_τ) on electrons due to the neutral currents. With diminishing f_B the suppression due to conversion should be relaxed, and therefore $\sin^2 2\theta$ should decrease according to (7) [29,30]. At $\Delta m^2 = 6 \cdot 10^{-6}$ eV^2 the best fit of the data for flavor mixing corresponds to the pairs of parameters [29]: $(f_B,\ \sin^2 2\theta) = (0.4,\ 1.0 \cdot 10^{-3}),\ (0.75,\ 4.3 \cdot 10^{-3}),\ (1.0,\ 6.2 \cdot 10^{-3}),\ (1.5,\ 9 \cdot 10^{-3}),\ (2.0,\ 10^{-2})$. The decrease of f_{Be} gives an additional shift of the allowed region to smaller values of $\sin^2 2\theta$. A consistent description of the data has been found for [29]

$$f_B \sim 0.4 - 2.0.$$

For unfixed values of the original fluxes, f_B, f_{Be} ..., the allowed region of neutrino parameters is controlled immediately by Gallium data and by the "double ratio". Namely, the mass squared difference

$$\Delta m^2 = (6^{+4}_{-3}) \cdot 10^{-6} eV^2, \tag{8}$$

is restricted by Gallium results which imply that the adiabatic edge of the suppression pit is in between the end point of the pp-neutrino spectrum and the Be-line. This bound does not depend on mixing angle in a wide region of θ. (For sterile neutrinos the bound is approximately the same). For fixed Δm^2 the mixing $\sin^2 2\theta$ is determined by the "double ratio"

$$R_{H/K} \equiv \frac{R_{Ar}}{R_{\nu e}},$$

where $R_{Ar} \equiv Q_{Ar}^{obs}/Q_{Ar}^{SSM}$ is the suppressions of signals in Cl–Ar experiment and Q_{Ar}^{SSM} is the predictions in the reference model [31]. The experimental value, $R_{H/K} = 0.65 \pm 0.11$, admits $\sin^2 2\theta = 1.0 \cdot 10^{-3} - 1.5 \cdot 10^{-2}$. Similar bound exists for the conversion to sterile neutrinos if one restricts the original boron neutrino flux by $\Phi_B \leq 1.5 \Phi_B^{SSM}$.

For very small mixing solution: $f_B \sim 0.5$, $\sin^2 2\theta_{es} \sim 10^{-3}$, all the effects of conversion in the high energy part of the boron neutrino spectrum (E5–6 MeV) become very weak. In particular, the distortion of the energy spectrum disappears, and the ratio charged-to-neutral currents $(CC/NC)^{exp}/(CC/NC)^{th}$ approaches 1. Thus studying just this part of spectrum it will be difficult to identify the solution (e.g., to distinguish the conversion and the astrophysical effects). Recent calculations in SSM with diffusion of heavy elements give larger boron neutrino flux [13,15], so that even with 25% decrease of nuclear cross-section and 1% decrease of central temperature of the Sun one still needs an appreciable conversion effect. This gives a hope that the problem can be resolved by SuperKamiokande/SNO experiments.

With increase of f_B the fit of the data in the large mixing domain becomes better [29]. Here the Kamiokande signal can be explained essentially by NC effect and mixing can be relatively small. Be- neutrino flux is sufficiently suppressed and suppression of the pp-neutrinos is rather weak. For $f_B = 2$ the values $\sin^2 2\theta = 0.2 - 0.3$ become allowed. Corresponding mass squared difference is $\Delta m^2 = 6 \cdot 10^{-6} - 10^{-4}$ eV^2.

4.3 The effect of third neutrino

The analysis of data in terms of two neutrino mixing is quite realistic, since in the most interesting cases (simultaneous solution of the solar and hot dark matter problems, or solar and atmospheric neutrino problems) third neutrino has large mass so that its Δm^2 is beyond the resonance triangle and its mixing to the electron neutrino is rather small. This reduces the three neutrino task to the case of two neutrino mixing. However, there is one interesting example where third neutrino could influence the solutions of the solar neutrino problem. It was considered previously [38,39,40] and reanalyzed recently in [41]: The third neutrino is in the region of the solution of the atmospheric neutrino problem: $m_3 \sim 0.1$ eV and it has an appreciable admixture to the electron neutrino state. Let us represent the ν_e as [10]

$$\nu_e = \cos\phi \nu' + \sin\phi \nu_3 \tag{9}$$

where

$$\nu' = \cos\theta \nu_1 + \sin\theta \nu_2$$

and ϕ is not small. In the case $m_3 \gg m_2$ ($m_2 < 10^{-2}$ eV) the third neutrino ν_3 "decouples" from the system (as far as we deal with the Sun) and its effect is reduced just to the averaged vacuum oscillations. In turn ν' converts resonantly to its orthogonal state. So that the survival probability can be written as

$$P = \cos^4\phi \; P_2 + \sin^4\phi, \tag{10}$$

where P_2 is two neutrino survival probability. Additional regions of the neutrino parameters $\Delta m^2 = (10^{-5} - 10^{-6})$ eV^2 and $\sin^2 2\theta = 3 \cdot 10^{-4} - 3 \cdot 10^{-3}$ are allowed for $\cos^4\phi \sim 0.5 - 0.7$. Now both pp- and Be- neutrinos can be outside the 2ν - suppression pit [39], where $P_2 \approx 1$ and according to (10) the suppression factor for them is $(\cos^4\phi + \sin^4\phi)$. This allows one to get about 1/2 suppression of the gallium production rate, and reconcile the Homestake and the Kamiokande at 2σ level. In [41] it is claimed that even the adiabatic solution (when the high energy part of the boron neutrino spectrum is on adiabatic edge) is not excluded. Indeed, now the distortion of the boron neutrino spectrum is weakened by factor $\cos^4\phi$ in comparison with two neutrino case. But even this is disfavored by the data. Large mixing with electron neutrinos (now $\sin^2 2\theta \sim 0.75$) is practically excluded by reactor experiment [42] and CHOOZ will finally check this possibility.

4.4 On the spin-flip effects

Resonance spin-flip precession can precisely reproduce the suppression profile [43,44,45], i.e. give very good description of averaged signals.

For values of the magnetic moment at the upper bound: $\mu \sim 3 \cdot 10^{-12}\mu_B$, where μ_B is the Bohr magneton, the strength of the magnetic field as big as 10^6 Gauss is needed. Traditional objection is that this field is much stronger than usually expected. There is another objection. In the most of calculations it was suggested that there is no *latitude* dependence of the field which is certainly incorrect: the toroidal field has different polarity in the southern and northern semispheres and there is the equatorial gap of the field. One can think that existing calculations correspond to the average field. However this means that there are regions with even stronger field than that mentioned above. Moreover, since the spin-flip effect is non-linear in the field one should calculate first the probabilities for different latitudes and then perform the averaging over the latitude rather than use the average field. In fact, it was shown [45] that for reasonable latitude distributions of the field the average suppression is too weak, moreover one expects an appreciable seasonal effect.

Time variations of signals are generic features of this solution. Where are these variations?

5 Standard and non-standard

The solution of the solar neutrino problem can be reconciled with solutions of (all ?) other neutrino anomalies like deficit of the atmospheric ν_μ- flux, possible signal of the $\bar{\nu}_\mu - \bar{\nu}_\tau$ oscillations, existence of hot component of dark matter. A number of schemes of the neutrino masses and mixing has been suggested in this context. Thinking in term "standard" and "non-standard" one can arrive at the following (at least the most popular) scenario.

5.1 "Standard" scenario of neutrino masses and mixing

(i). Neutrino masses are generated by the see-saw mechanism with masses of the RH components $M_R = 10^{11} - 10^{13}$ GeV. The mass scale 10^{13} GeV can originate e.g. from Grand Unification scale, M_{GU}, and the Planck scale, M_P,

as $M_R \sim M_{GU}^2/M_P$.
(ii) Second mass, m_2, is in the range

$$m_2 = (2-3) \cdot 10^{-3} eV, \tag{11}$$

so that the resonance flavor conversion $\nu_e \to \nu_\mu$ solves the solar neutrino problem. The desired mixing angle is consistent with

$$\theta_{e\mu} = \sqrt{\frac{m_e}{m_\mu}} - e^{i\phi}\theta_\nu, \tag{12}$$

where θ_ν comes from diagonalization of neutrino mass matrix. This relation is similar to the corresponding relation in quark sector which testify for certain quark-lepton symmetry (unification).
(iii) The third neutrino (for $m^D \sim 100$ GeV and $M \sim 3 \cdot 10^{12}$ GeV) has the mass about 5 eV. It composes the desired hot component of the dark matter.
(iv) The decays of the RH neutrinos with mass 10^{12} GeV can produce the lepton asymmetry of the Universe which can be transformed by sphalerons in to the baryon asymmetry [46].
(v) Large Yukawa coupling of neutrino from the third generation, e.g. $Y_\nu \sim Y_{top}$, gives appreciable renormalization effects in the region of momenta $M_R - M_{GU}$. In particular, the $b-\tau$ mass ratio increases by $(10-15)\%$ in the MSSM. In turn this disfavors the $b-\tau$ mass unification for low values of $\tan\beta$ [47].
(vi) Simplest schemes with quark - lepton symmetry lead to mixing angle for the e and τ generations: $\theta_{e\tau} \sim (0.3-3)V_{td}$ which is close to the bound from the nucleosynthesis of heavy elements (r-processes) in the inner part of the supernova: $\sin^2 2\theta_{e\tau} < 10^{-5}$ ($m_3 2$ eV) [48].
(vii) For $\mu-\tau$ mixing one expects [49] $\theta_{\mu\tau} \sim kV_{cb}\eta$, where $k = 1/3 - 3$ and $\eta \sim 0.6 - 0.7$ is the renormalization factor. If $m_3 3$ eV some part of expected region of mixing angles is already excluded by FNAL 531. Large part of the region can be studied by CHORUS and NOMAD. The rest (especially $m_3 < 2$ eV) could be covered by E 803.
(viii) The depth of $\bar\nu_\mu - \bar\nu_e$ oscillations with $\Delta m^2 \approx m_3^2$ turns out to be $4|U_{3\mu}|^2|U_{3e}|^2 \approx 4|\theta_{e\tau}|^2|\theta_{\mu\tau}|^2$. The existing experimental bounds on $\theta_{e\tau}$ and $\theta_{\mu\tau}$ give the upper bound on this depth: $< 10^{-3}$ [50] which is too small to explain the LSND result.

The standard scenario does not solve the atmospheric neutrino problem. One can consider scenario with three degenerate neutrinos or sacrifice the HDM suggesting that some other particles are responsible for the structure formation in the Universe. In the latter case $m_3 \sim 0.1$ eV and strong $\mu-\tau$ mixing allows to explain via $\nu_\mu - \nu_\tau$ oscillations the atmospheric neutrino deficit. Strong

$\mu - \tau$ mixing, could be related to relatively small mass splitting between m_2 and m_3 which implies the enhancement of the mixing in the neutrino Dirac mass matrix [51]. It could be related to the see-saw enhancement mechanism [52,53] endowed by renormalization group enhancement [53] or with strong mixing in the charge lepton sector [54].

5.2 More neutrino states?

Safe way to accommodate all the anomalies is to introduce one new neutrino state. As follows from LEP bound on the number of neutrino species this state should be sterile (singlet of standard group). Taking into account also strong bound on parameters of oscillations into sterile neutrino from Primordial Nucleosynthesis one can write the following "scenario" [55 - 62].

(i) Sterile neutrino has the mass $m_S \sim (2-3) \cdot 10^{-3}$ eV and mixes with ν_e, so that the resonance conversion $\nu_e - \nu_s$ solves the solar neutrino problem;
(ii) The masses of ν_μ and ν_τ are in the range 2 - 3 eV, they supply the hot component of the DM;
(iii) ν_μ and ν_τ form the pseudo Dirac neutrino with large (maximal) mixing and the oscillations $\nu_\mu - \nu_\tau$ explain the atmospheric neutrino problem;
(iv) ν_e is very light: $m_1 < 2 \cdot 10^{-3}$ eV. The $\bar{\nu}_\mu - \bar{\nu}_e$ mixing can be strong enough to explain the LSND result.
(v) However production of heavy elements in supernova via "r-processes" is problematic for this scenario.

Sterile neutrino can be used to explain the atmospheric neutrino problem in the context of standard scenario, if one ignores the Nucleosynthesis bound.

5.3 Sterile neutrino or light singlet fermion?

Introducing sterile neutrino S one encounters several questions:
What is the origin of this neutrino?
How it mixes with usual neutrinos?
How one can explain its small mass?

1. *Origin.* Natural candidate is if course, the RH neutrino component. However in this case the see-saw mechanism does not operate. S could be the component of the multiplet of extended gauge symmetry - like $SO(10)$- singlet from 27-plet of E_6 [63]. In [64] it was suggested that S is the mirror neutrino from mirror standard model. In all these cases one has three singlet fermions.

Let us consider another possibility [60,61]:
(i). S has an origin beyond usual fermionic structure, and in particular, beyond the see-saw mechanism. So that the see-saw explains the lightness of the active neutrinos in the usual way.
(ii). S has no generation structure and probably is generation blind. There is only one light singlet fermion (although this is not necessarily).
(iii). Supersymmetry may be a natural framework of the appearance of such a fermion. A number of singlet superfields was introduced for different purposes: to generate μ term, to realize PQ-symmetry breaking, to break spontaneously lepton number, etc.. String theory typically supplies some singlets. Fermionic components of these superfield could be identified with desired sterile neutrino.

2. *Mixing*. The standard see-saw structure

$$hL\nu^c H_2 + M\nu^c\nu^c \tag{13}$$

involves three fields: doublet neutrino, RH neutrino component, ν^c, and Higgs doublet, H_2. Correspondingly, there are three possible ways to mix S with active neutrinos:

(i) via direct coupling to the left handed neutrinos:

$$\epsilon LSH_2. \tag{14}$$

The parameter ϵ is of the order

$$\epsilon \sim \frac{m_{3/2}}{M_P}, \tag{15}$$

where M_P is the Planck mass. This means that S could be the field from hidden sector which is mixed with usual neutrinos via gravitational interactions.

(ii). mixing via interactions with RH neutrinos:

$$\lambda\nu^c Sy, \tag{16}$$

where y is an additional singlet field which acquires the VEV $< y \sim m_{3/2}$ as the result of SUSY breaking. The interactions (13) and (16) allow one to explain simultaneously both the mixing of S with neutrinos and the desired mass of S without introduction of new mass scales.

(iii). mixing via Higgsino:

$$\frac{\mu}{< S >} H_1 H_2 S + \epsilon' LH_2. \tag{17}$$

The first term which mixes S with Higgsino can be responsible also for generation of the μ-term . Second term mixes Higgsino with neutrino thus breaking

R-parity. It can be generated spontaneously by the interaction (13), if sneutrino acquires non-zero VEV, or explicitly by, e.g., gravitational interactions.

3. *Mass.* Spontaneous violation of global symmetry, ($U(1)_G$ in the simplest case) like Peccei-Quinn or lepton number symmetries or horizontal symmetry leads to the appearance of the massless boson. In the limit of exact supersymmetry the fermion partner is also massless. However, violation of supersymmetry results in generation of mass of S. In supergravity one has typically $m_S \sim m_{3/2}$. The mass m_S can be further suppressed by special choice of (i) the superpotential or (ii) Kähler potential. In the first case it is quite easy to get $m_S \sim m_{3/2}^2/M_G$ which leads to desired value of m_S for $M_G \sim 10^{16}$ GeV.

Using non-minimal kinetic terms one can suppress m_S at tree level, one loop or even two loops. In the case of strict no-scale supergravity the mass is generated in three loops which is sufficient to explain the smallness of m_S for rather natural values of parameters.

Another possibility [61] is to use R-symmetry to protect the mass, to forbid undesired mixing of S and to ensure that S does not acquires a VEV. (In this case R-parity can be conserved). For more details see [65].

Thus discovery of solar neutrinos conversion into sterile neutrinos (and future experiments will be able to do this) may give the hint to supersymmetry and to really very rich physics beyond the standard model.

6 Conclusion

Before Gallium experiments we had the following criteria: The counting rate much below 70 - 75 SNU is the proof of new neutrino physics (oscillations, conversion etc.). The counting rate much bigger 75 SNU testifies for astrophysical solution. Nature has chosen precisely 75 SNU, and has stayed us in uncertain situation for more than 5 years. Although present rather precise data strongly indicate new physics we have not passed through simple *a priori* criteria.

Now we have new chance. *A priori* criteria are: distortion of the energy spectrum of boron neutrinos, anomalous ratio of charged to neutral current number events, day-night effect, seasonal variations...

Following previous logic of Nature one can imagine that neither distortion nor time variations or CC/NC anomaly will be observed. This is quite possible situation, e.g. if the original boron neutrino flux is small, and the transitions of this flux is not needed. Is this the proof of Astrophysical solution? For boron neutrinos - Yes. Then to explain other experiments one should suggest strong suppression of the Beryllium neutrino flux. This is realized in very small mixing MSW - solution. To proof this one should wait again - wait for experiments which are sensitive to beryllium neutrinos and in this case BOREXINO results can be decisive...

Is Nature going again play with us? Let us see what happens ...

References

1. K. Nakamura, ICRR-report 309-94-4,
2. SNO-detector: Ed. E.W. Beier, D. Sinclair, Report SNO-89-15 (1989).
3. CHORUS Collaboration, N. Armenise et al, CERN-SPSC/90-42 (1992).
4. R. I. Steinberg, Proc. of the 5th Int.Workshop on Neutrino Telescopes, ed. by M. Baldo-Ceolin, (1993) 209.
5. K. Lande, Nucl. Phys. B38 (Proc. Suppl.) (1995) 47.
6. M. Nakahata, Talk given at 4th school on Non-accelerator particle physics, ICTP, Trieste July 18 - 28 (1995).
7. S. Eliott, Talk given at XXX Rencontres de Moriond, Les Arcs, Savoie - France, March 11 - 18 (1995).
8. GALLEX collaboration, P. Anselmann et al., Phys. Lett.,B357 (1995) 232.
9. T. Stanev, Talk given at the Int. Workshop: Solar Neutrino problem:Astrophysics or Oscillations, Feb.28 - March 1 (1994), ed. V. Berezinsky and E. Fiorentini.
10. A. Yu. Smirnov, Report DOE/ER/40561-136 -INT94-13-01.
11. GALLEX Collaboration, P. Anselmann, Phys. Lett. B342 (1995) 440.
12. P. Galeotti, Talk given at the conference on New projects of Gran-Sasso Laboratory, March 1995.
13. C. R. Proffitt, Astrophys. J. 425 (1994) 849;
14. A. Kovetz, G. Shaviv, Astrophys. J. 426 (1994) 787;
15. J. N. Bahcall and M. H. Pinsonneault, IASSNS-AST 95/24.
16. V. Castellani et al., Preprint INFNFE-04-95.
17. V. N. Tsytovich, Talk given at the Int. workshop: Solar Neutrino problem: Astrophysics or Oscillations, Feb.28 - March 1 (1994), ed. V. Berezinsky and E. Fiorentini, p.238.
18. I. Barabanov, (1988) unpublished.
19. J. N. Bahcall and H. A. Bethe, Phys. Rev. Lett., 65 (1990) 2233; Phys. Rev. D47 (1993) 1298.
20. M. Spiro and D. Vignaud, Phys. Lett. B 242 (1990) 279.
21. V. Castellani, S. Degl'Innocenti and G. Fiorentini, Phys. Lett. B 303 (1993) 68: Astron. Astroph. 271 (1993) 601.

22. H. Hata, S. Bludman and P. Langacker, Phys. Rev. D 49 (1994) 3622.
23. V. S. Berezinsky, Comm Nucl. Part. Phys. 21 (1994) 249.
24. W. Kwong and S. P. Rosen, Phys. Rev. Lett., 73 (1994) 369.
25. S. Parke, Phys. Rev. Lett. 74 (1995) 839.
26. S. Degl'Innocenti, G. Fiorentini and M. Lissia, Ferrara preprint INFNFE-10-94.
27. N. Hata and P. Langacker, Phys. Rev. D 50 (1994) 632
28. N. Hata and P. Langacker, Report UPR -0625T, hep-ph/9409372.
29. S. P. Krastev and A. Yu. Smirnov, Phys. Lett. B 338 (1994) 282.
30. V. S. Berezinsky, G. Fiorentini and M. Lissia, Phys. Lett., B 341 (1994) 38.
31. J. N. Bahcall and M. H. Pinsonneault, Rev. Mod. Phys. 64 (1992) 85.
32. A. Dar and G. Shaviv, Report Technion-Ph-94-5 (1994).
33. Z. G. Berezhiani and A. Rossi, Phys. Rev. D51 (1995) 5229. Preprint INFNFE 12-95, July 1995.
34. P. I. Krastev and S. T. Petcov, Nucl. Phys. B449 (1995) 605.
35. P. I. Krastev and S. T. Petcov, report SISSA-9/95/EP.
36. E. Calabresu et al., INFNFE AO-95, INFNCA-TH9512.
37. A. Yu. Smirnov, D. N. Spergel and J. N. Bahcall, Phys. Rev. D49 (1994) 1389.
38. see e.g. X. Shi and D. Schramm, Phys. Lett., B283 (1992) 305; X. Shi, D.Schramm, J. Bahcall, Phys. Rev. Lett. 69 (1992) 717.
39. A. Yu. Smirnov, Proc. of the Int Symposium on Neutrino Astrophysics, Takayama/Kamioka 19 - 22 October 1992, ed. by Y. Suzuki and K. Nakamura, p.105.
40. A. S. Joshipura and P. I. Krastev, Phys. Rev. D50 (1994) 3484.
41. E. Ma and J. Pantaleone, Preprint UCRHEP-T140 (March 1995); E. Ma, These Proceedings.
42. S. M. Bilenky, C. Giunti and C. W.Kim, Preprint hep-ph/9505301
43. E. Kh. Akhmedov A. Lanza and S. Petcov, Phys. Lett. B348 (1995) 124.
44. C.-S. Lim and H. Nunokawa, Preprint KEK-TH-416, KOBE-TH-94-02.
45. P.I. Krastev and A. Yu. Smirnov, Z. Phys. C49 (1991) 675.
46. M. Fukugita and T. Yanagida, Phys. Lett., B174 (1986) 45.
47. F. Vissani and A. Yu. Smirnov, Phys. Lett., B 341 (1994) 173; A. Brignole, H. Murayama and R. Rattazzi, Phys. Lett. B335 (1994) 345.
48. Y. -Z. Qian et al., Phys. Rev. Lett., 71 (1993) 1965.
49. S. Dimopoulos, L. J. Hall and S Raby, Phys. Rev. D47 (1993) R3697. H.-C. Cheng, M. S. Gill and L. J. Hall, LBL-33936, UCB-PTH-93/11.
50. K. S. Babu, J.C. Pati and F. Wilczek, Preprint IASSNS-HEP 95/37.

51. M. Fukugita, M. Tanimoto and T. Yanagida, Prog. Theor. Phys. 89 (1993) 263.
52. A. Yu. Smirnov, Phys. Rev D48 (1993) 3264.
53. M. Tanimoto, Phys. Lett., 345B (1995) 477; Preprint-EHU-95-8.
54. G.K. Leontaris, S. Lola and G.G. Ross, CERN-th-95-133, HD-HEP-95-8,IOA-95 323.
55. D. O. Caldwell and R. N. Mohapatra, Phys. Rev. D48 (1993) 3259.
56. J. Peltoniemi and J. W. F. Valle, Nucl. Phys. B406 (1993) 409.
57. J. Peltoniemi, D. Tommasini and J. W. F. Valle, Phys. Lett. B298 (1993) 383.
58. J. T. Peltoniemi, Mod. Phys. Lett. A8 (1993) 3593.
59. J. R. Primack, J. Holtzman, A. Klypin, and D. O. Caldwell, Phys. Rev. Lett. 74 (1995) 2160; and references therein.
60. E. J. Chun, A. S. Joshipura and A. Yu. Smirnov, IC/95/76, PRL-TH/95-7, hep-ph/9505275.
61. E. J. Chun, A. S. Joshipura and A. Yu. Smirnov, IC/95/164 , PRL-TH/95-11, hep-ph/9507371.
62. E. Ma and P. Roy, preprint UCRHEP-T145 (hep-ph/9504342).
63. E. Ma, Preprint UCRHEP-T149
64. R. Foot, R. Volkas, Preprint UM-P-95/49; Z.G. Berezhiani and R. N. Mohapatra INFN-FE-06-95.
65. E. J. Chun, These proceedings.

The Atmospheric Neutrino Anomaly and Long Baseline Neutrino Oscillation Experiments

Barry C. Barish
California Institute of Technology
Pasadena, California 91125 USA

September 20, 1995

Abstract

At this workshop I briefly review the evidence for an atmospheric neutrino anomaly and give a critical analysis of this result. In addition, I discuss the potential for resolving this anomaly with long baseline neutrino oscillation experiments using neutrino beams from a high energy accelerators directed toward a major underground detector at a distant location. These possibilities are especially topical due to the continuing evidence for the anomaly, the considerations by major laboratories for such efforts, and the possibility that the effect is due to neutrino oscillation phenomena.

1 Introduction

Neutrino experiments have been conducted at reactors for almost 40 years, at high energy accelerators for about 30 years, and in stand-alone underground facilities for almost as long. Now, there are serious proposals for hybrid experiments that will detect neutrinos from high energy accelerator produced neutrino beams hundreds of kilometers away. The motivation for this is to perform a definitive experiment for vacuum neutrino oscillations in the region of parameter space indicated by the atmospheric neutrino anomaly.

Large underground detectors have the capability to study neutrinos resulting from the collisions of high energy primary cosmic rays near the top of the atmosphere. Studies over the past several years in a number of detectors have produced strong indications of an anomaly at low energies in the ratio of detected muon neutrino to electron neutrino events. This has been interpreted as due to vacuum neutrino oscillations, either $\nu_\mu \rightarrow \nu_e, \nu_\mu \rightarrow \nu_\tau$ or possibly $\nu_\mu \rightarrow \nu_{sterile}$. A practical way to follow up these observations is by distant detection of reactor or accelerator neutrinos. I concentrate here on the case of high energy accelerator neutrinos. Improved accelerator neutrino beams and powerful next generation underground detectors make this a very attractive possibility.

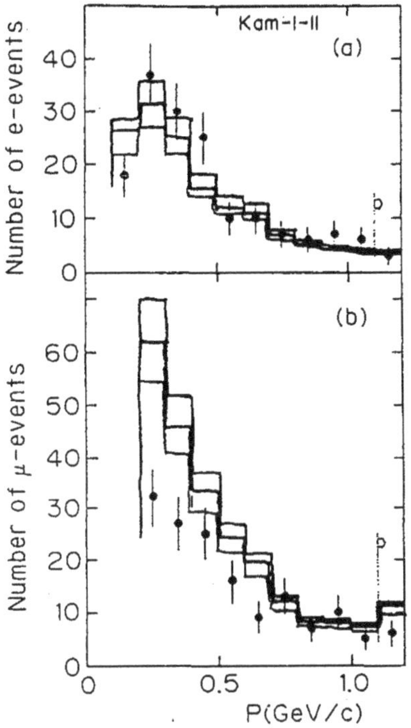

Figure 1: Kamiokamde results for contained events indicative of a deficit of ν_μ events.

In this presentation, I review the key information obtained from atmospheric neutrinos and then discuss the possibility for a long baseline experiment above τ threshold; directing a neutrino beam from the main injector of Fermilab to a new detector at Soudan (eg. MINOS).

2 The Atmospheric Neutrino Anomaly

Atmospheric neutrinos originate from the interactions and subsequent hadron showers produced by primary cosmic ray interactions near the top of the atmosphere. Most of the neutrinos come from the produced pions that decay and yield $\pi \rightarrow \mu + \nu_\mu$ and the subsequent muon decay $\mu \rightarrow e + \nu_e + \nu_\mu$. Although there is some model dependence and extrapolations to regions where no data exist in phenomenologically predicting these ν_μ and ν_e fluxes, the ratio of these fluxes is quite insensitive to these uncertainties.

Experiments on atmospheric neutrinos cover a large energy range and a large variation in path lengths. This makes these studies particularly attractive for investigating possible neutrino oscillation effects. The neutrino interactions

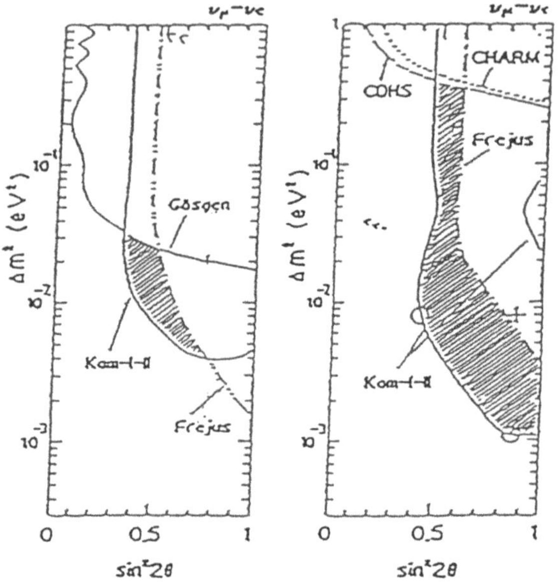

Figure 2: The interpretation of deficit of ν_μ interpreted as due to neutrino oscillations yields the allowed region which is shaded

are observed in three topologies of quite different neutrino energies - contained events (ν_μ and ν_e with $\overline{E}_\nu < 0.6$ GeV), stopping events (mostly ν_μ's with $\overline{E}_\nu \sim$ 5 GeV), and thru-going μ's ($\overline{E}_\nu \sim 100$ GeV).

Figure 1 shows the distributions of contained events from Kamiokande for the ν_μ and ν_e events. The results are usually presented in terms of a ratio of ν_μ/ν_e, in order to eliminate uncertainties concerning the 'beam' of neutrinos resulting from primary cosmic ray showers at the top of the atmoshere. This result can, in principle, be due to either an excess of ν_e's or a deficit of ν_μ's. However, estimates of the absolute neutrino flux indicate that the result is most likely due to a defict of ν_μ's. If so, this deficit could result from neutrino oscillations of the type $\nu_\mu \rightarrow \nu_e$ or $\nu_\mu \rightarrow \nu_\tau$ in the region of ΔM^2 vs $sin^2 2\theta$ shown in Figure 2. A negative search from the Frejus detector limits the allowed region as shown and confirming results from IMB and Soudan support the Kamiokande result, but are somewhat less restrictive in parameter space. A summary [1] of the main results is shown in Table 1.

Potentially, more information is available by studying the neutrinos at higher energy. Results from IMB on stopping muons are consistent with no oscillations

ATMOSPHERIC ν_μ/ν_e RATIOS

Experiment	ktonyr	$(\mu/e)_{obs}/(\mu/e)_{calc}$
Kamioka	6.1	0.60±0.07 ± 0.05
IMB3	7.7	0.54±0.05 ± 0.07
Frejus	1.56	0.87±0.21 ± 0.00
Nusex	0.40	0.99±0.40 ± 0.00
Soudan 2	1.00	0.69±0.19 ± 0.09

Table 1: The ratio ν_μ/ν_e for contained Neutrino Event Rates. [1]

and limit the low mass region, however, results from Kamiokande in a somewhat different analysis using partially contained events are consistent with oscillations, and in fact have a zenith angular distribution consistent with predictions from neutrino oscillations. Finally, at yet higher energies, data on through going muons agree within 10-15% with predictions. Unfortunately, at high energies no ratio is possible because of the short range of electrons and the results are limited by the systematic errors of the predictions of the absolute neutrino flux which is compared with the experimental observations for ν_μ's. The recent results from MACRO on thru-going events are consistent both with no effect, but favor an effect with the indicated parameters from Kamiokande contained events. Other experiments have similar results and only Baksan appears mildly inconsistent with the oscillation parameters.

3 Interpreting The Atmospheric Anomaly

The results for contained neutrino events indicate an anomalous ratio of observed $N(\mu)/N(e)$ while the events at higher energy, although consistent with that result, are not, in themselves, compelling. So, in reality, we are primarily confronted with the problem of whether the effect for contained events is real or instrumental. It is worth noting that despite the number of different experiments contributing results, the main statiscially significant results comes from two Cerenkov detectors (Kamiokande and IMB) and therefore the result is based primarily on a single technique. The Soudan II detector, which uses a fine grain sampling calorimeter, also has positive evidence, though statistically limited due to the small size of the detector. However, the future prospects for Soudan are statistically limited and the other result from a calorimetric detetor

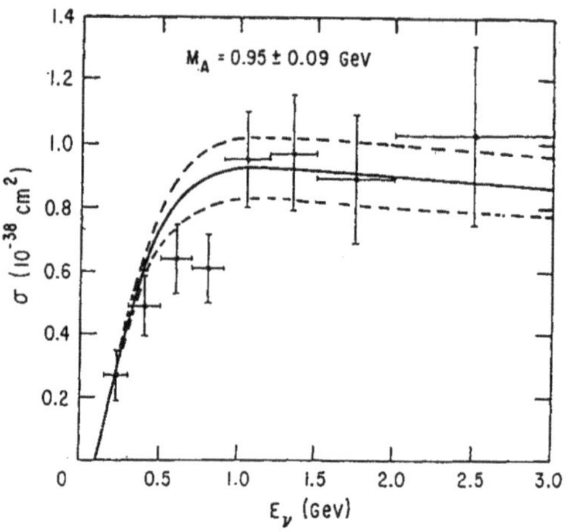

Figure 3: Total cross section $\nu + n \rightarrow \mu + p$ as a function of neutrino energy and a comparison with threshold and form factor behavior. The highest energy data point extends from 2.0 to 6.0 GeV.

(Frejus) is negative.

It appears that until there are long baseline experiments, which will have more constraints due to control over the neutrino beam, the experimental progress depends on experimental checks of the Cerenkov technique, better statistics when Super Kamiokande is operational, and extension of the search to higher energy in MACRO and other large detetors.

The main instrumental questions for the Kamiokande result are whether there are systematic problems in particle identification of electrons and muons, whether there are any systematic difference in vertex reconstruction ability between muons and electrons, and finally a question of possible neutron background has been raised. The Kamiokande group has undertaken experimental tests at KEK in a test beam using a prototype Cerenkov detector. Important results from those tests have been presented recently by the Kamiokande group. In particular, they used beam tests at KEK to establish their particle ID capabiltiies for low energy electrons and muons consistent with expectations of their Monte Carlo. In addition, they have established the ability to reconstruct the vertex. These results significantly strengthen the conclusion that the result is sound experimentally. The problem of neutron background is a subtle one and the Kamiokande experimentors must resolve these questions satisfactorily and present their conclusions.

Assuming the anamoly is real, the next question is whether there are phenomenological effects for low energy neutrinos that are the actual cause of the anomaly. Contained events are mostly for E < 1 GeV, and are dominated by the quasi-elastic processes $\nu_l + n \rightarrow l^- + p$ and $\overline{\nu}_l + p \rightarrow l^+ + n$. These reactions exhibit threshold effects and pseudoscalar form factor effects that must be

Events	Data	Monte Carlo predictions			
		BGS	HK	LK	BN
μ-like	151	261	234	205	179
e-like	159	165	146	128	110
Total	310	426	380	333	289

Table 2: Contained event rates for Kamioka: 4.92K ton year.

taken into account. Figure 3 shows the low energy data and the form factors [2], which appear consistent with each other. We conclude that the anomaly does not appear explainable from variations in the phenomenologs.

Table 2 shows a comparison of several Monte Carlo predictions [3] with the Kamiokande results. Note, that with the exception of Bugaev and Naumov (BN), all indicate that the anomaly is due to a deficit of μ events. If indeed the anomaly were due to an excess of electrons, Figure 4 shows that the result is interpretable as being due to proton decay with a lifetime of $\tau = 4\ 10^{31}$ years. Although this disagrees with proton decay limits, it is crucial to resolve the differences in the predictions. In fact, the primary difference between Barr et al (BGS) and Bugaer and Naumov (BN) appears to be due to differences in the extrapolations of the pion yields from known data to the small x region. To avoid this problem, Perkins [4] has used an alternate method to determine the ν_μ flux by using the results on the high altitude cosmic ray μ flux. He obtains the result shown in Figure 5, which is in good agreement with Barr et al and independently indicate a deficit of ν_μ.

It thus appears that the anomaly is neither instrumental in origin (with a remaining caveat concerning neutrons) nor due to phenomenological uncertainities in either low energy cross sections or form factor effects. Therefore, the basic evidence for the anomaly seems sound and provide good motivation for developing plans and preparations for definitive follow up experiments based on distant detection of high energy accelerator neutrino beams.

4 Long Baseline Neutrino Experiments

To summarize the results discussed above, the data on contained ν events strongly indicate an anomalous ration $N\mu/N_e$, where the result is approximately 0.6 of expectations. Experimental checks at KEK which address the instrumental concerns substantiate the result with a possible remaining question regarding meutrons. Also, the magnitude of the effect cannot be explained by uncertain-

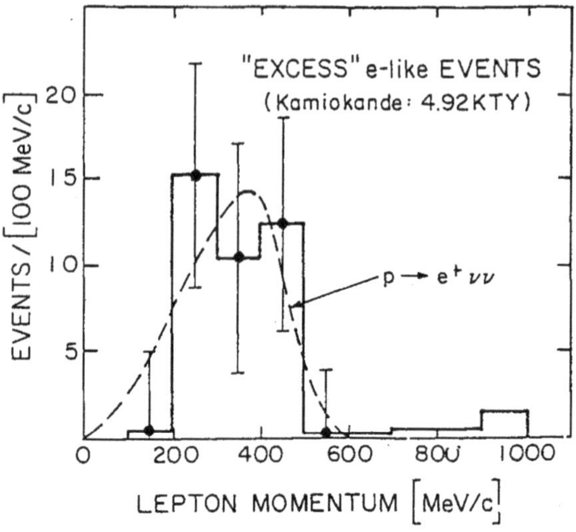

Figure 4: An interpretation of the effect as an excess of ν_e.

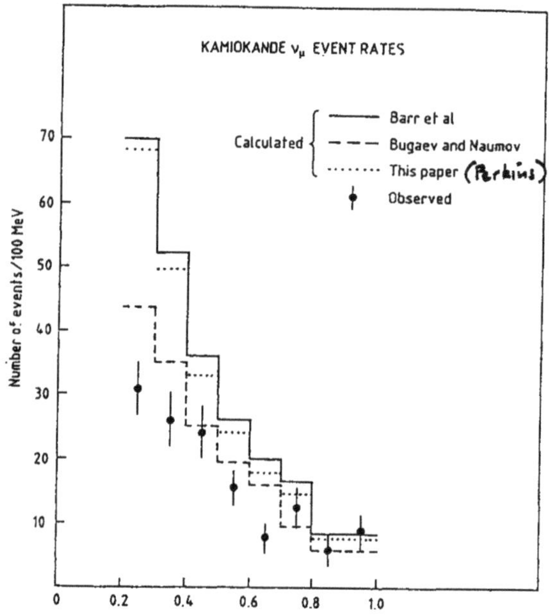

Figure 5: The dashed curve represents an independent analysis of the predocted muon neutrino flux by Perkins which support the interpretation of a ν_μ deficit.

ties in either the neutrino cross sections or form factors at low energies. Finally, phenomenological predictions of the fluxes are not in total agreement as to whether the reported effect represents an excess of electrons or a deficit of nuons. However, an alternate method by Perkins indicates that it is due to a deficit of ν_μ's.

We conclude that the anomaly appears real and that it is likely due to a deficit in the number of observed muon neutrinos. If so, we are led to the possible explanation that this is due to neutrinos oscillations. This intriging result provides strong motivation for the development of long baseline neutrino experiments where definitive experiments are possible.

Many ideas exist for long baseline neutrino experiments, but the most developed have been a proposal at BNL below τ threshold, and at Fermilab above τ threshold. Since they illustrate the class of experiments below τ threshold (BNL, KEK) and above τ threshold (CERN, Fermilab), I will describe these approaches.

4.1 The Brookhaven Proposal (low energy option)

The general approach of the experiments below τ threshold is to do a very well controlled disappearance experiment where a small reduction in the ν_μ flux with distance can be established. The proposed BNL experiment would use a new neutrino beam more intense than previous beams due to the improved RHIC booster injection (x 4) and a longer decay length (x 1.7). The estimated yield of neutrinos is given by $F \propto k/r^2$ with $k = 2.710^{11}$ $\nu_\mu km^2/\text{sec}$. The energy spectrum of this beam is below threshold for τ production and therefore dictate developing an experimental approach based on a disappearance experiment. The strategy is to compare a distant target with a near target testing for ν_μ disappearance. In order to check for consistency, three locations are used and the event rate per 10^{20} protons on target is $\mathrm{N_1} = 638\mathrm{K}$ (closest detector), and $\mathrm{N_3} = 18\mathrm{K}$ (for detector) for no neutrino oscillations (for an early version of this proposal).

In such an experiment, there are several possible sources of background with various checks. The beam contains a $\sim 1\%$ contamination from ν_e and these actually provide an important check described below. The $\overline{\nu}_\mu$ contamination is less than 4% and has no effect on the result for an oscillation experiment. Finally, there is a substantial background from cosmic rays, since this detector is on the surface. The needed rejection is achieved by a combination of using the machine bunch structure timing requirement and detailed pattern recognition.

The strategy is to take the ratios $N_3/N_1, N_{20}/N_1$, etc. and by detailed comparisons determine whether the observed rate at 20 km shows a deficit (disappearance), of $\nu_\mu \to \nu_\tau$. A check would be that in this case, the ν_e events should show no effect. If however, the effect is actually from $\nu_\mu \to \nu_e$, the observed number of ν_e events at 20 km will-increase tenfold!

The proposal at BNL received scientific approval by the Brookhaven Program Advisory Committee and was considered by a special DoE HEPAP subpanel, along with the Fermilab MINOS proposal, for funding. The panel has recently concluded that despite the attractiveness of the BNL proposal, they recommend the Fermilab experiment.

4.2 Fermilab and Gran Sasso (high energy options)

Fermilab neutrino beams are above τ threshold so that the approach of the experiments is to observe τ appearance from $\nu_\mu \to \nu_\tau$ oscillations. This is also true of the CERN - Gran Sasso option, where $\nu_\mu \to \nu_\tau$ can be observed in the electron mode in ICARUS. A new neutrino beam from the 120 GeV main injector will be developed and pointed at Soudan. The design proton intensity of 10^{14} protons every 1.5 sec will produce a large neutrino flux so that both wideband and narrow band beams can be used and are being designed into the facility.

A collaboration was formed from the proponents of three independent expressions of interest. A beam and a 'reference' detector design for MINOS was proposed in January 1995 in a proposal to Fermilab. The experiment will consist of both a near detector of the Fermilab site and a new far detector on the Soudan mine 730 km away.

To illustrate the potential physics capabilities and demonstrate feasibility a 'reference' detector has been proposed. Detector optimization and detailed design is expected over the next two years, followed by construction and initial operation in about 2000. The magnetized plate detector of 10Kton is based on 4 cm magnetized iron absorber plates. A sketch of this layout is shown in Figure 6 and the sensitivity of such a detector is shown in Figure 7. It is assumed that the iron is magnetized toroidally with a field of 20 Kgauss inside of the plates. For studies of the ability to separate different types of events a GEANT similuation of the detector has been used.

The detector makes use of topologies of contained events to identify different types of neutrino interactions on a statistical basis. Monte Carlo events are classified as 'muon-like' (containing a long muon track), 'electron-like' (a short, showering event primarily from ν_e interactions), 'hadron-like' (a long showering event with no clear muon track; primarily NC interactions) or 'other'.

Most interesting and convincing would be an experiment that observed an unambiguous signatures of ν_τ appearance. Preliminary studies suggest this should be possible, although very difficult. One possibility is to use quasi-elastic events and make kinematic cuts based on looking for anomalous transverse momentum in the resulting leptons. However, the best signatures are based on the use of a narrow band beam and the event rates are reasonable. Given knowledge of the neutrino energy, it is possible to make kiinematic cuts for events with τ decaying to μ which allows a statistical identification of the ν_τ events. There is a trade-off in use of these two approaches. The narrow-band beam gives a smaller flux but the knowledge of the beam energy gives a valuable kinematic constraint allowing for less stringent analysis cuts. The approach of using quasi-elastic events allows use of the broad-band beam but requires stringent analysis cuts to select the appropriate sample of events. Further study is required to determine the best strategy to pursue.

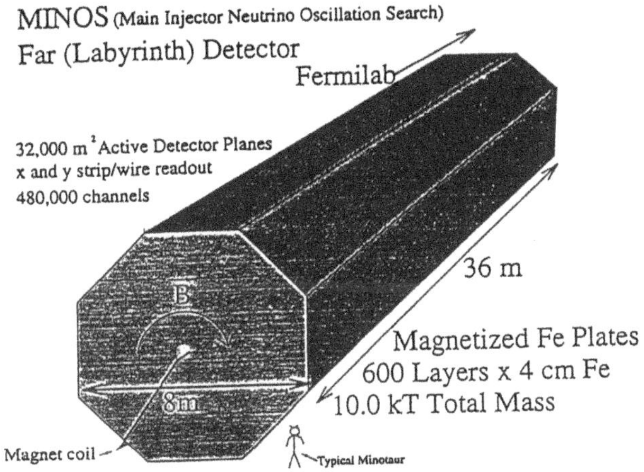

Figure 6: Sketch of one possible detector configuration.

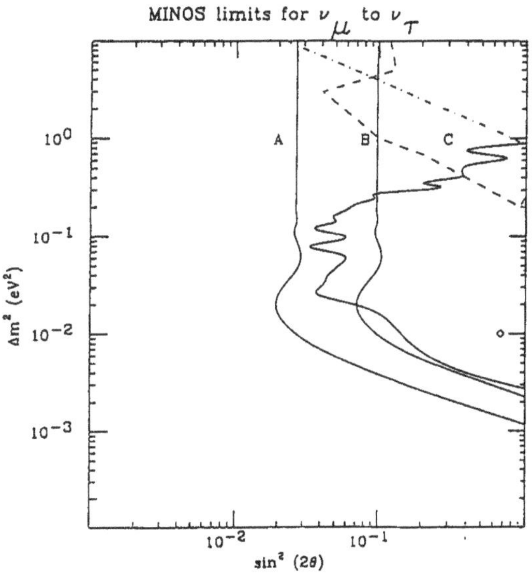

Figure 7: MINOS two year 90% CL limit curves for $\nu_\mu \rightarrow \nu_\tau$ oscillations. Curve "A" is for the $\frac{cc}{total}$ test. Curve "B" is for the near/far rate comparison and is cominated by 4% systematic error. Curve "C" comes from the total energy measurement. The dashed curve shows measured limits from the CDHS experiment, and the dot-dashed curve shows FNAL E-531 limits. The diamond is the Kamiokande bestfit point.

5 Conclusions

Evidence for an atmospheric neutrino anomaly may be indicative of neutrino oscillations in the region $\Delta M^2 \sim .01$ ev^2 and having large mixing angles. Assuming this data and interpretation are correct, it appears that a conclusive experiment can be done using a long baseline neutrino experiment. This may be done either from a careful dissappearance experiment below τ-threshold (BNL or KEK) or an appearance experiment above τ-threshold (Fermilab or CERN). Recently, the Fermilab MINOS experiment was chosen over the BNL experiment for the U.S program.

References

[1] Hirata et al Phys. Lett. 146 (1994); Casper et al Rev. Lett. 66 2561 (1991); Perdereau et al Proc. of Tests of Fundamental Conservation Laws in Physics (1991); Aglietta et al Europhys. Lett 8 611 (1989); Robak Thesis "Measurement of the Atmospheric Neutrino Flavor Ratio with the Soudan 2 Detector"; Univ. of Minn. (1992)

[2] S. Barish et al PR D16 3103 (1977)

[3] Barr, Gaisser, Stanev (GBS) PR D39 3532 (1989); Honda et al (HK) PL B248 193 (1990); Lee and Koh (LK) NC 105B 883 (1990); Bugaev an Naumov (BN) PL B232 391 (1989).

[4] Perkins, The Atmospheric Neutrino Problem: A Critique (unpublished).

PROBLEMS OF DARK MATTER

A.D. DOLGOV

Instituto de Fisica Corpuscular - C.S.I.C.
Departament de Fisica Teorica, Universitat de Valencia
46100 Burjassot, Valencia, SPAIN;
ITEP, 117259, Moscow, Russia

The data indicating existence of different forms of dark matter in the universe as well as the role of this matter in structure formation are briefly reviewed. It is argued that vacuum energy gives a dominant contribution into the total energy density of the universe. The model of structure formation with unstable tau-neutrino with MeV-mass and KeV-majoron is described.

It is commonly agreed now that the matter (or energy) in the universe is predominantly (90% or even more) invisible and that most probably this invisible matter is not the normal baryonic staff that we are familiar with. The agreement however does not go beyond this point and as for the nature of this dark matter the problem remains open and will probably be open for quite some time. There are some hints from cosmology regarding the properties of dark matter. They are given by direct astronomical observations and by the theory of large scale structure formation in the universe. From the other side the elementary particle theory gives plenty predictions about possible new stable particles which might be constituents of dark matter. Unfortunately these predictions do not directly follow from the well established minimal $SU(3) \times SU(2) \times U(1)$ - theory but from one or other higher energy extensions of this theory which have not yet been verified by experiment. So at the moment we are unable to chose different possibilities and have to rely only on aesthetic properties of the model which may be grossly misleading.

The fraction of a particular form of matter in the universe is given in terms of the dimensionless parameter

$$\Omega_j = \rho_j / \rho_c \tag{1}$$

where ρ_j is the mass (energy) density of this form of matter and ρ_c is the critical or closure density,

$$\rho_c = \frac{3H^2 m_{Pl}^2}{8\pi} = 1.88 h_{100}^2 \times 10^{-29}\, g/cm^3 = 10.5 h_{100}^2\, KeV/cm^3 \tag{2}$$

Here $m_{Pl} = 1.22 \times 10^{19}$GeV is the Planck mass (the Newtonian gravitational constant is $G_N = m_{Pl}^{-2}$). The Hubble constant H is parametrized in terms of dimensionless quantity h:

$$H = 100h\, km/sec/Mpc \tag{3}$$

It is usually assumed that $0.4 < h < 1$, but the recent data [1] favor the values in the upper part of this interval, $h = 0.75 - 0.85$. The larger values of h were rejected by many astronomers because of the universe age problem (see below) but now more and more evidence is accumulated in favor of that and the attitude seems to be gradually changing.

If the Hubble constant and Ω_j are known, the universe age can be expressed through them as

$$t_u = \frac{1}{H} \int_0^1 \frac{dx}{\sqrt{1 - \Omega_{tot} + \Omega_m x^{-1} + \Omega_{vac} x^2}} \tag{4}$$

where Ω_m and Ω_{vac} correspond respectively to the energy density of nonrelativistic matter and to the vacuum energy density (or, what is the same, to the cosmological constant); $\Omega_{tot} = \Omega_m + \Omega_{vac}$, and $1/H = 9.8h_{100}^{-1} \times 10^9$yr.

On the other hand the universe age can be found from nuclear chronology which uses measurements of the ratios of the long-lived isotopes, $^{187}Re/^{232}Th$ or $^{238}U/^{235}U$, and from the estimated ages of old stellar clusters. Both methods give close results in the range (12 - 20) Gyr. The recent analysis gives $t_u \approx 18$ Gyr, though somewhat smaller values are possibly not excluded (for the review and the list of references see e.g. paper [2]).

If the universe is dominated by nonrelativistic matter, as usually assumed, the age is approximately equal to $t_u \approx 9.8h^{-1}/(1+\sqrt{\Omega_m}/2)$ Gyr. Even with $\Omega = 0$ the large age, $t_U = 18\,Gyr$, and $h > 0.75$ are inconsistent. The observed tendency to large values of H and t_u presents a very strong argument in favor of nonvanishing vacuum energy. For $\Omega_{tot} = \Omega_m + \Omega_{vac} = 1$ eq.(4) gives

$$t_u = \frac{2}{3H\sqrt{\Omega_{vac}}} \ln \frac{1+\sqrt{\Omega_{vac}}}{\sqrt{1-\Omega_{vac}}} \tag{5}$$

With $h = 0.75$ and $\Omega_{vac} = 0.8$ we get $t_u = 14$ Gyr. If this is the case most of invisible energy in the universe is just the energy of empty space, of vacuum. Unfortunately our understanding of vacuum energy is very poor. Any reasonable estimate of it gives the result which is some 50-100 orders of magnitude higher than the observational limit $\rho_{vac} < 10^{-47}\,GeV^4$ (for the review see refs. [3,4]). For example there are contributions from quark and gluon condensates which are well established in QCD and which are about $10^{-4}\,GeV^4$. So there must exist a contribution into vacuum energy from something not related to quarks and gluons but with exactly the same magnitude and the opposite sign. It is hard to imagine an accidental cancellation with such an accuracy but no dynamical mechanism has yet been found. To my mind the best possibility for solving the cosmological constant problem is the adjustment mechanism [5,6] (for a recent paper and some references see [7] and reviews [3,4]), which ensures a cancellation of vacuum energy by an action of a new field coupled to gravity. This cancellation is generically not complete and a noncompensated amount of ρ_{vac} is always of the order of $\rho_c(t)$. In such models vacuum energy and the energy of the new field are essential at any stage of the universe evolution (in contrast to the models with normal time independent vacuum energy) and this might have an impact on primordial nucleosynthesis, structure formation, etc.

Astronomical data on the amount of different forms of matter in the universe can be summarized as following:

1. The directly seen luminous matter contributes very little to the total mass of the universe: $\Omega_{lum} = (3.5 \pm 1) \times 10^{-3} h^{-1}$.

2. Total amount of baryons found from primordial nucleosynthesis is $\Omega_{bar}^{(NS)} = 4 \times 10^{-3} \eta_{10} h^{-2}$ where $\eta_{10} = 10^{10}(N_B/N_\gamma)$ with $\eta_{10} = 1 - 7$. The recent analysis of the problem can be found e.g. in refs. [8,9,10]. At the moment there is no consensus whether η is relatively high, near the upper bound, or near the lower end of the interval. In the first case we should expect plenty of invisible baryons in the universe while in the second all baryons may be visible. To resolve this ambiguity it is very important in particular to find the amount of primordial deuterium. At the moment there are two conflicting sets of observations, one gives by number $D/H \approx 2 \times 10^{-4}$ [11,12] and requests a small η, $\eta_{10} \approx 1$, while the other gives $D/H = (1-2) \times 10^{-5}$ [13] and $\eta_{10} \approx 3$.

3. Flat rotational curves observed in gravitationally bound systems like gas around galaxies or galactic satellites give, depending on scale, $\Omega_{rot} = 0.1 - 0.3$. Rotational velocities are measured for hundreds of galaxies up to distances about 30 Kpc through HI hydrogen gas, up to 100 Kpc for hot X-ray gas around galaxies and up to 200 Kpc for galactic satellites [14,15,16]. These flat rotational curves mean

that the gravitating mass is not confined inside the luminous region with the size about 10 Kpc (for large galaxies) but linearly rises with the distance, $M(r) \sim r$. No cutoff for this behavior has yet been observed.

4. The recent analysis [17] of peculiar velocities of about 3000 galaxies gives $\Omega_{l.s.f} = [b(0.74 \pm 0.13)]^{5/3}$ where b is the biasing parameter, $b = (\delta\rho/\rho)_{vis}/(\delta\rho/\rho)_{tot}$. The value of the latter is not known but we believe that it is close to one and most probably $b \geq 1$.

5. The lower limits on the universe age, $t_U > 12\,Gyr$, and on the Hubble constant, $h > 0.7$, result in the bound $\Omega_{matter} < 0.2$, if cosmological constant is zero.

6. Recent observations [18,19] of hot X-ray gas in rich galactic clusters showed a surprisingly large fraction of baryons with respect to the total mass of the clusters. If the mass contained in the clusters represents a fair sample for the total mass in the universe then these data together with the nucleosynthesis constraint on Ω_B put the strong upper bound on the energy density in the universe: $\Omega_{clustered} \leq 0.15h_{100}^{-1/2}/(1+0.55h_{100}^{3/2})$. This bound is evidently valid for the clustered matter and is not applicable for the uniformly distributed one. These data are probably a good indication on nonzero vacuum energy $\Omega_{vac} \approx 0.8$. It is uniformly distributed and for the typical cluster size about 1 Mpc does not contribute much inside this radius.

7. Inflationary universe model predicts $\Omega = 1 \pm 10^{-4}$. Recently there appeared attempts [20] to reconcile inflation with $\Omega \neq 1$, which were stimulated by the new astronomical data. These attempts however do not look as natural as the original inflationary prediction since they request a tuning of inflationary and postinflationary stages in such a way that $\exp(2H_I\tau) = (T_R/T_0)^2(\Omega_0^{-1} - 1)^{-1}$, where T_R is the reheating temperature after inflation, $T_0 = 2.7\,K$ is the present-day temperature of the cosmic microwave radiation, and τ is the duration of inflation. This condition looks very strange because physically these epochs are not related. To my mind $\Omega = 1$ remains a very strong prediction of inflationary cosmology.

At the present stage one cannot definitely claim what is the correct value of Ω. I think that a nonzero value of ρ_{vac} should be seriously considered. To my mind the best choice is the following: $\Omega_{bar} \approx \Omega_{vis} \approx (0.3-0.5)\%$; $\Omega_{DM} = 0.1 - 0.2$, and $\Omega_{vac} = 0.8 - 0.9$ with $\Omega_{tot} = 1$.

Even if one believes that vacuum energy contributes (80-90)% to the total energy density in the universe, still there should be some other unknown form of invisible matter which provides flat rotational curves, large scale flows and contributes the remaining (10-20)% into Ω_{tot}. Nucleosynthesis constraint does not permit this matter to be baryonic and we do not know what it is. The only hints about its nature come from elementary particle theory and from the theory of large scale structure formation. Minimal supersymmetric extensions of the standard model predict existence of a massive stable particle with $m = O(100\,GeV)$. The calculated cosmic abundance of these particles naturally give Ω_{LSP} close to unity (here LSP stands for lightest supersymmetric particle). This coincidence is quite impressive because apriori one could expect Ω_{LSP} different from 1 by many orders of magnitude. This makes LSP a very good candidate for dark matter particle (for a recent review see [21]).

Still one may express some doubts regarding this possibility. Stability of LSP is ensured by R-parity which is conserved in simple supersymmetric models. However we know from the history of particle physics during last half of century that the only conservation laws survived which were protected by a well justified theoretical principle, like e.g. CPT-theorem or electric charge conservation (protected by gauge invariance in QCD). As for R-parity no such principle naturally appeared and thus one should expect that it is nonconserved. In that case LSP's are unstable and if their life-time is cosmologically small, we have to look for other possible candidates for dark matter particles. These could be axions, massive neutrinos, or maybe some other more exotic forms of matter. A classical field which probably compensates nonzero vacuum energy is a very interesting possibility. Models with broken R-parity which may kill stable LSP's, simultaneously give birth to another possible candidate for dark matter. R-parity breaking is realized through spontaneous breaking of leptonic charge conservation [22,23] associated with global symmetry group $U(1)_L$. Spontaneous breaking of a global symmetry group gives rise to appearance of a (pseudo)goldstone boson, majoron. If there is also an explicit symmetry breaking, the majorons would become massive and, if cosmologically long-lived, could be constituents of dark matter [24,25].

Some of dark matter candidates, like e.g. massive stable neutrinos, may be rejected on the basis of the theory of large scale structure formation. However not all the basic assumptions of this theory are solid enough to make really strong conclusions. While practically everybody agrees that the universe structure has been formed as a result of evolution of initially small fluctuations under the action of gravitational forces, there is no consensus about the shape of the spectrum of initial density perturbations. Introducing dimensionless quantity $\delta = \delta\rho/\rho$ we can present the power spectrum in the form

$$\langle \delta^2(x) \rangle = \int dk f^2(k) \tag{6}$$

(assuming that Fourier amplitudes are delta-correlated; one more assumption which may be questioned). Usually for $f(k)$, the most simple form is taken namely it is assumed that it does not contain any dimensional parameter, $f(k)^2 \sim 1/k$. This corresponds to flat or scale-free spectrum proposed by Harrison [26] and Zeldovich [27]. Inflationary models gave theoretical justification to this form of spectrum and now it is commonly used as the basic spectrum in calculations of structure formation. Sometimes as a simple generalization an arbitrary power law spectrum is considered, $f^2 \sim 1/k^n$. The spectrum with $n \neq 1$ may appear in some inflationary models [28] but with n not much different from unity. If we permit in principle an existence of primordial spectrum with an arbitrary n, introducing in this way a new scale to the theory, a combination of several terms with different n as well as a more complicated functions are also permitted but without a guiding principle for choosing a particular form of the spectrum, the theory would completely loose its predictive power. So we have to keep in mind that the judgement on the possible dark matter particles is made under assumption that the spectrum of perturbations has simple scale-free form.

The simplest model which was successfully used for description of structure formation until recently was the one with a single component cold dark matter particles (like LSP or axion) and flat spectrum of initial perturbations (for the review see e.g. refs. [29,30]). The model reasonably well described galaxy distribution at the scale of tens megaparsec with the only free parameter, the overall normalization of the spectrum. However the COBE measurements [31] has fixed the normalization at large scale end of the spectrum and with this normalization the predictions at smaller scales became approximately twice above the observations. An evident possible cure is to change the shape of the spectrum of primordial fluctuations. It seems to be a very interesting possibility but since this idea is not supported by the inflationary scenario, the major line of investigation goes along consideration of different forms of dark matter with the same flat spectrum of density perturbations or maybe an introduction of the cosmological constant. The basic idea of all these models was to suppress the power of the evolved spectrum at smaller scales relative to that at larger (COBE) scales.

One possibility is to shift the epoch of matter dominance (MD) to a later stage in a simple cold dark matter model. Since the characteristic scale at which perturbations started to rise in this model is determined by the horizon size at the onset of the MD stage, shifting it to a later moment gives less time for rising of the fluctuations and correspondingly less power at galactic and cluster scales. This goal can be achieved if one assumed that universe is open so that $h_{100}^2\Omega \approx 0.2$. However the low value of Ω is disfavoured by inflationary scenarios which (at least in simple versions) predict $\Omega = 1$. One can recover this prediction of inflation in the universe with low matter density if the cosmological constant Λ (or in other words, vacuum energy) is nonzero. As we have mentioned above the recent data indicating a rather high value of the Hubble constant, $h_{100} \approx 0.8$, support the idea of nonzero Λ with the fraction of the vacuum energy $\Omega_{vac} = 0.8$. The models with nonzero Λ give a satisfactory description of the observed structure [32] with flat spectrum and COBE normalization.

A mixed (hot+cold) dark matter scenario can also do the necessary job of diminishing the power at small scales because (initially flat) perturbations in hot dark matter are erased at scales smaller than $10^{14} M_{\odot}$ by free streaming if the dark matter is collisionless as is the case of neutrinos. A good description of the structure requests 70% of CDM and 30% of HDM [29]. It gives even better description if there are two equal mass neutrino species each with the mass 2.5 eV [33] as is suggested by the recent indications of neutrino oscillations by the Los Alamos group.

Recently there appeared a renewed interest to the idea of structure formation with unstable particles [34]. It is assumed in these models that there exists a massive long-lived particle, usually tau-neutrino with the mass in MeV range which decayed into massless species at the epoch when the mass density of the parent particles dominated the energy density of the universe. Correspondingly the present-day energy density of relativistic particles would be bigger than in the standard scenario and the onset of MD stage would take place later.

The common shortcoming of these models is that they all demand a certain amount of fine-tuning. Generally one would not expect that the contribution from hot and cold dark matter into the universe mass are about the same, they may differ by many orders of magnitude. One would also suspect that the vacuum energy which remains constant in the course of the universe expansion is by no means related to the critical energy now, which goes down with time as m_{Pl}^2/t^2. However adjustment mechanism [5,6] may possibly solve this problem. The models with unstable particles mentioned above are also based on the assumption of two independent components with a close contribution into Ω: the massive unstable particles themselves and unrelated cold dark matter. This is definitely unnatural and this shortcoming stimulated search for other models. Recently in ref. [35] a return to to the universe with a single cold dark matter component (of course except for baryons) was advocated. For successful description of the observed structure the authors need the power index of the spectrum $n = 0.8 - 0.9$, a low value of the Hubble constant, $h = 0.45 - 0.5$, and a large contribution of tensor perturbations (gravitational waves) into quadrupole fluctuation of background radiation temperature, $C_T/C_S = 0.7$.

Another attempt to overcome unnaturalness of multicomponent dark matter model has been done in paper [25]. In this paper a model is considered in which unstable particles and the particles of cold (or possibly warm) dark matter are closely connected. In fact the decay of the former produces particles of the present-day dark matter. A necessary background model of this kind in particle physics was proposed some time ago [22,23] as an attempt to find a phenomenologically acceptable description of R-parity breaking. The underlying mechanism is the spontaneous breaking of leptonic charge conservation at electroweak scale. The model contains a Majorana type tau-neutrino with the mass around MeV which decays into massive but light majoron J with mass in KeV region. Life-time with respect to this decay, as estimated in refs. [22], could be of order of years depending on the values of parameters, i.e. in the interval interesting for structure formation. It is worth noting that there is no stable SUSY particle in this model so the dark matter cannot be associated with it but the model itself produce a candidate for dark matter, namely a massive majoron. The model possesses some features like sufficiently large diagonal coupling of ν_τ to majorons or selfinteraction of majorons which rather naturally permit to resolve appearing cosmological problems in particular the problem of extra massless particle species during primordial nucleosynthesis. The cosmological properties of this model are rather unusual and are interesting by themselves. Dark matter particles (majorons) in this model are strongly self-interacting and thus the structure formation in this model is different from the traditional one with collisionless dark matter particles [36,37,38,25]. In particular the shape of galactic halo would depend on the dark matter selfinteraction. The recent data [39] might give an indication that collisionless dark matter gives a poor description of the shape of halo in dwarf galaxies.

At first sight the model of ref. [25] encounters very serious cosmological difficulties. Tau-neutrino with the mass in MeV range and stable at nucleosynthesis time scale would strongly destort primordial abundance of light elements [40,41] and should be excluded. However the Yukawa coupling of ν_τ to majorons helps to reduce the number density of tau-neutrinos down to a safe value through the reaction $2\nu_\tau \to 2J$. In this model the effective number of relativistic degrees of freedom during nucleosynthesis may be even below 3 in agreement with the recent claims [8,42].

However a strong reduction of the number density of ν_τ at nucleosynthesis through the annihilation into majorons produces the equilibrium amount of majorons such that the ratio of their number density to that of photons becomes $n_J/n_\gamma = 0.5(T_J/T_\gamma)^3$. By Gerstein-Zeldovich bound [43,44] the mass of such particles cannot exceed roughly speaking 30 eV. On the other hand for succesful structure formation we need $m_J = O(KeV)$ and it looks as if the model does not work. Fortunately this is not so because of relatively strong majoron selfcoupling: λJ^4 with $\lambda = 0.1 - 0.01$ [24]. This interaction gives rise to majoron cannibalism through the process: $4J \to 2J$. This process is fast enough to reduce the majoron number density by an order of magnitude [25] and to permit the majorons to have the mass appropriate for warm dark matter particles.

To conclude we definitely know that the world predominantly consists of matter which is different from what we directly see around. We do not know what kind of matter it is. Most probably this unknown matter consists of several different components. Why these components have similar energy densities in the present-day universe is also unknown. One of the component is very probably just energy of vacuum. The problem of vacuum energy is, to my mind, the central one in cosmology and particle physics in particular in connection with dark matter. A new massless field coupled only to gravity may solve the cosmoogical constant problem and to contribute to the dark matter. As for more traditional components which should be also present, even if vacuum energy is nonvanishing, the search for low energy supersymmetry is very important In particular it is essential to know if R-parity is indeed conserved and the lightest supersymmetric particle is stable. An improvement of accuracy in determination of tau-neutrino mass may have an important impact on cosmology. A related phenomenon is the primordial nucleosynthesis because the produced abundance of light element depend on ν_τ mass if it is in MeV range. Hopefully in the nearest future the ambiguity in determination of the amount of primordial deuterium, which is very sensitive to the total density of baryons will be setteld down. If one is optimistic one may hope that the mistery of building blocks of the universe will be resolved before the end of the next (or this, if one is superoptimistic) century.

Acknowledgments

This work was supported by DGICYT under grants PB92-0084 and SAB94-0089 (A. D.).

1. M.J. Pierse, D.L. Welch, R.D. McClure, et al, Nature, **371** (1984) 385; W.L. Freedman, B.F. Madore, P.B.Stetson, et al, Nature, **371** (1994) 757; P. Nugent, D. Branch, E. Baron, et al, Phys.Rev.Lett. (to be published).
2. S. van den Bergh, to be published in Dahlem Workshop Report (Chichester: John Wiley and Sons); astro-ph/9506027.
3. S. Weinberg, Rev. Mod. Phys. **61** (1989) 1.
4. A.D. Dolgov, Proc. of 24th Rencontre de Moriond Astrophys. Session, Les Arcs, France, 1989, p.227.
5. A.D. Dolgov, Proc. Very Early Universe, 1982, p. 449. Cambridge Univ. Press, ed. G. Gibbons, S.W. Hawking, and S.T. Siklos.
6. A.D. Dolgov, Pis'ma ZhETF, **41** (1985) 280.
7. Y. Fujii, gr-qc/9508029.
8. K.A. Olive and G. Steigman, Ap.J. Suppl. **97** (1995) 49.
9. K.A. Olive and S.T. Scully, Int.J.Mod.Phys. **A** (1985)(to be published); UMN-TH-1341/95.
10. C.J. Copi, D.N. Schramm, and M.S. Turner, FERMILAB-Pub-95/257-A, astro-ph/9508029.
11. R.F.Carswell, M. Rauch, R.J. Weynman, et al, MNRAS, **268** (1994) L1.
12. A. Songalia, L.L. Cowie, C. Hogan, and M. Rugers Nature, **368** (1994) 599.
13. D. Tytler and X.-M. Fan, BAAS (1995) 1424.

14. K.M. Ashman, PASB**104** (1992) 1109.
15. R. Cen and J.P.Ostriker Ap.J. **417** (1993) 415.
16. M. Persic, P. Salucci, Ap.J.,Suppl. (in press), astro-ph/95.
17. M.J. Hudson, A. Dekel, S. Courteau, et al. astro-ph/9501074; MNRAS (to be published).
18. S.D. White, J.F.Navarro, A.E. Evrard, and C.S. Frank, Nature, **366** (1993) 429.
19. D.A Buote and C.R Canizares Ap.J. (Feb., 1996) (to be published); astro-ph/9504049.
20. M. Kamionkowski, B. Ratra, D.N. Spergel, N. Siguyama, Ap.J. **434** (1994) L1; M. Bucher, A.S. Goldhaber, and N.Turok, PUPT-1507, hep-ph/9411206 (1994); L. Amendola, C. Baccigalupi, and F. Occhionero, astro-ph/9504097 (1995); A.D. Linde and A. Mezhlumian, SU-ITP-95-11, astro-ph/9506017 (1995).
21. E. Diehl, G.L. Kane, C. Kolda, and J.D. Wells, hep-ph/9502399, Phys.Rev. D (to be published (1995).
22. A Masiero, J. W. F. Valle, Phys.Lett. **B251** (1990) 273; J. C. Romão, C. A. Santos, and J. W. F. Valle, Phys.Lett. **B288** (1992) 311; J. C. Romão and J. W. F. Valle, Phys.Lett. **B272** (1991) 436; Nucl. Phys. **B381** (1992) 87.
23. Y. Chikashige, R. Mohapatra, R. Peccei, Phys.Lett. **98B** (1981) 265.
24. V. Berezinskii and J.W.F. Valle, Phys.Lett. **B318** (1993) 360.
25. A.D. Dolgov, S. Pastor, and J.W.F. Valle, FTUV/95-14, IFIC/95-14.
26. E.R. Harrison, Phys.Rev. **D1** (1970) 2726.
27. Ya.B. Zeldovich, MNRAS **160** (1972) 1p.
28. K. Freese, J.A. Frieman, and A.V. Olinto, Phys.Rev.Lett. **65** (1990) 3233; F.C. Adams, J.R. Bond, K. Freese, J.A. Frieman, and A.V. Olinto, Phys.Rev. **D47** (1993) 426.
29. J. Primack, in Proceedings of the International School on Cosmological Dark Matter, Valencia Oct.1993, (World Scientific, 1994), edited by J. W. F. Valle and A. Perez, p.81.
30. C. Frenk in Proceedings of the International School on Cosmological Dark Matter, Valencia Oct.1993, (World Scientific, 1994), edited by J. W. F. Valle and A. Perez, p.65.
31. G. Smoot et al, Ap.J. **396** (1992) L1; E. Wright et al, Astrophys. J. **420** (1994) 1.
32. G. Efstathiou, W.J. Sutherland, and S.J. Maddox, Nature **348** (1990) 705.
33. J. Primack, J. Holtzman, A. Klypin, and D. Caldwell, Phys. Rev. Lett. **74** (1995) 2160.
34. H. Kikuchi, E. Ma, Phys.Rev.**D51**(1995) 296; S. Dodelson, G. Gyuk, M. Turner, Phys.Rev.Lett. **72**(1994) 3754; H.B. Kim and J.E. Kim, Nucl.Phys. **B433**(1995) 421; J. Bardeen, J. Bond, G. Efstathiou, Ap.J. **321** (1987) 28; J. Bond, G. Efstathiou, Phys.Lett. **B265**(1991) 245; M. Davis et. al. Nature,**356**(1992) 489; A.S. Joshipura, J.W.F. Valle, Nucl.Phys. **B440** (1995) 647; M. White, G. Gelmini, J. Silk, Phys.Rev. **D51**(1995) 2669; G. Gelmini, D.N. Schramm, J.W.F. Valle, Phys.Lett.**B146** (1984) 311; P. Hut and S.D.M. White, Nature, **310** (1984) 637.
35. M. White, D. Scott, J. Silk, and M. Davis, astro-ph/9508009.
36. E.D. Carlson, M.E. Machacek, and L.J. Hall, Ap.J. **398** (1992) 43.
37. M.E. Machacek, Ap.J. **431** (1994)
38. A. A. de Laix, R.J. Scherrer, and R.K. Schaffer, Ohio preprint OSU-TA-9/94, BA-95-08, astro-ph/9502087.
39. B. Moore, Nature, **370** (1994) 629; R.A. Flores and J.R. Primack, Ap.J. **427** (1994); A. Burkert, astro-ph/9504041.
40. E. Kolb, M.S. Turner, A. Chakravorty, and D.N Schramm, Phys.Rev.Lett. **67** (1991) 563.
41. A.D. Dolgov and I.Z. Rothstein, Phys.Rev.Lett. **71** (1993) 476.
42. N. Hata, R.J. Scherrer, G. Steigman, et al, OSU-TA-6/95, UPR-0654T, hep-ph/9505319.
43. S.S. Gershtein and Y.B. Zeldovich, Pis'ma ZhETF, **4** (1966) 258.
44. R. Cowsik and J. McLeland, Phys.Rev.Lett. **29** (1972) 669.

MIXED DARK MATTER AND SUPERSYMMETRY

A. Masiero

Dipartimento di Fisica - Univ. di Perugia

and

Istituto Nazionale di Fisica Nucleare – Sez. di Perugia

06100 Perugia (Italy)

Abstract. The available data on large scale structures seem to favour models with mixed dark matter (MDM), i.e. with a hot and cold component in a rather well–defined amount, or with some form of "warm" dark matter. I discuss some prospects for these new scenarios for DM in the context of supersymmetric extensions of the electroweak standard model. In particular, I emphasize the intriguing link which exists between the present prospects of solution of the DM puzzle and the explicit or spontaneous breaking of baryon and/or lepton number symmetries. Some consequences on the issue of baryogenesis are worked out.

1. INTRODUCTION

There exists little doubt now that ordinary matter (baryons) is insufficient to provide the large amount of non–shining matter (dark matter) which has been experimentally proved to exist in galactic haloes and at the level of clusters of galaxies. In a sense, this might constitute the "largest" indication that we have that new physics beyond the electroweak standard model (SM) should be present. Indeed, if one takes the bound on the abundance on baryons coming from nucleosynthesis seriously, say $\Omega - b < 8 \cdot 10^{-3}$, one has to conclude that the SM fails to provide at least half of the whole energy density of the Universe (considering that we have direct indication that Ω should be at least $2 \cdot 10^{-2}$) [1]. Here Ω denotes the ratio of the energy density ρ to ρ_c, the critical energy density, $\rho_c = 3H_0^2/8\pi G$, with G the gravitational constant and H_0 the present value of the Hubble parameter. We could make these considerations on the insufficiency of the SM to provide an adequate Ω even more dramatic if we accept the theoretical input that the Universe underwent some inflationary era and that inflation produced $\Omega = 1$. In that case, at least 90% of the whole energy density of the Universe should be provided by some physics which is not contained in the SM.

The fact that some new source of energy density in addition to ρ_B should be envisaged to obtain Ω between 0.2 and 1 is only one of the two tasks which should be fulfilled by DM. The other crucial request is that DM must provide a suitable gravitational driving force for the primordial density fluctuations to evolve into the large scale structures (galaxies, clusters and superclusters of galaxies) that we observe today [1]. The constraints of DM coming from the evolution of the large scale structures have kept accumulating in these last few years [2]. Let me try to summarize here in a few words what are the major implications of these studies for a particle physicist.

We deal with two kinds of DM: cold dark matter (CDM) and hot dark matter (HDM). Although not entirely correct, we can roughly distinguish these two types of DM according to whether the DM candidate was relativistic or not at the moment it decoupled. Light neutrinos in the eV range are the most typical example of HDM, being their decoupling temperature of 0(1 MeV). On the other hand, the lightest supersymmetric (susy) particle (LSP) in the tens of GeV range (the most likely candidate is the lightest neutralino) represents a good example of CDM. When it decouples it is generally non–relativistic. For instance, in the case of the lightest neutralino the decoupling temperature is typically one order of magnitude smaller then its mass.

As long as the light HDM candidate remains relativistic there is no room for the growth of density fluctuations. Hence, in HDM scenarios first the superlarge structures form and they have to fragment into smaller structures (up–bottom scenario). With neutrinos in the eV range the problem is that we produce too many superlarge structures and we find a conflict in reproducing the experimental spectrum related to the distribution of structures on different scales. The opposite problem arises in the CDM case: in this case too much power in the spectrum is present at low mass scales (galactic scales). Hence the so–called CDM standard model [3], with $\Omega = 1$ CDM providing up to 90–95% of the whole energy density of the Universe and scale–invariant originary density fluctuation spectrum (Zeldovich spectrum), is today in trouble, although the trouble is not as severe as for the case of pure HDM. Several solutions have been proposed.

A general feature is that some amount of CDM should in any case be present. An interesting possibility is that Ω_{CDM} is ~ 0.2 and the remaining contribution to reach $\Omega = 1$ is provided by a non–vanishing value of the cosmological constant. Needless to say, we have not found so far any satisfactory reason to have $\Lambda =$

0; it becomes even more difficult to think of a possible solution which avoids a tremendous fine–tuning and yields some specific non–zero value of Λ.

Other possibilities are related to modifications of the original density fluctuation spectrum (tilted models). Another interesting suggestion is that the moment we have the transition from radiation to matter dominance is somewhat delayed, hence producing less power on small scales. Very slowly decaying relic particles into relativistic products might be a way of implementing this idea [4], although the particle physics models that have been proposed seem somewhat contrived.

Finally, the possibility which encounters most favour so far is the so–called mixed dark matter (MDM) [5], where a wise cocktail of HDM and CDM is present ($\Omega_{HDM} \sim 20 - -30\%$, $\Omega_{CDM} \sim 70\%$). This suggestion will constitute the case that is going to be discussed in this talk in the context of susy extensions of the SM.

In conclusion, I think that it is fair to say that there is still large uncertainty on the nature and amount of DM which is needed. Although the standard CDM scenario is now in severe trouble, it seems that some amount of CDM (with Ω_{DM} ranging from 0.2 up to 0.7) is required. The major challenge for a particle physicist in the MDM picture is to provide at the same time a source for H– and C–DM in the right amount, roughly $\Omega_{CDM} \sim 3\Omega_{HDM}$.

2. THE LIGHTEST SUPERSYMMETRIC PARTICLE (LSP)

In models where a discrete symmetry, matter R–parity [6] discriminates between ordinary and SUSY particles, the lightest SUSY particle (LSP) is absolutely stable. For several reasons the lightest neutralino is the favourite candidate to be the LSP fulfilling the role of CDM [7,8].

The neutralinos are the eigenvectors of the mass matrix of the four neutral fermions partners of the W_3, B, H^0 and H_2^0. There are four parameters entering this matrix: M_1, M_2, μ and $tg\beta$. The first two parameters denote the coefficient of the SUSY breaking mass terms $\bar{B}\bar{B}$ and $\bar{W}_3\bar{W}_3$ respectively, μ is the coupling of the $H_1 - H_2$ term the superpotential. Finally $tg\beta$ denotes the ratio of the $VEV's$ of the H_2 and H_1 scalar fields

In general M_1 and M_2 are two independent parameters, but if one assumes that a grand unification scale takes place, then at the grand unification $M_1 = M_2 = M_3$, where M_3 is the gluino mass at that scale. Then at M_w one obtains:

$$M_1 = \frac{5}{3} tg^2\theta_w M_2 \simeq \frac{M_2}{2}, \quad M_2 = \frac{g_2^2}{g_3^2} m_{\tilde{g}} \simeq m_{\tilde{g}}/3 \; , \tag{1}$$

where g_2 and g_3 are the $SU(2)$ and $SU(3)$ gauge coupling constants, respectively. The relation (1) between M_1 and M_2 reduces to three the number of independent parameters which determine the lightest neutralino composition and mass: $tg\beta, \mu$ and M_2. Hence, for fixed values of $tg\beta$ one can study the neutralino spectrum in the (μ, M_2) plane. The major experimental inputs to exclude regions in this plane are the request that the lightest chargino be heavier than $M_Z/2$ and the limits on the invisible width of the Z hence limiting the possible decays $Z \to \chi\chi, \chi\chi'$. Moreover if the GUT assumption is made, then the relation between M_2 and $m_{\tilde{g}}$ implies a severe bound on M_2 from the experimental lower bound on $m_{\tilde{g}}$ of CDF (roughly $m_{\tilde{g}} > 120$ GeV, hence implying $M_2 > 40$ GeV). The theoretical demand that the electroweak symmetry be broken radiatively, i.e. due to the renormalization effects on the Higgs masses when going from the superlarge scale of supergravity breaking down to M_W, further constrains the available (μ, M_2) region.

The first important outcome of this is that the lightest neutralino mass exhibits a lower bound of roughly 10 to 20 GeV [9]. The prospects for an improvement of this lower limit at LEP 200 crucially depends on the composition of χ [9]. If χ is mainly a gaugino, then it is difficult to go beyond 40 GeV for such a lower bound, whilst with a χ mainly higgsino the lower bound can jump up to $m_\chi > M_W$ at LEP 200.

Let us focus now on the role played by χ as a source of CDM. χ is kept in thermal equilibrium through its electroweak interactions not only for $T > m_\chi$, but even when T is below m_χ. However for $T < m_\chi$ the number of $\chi's$ rapidly decrease because of the appearance of the typical Boltzmann suppression factor $\exp(-m_\chi/T)$. When T is roughly $m_\chi/20$ the number of χ diminuished so much that they do not interact any longer, i.e. they decouple. Hence the contribution to Ω_{CDM} of χ is determined by two parameters: m_χ and the temperature at which χ decouples (T_D). T_D fixes the number of $\chi's$ which survive. As for the determination of T_D itself, one has to compute the χ annihilation rate and compare it with the cosmic expansion rate [7].

Several annihilation channels are possible with the exchange of different SUSY or ordinary particles, $\tilde{f}, H, Z$, etc. Obviously the relative importance of the channels depends on the composition of χ. For instance, having assumed χ to be a pure gaugino in the case discussed in the previous section, then the $\tilde{f}$ exchange represents the dominant annihilation mode.

Quantitatively [10], it turns out that if χ results from a large mixing of the gaugino ($\tilde{W}_3$ and $\tilde{B}$) and higgsino ($\tilde{H}_1^0$ and $\tilde{H}_2^0$) components, then the annihilation is too efficient to allow the surviving χ to provide Ω large enough. Typically in this case $\Omega < 10^{-2}$ and hence χ is not a good CDM candidate. On the contrary, if χ is either almost a pure higgsino or a pure gaugino then it can give a cospicuous contribution to Ω.

In the minimal SUSY standard model there are five new parameters in addition to those already present in the non–SUSY case. Imposing the electroweak radiative breaking further reduces this number to four. Finally, in simple supergravity realizations the soft parameters A and B are related. Hence we end up with only three new, independent parameters. One can use the constraint that the relic χ abundance provides a correct Ω_{CDM} to restrict the allowed area in this 3–dimensional space. Or, at least, one can eliminate points of this space which would lead to $\Omega_\chi > 1$, hence overclosing the Universe. For χ masses up to 150 GeV it is possible to find sizable regions in the SUSY parameter space where Ω_χ acquires intersting values for the DM problem. A detailed and updated analysis is presented in ref. [11] where one can compare the allowed SUSY parameters area with or without the constraint $0.1 < \Omega_\chi h^2 < 0.7$, where h is the Hubble parameter.

There is a further phenomenological constraint which helps in restricting even more severely the available regions of SUSY parameter space where $\Omega_\chi h^2$ can be relevant for the DM problem: it is the recent measurement of the decay $b \to s + \gamma$ at the inclusive level by the CLEO collaboration. Two papers [12] have recently thoroughly investigated the problem of the direct detection of relic neutralinos in processes of neutralino–nucleus scattering including the constraint arising from the experimental result of $BR(b \to s + \gamma)$. It turns out that large portions of the SUSY parameter space where it would be possible to have a neutralino–nucleus scattering rate high enough to be detectable in the next round of experiments predict very large values for $BR(b \to s + \gamma)$ vastly exceeding the experimental result. However, there still survive particular regions where rates as high as 10^{-1} events/kg/day for a ^{76}Ge detector are allowed. This is the case, for instance, for relatively large $\tan\beta$ ($\tan\beta \sim 20$) and moderate values of the SUSY parameters ($\tilde{m} = 200$ GeV, $\mu = -300$ GeV, $M_Z = 100$ GeV). For a complete discussions I refer the interested reader to the works of ref. [12].

I close this section with a remark concerning the possibility that gauginos are massless, i.e. $M_1 = M_2 = M_3 = 0$, to start with and that R-invariance (the continuous $U(1)$ symmetry associated with the fermionic partners of the gauge bosons, not to be confused with the discrete R-parity) is broken spontaneosuly by Higgs $VEV's$ or else explicitly by dimension 2 or 3 SUSY–breaking terms in the low energy effective lagrangian. Gluino and lightest neutralino masses then depend on only a few parameters. For a breaking scale of a few hundred GeV or less, the gluino and the lightest neutralino have masses typically in the range $10^{-1} - 2$ GeV. On the other hand, for a SUSY–breaking scale several TeV or larger, radiative contributions can yield gluino and lightest neutralino masses of O(50–300) GeV and O(10–30) GeV, respectively. As long as the Higgs $VEV's$ are the only source of R-invariance breaking, or if SUSY breaking only appears in dimension 2 terms in the effective lagrangian, the gluino is generically the lightest SUSY particle, hence modifying the usual phenomenology (and in particular the conventional view of the DM in SUSY) in interesting ways. For reasons of space I cannot deal more with this interesting (or at least curious) issue here and I recommend the works in ref.13 and 14 for a detailed discussion of the particle physics and cosmological aspects, respectively.

3. LEPTON NUMBER VIOLATION IN SUSY

In the previous section I discussed the more conventional SUSY schemes where R parity is imposed to avoid all the B and L violating operators in the superpotential. From the cosmological point of view the most important consequence of the presence of R is that there exists a stable SUSY particle which has good chances to constitute the CDM in an MDM scenario. As for the hot part of the MDM one can think of neutrinos getting a small mass (in the eV range). In some SUSY GUT's like SO(10) this is naturally achieved through a see–saw mechanism.

Let me comment now the alternative possibility that R–parity is replaced by some other symmetry, for instance B–parity, allowing for B or L explicit violation in the superpotential. The removal of R–parity has an unpleasant consequence for the DM problem: we lose our beloved CDM candidate represented by the stable LSP. In models with broken R–parity the LSP can decay into ordinary particles and, generally, these decays are much faster than what would be required to make the LSP survive until today.

The only exceptions are situations of extremely tiny violations of R–parity. An example is offered in ref. [15]. Not only can the lightest neutralino still be the CDM today, but its slow decays can have an experimental impact: for instance, we considered the possibility of the LSP radiative decays into a $\nu + \gamma$ with a possibly "visible" neutrino line. The negative result of a search performed at Kamiokande of such neutrinos led to a sharp improvement [16] on the bounds of the LSP lifetime (it turns out that τ_{LSP} must exceed the Universe lifetime by several orders of magnitude).

Although the absence of R parity carries the bad news that in general we lose the obvious SUSY candidate for CDM, it can have a positive impact on the other side of a mixed dark matter (MDM) scenario, i.e. it can yield a good amount of HDM. The point is that R violation is accompanied by L violation (for instance in schemes with B–parity), hence allowing for nonvanishing neutrino (Majorana) masses. In addition to the presence of m_ν there are several other important astrophysical implications: possibly large neutrino magnetic moments, new features in the implementation of the MSW mechanism for the solar neutrino problem, etc. [17].

The explicit violation of L through the presence of L violating operators in the superpotential is severely limited not so much by phenomenological constraints [18] , but rather by a powerful cosmological argument related to the survival of the cosmic matter–antimatter asymmetry [19].

The argument goes as follows. It is well–known that owing to the anomalous character of the L and B currents, these two numbers are violated at the quantum level. Only the combination $B - L$ is conserved. Although these violations are unlikely to produce any visible effect at zero temperature, they become quite relevant at high temperature [20]: the associated B and L violating processes have rates larger than the expansion rate of the Universe (at least for 100 GeV $< T <$ critical temperature of the electroweak phase transition, but, presumably, also for $T > T_c$), hence leading to an equal erasement of the pre–existing B and L asymmetries. Hence, if one starts with $\Delta B = \Delta L$, which is the case in GUT's with $B - L$ conservation, one ends up with $\Delta B = \Delta L = 0$ at the electroweak phase transition.

Whether these same quantum effects which are responsible for the cosmic ΔB erasement can be used to produce a new ΔB at the time of the electroweak phase transition is very doubtful. The survival of a lately produced ΔB seems to require an excessively light Higgs boson in the SM and also the amount of CP violation is unlikely to be sufficient to obtain a sizeable ΔB. However, both these objections are far from being settled and further work is needed to make some final assessment on this intriguing issue. A safer way to solve problem is represented by a different boundary condition at the GUT scale with $\Delta B \neq \Delta L$. If this is the case, given that quantum effects preserve $B - L$ it is never possible to reach a total erasement of ΔB. This is the reason why models like SO(10) where $B - L$ is violated (hence allowing for $\Delta B \neq \Delta L$) are certainly favoured with respect to GUT's with $B - L$ conservation (like SU(5)). Moreover SO(10) schemes can lead to neutrino masses in the convenient range to provide viable candidates for HDM.

All what I said above holds provided that during the interval time from the production of the cosmic ΔB (for example at the GUT time) down to the electroweak phase transition no other B or L violating interaction is in equilibrium apart from the abovementioned anomalous quantum effects. For instance, if R violating processes are present and are fast enough to be in equilibrium at some moment, since they violate either B or L they certainly violate $B - L$ and hence no combination of ΔB and ΔL can survive (independently from whether $\Delta B = \Delta L$ or $\Delta B \neq \Delta L$ to start with). Requiring the R–violating induced processes to be out of equilibrium places such a severe bound [19] on the strength of the R violation in the superpotential that certainly one could forget about any phenomenological implication of R breaking. As usual, however, this is not the end of the story concerning SUSY models without R parity. Several solutions have been pointed out to let ΔB survive even in the presence of non–negligible R–breaking effects.ref.21 Nervertheless the above cosmological observation represents a severe warning for the construction of consistent SUSY schemes which are alternative to those with the traditional matter R–parity.

One final comment on R–parity breaking is in order. We know that many continuous or global symmetries of the initial lagrangian can be spontaneously broken. One might wonder whether R–parity can undergo a similar destiny. Long ago it was pointed out [22] that there are regions of the SUSY parameter space where the minimization of the scalar potential leads to a nonvanishing VEV for the scalar partner of the neutrino, the sneutrino. This would correspond to the spontaneous breaking of L and R–parity. By now we know that such a breaking is phenomenologically forbidden. Indeed, the Z boson could decay into the Goldstone boson associated to the breaking of L and the scalar partner of it. The stringent bound on the invisible width of the Z excludes this possibility.*

* It was recently discussed the possibility that gravitational effects spoil any global symmetry [23]. If this is the case, L might be explicitly broken very tinily. The subsequent "spontaneous" breaking through a VEV of the sneutrino gives rise to a pseudo–Goldstone boson. Interestingly enough, even though the explicit break-

Alternatively one can supplement the usual particle spectrum of the minimal SUSY model with one or more gauge singlet scalar superfields which carry L and acquire a VEV [25,26]. In this case the Goldstone boson being a gauge singlet does not couple to the Z boson. In relation to the above considerations on baryogenesis and R–breaking, it is relevant to notice that the breaking of R can be induced radiatively, i.e. by the evolution of the singlet masses dictated by the renormalization group equations. It was recently shown [26] that this radiative breaking can delay the breaking of R down to temperature so low that the B violating quantum effects are no longer effective, i.e. typically $T < 100$ GeV.

4. MIXED AND WARM DM

As discussed in the Introduction, schemes with pure hot DM or pure cold DM seem disfavoured by recent (and also less recent) observations.

MDM [5] relies on a scenario where $\Omega_{CDM} \simeq 2\Omega_{HDM} \simeq 0.6$ and $\Omega_B \lesssim 0.1$. In principle one does not have to sweat so much to realize a scheme of this kind. Take a SUSY model with R—parity where neutrinos are massive. Then the lightest neutralino can play the role of CDM, while a neutrino of few eVs yields the HDM. Choosing the parameters conveniently one can obtain the prescribed cocktail of C- and H- DM. The problem that I see is just in this convenient choice of parameters. This is another way to say that one actually performs a fine–tuning to obtain the correct amount of Ω_{CDM} and Ω_{HDM} and this is certainly unsatisfactory. This is the reason which prompted some authors to investigate some possible common origin for HDM and CDM in order to justify close relation of their contributions to Ω. In the work of ref. [27] it was proposed to have the relative abundances of the HDM and CDM components set by the same scale. In their model, this is the scale of B-L spontaneous breaking of O(1 TeV). The HDM is given by the tau neutrino, while CDM is provided by the fermionic partner of the Goldstone boson associated to the B-L breaking.

Together with Bonometto and Gabbiani, we proposed [28] an example where one same particle may play the twofold role of HDM and CDM. In SUSY the axion possesses a fermionic partner, the axino ($\tilde{a}$). In fact, the $\tilde{a}$ is likely to be the lightest SUSY particle. Now, axinos can be produced via two entirely different

---

ing is very small, the mass of this particle can easily exceed the Z mass hence preventing the abovementioned decay which contributed to the Z invisible width [24].

different mechanisms in these models. First there are the axinos which are produced with the axions and were formerly in thermal equilibrium with the other components of the Universe, subsequently decoupling at a temperature $< V_{PQ}$ (the Peccei–Quinn scale) much higher than their mass. This $\tilde{a}$ component will be an effective CDM as only fluctuations involving masses $\lesssim 0.1 M_{\odot}$ will be erased at its derelativization. It was shown that they can account for Ω close to one [29]. This kind of "primordial" axinos are not the only axinos surviving today. Indeed if the $\tilde{a}$ is the lightest SUSY particle, all the SUSY particle must eventually decay into it.

Calling χ the lightest neutralino, we can expect the typical decay $\chi \rightarrow \tilde{a} + \gamma$ to occur through a supersymmetrization of the ordinary $a - \gamma\gamma$ coupling.

These "second hand" axinos can easily behave as hot dark matter, derelativizing at a redshift $z \sim 10^4$. Accordingly, fluctuations in such component will be erased up to a mass $\sim 10^{15} M_{\odot}$.

The detailed study of the conditions which make this scheme a viable MDM scenario is presented in ref. [28]. The major ingredients are a Peccei–Quinn scale of $O(10^{10}$ GeV), heavy sfermions in the TeV range and the lightest neutralino being a pure gaugino.

As interesting as these attempts to find some common origin of the hot and cold components of DM might be, it seems clear to me that at the moment we are lacking some particle physics model where it is compelling to have both kinds of DM and in the correct relative amount. An interesting prospect in this sense may be represented by models based on effective supergravities obtained from 4D super- strings. Here beside the usual lightest neutralino we can find other susy particles both in the bosonic and fermionic sector that can be rather light and produce neutralino decays. The gravitino and dilatinos are example of this class of new particles which are the low energy "relic" of the superlarge breaking of local susy in effective supergravities. Whether or not they can play an active role in this problem of producing "natural" candidates for both kinds of DM is a question presently under study.

Acknowledgements. I wish to thank my "astroparticle" collaborators, V. Berezinsky, S. Bonometto, S. Borgani, D. Comelli, F. Gabbiani, G. Giudice, M. Pietroni, A. Riotto and J.W.F. Valle. It is also a great pleasure to thank the organizers for the nice and fruitful atmosphere in which the meeting took place.

REFERENCES

[1] For a clear introduction to the "observational" aspects of the DM problem, see, for instance: Kolb and S. Turner, in The Early Universe, Addison–Wesley, New York, 1990;
Dark Matter, M. Srednicki ed. (North Holland, Amsterdam, 1989);
J.R. Primack, D. Seckel and B. Sadolet, Ann. Rev. Nucl. Part. Sci., 38 (1988) 751.

[2] For a review see J. Primack in Electroweak, Moriond 1995.

[3] For a recent review of CDM, see M. Davis, G. Efstathiou, C.S. Frenk and S.D.M. White, Nature 356 (1992) 489.

[4] H.B. Kim and J.E. Kim, preprint SNUTP 94–48, hep–ph/9405385 (1994).

[5] Q. Shafi and F.W. Stecker, Phys. Lett. 53 (1984) 1292; S.A. Bonometto and R. Valdarnini, Astroph. J. 299 (1985) L71; S. Achilli, F. Occhionero and R. Scaranella, Astroph. J. 299 (1985) L77; J.A. Holtzman, Astroph. J. Suppl. 71 (1981) 1; A.N. Taylor and M. Rowan–Robinson, Nature 59 (1992) 396; J.A. Holtzman and J. Primack, Astroph. J. 396 (1992) 113; D. Yu. Pogosyan and A.A. Starobinski, CUA preprint (1993); A. Klypin, J. Holtzman, J. Primack and E. Regoös, SCIPP 92/52.

[6] For a phenomenologically oriented introduction to SUSY, see: H.P. Nilles, Phys. Rep. 110C (1984) 1; H. Haber and G. Kane, Phys. Rep. 117C (1985) 1; For construction of the N=1 supergravity lagrangian, see E. Cremmer, S. Ferrara, L. Girardello and A. Van Proeyen, Phys. Lett. B116 (1982) 231; Nucl. Phys. B212 (1983) 413.

[7] J. Ellis, J.S. Hagelin, D.V. Nanopoulos, K. Olive and M. Srednicki, Nucl. Phys. B238 (1984) 453.

[8] For reviews, see, for instance: J. Ellis, Phil. Trans. R. Soc. Lond. A336 (1991) 247; L. Roszkowski, proceedings of the Joint International Lepton–Photon Symposium and Europhysics Conference on High Energy Physics (1991), Geneve, Switzerlans; K.A. Olive, Proc. of Ten Years on SUSY Confronting experiment, eds, J. Ellis, D.V. Nanopoulos and A. Savoy–Navarro CERN Sept. 1992.

[9] L. Roszkowski, Phys, Lett. B252 (1990) 471; L. Roszkowski, Proceedings of the UCLA International Conference on Trends in Astroparticle Physics, 1990.

[10] A. Bottino, V. de Alfaro, N. Fornengo, G. Mignola and M. Pignone – preprint DFTT 37/93 (1993); M. Drees, G. Jungman, M. Kamionkowski and M. Nojiri, Phys. Rev. D49 (1994) 636.

[11] P. Gondolo, M. Olechowski and S. Pokorski; Proc. of the XXVI International Conference on High Energy Physics, Dallas, August 1992.

[12] P. Nath and R. Arnowitt, preprint CERN–TH–7463/94 (1994); F. Borzumati, M. Drees and M. Nojiri, preprint DESY 94–096 (1994).

[13] G. Farrar and A. Masiero, preprint RU–94–38, hep–ph 9410401 (1994).

[14] G. Farrar and R. Kolb, preprint RU–95.

[15] V. Berezinski, A. Masiero and J.F.W. Valle, Phys. Lett. B266 (1991) 382.

[16] M. Mori et al., Phys. Lett. B287 (1992) 217.

[17] For a review, see A. Masiero, Nucl. Phys. B (Proc. Suppl.) 28A (1992) 130.

[18] V. Barger, G.F. Giudice and T.Y. Han, Phys. Rev. D40 (1989) 2987.

[19] B. Campbell, S. Davidson, J. Ellis and K.A. Olive, Phys. Lett. B256 (1991) 457; Astropart. Phys. 1 (1992) 77; W. Fischer, G.F. Giudice, R.G. Leigh and S. Paban, Phys. Lett. B258 (1991) 45,

[20] V.A. Kuzmin, V.A. Rubakov and M.E. Shaposhnikov, Phys. Lett. B155 (1985) 36; V.A. Matveev, V.A. Rubakov, A.N. Tavkhelidze and M.E. Shaposhnikov, Usp. Fiz. Nauk 156 (1988) 253.

[21] H. Dreiner and G.G. Ross, preprint OUTP–92–08; A.G. Cohen and A.E. Nelson, Phys. Lett. B297 (1992) 111; A. Masiero and A. Riotto, Phys. Lett. B289 (1992) 73; D. Comelli, M. Pietroni and A. Riotto, preprint SISSA 93/50/A.

[22] C. Aulakh and R.N. Mohapatra, Phys. Lett. B119 (1983) 136; J. Ellis, G. Gelmini, C. Jarlshog, G.G. Ross and J.F.W. Valle, Phys. Lett. 1985 142; G.G. Ross and J.W.F. Valle, Phys. Lett. B151 (1985) 375; R. Barbieri, D.E. Brahm, L.J. Hall and S.D.H. Hsu, Phys. Lett. B238 (1990) 86; B. Gato, J. Leòn, J. Perèz–Mercader and M. Quiros, Nucl. Phys. B260 (1985) 203; M.S. Carena and C.E.M. Wagner, Phys. Lett. B186 (1987) 361.

[23] S. Giddings and A. Strominger, Nucl. Phys. B307 (1988) 854; S. Coleman, Nucl. Phys. B310 (1988) 643; R. Holman, S.D.H. Hsu, T. Kephart, E. Kolb, R. Watkins and L. Widrow, Phys. Lett. B282 (1992) 132; M. Kamionkowski and J. March–Russel, Phys. Lett. B282 (1992) 137; S. Barr and D. Seckel, Phys. Rev. D46 (1992) 539; E. Kh. Akhamedov, Z.G. Berezhiani, R.N. Mohapatra and G. Senjanovic, Phys. Lett. B299 (1993) 90.

[24] D. Comelli, A. Masiero, M. Pietroni and A. Riotto, Phys. Lett. B324 (1994) 397.

[25] A. Masiero and J.F.W. Valle, Phys. Lett. B251 (1990) 273; J.C. Romao, C.A. Santos and J.F.W. Valle, preprint FTUV/91-06.

[26] G.F. Giudice, A. Masiero, M. Pietroni and A. Riotto, Nucl. Phys. B396 (1993) 243; M. Chaichian and R. Golzalez Felice, preprint BUTP-92/36 (1992); J. Umemura and K. Yamamoto, Phys., Phys. Lett. B313 (1993) 89.

[27] R.N. Mohapatra and A. Riotto, Phys. Rev. Lett. 73 (1994) 1324.

[28] S.A. Bonometto, F. Gabbiani and A. Masiero, Phys. Rev. D49 (1994) 3918.

[29] K. Rajgopal, M.S. Turner and F. Wilczeh, Nucl. Phys. B358 (1991) 447.

DOUBLE BETA DECAY SEARCHES

ANGEL MORALES

Laboratorio de Física Nuclear y Altas Energías, University of Zaragoza,
50009 Zaragoza, Spain

An overview of the double beta decay searches is presented and some of the achievements occurred recently in the field are summarized. The relevance of this topic as a window beyond the standard model is emphasized.

1 Introduction and Phenomenological Remarks

1.1 The Nuclear Double Beta Decay

The Nuclear Double Beta Decay [1] (DBD) is a low energy laboratory where some fundamental issues in Particle Physics and, more specifically, in Neutrino Physics, can be investigated. The DBD tests the lepton number conservation law and explores the nature of the neutrino. It addresses the question whether the neutrino is self-conjugated and whether it has a Majorana mass. On the other hand, it may reveal the existence of the majoron and, by extension, the presence of a category of scalar particles which play a fundamental role in the mass generation problem. So, DBD is a window beyond the Standard Model and this fact explains the great interest aroused in this field of research over the past years [1]. Particle Dark Matter searches have also benefitted from the low background techniques developed for double beta decay searches.

The double electron decay ($2\beta^-$) is a weak process where a nucleus gets transformed into another isobar through the emission of two electrons which may or may not be accompanied by their corresponding anti-neutrinos.

$$
\begin{aligned}
(A, Z) \rightarrow \quad & (A, Z+2) + 2e^- + 2\bar{\nu}_e & \therefore\ (2\beta)_{2\nu} \\
\rightarrow \quad & (A, Z+2) + 2e^- & \therefore\ (2\beta)_{0\nu} \\
\rightarrow \quad & (A, Z+2) + 2e^- + M & \therefore\ (2\beta)_{0\nu,M}
\end{aligned}
$$

The decay without neutrinos $(2\beta)_{0\nu}$ is of fundamental importance because it has no room within the Standard Model. The same trascendental consequences would have the existence of the neutrinoless decay accompanied by a Goldstone boson, the Majoron [2], sharing the available energy with the leptons to provide -like in the standard two-neutrino case- a continuous energy spectrum.

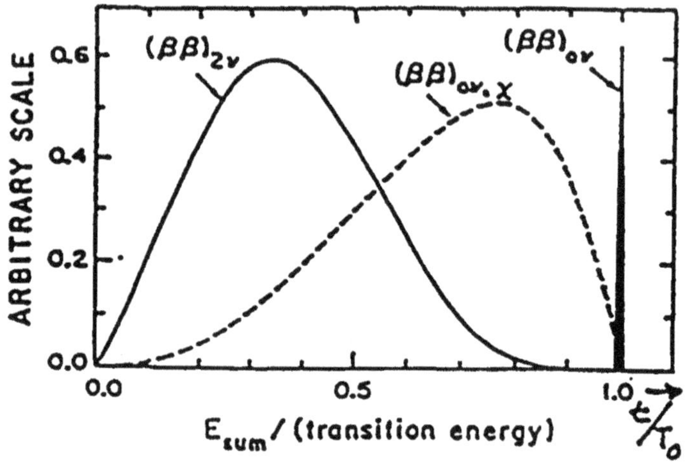

Fig. 1: Two electrons summed energy spectra in the cases of two-neutrino and neutrinoless double beta decay

The decay with neutrinos is a conventional, although rare, second order weak process. Direct counter experiments have observed this process in ^{76}Ge, ^{82}Se, ^{100}Mo, ^{116}Cd, ^{150}Nd and ^{238}U. Geochemical experiments have proved their existence also in ^{128}Te, ^{130}Te, ^{96}Zr and again in ^{82}Se. The half-lives of these rare process range from 10^{19} y up to 10^{24} y. Besides its intrinsic interest, these measurements have provided a test of the nuclear structure calculation [3], [4], [5], [6], of paramount importance in the reliability of the results on the neutrino mass limit derived from the fundamental, not yet seen, neutrinoless decay. The two-neutrino decay half-life is expressed customary as [7]

$$\left[T_{1/2}^{2\nu}\left(0^+ \rightarrow 0^+\right)\right]^{-1} = G_{2\nu}\left|M_{GT}^{2\nu}\right|^2 \tag{1}$$

where $G_{2\nu}$ is a phase space integral and $M_{GT}^{2\nu}$ the double Gamov Teller nuclear matrix element

The neutrinoless DBD, on the contrary, would imply the violation of the lepton number conservation law and the existence of a self-conjugated massive Majorana neutrino. In fact, for this process to happen, one should have i) a self-conjugate (Majorana) neutrino $\nu_m=\overline{\nu}_m$, and ii) the $\overline{\nu}_m$ emitted which is mainly R (right-handed) (strictly R if $m_\nu = 0$) must have a left-handed L impurity admixture to match the helicity needed to be absorbed. This admixture $\overline{\nu}_m^L$, of order m_ν/ω, can be provided by a Majorana mass or explicitely by a right-handed, V+A, interaction, but at the fundamental level, the Majorana neutrino needed for such a process should have a mass [8].

The transition probability in the case of a neutrinoless decay driven by a Majorana neutrino mass term.is expressed as [7]

$$\left[T_{1/2}^{0\nu}(0^+ \to 0^+)\right]^{-1} = G_{0\nu}\left|M^{0\nu}\right|^2 \frac{<m_\nu>^2}{m^2} \qquad (2)$$

with $G_{0\nu}\left|M^{0\nu}\right|^2 \equiv F_N$ (which will be called the Nuclear Factor-of-Merit) and where $M^{0\nu} = M_{GT}^{0\nu} - (g_V / g_A)^2 M_F^{0\nu}$ The expression $< m_\nu >= \sum_j \varpi_{em} m_m U_{em}^2$ is the so-called neutrino mass parameter. U_{em} is the unitary mixing matrix describing the coupling of the neutrinos to electrons, ω_{em} a phase factor and m_m the neutrino mass eigenvalue. $M_{F,GT}^{0\nu}$ are respectively the Fermi and Gamow Teller nuclear matrix elements, and $G_{0\nu}$ the neutrinoless phase space integral.

The most important experimental outcome in DBD decay searches would be the lower bound on the neutrino mass implied by the observation of a neutrinoless decay process. On the contrary, the non-appearance of $(2\beta)_{0\nu}$ will provide only an upper limit for the mass parameter $<m_\nu>$. The fundamental result of a $(2\beta)_{0\nu}$ discovery would be expressed as

$$< m_\nu >_{exp} = \frac{m_e}{\left(F_N T_{1/2}^{0\nu}\right)^{1/2}} \text{ eV} \qquad (3)$$

where, for instance, in the ^{76}Ge case (the best explored DBD candidate), some theoretical values of the nuclear factor-of-merit computed in various nuclear models are: F_N=1.56x10^{-13} y^{-1} (Haxton'84); 9.67x10^{-15} y^{-1} (Caltech'88); 1.21x10^{-13} y^{-1} (Tubingen'87); 1.12x10^{-13} y^{-1} (Heidelberg-Tokyo'89); 1.3x10^{-13} y^{-1} (Valencia'91) [3], [4], [5], [6], [11]. $T_{1/2}^{0\nu}$ would be the half-life of the measured neutrinoless DBD. Cancellations in the $<m_\nu>$ expression may occur, but $<m_\nu> \leq m_m^{heaviest}$,i.e., at least one neutrino will have a mass not smaller than $<m_\nu>_{exp.}$, above.

Up to now the neutrinoless DBD has not been observed and so, only an upper limit on. $<m_\nu>$ can be settled from the measured half-life lower limit. In fact, $<m_\nu>_{exp}$ can be much smaller than the actual neutrino masses.

To deduce a neutrino mass limit from these considerations, we need to have reliable matrix element calculations. Assuming, *bona fide*, that these calculations are correct, enriched isotopes = detectors are now measuring neutrino mass limit down to ~ one eV ($<m_\nu>$) as derived from current half-life lower limits of $\geq 10^{24}$y (^{76}Ge). There exists now a large body of nuclear matrix elements calculations, both in the shell model [3] and in the pnQRPA method [4], [5] and [6]. In this last approach, the $(2\beta)_{2\nu}$ theoretical nuclear matrix elements exhibit an extreme sensitivity upon the choice of the nuclear parameters (say the particle-particle force g_{pp}) but this shortcoming is being overcome by other methods, like the Operator Expansion Method [9], or through phenomenological approaches [10]. In the case of the $(2\beta)_{0\nu}$ nuclear matrix elements, where their reliability needs to be much more important, the uncertainty of the calculation is acceptable, and so the conclusions derived on the possible neutrino mass -or neutrino mass limit- can be taken on more firm grounds.

There exist several hints which point towards a non-zero neutrino mass. The atmospheric neutrino anomaly, the solar neutrino deficit and the LSND experiment could be explained within the frame of neutrino oscillations which require neutrino non-zero masses and mixing between the various neutrino flavours [12]. On the other hand, the recent COBE data on the microwave cosmic background radiation anisotropies, as well as small scale surveys data, imply that the simple cold dark matter model of cosmic structure formation should be implemented with a 25%~30% of hot dark matter for matching properly the observed spectral power both at large and small scales of the universe [13]. The light neutrino, archetypical hot dark matter candidate, could well do the job, and at the same time solve the neutrino oscillation problem with a mass of a few eV [14]. The level of sensitivity of the most advanced double beta decay experiments, which have already reached the 1eV value for the upper limit of the neutrino mass parameter (within the caution of the nuclear physics uncertainties), may be well close to the finding of the neutrinoless double beta decay.

2 Experimental Remarks

To investigate the 2β decays, three general approaches have been followed:

a) Geochemical experiments, where one looks for isotopic anomalies in noble gases daughter of 2β decaying nucleus over geological time scales. Determination of the gas retention age of the ore is crucial. Examples are the $^{82}Se \rightarrow {}^{82}Kr$ transition [15] and the celebrated case of the $^{128,130}Te \rightarrow {}^{128,130}Xe$ decays [16].
b) Radiochemical ("milking") experiments, where the daughter nuclei from large amount of a double beta emitter are accumulated, extracted and counted (like in solar ν experiments). The characteristic decay of the accumulated nuclei -which should be themselves radioactive- through a clear and distinct signal is crucial. A recent example of the method is provided by the 2β decay $^{238}U \rightarrow {}^{238}Pu$ [17].
c) Counter experiments, where the two electrons emitted by a 2β-decaying source are registered, and the features of the events recorded, according to the capabilities of the various types of the detectors employed. The two-electron energy spectrum, the single energies or, in various devices, the tracking of the particles have provided experimental evidence of the two-neutrino double beta decay in several cases, as well as half-life limits in others. The merits of the various types of counters are different as intended for diverse 2β emitters. They all share, however, the same constraint: to minimize the product background x energy resolution, which are the main ingredients of their sensitivities. Many

examples of direct, counter experiments are available. The best known are those of the 2ß emitters ^{76}Ge, ^{82}Se, ^{100}Mo, ^{116}Cd, ^{130}Te, ^{136}Xe, and ^{150}Nd.

Various different devices have been used in double beta decay direct searches: Germanium spectrometers, time projection chambers, stacks of silicon detectors, ionization and multielement proportional chambers, stacks of plastic scintillators, drift chambers plus plastic detectors, multiwire drift tubes in Geiger mode, crystal scintillators and a few more [1].

When looking for the neutrinoless decay, the issue of the choice of the emitter follows straightforward considerations. It is convenient for the emitter to be at the same time the detector itself. Moreover, it should have reasonable isotopic abundance or to be isotopically enriched. Emitters of large $Q_{2ß}$ have been currently the standard choice, to escape from the low energy region -where the background is larger- and to increase the expected $(2ß)_{0\nu}$ yield ($\sim Q^5$). Although the phase space for two-particle decays ($2ß_{0\nu}$) is much larger than that for four-particle decays ($2ß_{2\nu}$), a large $Q_{2ß}$ also increase, by phase space reasons, the relative (2ν)/(0ν) yield. And so, it may happen that in the case of those emitters which 2ν-decay very "fast", the 2ν events may become a background for 0ν searches when modest energy resolution devices are employed [1]. Although the phase space considerations are very useful in choosing the emitter, a favorable nuclear matrix element is obviously needed despite the possible uncertainty in its computation.

New techniques based on low temperature thermal detection are being applied to 2ß decays and other rare searches. The cryogenic detectors are supposed to provide better energy resolution and lower energy threshold, as well as a more effective absorption of the particle energy (thermal vs ionization). On the other hand, they will permit to enlarge the choice of suitable candidates for 2ß searches, in the modality of the source=detector devices. Noticeable examples are that of bolometers of ^{130}Te devoted to 2ß searches (Milano), or that of the projected Nd bolometer (UCSB/LBL/INR). Hybrid (redundant) detectors, i.e. low temperature calorimeters endowed with background rejection mechanisms through the simultaneous measurement of heat and ionization, are being developed by UCSB/LBL/CfPA/Stanford for dark matter searches with Ge crystals [18]. Also, phonon plus scintillation detection devices like CaF_2 (Milano), $CdWO_4$ (Milano/Kiev), NdF_3 (Zaragoza/USC) are either in progress or in project.

To weight the perspectives of a given experiment, it is useful to define the neutrinoless sensitivity or detector-factor-of-merit.

$$F_D^{0\nu} = \ln 2 \, N_{2ß} \sqrt{\frac{t}{R\Gamma}} \varepsilon_\Gamma = \ln 2 \frac{a}{A} 10^3 N_A \sqrt{\frac{Mt}{B\Gamma}} \varepsilon_\Gamma \tag{4}$$

where $N_{2ß}$ is the number of 2ß emitter nuclei in the M (kg) of source material, a the isotopic abundance, A the atomic weight, Γ the chosen energy bin (or energy resolution of the detector at the $Q_{2ß}$ energy value) and ε_Γ the detector efficiency including the probability of the events to fall in the bin $\Delta E = \Gamma$. On the other hand, R is the differential rate (c/keV y), B is the differential specific rate (in c/keV kg y), and t the counting time expressed in years.

The other prescription in the experimental searches refer to the nuclear physics qualities of the emitter, i.e. to its Factor-of-Merit $F_N^{0\nu}$. To increase the overall sensitivity, one should choose emitters and techniques optimizing $F_N F_D$ because $<m_\nu> \leq m / \left(F_N^{0\nu} F_D^{0\nu}\right)^{1/2}$. The values of $F_N^{0\nu}$ for various well-known emitters are given in Table 1, following -for sake of comparison- the estimates of MPI/Heidelberg [5].

Table 1. Neutrinoless Nuclear factor-of-merits computed according to the nuclear matrix elements of Max Planck Institut/Heidelberg [5] and the phase space integrals of Doi et al [7].

	^{76}Ge	^{82}Se	^{100}Mo	^{116}Cd	^{128}Te	^{130}Te	^{136}Xe	^{150}Nd	^{238}U
$F_N^{0\nu}(y^{-1})$	1.12×10^{-13}	4.33×10^{-13}	2.05×10^{-13}	5.36×10^{-13}	3.36×10^{-14}	5.34×10^{-13}	1.18×10^{-13}	7.74×10^{-12}	1.0×10^{-12}

3 Experimental Results

More than thirty experiments in progress or recently completed account for the large activity in the field. In the following, some comments on the current Germanium experiments will be given, whereas the results for the 2ν and 0ν decay modes of most of the double beta decay emitters explored until now, are summarized in Tables 2 and 3. A more completed description, as well as references to the previous experiments, can be found in the various review papers quoted in Ref. [1].

Three experiments already completed [19], [20], [21] reported a positive measurement of the two-neutrino decay mode of ^{76}Ge. Now there are two running experiments [22], [23], both using large quantities of isotopically enriched Germanium-76.

The Heidelberg-Moscow experiment [22] is operating a set of germanium detectors up to a total of five crystals which in its final set up have a total mass of 11.5 kg (10.9 Kg active mass) (85% enriched in ^{76}Ge) in the Gran Sasso Underground Laboratory. The background around the $Q_{2\beta}$ value of the expected neutrinoless transition Q = 2038.6 keV has been steadily decreasing, both because the deexcitation of cosmogenic induced activities and because of improvements in the shielding. For a combined exposure of 10.23 Kg y, the background obtained in the total accumulated spectrum (averaged in a 80 keV interval around the $Q_{2\beta}$-value), is B = 0.21 ± 0.02 c/kg keV y (excluding the peak region), and the overall energy resolution at this energy is Γ = 3.7 keV. The half-life lower limit obtained from such an exposure is $T_{1/2}^{0\nu} > 5.6 \times 10^{24}$ y (90% C.L). By using their estimation for the neutrinoless nuclear matrix element, that corresponds to an effective neutrino mass $<m_\nu>\leq 0.65$ GeV (90% C.L.).

In 1993, the H/M experiment had aroused a certain expectation because of the intriguing constant value of the neutrinoless half-life limits derived from their data along about eighteen months (from Mt=3.1 kg y to Mt =7.2 kg y) in spite of the decreasing background (respectively of B=0.36 c/keV kg y and B=0.27 c/keV kg y). The addition of new data has washed out this constant trend, returning the half-life limit evolution to the expected root square of t (or of the exposure).

The Heidelberg-Moscow experiment has also performed a measurement of the two-neutrino mode half-life. By using one of their detectors (M=2.76 kg), and an exposure of 19.27 mol y, after peak-removal, and a background sustraction [made by using a quantitative model built from measured quantities and a MC simulation (GEANT-3)], a best fit to the stripped spectrum gives a two-neutrino half-life

$$T_{1/2}^{2\nu} = [1.42 \pm 0.03 \text{ (stat.)} \pm 0.13 \text{ (syst.)}]\ 10^{21} \text{ y} \qquad (5)$$

Two new measurements of the two-neutrino mode ^{76}Ge half life are being reported by the International Germanium Experiment (IGEX) [23]. This consortium (USC, PNL, Zaragoza, INR and ITEP) is operating various germanium detectors (enriched up to 87.44% in ^{76}Ge) at Homestake, USA (two detectors of M = 2.1 kg and M = 2.2 kg); at Canfranc, Spain (two detectors of about 1 kg each and one of M=2.1 kg) and at Baksan, Russia (one detector of 1.1 kg), and is producing new large Ge crystals up to a total of about 18 kg of enriched germanium oxide. The detectors are to be placed at Canfranc at 2400 m.w.e. depth. The Phase-I of IGEX (which started three years after the H/M experiment), built three one-kilogram enriched detectors to test their performances and set the background conditions in a search for the best operation of a multikiligram of Ge-detectors. The crystals were grown at Tennelec/Oxford, USA, and the cryostats were made in electroformed copper at PNL. The standard shielding of IGEX detectors consists of very old (450 y or 2000 y) lead bricks in the inner part, surrounded by low activity lead and a neutron absorber. Active shieldings were also used in the sets installed in the underground sites of lesser depth.

As far as the neutrinoless decay half-life lower limit is concerned, by combining data from IGEX-I and IGEX-II detectors, the current limit of this experiment is $T_{1/2}^{0\nu} \geq 2.5\text{x}10^{24}$ y (90% C.L.) for a total exposure of Mt=18.1 mole years. The background in the region of 2000 keV-2500 keV is 0.2 c/keV kg y and the energy resolution at 2038 keV is Γ=3.5 keV.

As far as the two-neutrino decay half life is concerned, data from IGEX-I obtained in Canfranc from one enriched and one natural abundance Ge detector [of relative efficiency k(E)], have been used to derive a residual spectrum through the difference (after subtraction of the visible peaks) $N(E)=N_{enr}(E)-k(E)\ N_{nat}(E)$ leading to a half life of

$$T^{2\nu}_{1/2}=(1.05\pm0.20)\times10^{21}\ y \tag{6}$$

Another determination of the two-neutrino half-life is derived from the spectrum of two IGEX-Phase II detectors, (of 2.1 kg each one). After the various subtractions performed on the raw data (corresponding to an exposure of 1.135 kg·y), i.e. corrected for external gammas -8 peaks-, cosmogenic peaks -^{57}Co, ^{58}Co, ^{54}Mn, ^{68}Zn -, continuous cosmogenics -^{60}Co, ^{68}Ge- and ^{210}Bi Bremstralung, the best fitted to the residual spectrum corresponds to a two-neutrino half-life of

$$T^{2\nu}_{1/2}=(1.11\pm0.15)\times10^{21}\ y. \tag{7}$$

These two IGEX results are to be compared with other previous determination of this half-life, [19], [20], [21] and in particular with that obtained in the H/M experiment [22].

To summarize the results of the various experiments, we gather in Table 2 the best present limits on the neutrinoless half-lives. The theoretical nuclear matrix element from the Heidelberg group have been used to derive the neutrino mass parameter upper limits quoted in the last column. The 2ν decay half-lives values or limits measured are given in Table 3.

Table 2. Limits On Neutrinoless Decay Modes

Isotope	Experiment	$T^{0\nu}_{1/2}$	$<m_\nu>$ (eV)
^{48}Ca	HEP Beijing	$>1.1\times10^{22}$ y 68% C.L.	<8
^{76}Ge	MPIH/KIAE [22]	$>5.6\times10^{24}$ y 90% C.L.	<0.7
	IGEX [23] [37]	$>2.5\times10^{24}$ y 90% C.L.	<1
^{82}Se	UCI [24]	$>2.7\times10^{22}$ y 68% C.L.	<5
^{100}Mo	LBL/MHC/UNM [25]	$>3.5\times10^{22}$ y 68% C.L.	<6
	UCI [24]	$>2.6\times10^{21}$ y 90% C.L.	<24
	Osaka [26]	$>7\times10^{21}$ y 68% C.L.	<13
	NEMO [27]	$>6.4\times10^{21}$ y 90% C.L.	<13
^{116}Cd	Kiev [28]	$>1.5\times10^{22}$ y 90% C.L.	<5.7
	Osaka [26]	$>1.7\times10^{21}$ y 68% C.L.	<17
^{128}Te	Washington [16]	from $\frac{T_{1/2}(130)}{T_{1/2}(128)}=(3.52\pm0.11)\times10^{-4}$ $>7.7\times10^{24}$ y	<1.0
^{130}Te	Milano [29]	$>1.4\times10^{22}$ y 90% C.L.	<5.9
^{136}Xe	Caltech/UN/PSI [30]	$>4.2\times10^{23}$ y 90% C.L.	<2.3
	Milano [29]	$>1.2\times10^{22}$ y 95% C.L.	<13.6
^{150}Nd	UCI [24]	$>2.1\times10^{21}$ y 90% C.L.	<4
^{78}Kr(Kβ$^+$)	Zaragoza/INR [31]	$>5.8\times10^{21}$ y 68% C.L.	<850

from MPHI n.m.e.

Table 3. Two-neutrino decay results

Emitter	Group	$T_{1/2}^{2\nu}$
^{76}Ge	USC/PNL/ITEP [21]	$\left(0.92^{+0.07}_{-0.04}\right)x10^{21}$ y
	Heidelberg/Moscow [22]	$(1.42\pm0.03\pm0.13)x10^{21}$ y
	IGEX (Homestake) [37]	$(1.11\pm0.15)x10^{21}$ y
	IGEX (Canfranc) [37]	$(1.05\pm0.20)x10^{21}$ y
^{82}Se	UCI/ILL [24]	$\left(1.08^{+0.26}_{-0.06}\right)x10^{20}$ y.
	Heidelberg [15]	$(1.30\pm0.05)x10^{20}$ y
	Missouri [15]	$(1.0\pm0.4)x10^{20}$ y
^{78}Kr	Zaragoza/INR [31]	$T_{1/2}(K\beta^{+})_{2\nu}>2x10^{20}$ y (68% C.L.) $T_{1/2}(K\beta^{+})_{ind}>2x10^{21}$ y (68% C.L.)
^{96}Zr	Kawashima et al. [35]	$(2.9\pm0.9)x10^{19}$ y
^{100}Mo	Osaka[26]	$\left(1.15^{+0.30}_{-0.20}\right)x10^{19}$ y
	UCI [24]	$\left(1.16^{+0.34}_{-0.08}\right)x10^{19}$ y
	NEMO [27]	$(0.95\pm0.04)x10^{19}$ y
^{116}Cd	Osaka [26]	$\left(2.6^{+0.7}_{-0.5}\right)x10^{19}$ y
	Kiev [28]	$\left(2.7^{+0.5+0.9}_{-0.4-0.6}\right)x10^{19}$ y
	NEMO [27]	$\left(3.6^{+0.6}_{-0.5}\pm0.3\right)x10^{19}$ y
^{128,130}Te	Washington (St. Louis) '92 [16]	$T_{1/2}^{130}/T_{1/2}^{128}=(3.52\pm0.11)x10^{-4}$ y $T_{1/2}^{130}=(2.7\pm0.1)x10^{21}$ y $T_{1/2}^{128}=(7.7\pm0.4)x10^{24}$ y
	Missouri [16]	$T_{1/2}^{130}=(0.7\pm0.2)x10^{21}$ y $T_{1/2}^{128}=(1.4\pm0.4)x10^{24}$ y
	Heidelberg [16]	$T_{1/2}^{130}$=(1.5-2.75)$x10^{21}$y and $T_{1/2}^{128}>5x10^{24}$y
^{136}Xe	Caltech/Neuchatel/PSI [30]	$\geq2.3x10^{20}$ y (90% C.L.)
^{150}Nd	ITEP [36]	$\left(1.88^{+0.66}_{-0.39}\pm0.19\right)x10^{19}$ y
^{238}U	Chic./Sta.Fe/Los Alam.[17]	$(2.0\pm0.6)x10^{21}$ y

4 Hunting the Majoron

One of the most attractive theoretical approaches to the generation of neutrino masses is to assume that they are Majorana particles with masses arising as a result of the spontaneous symmetry breaking of the lepton number. The emerging massless Goldstone boson in these theories is called Majoron. This idea has been materialized in various models depending on the weak isospin properties of the Higgs field. Chicashiga, Mohapatra and Peccei (CMP) used a singlet (coupled extremely weak to neutrinos), whereas in the Gelmini and Roncadelli (GR) model, the non-zero of the vacuum expectation value of a Higgs triplet field spontaneously breaks the B-L global symmetry [2].

The GR triplet Majoron Model has been ruled out by the experimental measurement of the Z^0 width (LEP) because the Majoron of the GR model would contribute to the equivalent of two aditional neutrinos to the Z^0 width. So, $(2\beta)_{0\nu M}$. requires a massless or light scalar particle with couple to neutrino but not with Z. New Majoron models have been invented [32] to circumvent this difficulty and, at the same time, to allow, at an observable rate, double beta neutrinoless decays with Majoron emission, i.e. with a strong enough coupling <g>. It should be noted, however, that these new models predict energy spectra different from that derived in the popular GR model.

There has been, from time to time, hints pointing towards the existence of a majoron as revealed in the two-electron energy spectrum as a shoulder or an excess of counts near the end point. Nevertheless, these hints have disappeared, and the best fit to the spectrum has provided the $(2\beta)_{0\nu M}$. half lives quoted in Table (4), together with the corresponding coupling strength g, derived from them.

Table 4. Limits On Majoron Decay Modes

		$T_{1/2}^{0\nu,M}$ (y)	<g>
^{48}Ca	HEP Beijing	$>7.2\times10^{20}$ y 90% C.L.	$<5.1\times10^{-4}$
^{76}Ge	MPIH/KIAE	$>3.9\times10^{22}$ y 90% C.L.	$<1.1\times10^{-4}$
^{82}Se	UCI	$>4.4\times10^{20}$ y 90% C.L.	$<(2-3)\times10^{-4}$
^{100}Mo	Osaka	$>7.9\times10^{20}$ y 90% C.L.	$<3\times10^{-4}$
	NEMO	$>5.0\times10^{20}$ y 90% C.L.	2.4×10^{-4}
	UCI	$>3.9\times10^{20}$ y 90% C.L.	5.8×10^{-4}
^{116}Cd	Osaka	$>2.5\times10^{20}$ y 68% C.L.	
^{128}Te	Washington	$>7.7\times10^{24}$ y	$<2.4\times10^{-5}$
^{130}Te	Heidelberg	$>1.5\times10^{21}$ y	$<2.1\times10^{-4}$
^{136}Xe	CALTECH/UN/PSI	$>7.2\times10^{21}$ y 90% C.L.	$<2\times10^{-4}$
^{150}Nd	UCI	$>5.3\times10^{20}$ y 99% C.L.	$<7\times10^{-5}$
	INR Baksan	$>1.3\times10^{20}$ y 95% C.L.	$<1.4\times10^{-4}$

5 Nuclear Matrix Elements Calculations

The main uncertainties in the calculation of 2ß half-lives come, as it is well-known, from the nuclear matrix elements. Reliable calculations are of paramount importance to extract from experiments the neutrino properties we are looking for. Outstanding progress has been achieved over the last years, mainly due to the application of the Quasiparticle Random Phase Approximation. The strong reduction of the $(2\beta)_{2\nu}$ nuclear matrix element to cope with the large half-lives measured till now, has been understood and consolidated. The role of the particle-particle component of the spin-isospin force has been the main ingredient in understanding the suppression of the rates and various independent calculations essentially agree on such an effect. However, the suppression is extremely sensitive to the strength of the particle-particle interaction, which in fact may lead to null matrix elements for values of g_{pp} very close to (or within) its accepted range. The question of this extreme sensitivity of the QRPA predictions to the nuclear parameters has still to be settled. Small changes of the g_{pp} strength, produce dramatic changes in the half-lives. See for instance Ref. [4], [5], [6].

The lesser dependence on g_{pp} of the neutrinoless transition matrix elements has been confirmed by various groups (Caltech, Tubingen, Heidelberg, Tokyo...). In particular, they are rather insensitive to details of the nuclear structure (except for some cases). That makes the theoretical prediction of $T_{1/2}^{0\nu} < m_\nu >^2$ rather reliable. The QRPA calculation of MPI/Heidelberg [5] has been extended to all potential 2ß emitters and their results are being used now as a reference mark, although in a few cases, the results are disappointing. Other calculations (QRPA with some phenomenological ingredients) approach the correct experimental result in the 100Mo case (Caltech) [33].

The shell model calculations of 2ß nuclear matrix elements (Haxton-Stephenson [3], Vergados [34]) have been a current reference for may years and still are alive and well. The method has well-known advantages, but the severe truncation on the model space needed to handle the computations may exclude some configurations relevant for the final result. The weak coupling shell model has been applied with reasonable success to Ca, Ge and Se, but gives a rate of almost two orders of magnitude faster than observed in the case of ^{130}Te.

The Operator Expansion Method (OEM) has been applied by the Heidelberg group [9] to the calculation of nuclear matrix element with diverse success. In this method, the GT matrix element is converted into a matrix which is not explicitely dependent on intermediate states. Thus, OEM do not use explicitely the intermediate energy spectrum, reducing so the dependence of the p-p force -which affects the distribution of the ß strength in the intermediate nucleus. The energy denominators $1/E_N$-E_I of the nuclear matrix elements are expanded into an infinite series, and so nuclear energies of the intermediate and initial -or final states- are lifted to the numerators, and then replaced by the corresponding hamiltonian operators, leading to a series of conmutators. The nuclear matrix elements involve the bare nucleon-nucleon interaction without any adjustable parameter. The initial and final wave functions are provided by QRPA or shell model. The operator expansion method (Tokyo, Heidelberg) as well as other Green function methods (Seattle, Caltech) [9], have been applied to Ge, Se, Mo, Te, U and Cd double beta transitions.

The naive phenomenological approach [10]of the dominance of the lowest 1^+ intermediate state (Zaragoza '84), fits reasonably well the experimental results in those cases where the approach was applied. This trend has been recently emphasized also by other authors [33]. Other approaches -like the one based in SU(4) for double beta decay [11]-, to avoid the extreme fine tuning of the microscopic approaches, have been applied to various emitters with results similar to those of the QRPA and shell model.

Some of the more relevant results concerning the theoretical prediction of half-lives for various double beta decays are illustrated in Fig. 2 in comparison with the experimental results.

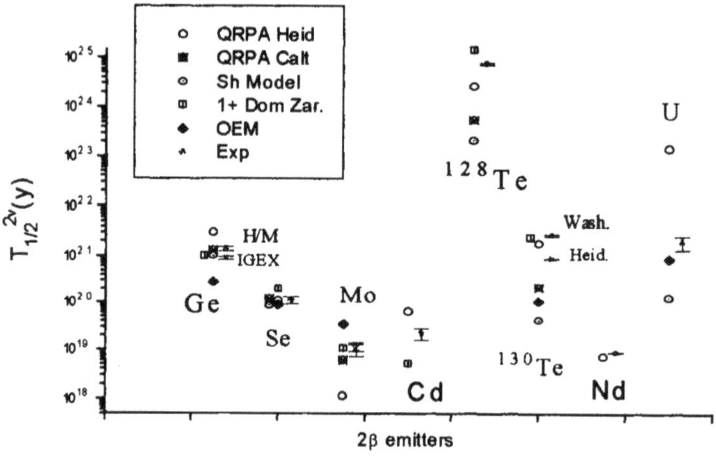

Fig. 2. (2ß)2ν Half-lives. Theory vs Experiment

5 Conclusions on recent 2ß results and perspectives

More than thirty double beta decay experiments are running or have been recently completed. A few more are planned, including large-scale projects and new detection techniques. The progress accomplished has been certainly impressive. As a result, the 2ß decay accompanied by neutrinos has been observed in ^{76}Ge, ^{82}Se, ^{93}Zr, ^{100}Mo, ^{116}Cd, ^{128}Te, ^{130}Te, ^{150}Nd and ^{238}U. Progress in the theoretical estimates of the nuclear matrix elements of the transitions, both in their reliability and in the number of 2ß emitters to which they now extend, have been steadily accompassing the experimental achievements.

In spite of the progress, the neutrinoless, lepton violating decay -if it exists- has escaped detection untill now. Some hints have evaporated but others might be starting to emerge from the ubiquous background.

Taken as mere half life limits, the experimental results in the neutrinoless ^{76}Ge decays would imply a neutrino mass upper limit of $<m_\nu>$ below one eV, which is, untill now, the main result of the searches on the double beta/neutrino mass connection.

The next future will see the extension of the $(2\beta)_{0\nu}$ searches exploring large amount of the qualified isotopes ^{100}Mo, ^{116}Cd, ^{130}Te, ^{150}Nd, ^{136}Xe, i.e. by using new and larger TPC, large tracking detectors, big liquid scintillators, a large-size NEMO detector and bolometers or other low temperature detectors. Also the use of hybrid detectors seems to be in the objective of various groups. For the time being, however, the classical double beta search with ^{76}Ge detectors seems to be the most sensitive. Nevertheless, in the case that the neutrinoless decay were found, to prove again its existence in a variety of isotopes will be imperative.

Acknowledgments

The author wishes to thank his colleagues of the IGEX Collaboration for useful conversations and to J. Valle for his invitation to lecture in this Workshop. The financial support of CICYT (Spain) is also acknowledged.

References

[1] See, for instance, A. Morales, "Double Beta Decay and the Neutrino Mass", Proc. *6th Workshop on Neutrino Telescopes*, Venice 1994, Ed. M. Baldo Ceolin, and references therein.

[2] V. Chicashige, R.N. Mohapatra, R.D. Peccei, *Phys. Lett.* **98B** (1981) 265.
G.B. Gelmini, M. Roncadelli, *Phys. Lett.* **99B** (1981) 411.
H.M. Georgi, S.L. Glashow, S. Nusinov, *Nucl. Phys.* **B193** (1981) 297.

[3] W.C. Haxton and G.J. Stephenson, *J. Prog. in Part. & Nucl. Phys.* **12** (1984) 409.

[4] P. Vogel and M.R. Zirnbauer, *Phys. Rev. Lett.* **57** (1986) 3148.
J. Engel et al., *Phys. Rev.* **C37** (1988) 731.

[5] A. Staudt et al., *Europhys. Lett.* **13** (1990) 31 and references therein.
A. Staudt et al., *Phys. Lett.* **B242** (1990) 17.

[6] T. Tomoda, *Rep. Prog. Phys.* **54** (1991) 53 and references therein.

[7] M. Doi et al., *Prog. Theor. Phys. Suppl.* **83** (1985) 1

[8] J. Schechter and J.W.F. Valle, *Phys. Rev.* **D25** (1982) 2951
B. Kayser et al. "New and exotic phenomena", edited by O. Facker, J. Tran Thanh Van, *Editions Frontieres*, 1987, pg. 349.

[9] Y.R. Wu et al., *Phys. Lett.* **B272** (1991) 169
Y.R. Wu et al., *Phys. Lett.* **B276** (1992) 274
W.C. Haxton, Proc. of *Neutrino'92*, Granada June 1992, edited by A. Morales, *Nucl. Phys. (Proc. Suppl.)* **31** (1993) 88
M. Hirsch et al., *Z. Phys.* **A 345** (1993) 163

[10] A. Morales et al., *J. Phys. C3* **Suppl. 45** (1984) 147 and *Anales de Física* **A80** (1984) 9 and **A80** (1984) 15.

[11] J. Bernabéu et al., *Z. Phys.* **C46** (1990) 323, Ibid 51 (1991) 499.

[12] An updated review of these questions can be seen in Proc. Neutrino 94, Eilat, Israel, *Nucl. Phys. B (Proc. Suppl.)* **38**(1995).
For the LSND experiment, see C. Athanassopoulos et al. *Phys. Rev. Lett.* **75**(1995)2650

[13] M. Turner, "Dark Matter: Theoretical Perspectives", Fermilab-Conf. 92/382.A

[14] D.O. Caldwell, *Nucl. Phys. B (Proc. Suppl.)* **38**(1995)390.

[15] T. Kirsten Science Underground, edited by M. Nuto et al., *AIP Conf Proc.* No. 96, N. York 1983, pg. 396.
O.K. Manuel et al., Nuclear Beta Decay and Neutrinos, edited by T. Kotani et al., *World Scientific*, Singapore 1987.

[16] T. Bernatowicz et al. *Phys. Rev. Lett.* **69** (1992) 2341, and *Phys. Rev.* **C47** (1993) 806.
O.K. Manuel, *J. Phys. G. Nucl. Part. Phys.* **17** (1991) S221 and *Nucl. Phys.* **A529** (1991) 29.
T. Kirsten et al. Phys. *Rev. Lett.* **50** (1983) 474 and *Z. Phys.* **C16** (1983) 189 and Nuclear Beta Decays and Neutrinos, edited by T. Kotani et al., *World Scientific*, Singapore 1986, pg. 81.
E.W. Hennecke et al. *Phys. Rev.* **C11** (1975) 1378
O.K. Manuel et al. Nuclear Beta Decay and Neutrinos edited by T. Kotani et al., *World Scientific*, Singapore 1987, pg. 71

[17] A.L. Turkevich et al., *Phys. Rev. Lett.* **67** (1991) 3211.

[18] B. Sadoulet, *Journ. Low T. Phys.* **93** (1993) 821.

[19] A.A. Vasenko et al., *Mod. Phys. Lett.* **A5** (1990) 1299

[20] H.S. Miley et al., *Phys. Rev. Lett.* **65** (1990) 3092.

[21] F.T. Avignone et al., *Phys. Lett.* **B256** (1991) 559.

[22] A. Balysh et al., *Phys. Lett.* **B283** (1992) 32, **B322** (1994) 176, **B356**(1995)450

[23] R. Brodzinski et al., Proc. *Neutrino 92*, Granada June 1992, edited by A. Morales, *Nucl. Phys. B (Proc. Suppl.)* **31** (1992) pg. 76
F.T. Avignone et al., Proc. TAUP'93, Gran Sasso, September 1993, edited by C. Arpesella, E. Bellotti and A. Bottino, *Nucl. Phys. B (Proc. Suppl.)* **35** (1993) 354.

[24] S.R. Elliot et al. *Phys. Rev. Lett.* **59** (1987) 2020
M.K. Moe et al., *Prog. Part. Nucl. Phys.* **32** (1994).
S.R. Elliot et al. *J. Phys. G.* **17** (1991) S145, *Nucl. Phys. (Proc. Suppl.)* **31B** (1993) 68. Phys. Rev. **C46**(1992)1535

[25] M. Alston-Garnjost et al., *Phys. Rev. Lett.* **71**(1993)831

[26] H. Ejiri et al., *Phys. Lett.* **B258** (1991) 17, *J. Phys. G., Nucl. Part. Phys.* **17** (1991) S155, *Nucl. Phys. B (Proc. Suppl.)* **28A** (1992) 219

[27] NEMO Collaboration, *Nucl. Phys. B (Proc. Suppl.)* **35** (1994) 369

[28] F. Danevich et al., Phys. Lett. **B344**(1995)72

[29] A. Alessandrello et al. *Phys. Lett.* **B285** (1992) 176, **B335** (1994) 519.

[30] J.L. Vuilleumier et al., *Phys. Rev. D.* **48**(1993)1009

[31] C. Sáenz et al., *Phys. Rev. C* **50**(1994)1170

[32] Z.G. Berezhiani, A.Yu. Smirnov, J.W.F. Valle, *Phys. Lett.* **B291** (1992) 99
C.P. Burgess and J.M. Cline, *Phys. Lett.* **B298** (1993) 141

[33] A. Griffiths and P. Vogel, *Phys. Rev.* **C47** (1993) 2910.

[34] J.D. Vergados, *Phys. Rep.* **133** (1986) 1.

[35] A. Kawashima et al. *Phys. Rev.* **C47**(1993)2452

[36] V.A. Artemiev et al. *Phys. Lett.* **B345**(1995)564

[37] IGEX Collaboration, Proc. TAUP 95, Toledo, Spain. Ed. by A. Morales, J. Morales, J.A. Villar, to appear in *Nucl. Phys. B (Proc. Suppl.)* 1996

FUTURE ν_τ OSCILLATION EXPERIMENTS AND PRESENT DATA

M. C. Gonzalez-Garcia

Theory Division, CERN, CH-1211 Geneva 23, Switzerland.

Our goal in this paper is to examine the discovery potential of laboratory experiments searching for the oscillation $\nu_\mu(\nu_e) \to \nu_\tau$, in the light of recent data on solar and atmospheric neutrino experiments, which we analyse together with the most restrictive results from laboratory experiments on neutrino oscillations in a four-neutrino framework.

1 Introduction

If neutrinos have a mass, a neutrino produced with flavour α, after travelling a distance L, can be detected in the charged-current (CC) reaction $\nu\ N' \to l_\beta\ N$ with a probability

$$\langle P_{\alpha\beta}\rangle \quad = \delta_{\alpha,\beta} - 4\sum_{i=1}^{n}\sum_{j=i+1}^{n} \mathrm{Re}[U_{\alpha i}U^\star_{\beta i}U^\star_{\alpha j}U_{\beta j}]\langle\sin^2\left(\frac{\Delta_{ij}}{2}\right)\rangle, \qquad (1)$$

where U is the mixing matrix. The average includes the dependence on the neutrino energy spectrum, the cross section for the process in which the neutrino is detected, and detection efficiency for the experiment. The probability, therefore, oscillates with oscillation lengths $\frac{\Delta_{ij}}{2} = 1.27\frac{|m_i^2-m_j^2|}{\mathrm{eV}^2}\frac{L/E}{\mathrm{m/MeV}}$ where E is the neutrino energy. For oscillation lengths such that $\Delta m^2_{ij} \gg 1/(L/E)$ the oscillating phase will have been over many cycles before the detection and therefore it will have averaged to 1/2. On the other hand, for $\Delta m^2_{ij} \ll 1/(L/E)$, the oscillation did not have time to give any effect.

Present data from solar and atmospheric neutrino experiments favour the hypothesis of neutrino oscillations. All solar neutrino experiments [1] find less ν_e than predicted theoretically. As for atmospheric neutrino experiments, two of them [2,3,4] measure a ratio ν_μ/ν_e smaller than expected from theoretical calculations. Nevertheless, this interpretation requires confirmation from further experiments, in particular from laboratory experiments, where the experimental conditions, in particular the shape, energy, and flux of the neutrino beam are under control.

At present all laboratory neutrino experiments report no evidence for neutrino oscillation [5] with the possible exception of LSND [6], which looks for

$\nu_\mu \to \nu_e$ oscillations. In addition a number of new experiments are in the process of starting taking data of being propossed both at CERN [7] and Fermilab [8]. Our goal is to examine the discovery potential of these experiments searching for the oscillation $\nu_\mu(\nu_e) \to \nu_\tau$, in the light of recent data on solar and atmospheric neutrino experiments, which we analyse together with the most restrictive results from laboratory experiments on neutrino oscillations in a four-neutrino framework [9].

2 Four-Flavour Models

Naive two-family counting shows that it is very difficult to fit all experimental information mentioned above with three neutrino flavours. In the spirit of Pauli, one is tempted to introduce a new neutrino as a "desperate solution" to understand all present data. The nature of such a particle is constrained by LEP results on the invisible Z width as well as data on the primordial ^{4}He abundance. Those rule out the existence of additional, light, active neutrinos. In consequence the fourth neutrino state must be sterile.

If one assumes a *natural* mass hierarchy with two light neutrinos with their main projection in the ν_s and ν_e directions and two heavy neutrinos with their largest component along the ν_μ and ν_τ flavours the mixing matrix can be parametrized in a general way as

$$U = \begin{array}{c|cccc} & \nu_s & \nu_e & \nu_\mu & \nu_\tau \\ \hline \nu_1 & c_{es} & s_{es}c_{e\mu}c_{e\tau} & s_{es}s_{e\mu} & s_{es}c_{e\mu}s_{e\tau} \\ \nu_2 & -s_{es} & c_{es}c_{e\mu}c_{e\tau} & c_{es}s_{e\mu} & c_{es}c_{e\mu}s_{e\tau} \\ \nu_3 & 0 & -c_{\mu\tau}s_{e\mu}c_{e\tau} - s_{\mu\tau}s_{e\tau} & c_{\mu\tau}c_{e\mu} & -c_{\mu\tau}s_{e\mu}s_{e\tau} + s_{\mu\tau}c_{e\tau} \\ \nu_4 & 0 & s_{\mu\tau}s_{e\mu}c_{e\tau} - c_{\mu\tau}s_{e\tau} & -s_{\mu\tau}c_{e\mu} & s_{\mu\tau}s_{e\mu}s_{e\tau} + c_{\mu\tau}c_{e\tau} \end{array} \quad (2)$$

with $c_i = \cos\theta_i$ and $s_i = \sin\theta_i$. For the sake of simplicity we have assumed no CP violation in the lepton sector. We also required that the sterile neutrino does not mix directly with the two heavy states to verify the constraints from big bang nucleosynthesis [10]. Such a hierarchy appears naturally, for instance if one advocates an $L_e \pm L_\mu \mp L_\tau$ discrete symmetry for the mass matrix [12]. In Ref. [11] a similar mass pattern is also generated via a combination of see-saw mechanism and loop mechanism. In this approximation $m_1, m_2 \ll m_3, m_4$ and $|m_3^2 - m_4^2| \ll m_3^2, m_4^2$.

The value of $m_3 \approx m_4$ is inferred from the dark matter data. Currently, the best scenario to explain the data considers a mixture of 70% cold plus 30% hot dark matter [13]. This implies $m_3 \simeq m_4 =$ 2–3.5 eV. Such a mass pattern has been argued to yield satisfactory results in Cold+Hot Dark Matter scenarios

[14]. We define

$$\Delta m^2_{solar} = |m_1^2 - m_2^2| \quad \Delta m^2_{AT} = |m_3^2 - m_4^2|$$
$$\Delta M^2_{DM} = |m_1^2 - m_3^2| \simeq |m_1^2 - m_4^2| \simeq |m_2^2 - m_3^2| \simeq |m_2^2 - m_4^2| \simeq (4-10)\ \text{eV}^2\ . \tag{3}$$

Transition probabilities between the different flavours will have therefore contributions from the three oscillation lengths due to the three different mass differences in the problem which we will denote $\sin^2(\Delta_{solar}/2)$, $\sin^2(\Delta_{AT}/2)$, and $\sin^2(\Delta_{DM}/2)$, respectively.

3 Global Analysis

At present the most precise laboratory experiments searching for neutrino oscillations are [5] the reactor experiment at Bugey, which looks for ν_e disappearance, and the CDHSW experiment at CERN, which searches for ν_μ disappearance. The E776 experiment at BNL searchs for the $\bar{\nu}_\mu \to \bar{\nu}_e$ appearance channel and the E531 experiment at Fermilab for the $\nu_\mu \to \nu_\tau$ channel. Neither of these experiments shows evidence for neutrino oscillation on those channels. Recently the Liquid Scintillator Neutrino Detector (LSND) experiment [6] has announced the observation of an anomaly that can be interpreted as neutrino oscillations in the channel $\bar{\nu_\mu} \to \bar{\nu_e}$. Most of the oscillation parameters required as explanation are already excluded by the E776 and KARMEN experiments.

For the Bugey reactor experiment the relevant transition probability is the ν_e survival probability. For any value of the atmospheric mass difference this probability will always verify

$$0.93 \lesssim P_{ee}^{Bugey} \leq 1 - 2c_{e\mu}^2 c_{e\tau}^2 (1 - c_{e\mu}^2 c_{e\tau}^2) \ \Rightarrow\ c_{e\mu}^2 c_{e\tau}^2 \geq 0.96\ . \tag{4}$$

For CDHSW the relevant probability is the ν_μ survival probability

$$0.95 \lesssim P_{\mu\mu}^{CDHSW} \leq 1 - \frac{1}{2}\sin^2(2\theta_{e\mu}) \ \Rightarrow\ \sin^2(2\theta_{e\mu}) \lesssim 0.1\ . \tag{5}$$

For E776 the situation is somehow more involved, since the value of the oscillating phase $\langle \sin^2(\Delta_{DM}/2) \rangle$ varies in the range $\Delta_{DM} = 4$–$10\ \text{eV}^2$ due to the wiggles of the resolution function of the experiment. Also, the experiment is sensitive to the atmospheric mass difference. We find that the limit $1.5 \times 10^{-3} \geq P_{e\mu}^{E776}$ is verified for any value of the atmospheric mass difference and the $\mu\tau$ mixing angle, if

$$\sin^2(2\theta_{e\mu}) c_{e\tau}^2 \leq (2\text{–}5) \times 10^{-3} \tag{6}$$

The limit from E531 on the mixings $e\mu$ and $e\tau$ is always less restrictive than the previous ones for any value of Δm^2_{AT} and ΔM^2_{DM}.

Combining these constraints we obtain that $e\mu$ and $e\tau$ mixings are constrained to

$$\sin^2(2\theta_{e\mu}) \leq (2\text{–}5)\times 10^{-3} \qquad \sin^2(2\theta_{e\tau}) \leq 0.16 \tag{7}$$

where the range of $\sin^2(2\theta_{e\mu})$ depends on the specific value of ΔM^2_{DM}.

If we now turn to the effect due to the oscillation with Δ_{AT}, we can rewrite the relevant probabilities for the different experiments expanding in the small angles $e\mu$ and $e\tau$:

$$\begin{aligned} P^{CDHSW}_{\mu\mu} &\simeq 1 - \tfrac{1}{2}\sin^2(2\theta_{e\mu}) - \sin^2(2\theta_{\mu\tau})\sin^2(\tfrac{\Delta_{AT}}{2}) \\ P^{E531}_{\mu\tau} &\simeq \sin^2(2\theta_{\mu\tau})c^2_{e\tau}\sin^2(\tfrac{\Delta_{AT}}{2}) \\ P^{776}_{e\mu} &\simeq \sin^2(2\theta_{e\mu})c^2_{e\tau}\sin^2(\tfrac{\Delta_{DM}}{2}) + \sin^2(2\theta_{\mu\tau})s^2_{e\tau}\sin^2(\tfrac{\Delta_{AT}}{2})\,. \end{aligned} \tag{8}$$

With the constraints in Eq. (7), the Bugey experiment is not sensitive to oscillations with Δ_{AT}. The relevant exclusion contours for each channel are shown in the Figures.

We now turn to the atmospheric neutrino data. Neutrinos are produced when cosmic rays hit the atmosphere and initiate atmospheric cascades. The mesons present in the cascade decay leading to a flux of ν_e and ν_μ which reach the Earth and interact in the different neutrino detectors. Naively the expected ratio of ν_μ to ν_e is in the proportion 2 : 1, since the main reaction is $\pi \to \mu\nu_\mu$ followed by $\mu \to e\nu_\mu\nu_e$. However, the expected ratio of muon-like interactions to electron-like interactions in each experiment depends on the detector thresholds and efficiencies as well as on the expected neutrino fluxes. Currently four experiments have observed atmospheric neutrino interactions. Two experiments, Kamiokande [2,3] and IMB [4], have observed a ratio of ν_μ-induced events to ν_e-induced events smaller than the expected one. In particular Kamiokande has performed two different analyses for both sub-GeV neutrinos [2] and multi-GeV neutrinos [3], which show the same deficit. On the other hand, the results from Fréjus and NUSEX [15] appear to be in agreement with the predictions.

The results of the three most precise experiments in terms of the double ratio $R_{\mu/e}/R^{MC}_{\mu/e}$ of experimental-to-expected ratio of muon-like to electron-like events are

$$\begin{aligned} R_{\mu/e}/R^{MC}_{\mu/e} &= 0.55 \pm 0.11 \quad \text{for IMB} \\ R_{\mu/e}/R^{MC}_{\mu/e} &= 0.6 \pm 0.09 \quad \text{for Kamiokande sub-GeV} \\ R_{\mu/e}/R^{MC}_{\mu/e} &= 0.59 \pm 0.1 \quad \text{for Kamiokande multi-GeV} \\ R_{\mu/e}/R^{MC}_{\mu/e} &= 1.06 \pm 0.23 \quad \text{for Fréjus} \end{aligned} \tag{9}$$

The statistical and systematic errors have been added in quadrature. The systematic error contains a 5 % contribution due to the neutrino flux uncertainties. For the Montecarlo prediction we have used the expected fluxes from [16] depending on the neutrino energies. Use of other flux calculations would yield similar numbers.

In each experiment the number of μ events, N_μ, and of e events, N_e, in the presence of oscillations will be

$$N_\mu = N^0_{\mu\mu}\langle P_{\mu\mu}\rangle + N^0_{e\mu}\,\langle P_{e\mu}\rangle\,, \qquad N_e = N^0_{ee}\langle P_{ee}\rangle + N^0_{\mu e}\langle P_{\mu e}\rangle\,, \tag{10}$$

where

$$N^0_{\alpha\beta} = \int \frac{d^2\Phi_\alpha}{dE_\nu d\cos\theta_\nu}\frac{d\sigma}{dE_\beta}\epsilon(E_\beta)dE_\nu dE_\beta d(\cos\theta_\nu) \tag{11}$$

and

$$\langle P_{\alpha\beta}\rangle = \frac{1}{N^0_{\alpha\beta}}\int \frac{d^2\Phi_\alpha}{dE_\nu d\cos\theta_\nu}P_{\alpha\beta}\frac{d\sigma}{dE_\beta}\epsilon(E_\beta)dE_\nu dE_\beta d(\cos\theta_\nu)\,. \tag{12}$$

Here E_ν is the neutrino energy and Φ_α is the flux of atmospheric neutrinos ν_α; E_β is the final charged lepton energy and $\epsilon(E_\beta)$ is the detection efficiency for such charged lepton; σ is the interaction cross section $\nu\ N \to N'\ l$. The expected rate with no oscillation would be $R^{MC}_{\mu/e} = N^0_{\mu\mu}/N^0_{ee}$. The double ratio $R_{\mu/e}/R^{MC}_{\mu/e}$ is then given by

$$\frac{R_{\mu/e}}{R^{MC}_{\mu/e}} = \frac{\langle P_{\mu\mu}\rangle + \frac{N^0_{e\mu}}{N^0_{\mu\mu}}\langle P_{e\mu}\rangle}{\langle P_{ee}\rangle + \frac{N^0_{\mu e}}{N^0_{ee}}\langle P_{\mu e}\rangle}\,. \tag{13}$$

We perform a global fit to the data in Eq. (9). In Fig. 3 the results are shown for zero mixings $e\mu$ and $e\tau$ as in a two-family scenario. Figure 4 shows the effect of the inclusion of the mixings. As seen in the figure the inclusion of the $e\tau$ mixing leads to a more constrained area for the oscillation parameters. The effect of the $e\tau$ mixing is to increase the value of the double ratio since there is a decrease on the number of ν_e. Therefore a larger amount of $\mu\tau$ oscillation is needed to account for the deficit. Due to the small values allowed, a non-zero mixing $e\mu$ does not modify the analysis of the atmospheric neutrino data.

Finally just to coment that for solar neutrino experiments the presence of new mixings affect very little the analysis performed in the two family scenario, the two MSW [17] solutions are still valid. The effect of the mixings appears to go in the direction of favouring the small mixing solution [18]. The large mixing solution is also in conflict with the constraints from big bang nucleosynthesis [10].

Table 1: Summary of the performance parameters for CHORUS and NOMAD.

	NOMAD	CHORUS	E803/ NAUSICAA (CERN)	MINOS
Sensitivity:				
$P_{e\tau}$	1.3×10^{-2}	0.8×10^{-2}	7×10^{-4}	–
$P_{\mu\tau}$	2×10^{-4}	1.4×10^{-4}	1.4×10^{-5}	0.012

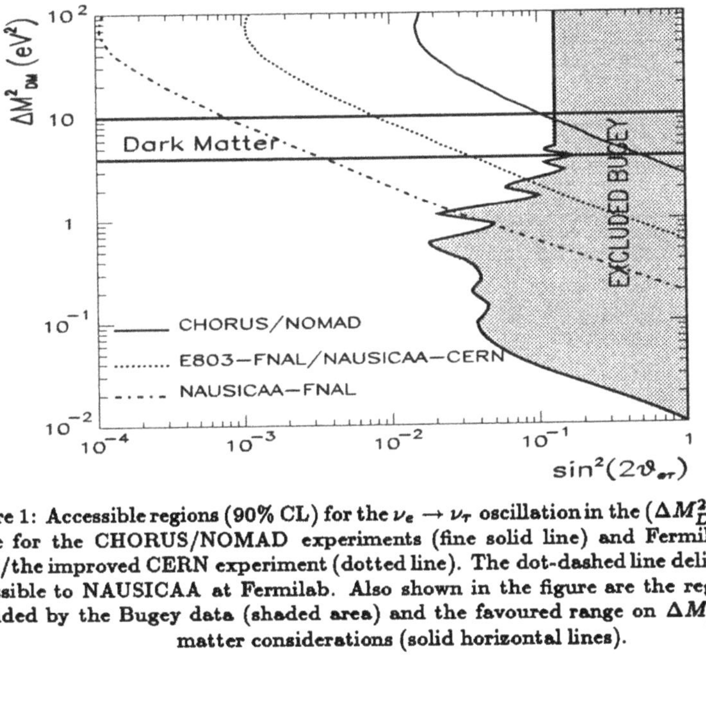

Figure 1: Accessible regions (90% CL) for the $\nu_e \to \nu_\tau$ oscillation in the $(\Delta M^2_{DM}, \sin^2(2\theta_{e\tau}))$ plane for the CHORUS/NOMAD experiments (fine solid line) and Fermilab experiment E803/the improved CERN experiment (dotted line). The dot-dashed line delimits the region accessible to NAUSICAA at Fermilab. Also shown in the figure are the region at present excluded by the Bugey data (shaded area) and the favoured range on ΔM^2_{DM} from dark matter considerations (solid horizontal lines).

4 ν_τ Oscillation Experiments:Discovery Potential

The two upcoming $\nu_\mu(\nu_e) \to \nu_\tau$ experiments, CHORUS and NOMAD [7], are ν_τ appearance experiments, i.e., they search for the appearance of ν_τ's in the CERN SPS beam consisting primarily of ν_μ's, with about 1% ν_e's. The mean energy of the ν_μ beam is around 30 GeV and the detectors are located approximately 800 m away from the beam source. Their expected performance are summarized in Table 1. There are a number of future $\nu_\mu(\nu_e) \to \nu_\tau$ experiments being discussed at present. As a specific example of these experiments we have considered a suggestion to upgrade the NOMAD detector [19]. We will refer to this future detector with the generic name of Neutrino ApparatUS with Im-

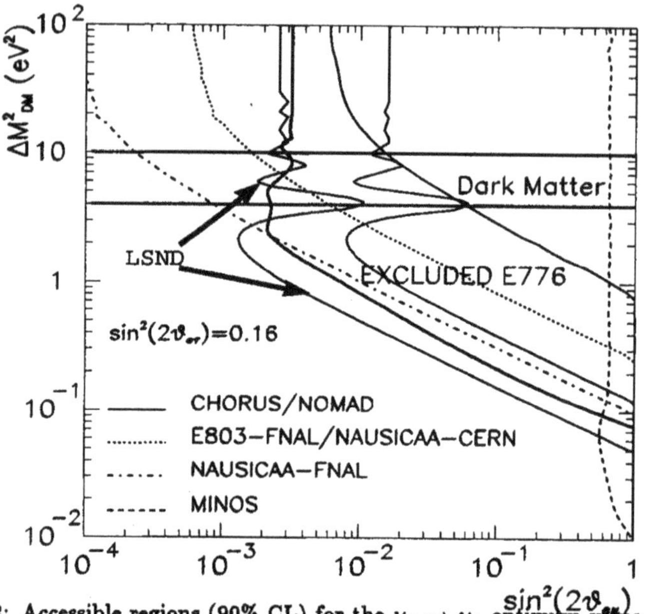

Figure 2: Accessible regions (90% CL) for the $\nu_\mu \to \nu_\tau$ optimum value of the mixing $e\tau$. The fine solid line corresponds to the CHORUS/NOMAD experiments and the dotted line corresponds to the Fermilab experiment E803 and to the improved CERN experiment. The dot-dashed line corresponds to NAUSICAA at Fermilab and the dashed line corresponds to MINOS. Also shown in the figure are the regions at present excluded by E776 data (shaded area) and the favoured range on ΔM^2_{DM} from dark matter considerations (solid horizontal lines). For comparison also the LSND data are shown.

proved CApAbilities (NAUSICAA). The detector performance is summarized in Table 1.

At Fermilab, a neutrino beam will be available when the main injector becomes operational, around the year 2001. Compared with the CERN SPS beam, the main injector will deliver a beam 50 times more intense, but with an average energy around one third of that of the SPS neutrinos. There are currently two experiments proposed to operate in this beam [8]. One is a short-baseline experiment, E803, and MINOS a long-baseline experiment, which proposes two detectors, separated by 732 km. This experiment can perform several tests to look for a possible oscillation $\nu_\mu \to \nu_\tau$ in the small mass difference range. The expected performance of E803 and MINOS are summarized in Table 1. Finally we will consider the possibility of installing the NAUSICAA detector as an alternative or a successor to E803 in the Fermilab beam which will improve by one order of magnitude its sensitivity.

After implementing the limits derived in Sec. 3 and considering the sensitivity of the experiments, one sees that for all facilities the only observable $\nu_e \to \nu_\tau$ transition oscillates with an oscillation length Δ_{DM} such that

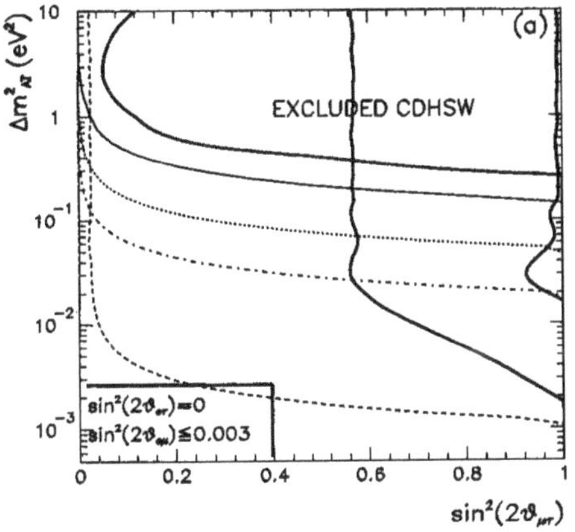

Figure 3: Accessible regions (90% CL) for the $\nu_\mu \to \nu_\tau$ oscillation in the $(\Delta m^2_{AT}, \sin^2(2\theta_{\mu\tau}))$ plane for zero value of the mixing $e\tau$. The fine solid line corresponds to the CHORUS/NOMAD experiments and the dotted line corresponds to the Fermilab experiment E803 and to the improved CERN experiment. The dot-dashed line corresponds to NAUSICAA at Fermilab and the dashed line to MINOS. Also shown in the figure are the regions at present excluded by CDHSW data (dark shaded area) and the atmospheric neutrino analysis (light shaded area).

$P_{e\tau} \simeq \sin^2(2\theta_{e\tau})\sin^2(\frac{\Delta_{DM}}{2})$. Figure 1 shows the regions accessible to the experiments in the $(\sin^2(2\theta_{e\tau}), \Delta M^2_{DM})$ plane. For transitions $\nu_\mu \to \nu_\tau$ a four-neutrino framework predicts (unlike the naive two-family framework) *two* oscillations, dominated by the characteristic lengths Δ_{DM} and Δ_{AT}. All experiments are in principle sensitive to both oscillations depending on the values of the mixing angles

$$P^{DM}_{\mu\tau} \simeq \sin^2(2\theta_{e\mu})\sin^2(\theta_{e\tau})\sin^2(\tfrac{\Delta_{DM}}{2}) \qquad P^{AT}_{\mu\tau} \simeq \sin^2(2\theta_{\mu\tau})\cos^2(\theta_{e\tau})\sin^2(\tfrac{\Delta_{AT}}{2}) \tag{14}$$

In Fig. 2 we show the regions accessible for the oscillation with oscillation lenght ΔM^2_{DM}) to the different experiments for an optimum value of the $e\tau$ mixing.

Figures 3 and 4 show the region accessible to the experiments for the oscillation in Δm^2_{AT}) for different values of the other mixings.

References

1. GALLEX Collab., Phys. Lett. **B327**, 377 (1994); SAGE Collab., Phys. Lett. **B328**, 234 (1994); Homestake Collab., Nucl. Phys. **B38** (Proc. Suppl.), 47 (1995); Kamiokande Collab., Nucl. Phys. **B38** (Proc. Suppl.), 55 (1995).

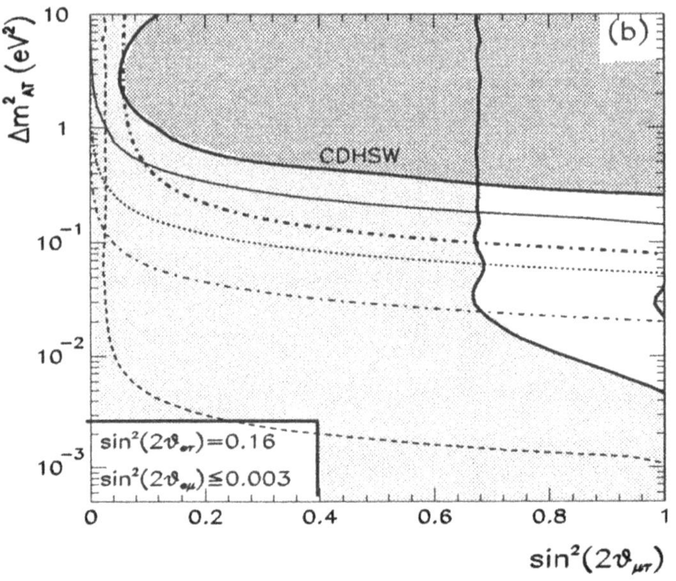

Figure 4: Same as previous figure for maximum mixing $e\tau$. Also shown in the figure are the region at present excluded by CDHSW data (dark shaded area) and the atmospheric neutrino analysis (light shaded area). These are the only constraints for zero or values of the $e\mu$ mixing $\sin^2(2\theta_{e\mu}) \ll 0.003$. For maximum value $\sin^2(2\theta_{e\mu}) \simeq 0.003$ the E776 limit (thick dot-dashed line) is relevant.

2. Kamiokande Collab., H. S. Hirata *et al.*, Phys. Lett. bf B205, 416 (1988) and Phys. Lett. **B280**, 146 (1992).
3. Kamiokande Collab., Y. Fukuda *et al.*, Phys. Lett. **B335**, 237 (1994).
4. IMB Collab., D. Casper *et al.*, Phys. Rev. Lett. **66**, 2561 (1991); R. Becker-Szendy *et al.*, Phys. Rev. **D46**, 3720 (1992).
5. BUGEY Collab. B. Achkar *et al.*, Nucl. Phys. **B424**, 503 (1995); CDHSW Collab., F. Didak *et al.*, Phys. Lett. **B134**, 281 (1984); E776 Collab., L. Borodvsky *et al.*, Phys. Rev. Lett. **68**, 274 (1992); E531 Collab., Phys. Rev. Lett. **57**, 2898 (1986); KARMEN Collab., B. Armbruster *et al.*, Nucl. Phys. **B38** (Proc. Suppl.), 235 (1995).
6. See for example W. C. Louis representing the LSND Collab., Nucl. Phys. **B38** (Proc. Suppl.), 229 (1995).
7. CHORUS Collab., N. Armenise et al., CERN-SPSC/90-42 (1990); NOMAD Collab., P. Astier et al., CERN-SPSLC/91-21 (1991), CERN-SPSLC/91-48 (1991), SPSLC/P261 Add. 1 (1991).
8. E803 Experiment, P-803 Proposal, Fermilab, October 1993; MINOS Experiment P-875 Proposal, Fermilab, March, 1995.
9. For more details see J. J. Gomez-Cadenas, and M. C. Gonzalez-Garcia,

CERN preprint CERN-TH/95-80, hep-ph/9504246.
10. X. Shi, D. N. Schramm and B. D. Fields, Phys. Rev. **D48**, 2563 (1993).
11. D. O. Caldwell and R. N. Mohapatra, Phys. Rev. **D48**, 3259 (1993);
12. J. Peltoniemi and J. W. F. Valle, Nucl. Phys. **B406**, 409 (1993)
13. E. L. Wright *et al.*, Astrophys. J. **396**, L13 (1992); M. Davis, F. J. Summers and D. Schlegel, Nature **359**, 393 (1992); A. N. Taylor and R. Rowan-Robinson, *ibid* **359**, 396 (1992); J. A. Holtzman, and J. R. Primack, Astrophys. J. **405**, 428 (1993).
14. J. R. Primack, J. Holtzman, A. Klypin, and D. O. Caldwell, UC Santa Cruz preprint SCIPP 94/28 and references therein.
15. Fréjus Collab., Ch. Berger *et al.*, Phys. Lett. **B227**, 489 (1989);NUSEX Collab., M. Aglietta *et al.*, Europhys. Lett. **8**, 611 (1989).
16. T. K. Gaisser, T. Stanev and G. Barr, Phys. Rev. **D38**, 85 (1988) and Phys. Rev. **D39**, 3532 (1989); V. Volkova, Sov. J. Nucl. Phys. **31**, 784 (1980) .
17. S. P. Mikheyev and A. Yu. Smirnov, Yad. Fiz. **42**, 1441 (1985); L. Wolfenstein, Phys. Rev. **D17**, 2369 (1985).
18. G. L. Fogli, E. Lisi and D. Montanino, Phys. Rev. **D49**, 3626 (1994); CERN preprint CERN-TH.7491/94.
19. J. J. Gomez-Cadenas, A. Bueno and J. A. Hernando, NOMAD/Memo 95-013(1995)

CORRELATED NEUTRINO OSCILLATIONS

ERNEST MA
Department of Physics, University of California,
Riverside, California 92521, USA

ABSTRACT

In solar neutrino oscillations, if ν_e has a significant third massive component, the allowed parameter space in Δm^2 and $\sin^2 2\theta$ for the first two components is shown to be greatly increased. This third component may be correlated to atmospheric neutrino oscillations, as shown in a specific predictive seesaw model of the 3×3 neutrino mass matrix. Possible variations to include the recent LSND results are briefly discussed.

1. Introduction

There are now three categories of data which show evidence of neutrino oscillations. (a) In experiments which detect neutrinos from the sun, there appears to be a deficit. Hence ν_e has apparently disappeared. (b) In experiments which measure the ratio ν_μ/ν_e in the atmosphere, there also appears to be a deficit. Hence a combination of ν_e and ν_μ diappearance and appearance may have also occurred. (c) The recent results of the LSND (Liquid Scintillator Neutrino Detector) experiment[1] seem to indicate that ν_e has appeared where there is originally only ν_μ.

Conventional interpretations of the above as neutrino oscillations always plot Δm^2 versus $\sin^2 2\theta$, assuming implicitly that only two neutrinos are involved in each case. As long as all mixing angles are small, this is a good approximation because the Δm^2 in each case is very different from one another. However, the atmospheric data[2] are strongly indicative of a large mixing between ν_μ and ν_e or ν_τ or both. Hence ν_e may well be composed of three mass eigenstates with a masive third component to account for all or part of the atmospheric oscillations, whereas the solar oscillations are explained by the first two components with a small Δm^2 together with a nonnegligible contribution from the massive third component. In the following it will be shown that this has the important consequence of enlarging the parameter space of Δm^2 and $\sin^2 2\theta$ for the first two components that is allowed by the present solar data.[3]

Since the LSND results are indicative of a much larger Δm^2, of order eV^2, a fourth neutrino is required if all of the data are to be explained by neutrino oscillations. This possibility will also be discussed.

2. Three-Neutrino Analysis of Solar Data

The work that I will describe in this section was done in collaboration with J. Pantaleone,[4] who has been considering neutrino oscillations of all three flavors for many years.[5] Other authors are now beginning to follow suit.[6]

Let the electron neutrino be a linear combination of three mass eigenstates:

$$\nu_e = \cos\theta\ \nu_1 - \sin\phi \sin\theta\ \nu_2 + \cos\phi \sin\theta\ \nu_3, \tag{1}$$

and assume*

$$\Delta m_{13}^2 \simeq \Delta m_{23}^2 \sim 10^{-2}\ \text{eV}^2. \tag{2}$$

Then for a given value of θ, one may use the solar data to find the allowed region in Δm_{12}^2 and $\sin^2\theta_{e2} \equiv 4|U_{e2}|^2(1-|U_{e2}|^2)$. The results for $\sin^2 2\theta = 0.35$ and 0.75 are shown below.

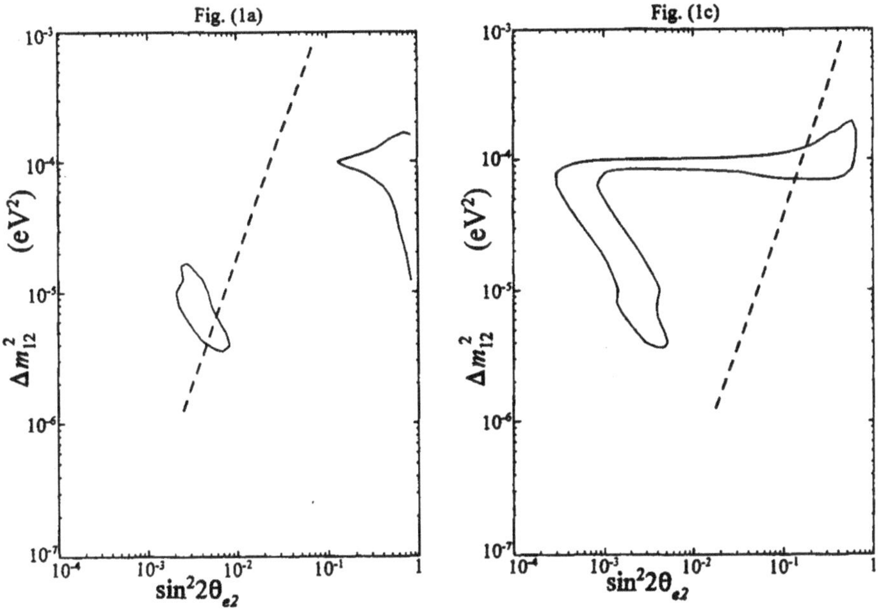

On the left is the allowed region for $\sin^2 2\theta = 0.35$ which differs from that of the two-neutrino analysis, *i.e.* $\theta = 0$, by only a little. However, there is already a firm indication that it has enlarged. On the right is the allowed region for $\sin^2 2\theta = 0.75$ which shows dramatically that it has greatly increased and that the adiabatic branch of the solution at around $\Delta m^2 = 10^{-4}$ eV^2 is now allowed. The dashed lines are theoretical predictions to be discussed in the next section.

*Note that if m_3 were of order a few eV, reactor data would require that ν_3 overlaps very little with ν_e. See the talk by M. C. Gonzalez-Garcia, these proceedings.

3. Seesaw Structure Revealed

Recall that the well-known empirical relationship for the Cabbibo angle in terms of the ratio of the d and s quarks, *i.e.* $\sin^2\theta_C \simeq m_d/m_s$, has led to the suggestion[7] that

$$\mathcal{M}_{ds} = \begin{bmatrix} 0 & a \\ a & b \end{bmatrix}. \tag{3}$$

This simple observation has generated over the years an enormous literature on quark mass matrices. It is an especially active field of research in the past two or three years. Consider now a trivial extension of this seesaw structure and apply it to the neutrino mass matrix, namely

$$\mathcal{M}_\nu = \begin{bmatrix} 0 & 0 & 0 \\ 0 & 0 & a \\ 0 & a & b \end{bmatrix}, \tag{4}$$

but in the basis $\cos\theta\, \nu_e - \sin\theta\, \nu_\mu$, ν_τ, and $\cos\theta\, \nu_\mu + \sin\theta\, \nu_e$. For small a/b, the mass eigenvalues are simply 0, $-a^2/b$, and b. The usual three neutrinos are related to the mass eigenstates by

$$\begin{pmatrix} \nu_e \\ \nu_\mu \\ \nu_\tau \end{pmatrix} = \begin{pmatrix} \cos\theta & -\sin\phi\sin\theta & \cos\phi\sin\theta \\ -\sin\theta & -\sin\phi\cos\theta & \cos\phi\cos\theta \\ 0 & \cos\phi & \sin\phi \end{pmatrix} \begin{pmatrix} \nu_1 \\ \nu_2 \\ \nu_3 \end{pmatrix}, \tag{5}$$

where $\sin\phi \simeq a/b \simeq \sqrt{m_2/m_3}$.

The electron neutrino is then as given by Eq. (1) and the discussion of the previous section applies. However, $\Delta m^2_{12} = m_2^2$ is now correlated with $\sin^2 2\theta_{e2}$ for a given choice of m_3 which is of course constrained by atmospheric data. In the figures above, the dashed lines represent the predictions of Eq. (4) for $m_3 \simeq 0.17$ eV (left) and 0.063 eV (right). They do indeed intersect the allowed regions.

The atmospheric neutrino oscillations are given by

$$P(\nu_\mu \to \nu_e) = \frac{1}{2}\cos^4\phi\sin^2 2\theta \left(1 - \cos\frac{t\Delta m^2_{23}}{2p}\right), \tag{6}$$

$$P(\nu_\mu \to \nu_\tau) = \frac{1}{2}\sin^2 2\phi\cos^2\theta \left(1 - \cos\frac{t\Delta m^2_{23}}{2p}\right). \tag{7}$$

Since the angle ϕ is small, ν_μ oscillates mainly into ν_e in this simplest realization of the seesaw ansatz. On the other hand, the $\nu_\mu - \nu_\tau$ submatrix may be rotated without affecting ν_e, in which case a better fit to the atmospheric data can be obtained.

4. A Specific Model

To obtain Eq. (4), start with ν_e, ν_μ, ν_τ, and four singlets: ν_S, N_1, N_2, N_3. Assume a discrete Z_6 symmetry $[\omega^6 = 1]$ and assign

$$(\nu_e, \nu_\mu, \nu_\tau) \sim (\omega, \omega^{-2}, 1); \quad (\nu_S, N_1, N_2, N_3) \sim (\omega, 1, \omega^2, \omega^{-2}). \tag{8}$$

The Higgs sector is taken to consist of two doublets $(\Phi_1, \Phi_2) \sim (1, \omega^{-3})$ and one singlet $\chi \sim \omega$. The resulting 7×7 mass matrix is then given by

$$\mathcal{M}_7 = \begin{bmatrix} 0 & 0 & 0 & 0 & 0 & m_1 & 0 \\ 0 & 0 & 0 & 0 & 0 & m_2 & 0 \\ 0 & 0 & 0 & 0 & m_3 & 0 & 0 \\ 0 & 0 & 0 & 0 & m_4 & 0 & m_5 \\ 0 & 0 & m_3 & m_4 & M_1 & 0 & 0 \\ m_1 & m_2 & 0 & 0 & 0 & 0 & M_2 \\ 0 & 0 & 0 & m_5 & 0 & M_2 & 0 \end{bmatrix}, \tag{9}$$

where m_1 comes from $\langle\phi_2^0\rangle$, $m_{2,3}$ from $\langle\phi_1^0\rangle$, and $m_{4,5}$ from $\langle\chi\rangle$. Large $M_{1,2}$ reduce the above to a 4×4 mass matrix

$$\mathcal{M}_4 = \begin{bmatrix} 0 & 0 & 0 & m_1 m_5/M_2 \\ 0 & 0 & 0 & m_2 m_5/M_2 \\ 0 & 0 & m_3^2/M_1 & m_3 m_4/M_1 \\ m_1 m_5/M_2 & m_2 m_5/M_2 & m_3 m_4/M_1 & m_4^2/M_1 \end{bmatrix}. \tag{10}$$

Assume now that m_4^2/M_1 dominates, then

$$\begin{aligned} \mathcal{M}_3 &= \begin{pmatrix} bs^2 & bsc & as \\ bsc & bc^2 & ac \\ as & ac & 0 \end{pmatrix} \\ &= \begin{pmatrix} c & 0 & s \\ -s & 0 & c \\ 0 & 1 & 0 \end{pmatrix} \begin{pmatrix} 0 & 0 & 0 \\ 0 & 0 & a \\ 0 & a & b \end{pmatrix} \begin{pmatrix} c & -s & 0 \\ 0 & 0 & 1 \\ s & c & 0 \end{pmatrix}, \end{aligned} \tag{11}$$

where $b = (m_1^2 + m_2^2) m_5^2 M_1 / m_4^2 M_2^2$, $a = m_3 m_5 \sqrt{m_1^2 + m_2^2} / m_4 M_2$, $s = m_1/\sqrt{m_1^2 + m_2^2}$, $c = m_2/\sqrt{m_1^2 + m_2^2}$. The desired seesaw structure is thus obtained.

5. Addition of a Fourth Neutrino

Since Eq. (10) contains a fourth neutrino which couples to both ν_e and ν_μ, it may be considered as a candidate for explaining the recent LSND results.[1] However, the required mass and mixing of this singlet neutrino with ν_e are then too large to be consistent with the nucleosynthesis bound on the number of light neutrinos.[8] To avoid this problem, the most natural thing to do is to use the singlet neutrino to explain the solar data in the matter-enhanced small-angle nonadiabatic solution, as has been pointed out by many authors.[9] In that case, ν_μ and ν_τ may be assumed to have masses of a few eV, but a small enough mass difference and large enough mixing to account for the atmospheric data. A small mixing between ν_e and ν_μ may then be invoked to explain the LSND results. A recently proposed model[10] uses a discrete Z_5 symmetry and the seesaw reduction of a 7×7 mass matrix to obtain four approximate light neutrino mass eigenstates $\cos\theta\ \nu_e - \sin\theta\ \nu_S$, $\cos\theta\ \nu_S + \sin\theta\ \nu_e$, $(\nu_\mu + \nu_\tau)/\sqrt{2}$, and $(\nu_\mu - \nu_\tau)/\sqrt{2}$, with eigenvalues 0, m_1, m_2, and $-m_2$ respectively. In addition, mixing occurs between ν_e and ν_μ, as well as between ν_S and ν_τ. Note that ν_μ and ν_τ are pseudo-Dirac partners, hence $\sin^2 2\theta = 1$ is required for atmospheric neutrino oscillations.

To accommodate a fourth neutrino in the present context, a possible variation is to double Eq. (3) and consider the 4×4 mass matrix

$$\mathcal{M}'_\nu = \begin{bmatrix} 0 & a & 0 & 0 \\ a & b & 0 & 0 \\ 0 & 0 & 0 & c \\ 0 & 0 & c & d \end{bmatrix} \tag{12}$$

in the basis $\cos\theta\ \nu_e - \sin\theta\ \nu_\mu$, ν_S, $\cos\theta\ \nu_\mu + \sin\theta\ \nu_e$, and ν_τ. Solar neutrino oscillations are as given before, but now the second mass eigenstate is mostly inert and there is no phenomenological constraint on the ratio a/b as in the case of Eq. (4). Atmospheric neutrino oscillations are mostly between ν_μ and ν_e, whereas the LSND results are explained by the fact that both ν_e and ν_μ mix with ν_τ. However, because of the seesaw ansatz, the latter is correlated with the former. Numerically, they are indeed consistent with both sets of data, although the value of Δm^2 in the LSND case is required to be less than about 3 eV^2.

6. Conclusions

As more neutrino experiments accummulate more data, there are two important messages for phenomenologists and model builders. First, the naive assumption that each case of neutrino oscillations is to be interpreted as between only two mass eigenstates must be abandoned. Atmospheric data tell us that a large mixing angle exists between ν_μ and ν_e or ν_τ or both. In this talk it has been shown that if ν_e has a significant third massive component, the analysis of solar data allows a much larger parameter space in Δm^2 and $\sin^2 2\theta$ for the first two components.

Second, the structure of the neutrino mass matrix is beginning to reveal itself. It is time to look for possible empirical relationships such as the well-known $\sin^2\theta_C \simeq m_d/m_s$ for quarks which may give us a glimpse of the underlying theory of the origin of masses. In this talk a first attempt, *i.e* Eqs. (4) and (12), has been noted.

7. Acknowledgements

I thank Prof. J. W. F. Valle and the other organizers of the International Workshop on Elementary Particle Physics: Present and Future (Valencia 95) for their great hospitality and a very stimulating program. This work was supported in part by the U. S. Department of Energy under Grant No. DE-FG03-94ER40837.

8. References

1. C. Athanassopoulos *et al.*, Los Alamos National Laboratory Report No. LA-UR-95-1238 (April 1995).
2. Y. Fukuda *et al.*, Phys. Lett. **B335**, 237 (1994).
3. For a review, see for example A. Yu. Smirnov, these proceedings.
4. E. Ma and J. Pantaleone, UCRHEP-T140 (March 1995).
5. T. K. Kuo and J. Pantaleone, Rev. Mod. Phys. **61**, 937 (1989); D. Harley, T. K. Kuo, and J. Pantaleone, Phys. Rev. D **47**, 4059 (1993); J. Pantaleone, Phys. Rev. D **49**, R2152 (1994).
6. A. S. Joshipura and P. I. Krastev, Phys. Rev. D **50**, 3484 (1994); M. Narayan, M. V. N. Murthy, G. Rajasekaran, and S. Uma Sankar, hep-ph/9505281; S. M. Bilenkii, C. Giunti, and C. W. Kim, hep-ph/9505301.
7. S. Weinberg, Ann. N. Y. Acad. Sci. **38**, 185 (1977).
8. See for example X. Shi *et al.*, Phys. Rev. D **48**, 2563 (1993); K. Enqvist *et al.*, Nucl. Phys. **B373**, 498 (1992).
9. J. T. Peltoniemi and J. W. F. Valle, Nucl. Phys. **B406**, 409 (1993); D. O. Caldwell and R. N. Mohapatra, Phys. Rev. D **48**, 3259 (1993); E. J. Chun, A. S. Joshipura, and A. Yu. Smirnov, hep-ph/9505275 and these proceedings; Z. G. Berezhiani and R. N. Mohapatra, hep-ph/9505385 and these proceedings.
10. E. Ma and P. Roy, UCRHEP-T145 (April 1995).

SUPERSYMMETRIC EFFECTS ON NEUTRINO OSCILLATIONS

E. ROULET

SISSA, Via Beirut 2/4,
34014 Trieste, Italy

It is shown that, in the minimal supersymmetric version of the Standard Model, the one-loop induced neutrino indices of refraction can affect the ν_μ–ν_τ oscillations significantly. This happens if one chargino is light ($m_{\chi^+} < 100$ GeV) and sleptons of the second and third families are non-degenerate and not too heavy. This effect can manifest in the indirect underground searches of supersymmetric dark matter, since matter enhanced oscillations of ν_τ's can modify the rates and spectrum of the energetic ν_μ's arising from the annihilation of neutralinos trapped in the sun.

Oscillations of neutrinos provide the best tool to test flavour-dependent neutrino properties, such as differences in the neutrino masses and, when the propagation takes place in a medium, also differences in their interactions with matter. In particular, this last situation is most clearly manifest in the MSW effect, arising from the fact that ν_e, unlike ν_μ and ν_τ, interact with electrons via W-boson exchange. This interaction modifies the ν_e index of refraction and can produce a resonant amplification of neutrino oscillations, leading to large flavour conversions even for small neutrino mixings.

On the other hand, at small momentum transfer muon and tau neutrinos interact indistinguishably with ordinary matter at the tree level in the Standard Model. Only at one-loop (via Z penguins and W box diagrams) a difference among the ν_μ and ν_τ indices of refraction n_{ν_μ,ν_τ} appears [1], due to the different masses of the corresponding leptons that run inside the loops, leading to $\Delta n_{\tau\mu} \equiv n_{\nu_\tau} - n_{\nu_\mu} \propto (m_\tau/M_W)^2$. The resulting effect is then just that $\epsilon \equiv \Delta n_{\tau\mu}/\Delta n_{\mu e} \simeq -5 \times 10^{-5}$, and is then quite small.

Being this effect absent at tree level and 'accidentally' small at one-loop in the SM, it is worth to consider it in the framework of the best motivated extension of the SM, i.e. in its minimal supersymmetric version. There are in this model new diagrams, involving the partners of the standard particles, which also contribute to the neutrino interactions with matter at one-loop (fig. 1). The value of $\Delta n_{\tau\mu}$ now turns out to depend mainly on the splitting among $\tilde{\tau}$ and $\tilde{\mu}$ slepton masses. This splitting can be the result of a large $\tilde{\tau}_L$–$\tilde{\tau}_R$ mixing or can just arise from non-universal initial conditions for the scalar masses in the soft-breaking terms. The result is that[3], for light charginos and

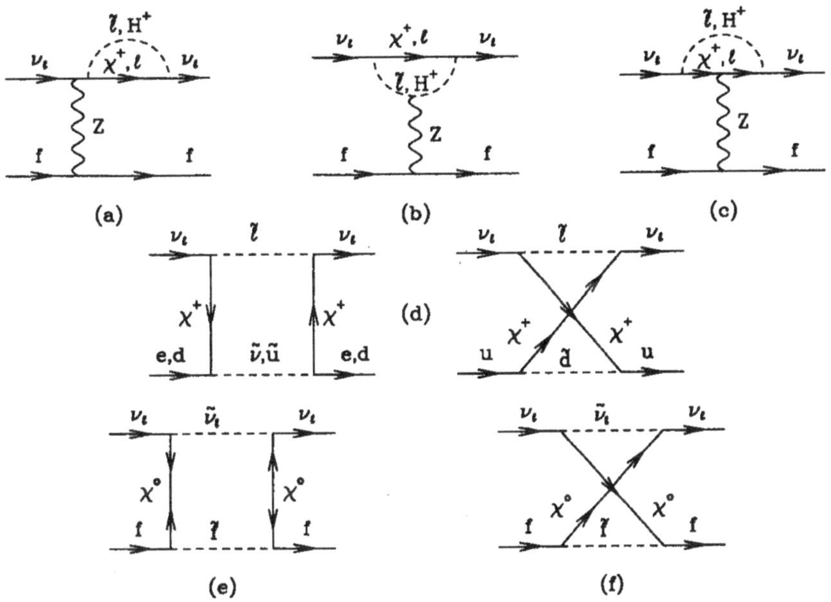

Fig. 1

Figure 1: Feynman diagrams describing the supersymmetric contribution to ν_ℓ–f forward scattering ($\ell = \mu,\ \tau$). The blob in fig. 1.a represents corrections to both ν external legs.

sleptons, a significant $\tilde{\mu}$–$\tilde{\tau}$ splitting can make the SUSY contribution to $\Delta n_{\tau\mu}$ dominant over the SM one, allowing for values of ϵ as large as $\sim 10^{-3}$.

Turning now to the possible physical applications of this effect, a first difficulty to observe any conversion among ν_μ and ν_τ is that for low energies, $E_\nu < 0.1$ GeV, these neutrinos are only detected by means of their neutral current interactions and are hence essentially indistinguishable. Furthermore, only e-type neutrinos are produced in the sun, and equal amounts of ν_μ and ν_τ are produced in supernovae (where the high densities could have enhanced the matter effects), so that oscillations among them do not give actually any overall result. These problems are not present in long-baseline ν_μ oscillation experiments on earth, but the resonance oscillation length in terrestrial matter (inversely proportional to $\Delta n_{\tau\mu}$) is typically much larger than the earth diameter, and hence oscillation effects are negligible also in this case. The only scenario we are left with in which the effect could be relevant is then in the propagation of energetic neutrinos in a stellar environement.

The situation that we will describe is actually directly related to the supersymmetric framework under consideration. It is well known that a nice feature of supersymmetry, once R-parity conservation is adopted to avoid B and L violation, is that the LSP, usually a neutralino χ, is stable and naturally becomes a good dark matter (DM) candidate. One of the main strategies that are being pursued at present to experimentally search for SUSY DM is through the observation of energetic neutrinos produced in the annihilation of DM trapped in the interior of the sun or the earth [4]. In particular, upward-going muons produced in the rock (or ice) just below underground detectors by energetic ν_μ and $\bar{\nu}_\mu$ (with $E_\nu >$ few GeV) may allow to probe significant regions of the supersymmetric parameter space in new installations such as Superkamiokande or Amanda. As has been shown[5], the usual MSW effect between ν_e and $\nu_{\mu,\tau}$ can affect the detection rate predictions. The ν_μ–ν_τ oscillations here discussed may also have important implications for the detection of neutralino annihilation signals, because the ν_μ and ν_τ fluxes from neutralino annihilation are generally quite different, so that oscillations among them modify the expected ν_μ flux to which the underground detectors are most sensitive.

The difference among ν_μ and ν_τ fluxes has its origin in the fact that the non-relativistic neutralino annihilation cross section into fermion pairs $f\bar{f}$ is proportional to m_f^2. Taking for instance the quite plausible situation in which $m_\chi < M_W$, that squarks are much heavier than sleptons, and that χ is mainly gaugino, neutralinos will then annihilate dominantly into τ pairs, which after decaying lead to much larger fluxes of ν_τ's than of ν_μ's.

To show the possible effects of energetic ν_μ–ν_τ matter enhanced oscillations in the solar interior, let us remind that, in the case of two flavour neutrino mixing, the resonance condition reads

$$\Delta n_{\beta\alpha} \equiv n_{\nu_\beta} - n_{\nu_\alpha} = \frac{\cos 2\theta \Delta m^2_{ji}}{2E^2}, \tag{1}$$

where E is the neutrino energy, θ is the angle describing the mixing of the flavour eigenstates $\nu_{\alpha,\beta}$ into mass eigenstates $\nu_{i,j}$ and $\Delta m^2_{ji} \equiv m^2_{\nu_j} - m^2_{\nu_i}$. We recall that for the ordinary MSW effect involving oscillations of ν_e into ν_α ($\alpha = \mu$ or τ), $\Delta n_{\alpha e} = \sqrt{2} G_F N_e/E$. The resonant condition can then be written as[2]

$$\Delta m^2_{ji} \cos 2\theta = \left(\frac{\Delta n_{\beta\alpha}}{\Delta n_{\mu e}}\right) \frac{E}{10 \text{ MeV}} \frac{\rho}{\rho_0}\, 10^{-4}\text{eV}^2, \tag{2}$$

where $\rho_0 \simeq 150$ g/cm^3 is the central density of the sun. Another requirement for the flavour conversion to be efficient is that the transition not be extremely non-adiabatic, which implies

$$\Delta m^2 \sin^2 2\theta > 2E \left|\frac{1}{N_e}\frac{dN_e}{dr}\right|_{res} \simeq 6 \times 10^{-8} \frac{E}{10 \text{ MeV}}\ \text{eV}^2. \tag{3}$$

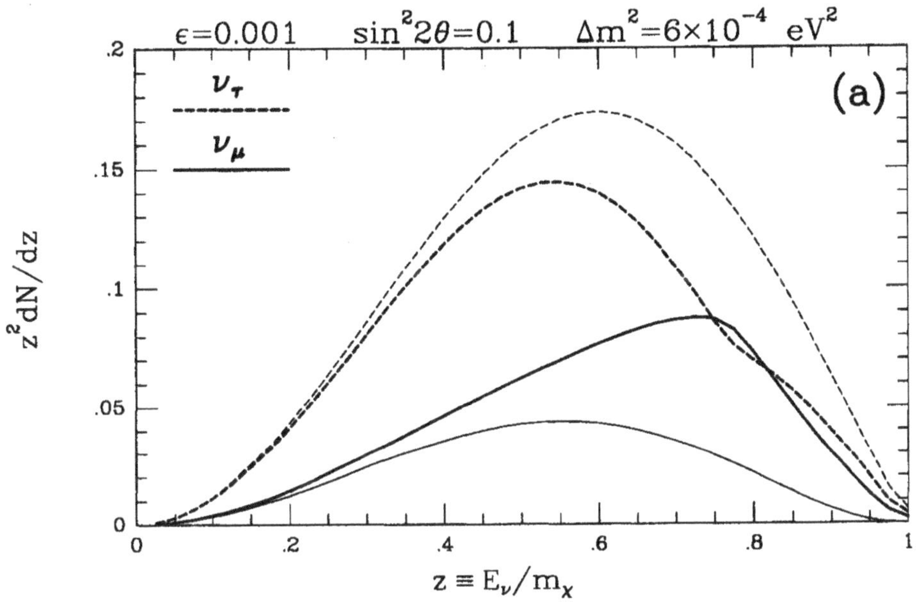

Figure 2: Differential ν_μ (solid lines) and ν_τ (dashed lines) yields ($\times z^2$) vs. $z \equiv E_\nu/m_\chi$, for χ annihilations into τ pairs. Thin lines describe the original spectra while thick lines include the matter effects, assuming $m_\chi = 50$ GeV.

In the usual MSW effect for $1 \div 10$ MeV neutrinos propagating in the sun, these last two conditions can be met in a 'triangle' in the Δm^2–$\sin^2 2\theta$ plane corresponding to $\Delta m^2 < 10^{-4}$ eV^2 and $\Delta m^2 \sin^2 2\theta > 10^{-8}$ eV^2. In the case of ν_μ–ν_τ oscillations under discussion, the r.h.s. of eq. (2) is suppressed by a factor $\epsilon < 10^{-3}$. Hence, for $E = 10 \div 100$ GeV, eq. (2) implies that $\Delta m^2_{\tau\mu} \cos 2\theta < 10^{-3}$ eV^2, while eq. (3) requires $\Delta m^2_{\tau\mu} \sin^2 2\theta > 6 \times 10^{-5}$ eV^2, leaving a smaller 'triangle' in the Δm^2–$\sin^2 2\theta$ plane than in the ordinary MSW case. This is mainly because the higher neutrino energies make the adiabatic condition harder to achieve. For values of $\epsilon < 10^{-4}$, the values of Δm^2 for which a resonance is met in the sun become too low to allow for an adiabatic transition and the triangles shrink towards large mixing becoming eventually negligible.

Hence, if $\epsilon > 10^{-4}$, there is a significant range in the parameter space for which the ν_μ–ν_τ matter oscillations in the sun can be important for energetic neutrinos, and in particular the parameters suggested by the atmospheric neutrino anomaly lie near the boundary of this region. To demonstrate the effects on the ν_μ spectra from DM annihilation of these flavour oscillations, we show in fig. 2 the ν_τ and ν_μ differential neutrino yields (with thin dashed and solid lines respectively) produced by the annihilation of neutralinos into a τ pair (the main annihilation channel in our example). What is actually plotted is z^2dN/dz (where $z \equiv E_\nu/m_\chi$), which is the relevant quantity for underground ν_μ detection, because both the CC ν_μ cross section and the muon range in the rock (or ice) are proportional to the neutrino energy.

With thick lines we show how the fluxes get modified after traversing the solar interior, assuming $m_\chi = 50$ GeV and taking $\epsilon = 10^{-3}$, $\Delta m^2 = 6 \times 10^{-4}$ eV^2 and $\sin^2 2\theta = 0.1$.

From the results, one can see that the detection rates may be sizeably modified by the matter enhanced oscillations here described, providing an interesting physical manifestation of the radiative supersymmetric effects studied. The uncertainties involved in the theoretical predictions of the DM annihilation signal (unknown neutralino mass and composition, uncertainties in the local halo density and DM velocity distribution, etc.) will however complicate the interpretation of any positive detection, so that these effects should actually be considered as providing a further spread in the theoretical predictions until these parameters become more constrained by accelerators and direct DM searches. The shape of the neutrino spectrum gives probably a clearer signature of the matter oscillations, and may provide a useful handle to identify them.

References

1. Botella F. J., Lim C. S. and Marciano W. J., Phys. Rev. D35 (1987) 896.
2. Kuo T. K. and Pantaleone J., Rev. Mod. Phys. 61 (1989) 937; Gelmini G. and Roulet E., preprint CERN-TH.7541/94. [6
3. Roulet E., Phys. Lett. in press (1995).
4. Primack J., Sadoulet B. and Seckel D., Ann. Rev. Nucl. Part. Sci. B38 (1988) 751.
5. Ellis J., Flores R. and Masood S., Phys. Lett. B294 (1992) 229.

Neutrino propagation in medium with random magnetic field

V.B. Semikoz

Instituto de Física Corpuscular - C.S.I.C.,
Departament de Física Teòrica, Universitat de València
46100 Burjassot, València, SPAIN

We consider active-sterile Majorana neutrino conversions and the Dirac neutrino spin-flip in the early universe hot plasma with strong random magnetic fields generated at electroweak phase transition time. Within a random magnetic field domain the magnetization asymmetry of the hot lepton-antilepton plasma produced by a uniform constant magnetic field is huge in contrast to their small density asymmetry, leading to a drastic change in the active-sterile conversion rates. Under the same conditions the Dirac neutrino spin-flip does not depend on any two-level energy splitting including vector and axial vector neutrino interaction potentials which contribute to the left-handed (active) neutrino spectrum in medium with magnetic field. Assuming that the random field provides the seed for the galactic field one can estimate the restrictions from primordial nucleosynthesis. Requiring that extra sterile neutrino (or a right-handed Dirac neutrino) does not enter in equilibrium with the active ones before nucleosynthesis, we find limits on the oscillation parameters (or on the Dirac neutrino magnetic moment) which are more stringent than in absence of random magnetic fields.

1 Nucleosynthesis constraints on primordial magnetic field and astrophysical bounds on seed for galactic magnetic field

If a strong magnetic field exists in early universe it should be a random small-scale one because the universe is isotropic at large scales at the same expansion time.

There are some generation mechanisms of the primordial random small-scale magnetic fields in early universe. The gauge invariance principle is not valid for small-scale domains separated by distances larger than the horizon scale l_H that allowed Vachaspati [2] to find in standard model an unusual form of the electromagnetic field which is proportional to gradients of the Higgs field VEV $F_{\mu\nu} \sim (\partial_\mu \phi^+ \partial_\nu \phi - \partial_\nu \phi^+ \partial_\mu \phi)$. However, how these domains enter under the horizon and what is happened with embedding of them into co-moving plasma with finite conductivity remain open questions in such model.[a]

In context of an Abelian gauge symmetry, spontaneously broken at the first-order transition, Kibble and Vilenkin have recently found another mech-

[a]Author is indebted to H. Rubinstein and D. Grasso for discussions around generation mechanisms of primordial random magnetic fields.

anism of the random magnetic field generation in bubble collisions [3] where a magnetic field flux loop arises along the interseption line of bubbles with the magnetic field value $B \sim (eR^2)^{-1} \sim \eta^2/e \sim 10^{24}\, G$. Here we estimate the loop radius at the phase transition time as the bubble radius R or the inverse VEV of the Higgs field, roughly, $R \sim \eta^{-1} \sim M_W^{-1}$.

Assuming any generation mechanism of a strong random magnetic field in early universe we put in the following the initial field amplitude $B_0 \sim T_0^2$, where T_0 is the phase transition temperature.

Using the rule for uncorrelated random magnetic field domains,

$$\langle B_i(x)B_j(y)\rangle = \frac{\delta_{ij}}{2\lambda}\delta^{(3)}(x-y), \tag{1}$$

we obtain the mean values

$$\begin{aligned} \langle B_{\|}\rangle = \langle B_{\perp}\rangle = \langle B_{\|}B_{\perp}\rangle = 0, \\ \langle B_{\|}B_{\|}\rangle = \langle B_{\|}^2 L_0\delta(t-t_0), \\ \langle B_{\perp}B_{\perp}\rangle = \langle B_{\perp}^2 L_0\delta(t-t_0). \end{aligned} \tag{2}$$

The correlation length λ is determined by the domain size L_0 and the value of root mean squared field $B \equiv B_{rms} = \sqrt{\langle B^2\rangle}$ at the horizon scale $L = l_H$ [1]:

$$\frac{1}{\lambda} = \frac{3}{\pi(3-2p)}B_{rms}^2(l_H)L_0^3, \tag{3}$$

for $p \neq 3/2$, and

$$\frac{1}{\lambda} = \frac{3}{\pi}\ln\frac{l_H}{L_0}B_{rms}^2(l_H)L_0^3, \tag{4}$$

for $p = 3/2$. In Eq.(2) the mean squared fields $\langle B_{\|}^2\rangle = \langle B^2\rangle/3$, $\langle B_{\perp}^2\rangle = 2\langle B^2\rangle/3$ are given by the r.m.s. field $B \equiv \sqrt{\langle B^2\rangle}$.

Different mechanisms of the magnetic field generation in the early universe give the following main phenomenological formula for r.m.s. field

$$B = B_0\left(\frac{T}{T_0}\right)^2\left(\frac{L_0}{L}\right)^p, \tag{5}$$

where we should put in correlators the maximal scale chosen at the horizon scale [b] $L = l_H(t)$.

[b] As explained in [1], this follows from the definition of the parameter λ for the initial δ-correlator in Eq.(1).

Note that if we put in Eq.(5) $B_0 \sim T_0^2$ (as in Vachaspati's mechanism, where $B_0 \sim T_{EW}^2$ [2]) the r.m.s. field with our constraints on the Dirac neutrino magnetic moment (in Eq.(15)), or on the neutrino mixing parameters (in Eq.(14)) depends only on the temperature of the universe T ($T \simeq T_{QCD}$ in Eq.(15)) and on two parameters describing the geometry of random field: (a) the scaling parameter p, and (b) the size of random domain $L_0^{min} = a(MeV/T)$.

There is a bound from primordial nucleosynthesis on the strength of magnetic field. It was shown [4,5] that a too large magnetic field strength would enhance the rates for the relevant weak processes at $T_{NS} \sim 0.1\ MeV$, resulting in additional helium production. This leads to $B \lesssim 3 \times 10^{10}\ Gauss$ [4,5] which implies, at present time [1],

$$B_{NS} \lesssim 1.8 \times (2.4)^p \times 10^{-7-11p} Gauss. \tag{6}$$

Now we can put some bounds on the magnetic field parameters p and L_0^{min} when we assume the hypothesis of a primordial origin of the observed galactic magnetic fields. In this case we require

$$L_0^{min} \gtrsim 10^3 (MeV/T) cm, \tag{7}$$

and $p \lesssim 1$. The first restriction comes from requiring that primordial fields survive against ohmical dissipation after recombination time and could seed the observed galactic magnetic field [4]. On the other hand, the limit $p \lesssim 1$ follows from Eq.(6) and the lower dynamo theory astrophysical bound on the strength of the galactic magnetic field [6] $B_{seed}(T_{now}) \gtrsim 10^{-18}\ G$. Moreover, in order to be a seed field of the Milky Way or Andromeda, one would require [6] $B_{seed} \gtrsim 10^{-13}\ G$ which corresponds to $p \lesssim 0.5$.

2 Neutrino conversion probability in medium with random magnetic field

In ultrarelativistic limit, within a domain located at $r = t$, ($B(t) \approx constant$), the active neutrino spectrum in medium with a strong constant magnetic field $E_a(q) = q + V + A(t)$ includes an additional axial interaction potential $A(t)$ which is proportional to the mean axial lepton current, $A(t) \sim \langle \bar{\psi}_e \gamma_z \gamma_5 \psi_e \rangle_0 \sim B_{\|}(t)$. Latter one is caused by magnetization of the lepton-antilepton gas in an external magnetic field. Here the magnetic field $B_{\|}(t)$ is aligned along the z-axis, and we denoted by the symbol $\langle ... \rangle_0$ the statistical averaging of the second-quantized charged lepton fields which describe many-particle system of the background plasma.

Using such modified spectrum in Hamiltonian of the Schrödinger equation for the neutrino doublet $(\nu_a \nu_s)$, we derive two types of evolution equations for the conversion probability which depend: in the case of the flavour transition $\nu_a \to \nu_s$ (without helicity change, or neglecting neutrino transition magnetic moment, $\mu = 0$) on the square of random magnetic field values $A(t) \sim B_{\parallel}(t)$, or $\sim B_{\parallel}(t)B_{\parallel}(t_1)$, and in the case of the spin-flavour transitions $\nu_{aL} \to \tilde{\nu}_{sR}$ (in the presence of magnetic moment, $\mu \neq 0$, but neglecting mixing, $s = 0$) on the square of the transversal magnetic field $\sim \mu^2 B_{\perp}(t)B_{\perp}(t_1)$.

This allows us to average with use of Eq.(2) exact dynamical equations over a random magnetic field distribution assuming a small scale of magnetic field, $L_0 \ll l_{conv}, V^{-1}$. Here V is the vector neutrino interaction potential and $l_{conv} \sim \Gamma^{-1}$ is the neutrino conversion length given by the damping parameter determined below.

For the flavour transitions we find the conversion probability [7]

$$P(t) = \frac{\Delta^2 \sin^2 2\theta}{2\Delta_m^2}\Big[1 - \exp(-\Gamma_{\parallel}t)\times$$
$$\times\Big[\cosh(\sqrt{\Gamma_{\parallel}^2 - \Delta_m^2}t) + \frac{\Gamma_{\parallel}}{\sqrt{\Gamma_{\parallel}^2 - \Delta_m^2}}\sinh(\sqrt{\Gamma_{\parallel}^2 - \Delta_m^2}t)\Big]\Big], \tag{8}$$

where the damping parameter $\Gamma_{\parallel} = \mu_{eff}^2 \langle B_{\parallel}^2\rangle L_0/2 = \mu_{eff}^2\langle B^2\rangle L_0/6$ is given by the r.m.s. field Eq.(5) and the coefficient

$$\mu_{eff} = -12c_A \times 10^{-13}\mu_B\frac{T}{MeV}, \tag{9}$$

which has no relation to a real magnetic moment and does not lead to a change of neutrino helicity. Here $c_A = \mp 0.5$ is the axial constant (upper sign for ν_e, lower one for $\nu_{\mu,\tau}$); $\Delta_m = \sqrt{(V - \Delta\cos 2\theta)^2 + \Delta^2 \sin^2 2\theta}$ is the standard MSW oscillation frequency. One can easily see that Eq.(8) reduces in absence of magnetic field, $\Gamma_{\parallel} = B = 0$, to the MSW solution.

For the strong magnetic fields $\Gamma_{\parallel} \gg \Delta_m$, Eq.(8) takes the aperiodic form

$$P(t) \simeq \frac{\Delta^2 \sin^2 2\theta}{2\Delta_m^2}\Big(1 - \exp(-\Gamma_{\parallel}t)\cos\Delta_m t\Big), \tag{10}$$

which after averaging (before neutrino decoupling) over fast neutrino collisions with matter is given by

$$P(t) = \frac{\Delta^2 \sin^2 2\theta}{8\Gamma_{\parallel}\Gamma_W}. \tag{11}$$

For the Dirac neutrino spin-flip ($\nu_L \to \nu_R$), or for the spin-flavour transitions ($\nu_{aL} \to \tilde{\nu}_{sR}$) neglecting neutrino mixing, $s = 0$, we derive another form of the conversion probability [1,8]

$$P(t) = \frac{1}{2}\Big(1 - \exp(-\Gamma_{\perp} t)\Big), \tag{12}$$

which does not depend on matter effects, collision frequency [1] or the mass difference Δm^2 (in Majorana case [8]) and given by the r.m.s. transversal magnetic field via $\Gamma_{\perp} = 4\mu^2 \langle B_{\perp}^2 \rangle L_0 = 8\mu^2 \langle B^2 \rangle /3$, or Eq.(5).

3 Nucleosynthesis constraints on active-sterile neutrino conversions

Now we move to the nucleosynthesis bounds on mixing parameters and μ. The rate for producing sterile neutrinos in the early universe hot plasma is given as the product of a typical weak interaction rate $\Gamma_W = 4.0 G_F^2 T^5$ times the conversion probability either Eq.(11) or Eq.(12). The constraint that follows from nucleosynthesis may be simply estimated as

$$\Gamma_s = \Gamma_W P \lesssim H, \tag{13}$$

where $H = 4.46 \times 10^{-22} (T/MeV)^2 MeV$ is the Hubble parameter.

For the seeding scenario with the particular choice of the magnetic field scaling parameter $p = 0.5$ in Eq. (5), and using the probability Eq.(11) we obtain the nucleosynthesis bound on active-sterile neutrino mixing parameters[7]

$$\mid \Delta m^2 \sin 2\theta \mid \lesssim 10^{-4} eV^2, \tag{14}$$

which is more stringent than in absence of magnetic fields [9]. More correct kinetic approach for the same problem and the variation of acceptable parameters p is given in Ref [[10]].

Using kinetic approach [11] for the Dirac neutrino spin-flip ($\nu_L \to \nu_R$) [1] or the Schrödinger equation for the Majorana neutrino spin-flavour transitions[8] we derive from nucleosynthesis constraint Eq.(13) the bound on neutrino magnetic moment

$$\mu \lesssim \frac{1.7 \times 10^{-21} \mu_B}{(L_0^{min})^{p+1/2} \Gamma_W^{1/2} H^p}, \tag{15}$$

where the scale parameter L_0^{min} given by Eq.(7), the weak process rate Γ_W and the Hubble parameter H are defined for the crucial QCD phase transition temperature T_{QCD} [1].

From last inequality we find the constraint on the neutrino magnetic moment, $\mu \lesssim 10^{-16} \mu_B$, for the seed field of the galactic magnetic field, $p \lesssim 1$, $B_{seed} \gtrsim 10^{-18}$ G [6], or even more, $\mu \lesssim 10^{-22} \mu_B$, for case of the seeding of magnetic field in MIlky Way or in Andromeda ($B_{seed} \gtrsim 10^{-13}$ G, $p \lesssim 0.5$, [6]).

We conclude that presence of a strong random primordial magnetic field allows to connect different fields of Physics in order to put from nucleosynthesis the stringent cosmological constraints on some fundamental Particle Physics parameters and to predict possible strength and structure of the seed magnetic field for the following dynamo enhancement of galactic magnetic fields.

The possible generation of small-scale magnetic fields in a supernova [13] allows us to apply the same approach for the Neutrino Astrophysics and put bounds on neutrino parameters [12,8].

Acknowledgments

This work was supported by DGICYT under grant SAB94-0325. I would like to thank the organizers for the invitation to participate in the Workshop and J.F.W. Valle and A.D. Dolgov for numerous fruitfull discussions.

References

1. T. Vachaspati, *Phys. Lett.* B **265**, 258 (1991).
2. , T.W.B. Kibble and Alexander Vilenkin, Preprint IMPERIAL/TP /94-95/11, TUTP- 95-2, NI94039 (1995).
3. K. Enqvist, A. Rez and V. Semikoz, *Nucl. Phys.* B **436**, 49 (1995).
4. B. Cheng, D.N. Schramm and J.W. Truran, *Phys. Rev.* D **49**, 5006 (1994).
5. D. Grasso and H. Rubinstein, *Astroparticle Physics* **3** (1995).
6. K. Enqvist, V. Semikoz, A. Shukurov and D. Sokoloff, *Phys. Rev.* D **48**, 4557 (1993).
7. V. Semikoz and J.F.W. Valle, *Nucl. Phys.* B **425**, 651 (1994).
8. S. Pastor, V. Semikoz and J.F.W. Valle, Valencia preprint FTUV/95-39, IFIC/95-41.
9. K. Kainulainen, *Phys. Lett.* B **224**, 191 (1990). R. Barbieri and A. Dolgov, *Phys. Lett.* B **237**, 440 (1990).
   K. Enqvist, K. Kainulainen and M.J. Thomson, *Nucl. Phys.* B **373**, 498 (1992).
10. K. Enqvist, J. Maalampi and V. Semikoz, Preprint HU-TFT-95728 May (1995), (accepted Nucl. Phys. B).
11. V. Semikoz, *Phys. Rev.* D **48**, 5264 (1993).
12. S. Pastor, V. Semikoz and J.F.W. Valle, *Astroparticle Physics* **3** 87 (1995).
13. C. Thomson and R.C. Dunkan, *Astrophys. J.* **408** 194 (1993).

STATUS OF THE CHORUS EXPERIMENT

The CHORUS collaboration

presented by C. Mommaert

IIHE-Vrije Universiteit Brussel, Pleinlaan 2,
1050 Brussels, Belgium

The CHORUS experiment searches for ν_μ-ν_τ oscillations in the CERN Wide Band neutrino Beam at the SPS accelerator. ν_τ charged current interactions are characterized by the short-lived τ-lepton. Its detection is possible in the 800kg emulsion target, which produces 30 grains of 0.5 μm diameter per 100 μm track length. An electronic detector provides the tracking back to the interaction point, which is used for automatic computer-assisted microscopes and a kinematical event selection, to reduce the sample of events to be scanned in the emulsion. The detection technique and the performance of the apparatus will be discussed. The vertex prediction by the electronic detector and the subsequent vertex finding in the emulsion is operating successfully.

1 Introduction

Observation of neutrino oscillations implies massive neutrinos. If the oscillation phenomenon in the ν_μ-ν_τ channel will be observed by the CHORUS collaboration [1], the ν_τ will be a candidate for the Hot Dark Matter of the Universe. At the same time, based on the quadratic see-saw mechanism, the ν_μ would then have a mass that could explain the observed solar neutrino deficit due to matter-enhanced neutrino oscillations (MSW effect) in the Sun.

2 Concept of the experiment

The muon neutrino energy at the CERN SPS Wide Band Beam peaks around $30\,GeV$. Combined with the mean distance of $\sim 600\,m$ neutrinos travel to reach the detector, this provides the best sensitivity for observing neutrino oscillation for $\Delta m^2 = m^2_{\nu_\tau} - m^2_{\nu_\mu} \sim 27\,eV^2$, a value favoured for dark matter. The situation is favourable for a ν_τ appearance experiment because the original contamination in the beam is of the order of 10^{-7}. The oscillation would be identified by a ν_τ-interaction, whose signature is given by detection of the short-lived tau-lepton. The decay modes searched for in CHORUS are the following :

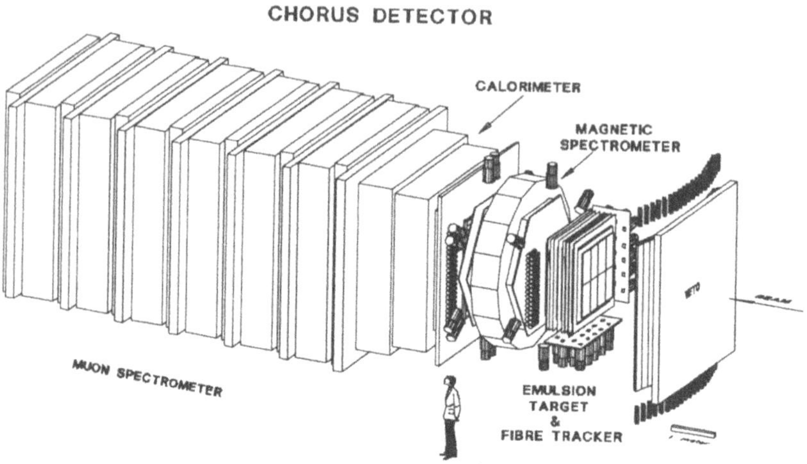

Figure 1: CHORUS detector layout.

$$\nu_\tau N \to \tau^-(\to \mu^- \ \overline{\nu}_\mu \ \nu_\tau)X. \quad (1)$$

$$\nu_\tau N \to \tau^-(\to h^- \ (n\pi^0) \ \nu_\tau)X. \quad (2)$$

$$\nu_\tau N \to \tau^-(\to \pi^+\pi^-\pi^-(n\pi^0)\nu_\tau)X. \quad (3)$$

These decays have to be discovered in a background of charged current and neutral current ν_μ interactions. A hybrid detector approach is based on the combined use of an electronic detector and a nuclear emulsion target. The electronic part with a pulsed magnet, a calorimeter and a muon spectrometer is used for kinematical event selection. A scintillating fiber tracking system allows to predict the location of tracks of interesting events inside the emulsion. The full detector is depicted in Fig. 1.

As an illustration of the kinematical event selection, one can consider the τ^--decay in (1). The kinematical properties of the final state particles should be contrasted with those corresponding to the dominant background process :

$$\nu_\mu N \to \mu^- X. \quad (4)$$

In the plane transverse to the ν beam, the hadronic shower X is back-to-back with the τ in (1), with the μ in (4) and the two neutrinos carry away transverse momentum in the first case, the so-called missing transverse momentum.

Kinematical event selection requires a high resolution calorimeter to measure the energy and direction of the hadronic shower and a magnetic muon spectrometer for the muon momentum vector. Events with a transverse momentum unbalance larger than 500 MeV/c are selected as candidates for reaction(1). The calorimeter technique used is based on lead-scintillator fiber sandwich, known as spaghetti calorimeter. The hadronic resolution obtained in a pion test beam is [2]

$$\frac{\sigma(E)}{E} = \frac{30.3 \pm 2.3\%}{\sqrt{E(GeV)}} + (1.8 \pm 0.7)\% \qquad (5)$$

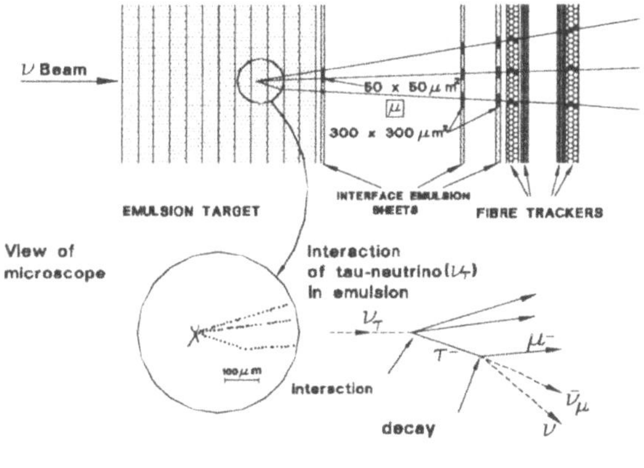

Figure 2: Vertex topology reconstruction.

Spectrometers are used for charge and momentum determination. A dedicated muon spectrometer is used for muon identification. A magnetic spectrometer, based on an air core magnet, without strayfield [3], determines charge and momentum of particles between 2 and 10 GeV before they enter the calorimeter. The momentum resolution is given by :

$$\frac{\Delta(p)}{p} = 1.6\%\, p + 16\%_{mult.scat} \qquad (6)$$

and is limited by multiple scattering (16%). The charge of τ-decay particles is determined with an accuracy of 8 σ for $p \leq 10 GeV$.

A massive target is necessary to provide a high rate of neutrino interactions and in order to be able to visualize the τ-decay, a high resolution of the order of $1\mu m$ is required. A nuclear emulsion target of 800kg was subdivided into four stacks. Since this technique integrates data over the exposure time, a track prediction system is needed to direct the computer-assisted microscopes for scanning. Scintillating fiber tracker planes and emulsion interface layers are interspersed with the bulk emulsion stacks as shown in Fig. 2.

With fibers of $500\mu m$ diameter, staggered in 7 layers and oriented in 4 projections per plane, we obtain a track prediction at the level of the emulsion interface layers with a Gaussian standard deviation of $175\mu m$. These interface layers are exchanged every 3 weeks for favourable background conditions to allow an accurate extrapolation onto the bulk emulsion with an accuracy of $\sim 50\mu m$.

3 The 1994-1995 data taking and perspectives

The 1994 CHORUS data taking lasted about 150 days. The data taking continued in 1995 and is expected to provide in total 2×10^{19} protons on target. On average, 0.7 events per 10^{13} protons per spill were recorded, out of which $\sim \frac{1}{3}$ neutrino interactions in the emulsion target. The proposal presented by the CHORUS collaboration [1] aimed at exploring the domain of small mixing angles, $\sin^2(2\Theta_{\mu\tau}) \sim 3 \times 10^{-4}$ for mass parameters $\Delta m^2 > 10\, eV^2$ (see Fig. 3).

This sensitivity corresponds to the observation of ~ 64 ν_τ events if oscillation occurs at the present 90% confidence limit [4]. Taking advantage of a suitable kinematical event selection, the corresponding scanning effort would be reduced to 50000 events, i.e. $\sim 10\%$ of the full event sample.

At the end of the 1994 run, one of the four emulsion stacks was removed, developed and replaced. The scanning of this stack has started. The four stacks will be replaced for continuing data taking in 1996/7 bringing the number of protons on target to 5×10^{19}.

4 First emulsion scanning

4.1 Emulsion interface layers

The scanning of interface emulsion layers yields an offset distribution of the difference between the predicted track position and the found track position with a standard deviation of $175\, \mu m$. The angular distribution between predicted

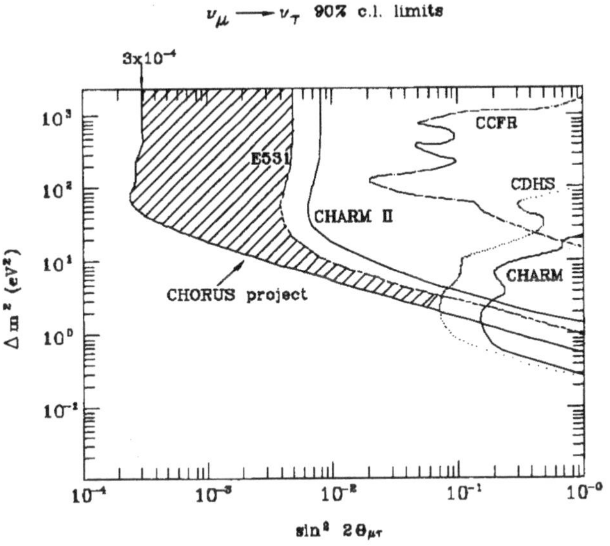

Figure 3: CHORUS sensitivity to oscillation parameters.

track and scanned track has a standard deviation of $3.2\,mrad$. These values are compatible with the design value [1]. At the end of June 1995, 4000 events have been measured, more than in the previous best experiment at FNAL [4].

4.2 Bulk emulsion events

The bulk emulsion scanning has confirmed the validity of the proposed detection technique. Predicted vertices are found in the bulk emulsion and the scanning predicts exiting tracks from the vertex region that coincide well with the present tracks in the fiber tracker, used for the vertex prediction. This is shown in Fig. 4, where a zoom on the target region displays the two types of tracks. The longer lines leaving the vertex are the track feedback from the scanning.

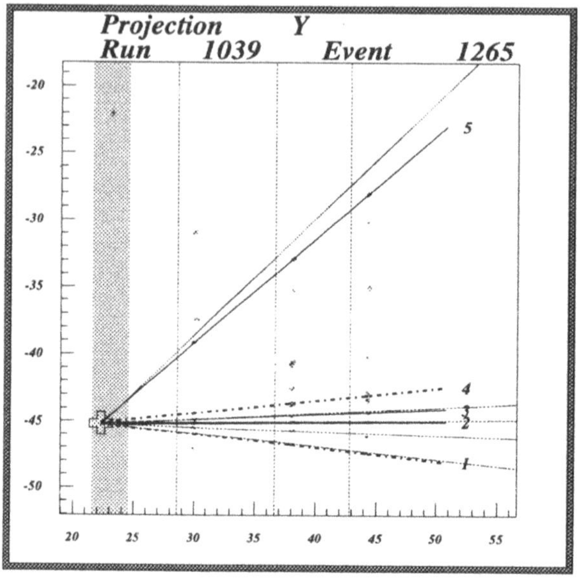

Figure 4: Vertex found in an emulsion stack, followed by 3 fiber tracker planes.The vertex was predicted based on the tracker information (shorter, numbered tracks) and found in the emulsion, from where exiting tracks (longer tracks without number) were predicted back.

References

1. N.Armenise et al.,CHORUS collaboration,CERN-SPSC/90-42,SPSC P254, and M.de Jong et al.,CHORUS collaboration,CERN-PPE/93-131.
2. PRoceedings of the 4th International Conference on Advanced Technology and PArticle Physics, Como, Italy 3-7 Oct.1994, to be published.
3. F.Bergsma et al., Nucl. Instr. and Methods A357(1995)243-248.
4. N.Ushida et al.,Phys.Rev.Lett.,57(23):2897-2900,1986.

THE SAN ONOFRE NEUTRINO OSCILLATION EXPERIMENT

N. MASCARENHAS, F. BOEHM, M. CHEN, B. COOK, H. HENRIKSON, R. HERTENBERGER, K. LOU, V. NOVIKOV, A. PIEPKE AND P. VOGEL

Norman Bridge Laboratory, California Institute of Technology, Pasadena CA 91125, USA

G. GRATTA, J. HANSON, D. MICHAEL

Lauritsen Laboratory, California Institute of Technology, Pasadena CA 91125, USA

We describe a reactor-based neutrino disappearance experiment with a sensitivity of $\Delta m^2 = 10^{-3}$ eV^2 and $sin^2 2\theta$=0.1 covering the region suggested by the atmospheric neutrino puzzle. We discuss the detection scheme, sensitivity and estimates for the expected signal and background. After two years of running a 12 ton liquid scintillation target placed 740 m from 2 x 3300 MW nuclear reactors will completely explore the parameter space for the channel $\bar{\nu_e} \rightarrow \bar{\nu_\mu}$ which may be responsible for the atmospheric neutrino anomaly. We will extend the explored region of Δm^2 by about an order of magnitude from the present limits

1 Motivation

There are strong indications of a solar neutrino anomaly[1] and recently an anomaly has been reported in the atmospheric neutrino flux by three independent experiments (IMB[2], Soudan[3] and Kamiokande[4]). An explanation for the atmospheric neutrino anomaly is a deficit of ν_μ via oscillations to either ν_e or ν_τ. We propose to explore $\bar{\nu_e}$ disappearance at a power reactor. The region to be explored is shown in Figure 1.

Reactors are powerful sources (see Table 1) of low energy (0 - 10 MeV) $\bar{\nu_e}$'s making them suitable for searches of oscillations in a region of small Δm^2. A 12 ton detector (cost $\approx$ \$1M) placed 740 meters from 2 x 3300 MW reactors is sensitive to a Δm^2 down to 10^{-3}eV^2 and a mixing angle $\sin^2 2\theta$ of about 0.1. Our experiment investigates $\bar{\nu_e}$ disappearance which includes the channel $\bar{\nu_e} \rightarrow \bar{\nu_\mu}$, a candidate that may be responsible for the atmospheric neutrino anomaly. Several long baseline experiments [5,6,7] (cost> \$50M) are being proposed at accelerators to investigate both the $\nu_\mu \rightarrow \nu_e$ and the $\nu_\mu \rightarrow \nu_\tau$ channels, the latter via τ appearance. Our experiment will extend the explored region of Δm^2 by about an order of magnitude from the present limits[8,9]. Recently,

Harrison et al.[10] have suggested that if maximal mixing exists between all three neutrino families, and if neutrino masses follow the Lipton mass hierarchy, namely $m(\nu_\tau) >> m(\nu_e), m(\nu_\mu)$, then the atmospheric and solar neutrinos results can be fit with a single $\Delta m^2 = 0.72 \times 10^{-2} eV^2$. Figure 2 shows the positron spectrum at San Onofre for these parameters. A significant deficit is expected. Thus, long baseline oscillation experiments at reactors[11,12] are a powerful, inexpensive and expedient way to explore neutrino oscillations.

Table 1: Parameters of the San Onofre Reactors

Thermal Power	neutrino source	neutrino energy	distance to target
2X3.3GW	$1.4 \times 10^{21} \nu s^{-1}$	0-10 MeV	740 m

2 The Detector

In the proposed low-energy reactor-neutrino oscillation experiment the reaction

$$\bar{\nu_e} + p \rightarrow e^+ + n, \quad (1)$$

calls for a hydrogen-rich scintillator target. The scintillator simultaneously detects the positron and neutron. The time correlated (e^+, n) signal provides a powerful signature to enhance the signal to noise.

The San Onofre detector will be situated at a shallow depth of 25 mwe. The cosmic ray hadronic component is completely negligible at this depth, however the muon flux, reduced from 130 $m^{-2}s^{-1}$ to 28 $m^{-2}s^{-1}$ remains an important source of background as the muons induce fast neutrons and radio-nuclei. The background from the latter would be overwhelming for a homogeneous detector. Our detector, therefore, is a fine grain segmented detector, where we identify the prompt positron as well as the two annihilation gamma rays in three adjacent detector cells, as shown in Figure 3.

In our detector we require that a positron ($E \geq 1$ MeV) form a fast triple coincidence with two annihilation gammas (energy window 50 - 600 keV). The neutrons are detected via Gd capture gamma rays (strong lines at E = 5.9 and 6.8MeV; capture time for a 0.1% Gd loaded scintillator = 27μs) in a time window of 100μs. With an estimated detection efficiency of 20%, we expect a signal of 24 neutrino events/day. The above-mentioned fast triple coincidence suppresses the neutron related background by 6×10^{-3} and radioactive decays leading to neutron-unstable states (such as ^{9}Li) by 2×10^{-3}.

The San Onofre detector consists of 66 cells filled with Gd-loaded mineral-oil-based scintillator. The fiducial volume is 12 tons. Each cell is made from 0.6 cm thick acrylic with dimensions 9m x 0.25m x 0.13m. The end layer of 0.8 m of each cell is inactive. The detector is surrounded by a one-meter-thick boron-loaded water buffer. Neutrons and gamma rays originating from the lab walls are attenuated by the buffer and the end layer of each cell. The buffer tanks are made of low radioactivity steel.

Outside the water buffer there is a hermetic muon veto. The veto is designed to tag through-going and stopping muons. Stacks of surplus liquid scintillator tanks from the MACRO experiment will constitute the muon veto. At the veto the muon flux through a horizontal area is 28 $m^{-2}s^{-1}$ at 25 mwe. Muons through the veto but not entering the fiducial volume open a 10μs veto. Through-going muons entering the fiducial volume initiate a 100μs veto, while muons stopping in the fiducial volume are vetoed for 300 μs. The dead time from vetoing is about 10%. The inherent design of the veto results in a veto inefficiency of less than 10^{-3}.

3 Backgrounds

3.1 Muon-Induced Radioactivity

The Kamiokande experiment[13] has reported on production of ^{16}N, ^{12}B, and ^{12}N. These rates have been scaled to San Onofre. ^{9}Li with a 35% branch of the beta delayed neutron emission is a possible radioactivity. While this is a serious background for a homogeneous detector it is of no practical concern in a segmented detector (where one requires a positron coincident with 2 annihilation gamma rays).

3.2 Fast Neutron Correlated Background

Muon interactions via nuclear spallation and negative muon capture lead to fast neutrons with energies up to 100 MeV. We have studied muon spallation theoretically[14] and have used the best known production rates to estimate the background for the San Onofre detector. As there remains considerable uncertainty in the neutron production yield and energy spectra we recently proposed[15] to study muon spallation in detail at the CERN SPS. By using a 100 GeV muon beam, targets of differing atomic number and thickness and neutron time-of-flight counters, we intend to study fast neutron production rates, angular distributions and energy spectra. This information will be used to improve our estimates for the neutron-related background.

3.3 Neutrons produced outside the veto

Neutrons produced outside the veto cannot be tagged by the associated muon but are largely attenuated by the 1m thick water buffer. The muon stop rate at 25mwe is about $50\text{kg}^{-1}\text{d}^{-1}$. From the capture-neutron production rate in rock at 25mwe of 11 $\text{kg}^{-1}\text{d}^{-1}$ and in oil of $3\text{kg}^{-1}\text{d}^{-1}$, we find that the number of capture neutrons produced outside the detector and entering the detector volume (after attenuation in the buffer) is about 160d^{-1}.

The neutrons from muon spallation are based on estimated neutron production yields at 25mwe of 3.5×10^{-5} g^{-1}cm^2 per muon, corresponding to a neutron production rate from muon spallation of about $13\text{kg}^{-1}\text{d}^{-1}$. An estimate of the number of spallation neutrons from rock reaching the detector yields 1300d^{-1}. The (non-vetoable) fast (> 5MeV) neutron flux in the fiducial volume of the detector originating from neutrons produced outside the veto, thus, is estimated to be about 1500d^{-1}. With a rejection factor of 6×10^{-3} from the 4 fold coincidence, this results in a background of 9d^{-1}.

3.4 Neutrons produced inside the veto

The veto suppresses the correlated background from fast neutrons which are created within the veto. The veto inefficiency of $< 10^{-3}$, allows only 0.6d^{-1} to the correlated 4-fold background. The fiducial volume acts as a veto for high energy (> 10MeV) muons. Only those muons that are stopped in the outer layer (5cm) of the detector, or about 10^{-3} of the 1s^{-1} unvetoed muons survive. With the same triple coincidence cut, this gives a negligible correlated 4-fold contribution from neutrons produced by muons in the fiducial volume.

The total correlated background from the above three sources is estimated at 9.6d^{-1}.

3.5 Correlated Multiple Neutrons

Multiple neutron arise in both muon capture and muon spallation. Multiple neutrons are born simultaneously and their capture events are thus correlated in time. Some of theses neutrons will exit the detector and become benign, while those capturing on Gd will result in a correlated gamma ray signature.

Muon capture multiple neutrons are not of major concern as they are tagged and suppressed (by a factor of about 10^{-7}) by the 300μs veto. But neutrons from muon spallation are a potential problem. For the 0.1% Gd loaded detector, spallation neutrons are suppressed, with the 100μs veto, by a factor of more than 900.

Assuming a double-to-single neutron ratio of 0.25, we estimate the double neutron background. Here, we neglect a small contribution from double neutrons from the buffer. After applying the factor of about 900 for a 100μs veto, the number of non-vetoable double neutrons which are candidates for correlated backgrounds is about 36d^{-1}. Requiring both neutrons to capture, resulting in two 8 MeV gamma bursts and also requiring for these gammas to produce a 4-fold coincidence signal, we find a reduction factor of 2.3$\times$10^{-2}. The rate due to the two-neutron background is less than 1d^{-1}.

3.6 Uncorrelated Backgrounds

There are two components of the uncorrelated accidental background: natural radioactivities in the detector and the surrounding walls, and muon-induced radioactivities including muon-produced radio isotopes and "neutron soup".

The passive shielding, the photo-multipliers, the scintillators, and the acrylic, contain small amounts of radioactive contaminants. We have studied the backgrounds due to contaminations from ^{40}K, Rn, U and Th. Concentrations of Th and U in the scintillation liquid have been found to be about 0.3 $\times$10^{-12}g/g[16] and a few times 10^{-12}g/g for Gd loaded samples. ^{40}K in the scintillator has been measured and is negligible, while ^{40}K from the PMTs will be reduced sufficiently by an 80 cm dead layer.

Gamma rays from radioactivities give rise to triple coincidences via Compton scattering or pair production. Our model estimates give a triple coincidence rate from all natural radioactive contributions of 4.5s^{-1} and a singles rate of 0.28s^{-1}, of which the largest triple contribution ($3.2s^{-1}$) is from the surrounding walls. This results in a total accidental 4-fold rate of 11/d.

We note that the accidental background can be obtained with good statistical accuracy between real positron triggers by inducing artificial triggers in the detector.

3.7 Summary of Backgrounds

The contribution from different sources of the background are shown in Table 2. The total background is 21/d compared with the expected signal of 24/d. Data collection will encompass 2 refuelling cycles, with 90 days background when each of the two reactors is refuelled and 400 days data with both reactors running.

Table 2: Summary of correlated and uncorrelated backgrounds.

Scintillator	Time [μs]			Background Counts [d^{-1}]			
	Window	Muon Veto		Correlated		Uncorrelated	Total
		Stop	Through	1 n	2 n	All	
Gd-loaded	100	300	100	8.6	1	11	20.6

4 A Prototype Segmented Detector

We conducted a prototype experiment from April 1993 to December 1993 at San Onofre (see[11]). 18 cells (0.13 tons fiducial, with unloaded liquid scintillator) from the Goesgen experiment were installed at the San Onofre Unit 2 reactor. The apparatus was at a depth of 20 mwe, 24.5m from the reactor core. A muon veto surrounded the detector but no water buffer could be implemented.

The experiment utilized the correlated (e^+,n) signature mentioned above. The analysis gave an excess over background of 12 (+/- 5.5) events/day (9 events/day expected). Reducing the threshold for the 511 keV gammas to 35 keV gave an excess of 22 (+/- 6.3) events/day (15 events/day expected), confirming the triple coincidence scheme.

5 Special Features and Custom Developments

Experience with the test detector led us to pursue many detector enhancements. We developed a 0.1% Gd loaded scintillator in collaboration with NE Technology. Neutron capture on Gd instead of protons reduces the capture time to 27 μs and the window to 100 μs (instead of 500 μs). The higher energy (8 MeV) neutron capture gamma rays allows a 3.5 MeV threshold for capture gamma detection, well above the natural radioactivity background. The NE scintillator has been tested in a full-size acrylic cell for 6 months. The light output, attenuation length and Gd concentration have been measured to be constant over time. The scintillator is measured to yield 4 photoelectrons per 4 PMTs for 50 keV deposited in the middle of the cell.

The detection efficiency of a segmented detector is greatly enhanced by including a large number of trigger patterns involving adjacent cells to detect the 511 KeV gamma rays from positron annihilation. The prompt triple coincidence must be performed in hardware which would require an enormous number of coincidence modules for the 66-cell detector. Instead, we developed a custom chip to perform this task using Xilinx Field Programmable Gate Array Logic (FPGA). The FPGA solution is inexpensive and allows the coincidence pattern to be changed by reprogramming the FPGA.

A custom DAQ system was developed. A VXI board (SODAQ) was designed to read out LeCroy FERA ADC's and TDC's. SODAQ reads the ADC's and TDC's via the high speed FERA port. Data is double buffered and readout with a workstation on the VXI bus. This system enables us to read data with minimal DAQ dead time. Gating logic for the time correlated (e^+,n) pattern is also included on the SODAQ board, eliminating the need for external gate generators.

Special photo-multipliers were developed for the detector by Philips. The XP4532 P/A is a 5-inch-diameter PMT, with 1.1 nsec transit time jitter, a resolved single photoelectron peak, minimum gain of 2 x 10^7 and low radioactivity (^{40}K = 0.08 ppm, ^{232}U = 0.29 ppm, ^{238}Th = 0.12 ppm).

6 Conclusion

Construction of the underground vault is scheduled to commence Fall of 1995. The complete detector will be installed by late 1996. Following 2 years of data taking, preliminary results are expected in 1998.

7 Acknowledgements

We thank the staff of Southern California Edison Co. (SCE) for support provided during installation and running of the prototype detector and Cadence Design Systems for the CAD design tools. W. Waller of the Caltech CCO helped with the figures. This work was supported by the U.S DOE Contract DE-FG03-88ER40397.

References

1. T. Kirsten, in *Neutrino 94*, Proceedings of the 16th International Conference on Neutrino Physics and Astrophysics Eilat, Israel (1994), p. 83, Ed. A. Dar, G. Eilam and M. Gronau, Nuclear Physics B (Proc. Suppl.) 38, January 1995.
2. R. Becker-Szendy *et al.* ibidem, p. 331.
3. M.C. Goodman (The Soudan 2 Collaboration). ibidem, p. 337.
4. Y. Fukuda *et al.* (Kamioka Collaboration), *Phys. Lett.* B **335**, 237 (1994)
5. E. Ables *et al.* (MINOS Collaboration), Fermilab Proposal E-875, (Feb 1995).
6. D. Beavis *et al.* BNL-PROPOSAL-889, (Jan 1995).
7. K. Nishikawa *et al.*, Letter of Intent for Long Baseline Neutrino Oscillation Experiment, using KEK-PS and Super-Kamiokande, (November 28, 1994).
8. G. Zacek *et al.*, *Phys. Rev.* D **34**, 2612 (1986).
9. B. Achkar *et al.*, *Nucl. Phys.* B **434**, 503 (1995).
10. P.F. Harrison *et al. Phys. Lett.* B **349**, 137 (1995).
11. F. Boehm *et al.*, Proposal for the San Onofre Neutrino Oscillation Experiment (1994).
12. Y. Declais *et al.*, Chooz Proposal, (1993).
13. K. Inoue, Thesis, Kamiokande (1993).
14. J. Delorme *et al.*, in press *Phys. Rev.* C, (1995).
15. F. Boehm *et al.*, CERN SPS Proposal P-293,CERN/SPSLC 95-52, (May 1995).
16. F. Boehm *et al.*, *Nucl. Instrum. Methods* A **300**, 395 (1991).

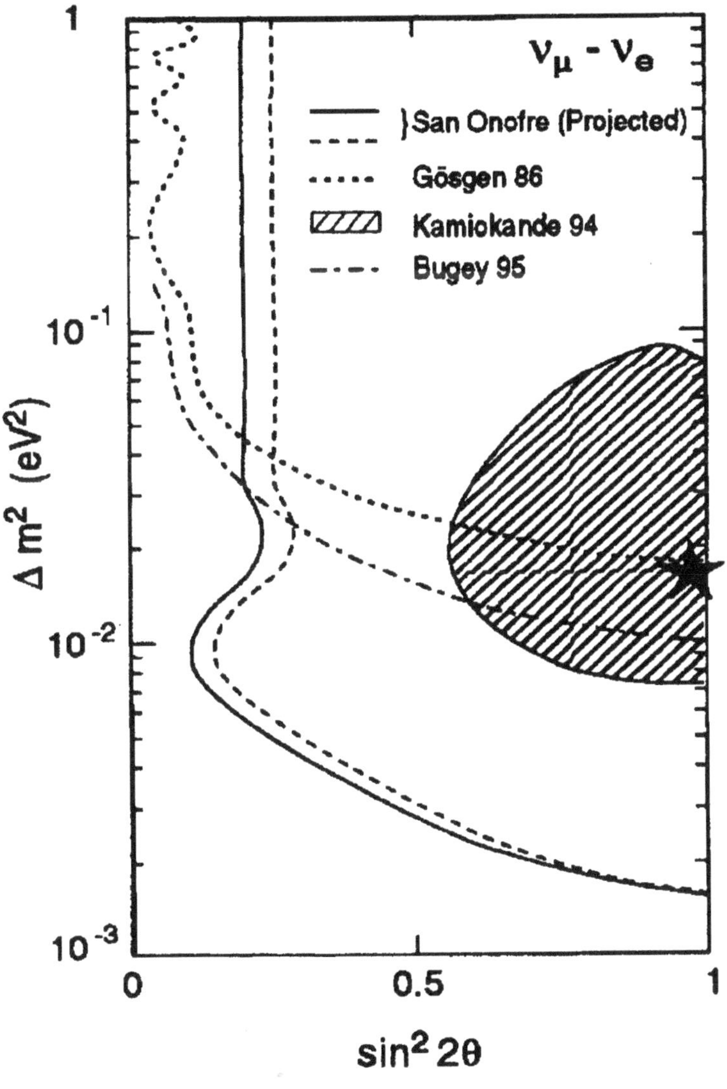

Figure 1: The region of parameter space that can be excluded at the 90 % CL by the San Onofre experiment. The outer curve is for an S/N = 1:0 and the inner curve for S/N = 1:1. We have included statistical and systematic errors. The latest region allowed by the atmospheric neutrino results[4] are shown by the hatched area. The star indicates the best bit for the Kamiokande data.

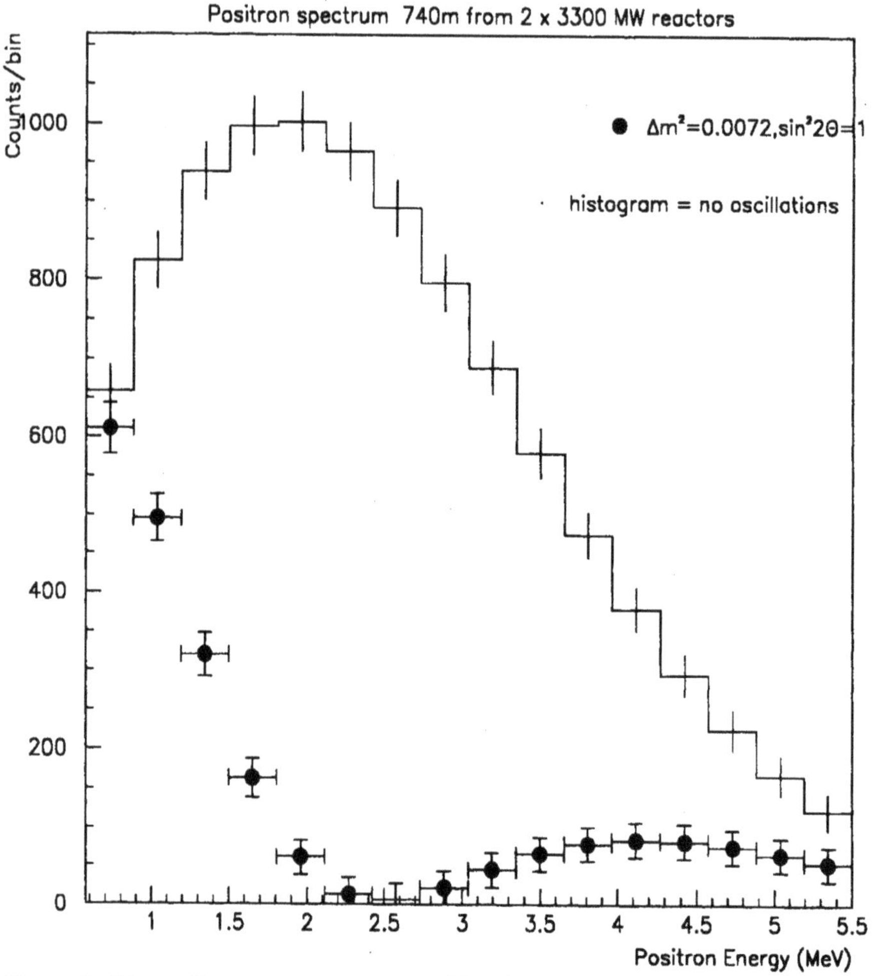

Figure 2: The positron spectrum measured by the San Onofre detector at 740 m from the 2 X 3300 MW reactors. This plot assumes 400 days of data with reactor ON and background subtraction with 90 days reactor OFF data. We assume an event rate of 24 counts/day with no oscillations (solid histogram) and a background of 21 counts/day. The points are with full mixing and $\Delta m^2 = 0.72 \times 10^{-2}\text{eV}^2$ which are a best fit to both solar and atmospheric neutrino results as suggested by Harrison[10] et al.

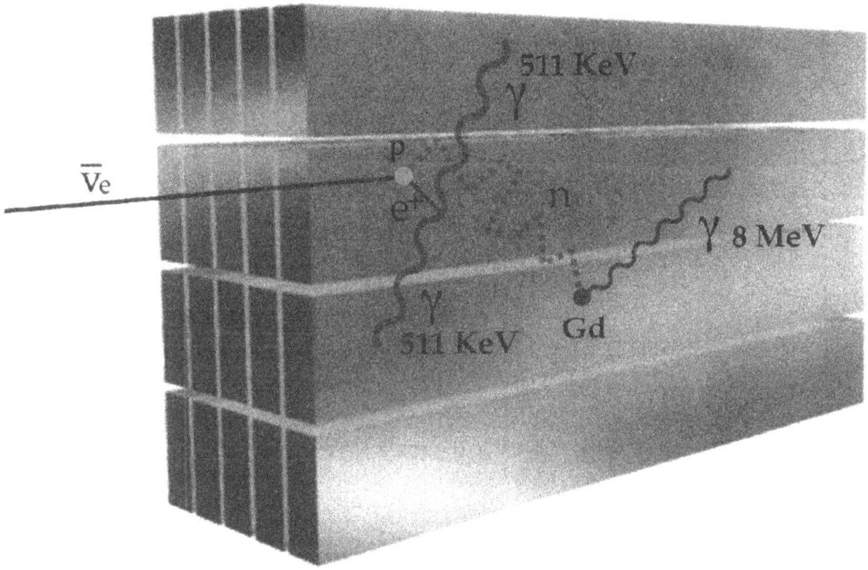

Figure 3: A typical neutrino signature in the San Onofre detector, a positron (E≥1 MeV) forms a fast triple coincidence with two annihilation gammas (energy window 50-600 keV). The neutrons are detected via Gd capture of gamma rays (strong lines at E=5.9 and 6.8 MeV; capture time for a 0.1% Gd loaded scintillator ≈ 27μs) in a time window of 100μs.

COSMOLOGY OF THE NEXT-TO-MINIMAL SUPERSYMMETRIC STANDARD MODEL

S.A. ABEL

Rutherford Appleton Laboratory, Chilton, Didcot, OX11 0QX, U.K.

S. SARKAR, P.L. WHITE[a]

Theoretical Physics, University of Oxford, 1 Keble Road, Oxford OX1 3NP, U.K.

We discuss the domain wall problem in the Next-to-Minimal Supersymmetric Standard Model, with particular attention to the usual solution of explicit breaking of the discrete symmetry by non-renormalisable operators. This "solution" leads to a contradiction between the requirements of cosmology and those of avoiding the destabilisation of the hierarchy.

1 Introduction

One of the most popular extensions of the Minimal Supersymmetric Standard Model (MSSM) is that of the Next-to-Minimal Supersymmetric Standard Model (NMSSM)[1] where the usual Higgs sector of two doublets is supplemented by a gauge singlet superfield coupling only through the Higgs sector superpotential. Apart from the important phenomenological consequences, this allows the elimination of the μ term from the MSSM superpotential by invoking a $\mathbf{Z}_3$ symmetry. The μ term presents a problem because one might expect it to be of order the Planck mass or at least the GUT scale, and yet reasonable phenomenology forces it to be of order the supersymmetry breaking scale $m_{3/2}$.[2]

The purpose of this talk is to briefly review one of the most important cosmological implications of the NMSSM, namely that of the domain walls which inevitably form at the weak scale when the $\mathbf{Z}_3$ is spontaneously broken[3]. The requirements that these walls do not destroy the predictions of standard cosmology forces us to explicitly break the $\mathbf{Z}_3$, which can be done with non-renormalisable operators (NROs). This in turn leads to problems with the destabilisation of the hierarchy, and we shall discuss how the relative sizes

[a]Talk presented by P.L. White

of the constraints affect the model. A more complete version of this work is contained in Reference 4.

2 The NMSSM

In the MSSM the Higgs sector contains two doublets H_1 and H_2 coupling to down- and up-type quarks respectively, with a superpotential containing a term $\mu H_1 H_2$. The origin of the μ term is obscure, but we must avoid taking it much larger than the supersymmetry breaking scale, which is typically assumed to be less than around 1TeV.

One possible solution to the μ problem, which has the additional advantage of giving a radically different low energy Higgs phenomenology, is that of the NMSSM[1]. Here we add a gauge singlet superfield N, and replace the μ term in the superpotential with the Higgs superpotential

$$W_{Higgs} = \lambda N H_1 H_2 - \frac{k}{3} N^3 \tag{1}$$

while the usual soft breaking terms are supplemented by another mass and two more trilinear terms instead of the bilinear term $B\mu H_1 H_2$ of the MSSM. It is then possible to arrange the parameters of the model in a natural way so that the singlet gets a vacuum expectation value (VEV) which is of the same order as those of the other Higgs fields, generating an effective μ parameter of form λx where $x = < N >$. The terms which we have left out of the potential consistent with gauge symmetry, namely $\mu H_1 H_2$, $\mu' N^2$, and $\mu'' N$, can all be banned by invoking a $\mathbf{Z}_3$ symmetry under which all chiral superfields have the same charge so that only trilinear terms are allowed in the superpotential. The inclusion of the N^3 term is necessary both because it is consistent with $\mathbf{Z}_3$ and because without it there is a Peccei-Quinn symmetry which gives a phenomenologically unacceptable axion.

3 Domain Walls

3.1 Formation and Structure

When a discrete symmetry is spontaneously broken, it generates domain walls[3]. This is simple to understand, since the discrete symmetry imposes that there are multiple degenerate vacua, and there is no way in which causally disconnected regions of the universe can conspire to all undergo a phase transition to the same vacuum (so long as the fields acquiring VEVs are in thermal equilibrium, which will always be true for the NMSSM). The universe must thus evolve to a state in which there are domains of different vacua, separated by "domain walls".

On purely dimensional grounds, or by analogy with analytically solvable toy models, we expect the domain walls to have a thickness δ and surface energy σ given by

$$\delta \sim \frac{1}{\nu} \qquad \sigma \sim \nu^3 \tag{2}$$

where ν is a typical VEV of the fields. In the NMSSM wall solutions may be found numerically[5], and we find that the above equations hold very well, with ν being replaced by some typical VEV of the singlet which is of roughly the same magnitude as that of the Higgs which gives the W and Z bosons their masses, *i.e.* 174GeV.

3.2 Dynamics

Once a wall network has formed, it will not remain static. There are three important forces acting on the network of domain walls, namely surface tension, friction, and pressure. The first of these is simply the effect of the constant surface energy density, which makes it energetically favourable for small bubbles of wall to disappear, for walls to smooth themselves out, and so on. This removes the smaller scale structure, and given enough time will remove the wall network entirely. However, by causality it is clear that the largest possible correlation length, by which we mean typical domain size, cannot be larger than the horizon scale, and so the best we can expect is that there is typically one wall per horizon volume throughout the evolution of the universe. As will be discussed later, even if the entire visible universe were to consist of only part of one domain, this still has unwelcome cosmological consequences.

The second force is that of friction. As the wall moves through space it will interact with the thermalised plasma, dissipating its energy and slowing its evolution. This effect can be shown to be insignificant for very high temperatures, when the particles in the plasma have very small reflection coeffecients and so do not exert a large force, and also at low temperatures when the density of the plasma is low[4]. We can thus neglect it, as it will only slow down the wall evolution in its early stages.

A final force which we may consider is that of pressure. If one of the vacua is slightly deeper than the others by some amount ε, as a consequence of some slight *explicit* violation of the $\mathbf{Z}_3$ symmetry in the potential, then there will be a force per unit area on the walls of order ε. Since the magnitude of the force due to surface tension is given by σ/R, where R is the curvature scale (effectively the correlation length), we see that the pressure will dominate the dynamics when

$$\varepsilon > \frac{\sigma}{R} \tag{3}$$

Since R is steadily increasing as a result of surface tension (and the expansion of the universe), we see that for large enough values of ε the pressure will come to dominate the dynamics before the present day. In this case the evolution of the wall network will begin with the curvature scale increasing such that it is always of order the horizon size, until ultimately pressure takes over, the true vacuum comes to dominate, and the wall network is completely eliminated. Numerical simulations suggest that once the pressure comes to dominate, disappearance of the walls is almost as fast as causality will allow[6].

3.3 Cosmological Implications

We now turn to the cosmological implications of a domain wall network. As the universe expands, the energy density due to the walls will fall as a^{-1} (for static walls) or a^{-2} (for ultra-relativistic walls in a radiation dominated universe), where a is the cosmological scale parameter. Since the energy densities of matter and radiation fall as a^{-3} and a^{-4} respectively, it is clear that the walls will ultimately dominate the dynamics of the universe, and we can check that this will have happened long before the present day for weak scale walls, and so the wall network must have long since disappeared.

In fact we may draw tighter constraints from other cosmological observations. If the walls decay after nucleosynthesis, then they will release a vast amount of energy which can be shown to cause photodestruction of the light elements whose abundances can be accurately measured[7]. Thus we require that the walls decay before a temperature of around 0.1 to 1MeV, giving a minimum value of ε to be

$$\varepsilon \gtrsim \lambda' \sigma M_W^2 / M_{Pl} \tag{4}$$

where λ' is of order 10^{-7}. We conclude that a tiny contribution to the superpotential from an NRO of form, say, $\lambda' N^4/M_{Pl}$ is sufficient to evade the cosmological constraints. Such an NRO will have negligible effect on the low energy phenomenology, and so we conclude that an NRO suppressed by at most one power of the Planck mass is sufficient to eliminate the walls in time to avoid cosmological problems.

4 Destabilising Divergences

The primary motivation for supersymmetry is the hierarchy problem. If the standard model is an effective theory valid to a large scale Λ then we expect the hierarchy to be destabilised by quadratic divergences and the standard model masses to be driven up to order Λ. Supersymmetry prevents this happening, since softly broken supersymmetry does not have quadratic divergences.

There is one important exception to this statement, in that tadpole diagrams in supersymmetry may be quadratically divergent[8] so long as we have NROs in our theory. These diagrams may be ruled out for one of three reasons: gauge invariance, which cannot save us from singlet tadpoles; $\mathbf{Z}_3$ symmetry, which our NROs must break to save us from the cosmological implications; and supersymmetry which is obviously broken for phenomenological reasons.

It is straightforward to check that every dimension five NRO which we can introduce in our superpotential allows the construction of at least one diagram at three loops or less which will be quadratically divergent. For example, the least dangerous such operator is $\lambda'(H_1H_2)^2/M_{Pl}$, where the first such diagram occurs at three loops. These diagrams generate a term in the superpotential of form $\mu''N$, where

$$\mu'' \sim \frac{\lambda'\lambda}{(16\pi^2)^2} m_{3/2} M_{Pl} \tag{5}$$

The presence of one factor of $m_{3/2}$ is guaranteed by the remark above that in the absence on supersymmetry breaking the non-renormalisation theorem prevents the generation of such terms, while the factor M_{Pl} is generated by taking the cut-off scale $\Lambda \sim M_{Pl}$.

The effects of such terms on the hierarchy is catastrophic, since they generate terms in the low energy Higgs potential which may be as large as those from radiative breaking at each order, and so destroy any reason for us to expect the electroweak scale to be very much less than the GUT or Planck scales. Requiring that the electroweak scale is not driven up by orders of magnitude in scale requires $\lambda' \lesssim 10^{-11}$, in clear contrast to the constraint from cosmology given above.

5 Conclusions

We have thus seen that the standard solution to the domain wall problem in the NMSSM must always introduce destabilising divergences which will destroy all predictive power in the Higgs sector and drive the electroweak scale to be many orders of magnitude too large. One may still solve the domain wall problem by introducing renormalisable terms which break the $\mathbf{Z}_3$, such as a μ term, but one then has a far more severe naturalness problem than is usually the case in the MSSM, since we must somehow ban all the dangerous NROs which can cause quadratically divergent singlet tadpoles, while simultaneously breaking every symmetry which could ban them.

References

1. P. Fayet, *Nucl. Phys.* **B90**, 104 (1975);
   H.-P. Nilles, M. Srednicki and D. Wyler, *Phys. Lett.* **B120**, 346 (1983);
   J.-M. Frere, D. R. T. Jones and S. Raby, *Nucl. Phys.* **B222**, 11 (1983);
   J.-P. Derendinger and C. A. Savoy, *Nucl. Phys.* **B237**, 307 (1984);
   L. Durand and J. L. Lopez, *Phys. Lett.* **B217**, 463 (1989);
   M. Drees, *Intern. J. Mod. Phys.* **A4**, 3645 (1989);
   J. Ellis, J. Gunion, H. Haber, L. Roszkowski and F. Zwirner, *Phys. Rev.* **D39**, 844 (1989).
2. L. Hall, J. Lykken and S. Weinberg, *Phys. Rev.* **D27**, 2359 (1983);
   J.E. Kim and H.P. Nilles, *Phys. Lett.* **B138**, 150 (1984);
   K. Inoue, A. Kakuto and T. Takano, *Prog. Theor. Phys.* **75**, 664 (1986);
   A.A. Ansel'm and A.A. Johansen, *Phys. Lett.* **B200**, 331 (1988);
   G. Giudice and A. Masiero, *Phys. Lett.* **B206**, 480 (1988).
3. Ya.B. Zel'dovich, I.Yu. Kobzarev and L.B. Okun, *Sov. Phys. JETP* **40**, 1 (1975);
   T.W.B. Kibble, J. Phys. **A9**, 1387 (1976); Phys. Rep.**67**, 183 (1980);
   A. Vilenkin, Phys. Rep. **121**, 263 (1985);
   A. Vilenkin and E.P.S. Shellard, *Cosmic Strings and Other Topological Defects* (Cambridge University Press, 1994)
4. S.A. Abel, S. Sarkar, P.L. White, hep-ph/9506359.
5. S.A. Abel, P.L. White, hep-ph/9505241.
6. W.H. Press, B.S. Ryden, D.N. Spergel, *Astrophys. J.* **347**, 590 (1989); *Astrophys. J.* **357**, 293 (1990);
   L. Kawano, *Phys. Rev.* **D41**, 1013 (1990).
7. J. Ellis, G. Gelmini, J. Lopez, D. Nanopoulos, S. Sarkar, *Nucl. Phys.* **B373**, 399 (1992);
   W. Hu and J. Silk, *Phys. Rev. Lett.* **70**, 2661 (1993);
   S. Sarkar, Oxford preprint OUTP-95-16P (submitted to *Rep. Prog. Phys.*);
   M.H. Reno and D. Seckel, *Phys. Rev.* **D37**, 3441 (1988);
   S. Dimopoulos, R. Esmailzadeh, L.J. Hall and G.D. Starkman, *Nucl. Phys.* **B311**, 699 (1989).
8. U. Ellwanger, *Phys. Lett.* **B133**, 187 (1983);
   J. Bagger and E. Poppitz, *Phys. Rev. Lett.* **71**, 2380 (1993);
   J. Bagger, E. Poppitz and L. Randall, hep-ph/9505244;
   V. Jain, hep-ph/9407382

ELECTROWEAK BARYOGENESIS WITH NEUTRAL HEAVY LEPTONS

N. RIUS

Theory Division, CERN
CH-1211 Geneva 23, Switzerland

We consider the possibility that baryogenesis occurs during the weak phase transition in a model with new neutral leptons which conserves total lepton number. Using present experimental bounds on the mixing between the neutral heavy leptons and the standard neutrinos we estimate that the model can account for the observed baryon to entropy ratio if the heavy lepton masses are $\geq M_Z$.

1 Baryon Asymmetry Puzzle

The baryon number to entropy ratio in the observed part of the universe is required to be $n_B/s \sim (4-6)10^{-11}$ by nucleosynthesis constraints [1]. In 1967, Sakharov [2] established the three basic requirements for obtaining this baryon asymmetry as a result of particle interactions in the early universe: a) Baryon number violation b) C and CP violation c) Departure from thermal equilibrium. These conditions may be fulfilled at weak scale temperatures [3], if the electroweak phase transition is first order.

In principle the Standard Model (SM) contains all the necessary ingredients for baryogenesis, but the effects of the CP violation from the Kobayashi-Maskawa phase are much too small to account for the observed n_B/s ratio [4]. However, electroweak baryogenesis is feasible in several simple extensions of the SM which contain additional sources of CP violation, namely two Higgs, singlet Majoron, and supersymmetric models [5].

The model we consider here provides a viable alternative to the see-saw mechanism for explaining the lightness of the known neutrinos. It arises in several contexts, such as E(6) superstring inspired models and Extended Technicolor with a GIM mechanism [6].

2 The Model

The gauge group is the standard $SU(2)\times U(1)$, with minimal quark sector. The lepton sector is extended with two electroweak singlet two-component leptons in each generation ν_R, s_L, and there is a new singlet scalar field, σ. In addition to the standard model lagrangian, the model is described by

$$\Delta\mathcal{L} = \bar{\Psi} f H l_R + \bar{\Psi} f_D H \nu_R + \bar{s}_L f_M \sigma \nu_R + h.c. - V(|\sigma|^2, |H|^2) \tag{1}$$

The Yukawa terms that lead to the neutral fermions masses are restricted by imposing total lepton number conservation.

We will assume that the the singlet σ acquires a vacuum expectation value at the same sçale as the standard Higgs doublet, and the theory undergoes a single phase transition at a critical temperature near the weak scale. In the broken phase the mass terms for the neutral leptons are

$$\bar{\nu}_R D \nu_L + \bar{\nu}_R S s_L \tag{2}$$

The mass matrix in (2) is singular, and ensures the existence of three massless Weyl neutrinos ν_a, regardless of the relative value of D and S. The other six Weyl fermions pair up into three heavy neutral Dirac fermions n_a. Note that in this model the ratio D/S can be much larger than the corresponding parameter of any model which invokes the see-saw mechanism to understand the smallness of neutrino masses.

After identifying the physical mass-eigenstates leptons by diagonalizing the mass matrices, we get the following form for the charged current leptonic weak interaction:

$$\mathcal{L}_{cc} = \frac{g}{2} W^\mu \sum_{a=1}^{3} \sum_{\alpha=1}^{6} \bar{e}_a \gamma_\mu L \left[K_L \nu'_L + K_H n_L \right] + h.c. \tag{3}$$

where ν'_L are the massless neutrinos, n_L are the left-handed part of the four component neutral heavy leptons (NHL) and

$$K = (K_L, K_H) \tag{4}$$

is a 3×6 mixing matrix, with $K_L \sim \mathcal{O}(1)$ and $K_H \sim \mathcal{O}(D/S)$. One can see that the charged current coupling of the mass eigenstates charged leptons to the massless as well as the heavy neutrinos is nontrivial, making possible the violation of individual lepton numbers L_e, L_μ and L_τ.

Due to the admixture of fermions of different weak isospin, there is no GIM mechanism in the neutral fermion couplings to the Z boson, which are given by

$$\mathcal{L}_{nc} = \frac{g}{2c_w} Z^\mu \sum_{\alpha,\beta} \bar{N}_\alpha \gamma_\mu L P_{\alpha\beta} N_\beta \tag{5}$$

where $N_\alpha = (\nu'_a, n_a)$, and $P = K^\dagger K$ is in general a nondiagonal projection matrix. The neutral couplings involving the massless neutrinos are diagonal but flavour dependent.

It can be shown [7] that for n generations the total number of physical parameters describing K is n^2 angles and $(n-1)^2$ phases. As a result, in this model CP violation can arise with only two generations of strictly massless neutrinos.

3 Electroweak Baryogenesis and Charge Transport Mechanism

If the electroweak transition is first order with supercooling, bubbles of true ground state (broken phase) nucleate and expand until they fill the universe; local departure from thermal equilibrium occurs in the vicinity of the expanding bubble walls. The two other Sakharov's conditions are also satisfied, since C and CP are known to be violated by the electroweak interactions and anomalous baryon number violation is fast at high temperatures.

The scenario that we consider for generating the baryon asymmetry is the charge transport mechanism [8]. As a bubble expands, particles in the unbroken phase which acquire masses in the true vacuum will tend to reflect off the advancing wall. Electroweak interactions are active throughout the trajectory and their CP-violating effects induce a different reflection probability for leptons and antileptons. This leads to a CP asymmetry in the reflected flux. In the symmetric phase anomalous $B+L$ violating interactions are in thermal equilibrium, therefore some of the anti-leptons radiating off of the expanding bubble will be turned into baryons. These baryons pass through the bubble wall into the broken phase, where baryon number is conserved.

An important survival requirement for the produced baryon asymmetry is that the sphaleron processes inside the bubble are slow enough so as not to wash it out. We don't consider the problem of whether a first order phase transition did take place; we assume that it did, and also that an optimal sphaleron rate is present. With a very optimistic estimation of the non-CP effects, the induced baryon asymmetry is at most $n_B/s \sim 10^{-2}\Delta_{CP}$ [9,10]. Therefore, $\Delta_{CP} \geq 10^{-8}$ is required.

4 CP Asymmetry

The observable CP asymmetry results from the interference of pure CP-violating phases in the mixing matrix K with CP-even phases, equal for particles and antiparticles. These are the reflection coefficients, which become complex when the particle energy is smaller than its mass in the true vacuum. We consider a two generation model for simplicity. Details can be found in [11].

4.1 Charged Leptons

The calculation of the charged lepton CP asymmetry is completely analogous to the computation done for quarks in the SM [10,4]. However, since the damping rate of leptons is dominated by weak interactions, it is possible that the resulting CP asymmetry is not as suppressed as in the case of strongly interacting quarks.

At finite temperature, T, the correct asymptotic states are quasiparticles instead of particles, built up of a resummation of the thermal self-energies of the particles [12]. In the determination of the quasiparticle dispersion relations we include only diagonal weak interaction corrections at leading T. The remaining corrections, *i.e.*, the non-diagonal weak corrections and those subleading in T, are essential ingredients to obtain an observable asymmetry, and we include them as a perturbation.

We are interested in the low momentum region, $p \ll \omega$, which is relevant for the total reflection of the charged leptons. Therefore we keep only linear terms in momentum. The effective unperturbed Hamiltonian is then

$$H^0_{eff} = \begin{pmatrix} -\frac{1}{3}i\sigma_z\partial_z - i\gamma + \omega^0_R & \frac{m}{2}\theta(z) \\ \frac{m}{2}\theta(z) & \frac{1}{3}i\sigma_z\partial_z - i\gamma + \omega^0_L \end{pmatrix} \tag{6}$$

where ω^0_R and ω^0_L are 3×3 diagonal matrices and γ is the damping rate of quasiparticles in a plasma.

Assume first $\gamma = 0$ and thus H^0_{eff} hermitean. We solve the Dirac equation resulting from the Hamiltonian (6) for eigenstates of the energy which are superpositions of an incoming, a reflected and a transmitted plane wave. It is possible to obtain an implicit equation for the reflection coefficients r_{ij}, which can be solved perturbatively by expanding the matrix r in powers of α_W. $r^{(0)}$ is complex in the region of total reflection, which is the same for particles and antiparticles. This gives rise to CP-even phases. To generate a CP-asymmetry they must interfere with the mixing matrix CP-odd phases, contained in the non-diagonal weak corrections.

The first effect in the asymmetry appears at $\mathcal{O}(\alpha_W^2)$ and it has the following structure:

$$\Delta_{CP}(\omega) = \alpha_W^2 Im(K_{Hi1}K^*_{Hj1}K^*_{Hi2}K_{Hj2})F(M_1, M_2)\lambda_i\lambda_j Im A(m_i, m_j) \tag{7}$$

where $F(M_1, M_2)$ is an antisymmetric function of the internal NHL masses which vanishes when one of the masses is zero, $A(m_i, m_j)$ is an antisymmetric function of the external charged lepton masses, and $\lambda_i = m_i/M_W$. Eq. (7) shows explicitly the GIM cancellation for both external and internal lepton masses.

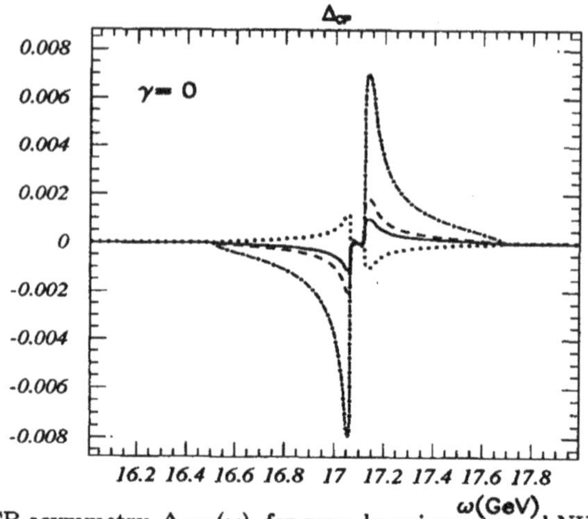

Figure 1: CP asymmetry, $\Delta_{CP}(\omega)$, for zero damping rate and NHL masses $M_2 = 300$ GeV and $M_1 = 100$ GeV (solid), 150 GeV (dash-dotted), 200 GeV (dashed) and 250 GeV (dotted).

The experimental bounds on the elements of the submatrix K_H depend on the NHL mass [13]. For 3 GeV $\leq M \leq M_Z$ the strongest limits come from LEP and they are too stringent to yield an observable CP asymmetry, since $Im(K_{Hi1}K^*_{Hj1}K^*_{Hi2}K_{Hj2}) \leq 3 \times (10^{-9} - 10^{-7})$. If $M \geq M_Z$, there are low-energy constraints which arise both from the non-observation of lepton flavour violation and from universality. Those limits are slightly weaker for the mixing of the third family with any of the first two, so we consider this case and we obtain

$$Im(K_{Hi1}K^*_{H\tau 1}K^*_{Hi2}K_{H\tau 2}) \leq 10^{-5} \quad i = e, \mu \tag{8}$$

In fig. 1 we show the CP asymmetry as a function of the energy of the quasiparticle, ω. We have fixed the mass of the heavier NHL to $M_2 = 300$ GeV and we plot the result for different values of the lighter NHL mass.

Consider now a non vanishing damping rate, γ. Due to incoherent thermal scattering with the medium the quasiparticle has a finite life-time $\sim 1/2\gamma$ and it turns eventually into a new state, out of phase with the initial one. The damping rate for leptons is essentially dominated by weak interaction processes and at zero momentum is of order $\gamma \sim \alpha_W T$ [14], *i.e.*~ 4 GeV at $T = 100$ GeV. Although the life-time is larger than the overall time scale of the problem, given roughly by the inverse of the thermal mass or plasma frequency $\sim gT \sim 20$ GeV for $T = 100$ GeV, it is small in comparison with the tree-level reflection

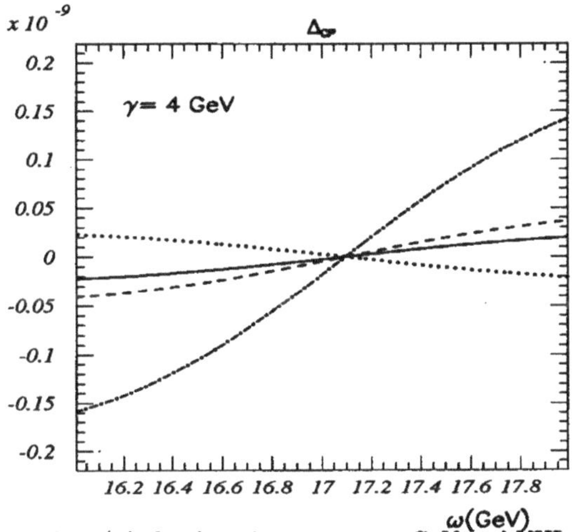

Figure 2: $\Delta_{CP}(\omega)$, for damping rate $\gamma = 4$ GeV and NHL masses as in fig. 1.

times $\sim 1/m$ for the charged quasi-leptons. In other words, the quasiparticles have a mean free path of $\sim 1/6\gamma$ and over larger scales, such as those needed for total reflection, the quantum coherence is lost, leading to a suppression of any pure quantum effect such as CP violation [4].

The result for the CP asymmetry including the damping rate effect is shown in fig. 2 as a function of the energy. It is suppressed with respect to fig. 1 by a factor $\sim 10^{-7}$. It is clear that the CP asymmetry in the reflected flux of charged leptons is several orders of magnitude below what observation requires.

4.2 Neutral Heavy Leptons

The NHL can be much heavier than the charged leptons, so in principle one expects that the suppression due to the damping rate will not be so important in this case. The dominant contribution to the CP asymmetry comes from the energy region in which one of the NHL is totally reflected and the other is partially reflected and partially transmitted, $w \sim M$. However, the effective Hamiltonian (6) describes the quasiparticles in the low momentum region, and thus the approach used in the previous section is not valid now.

Since the thermal masses of the NHL are small, $\omega^0_{L,R} \sim 0.1M$, as a first approximation we have estimated the CP asymmetry in the $T = 0$ limit. We follow the approach of ref. [15], where a similar calculation is done for SM quarks. The essential non-perturbative effect is the wall which separates the two symmetry phases, unbroken and spontaneously broken, during the first order phase transition. The exact Feynman propagator of a free fermion in the presence of a thin wall has been computed in [15]. It contains both massless and massive poles, and due to the wall the initial and final z components of the momentum need not be equal. Then, one can use perturbation theory in the gauge and Yukawa couplings of fermions to bosons and perform one loop calculations. As NHL masses can not be neglected the expressions are more cumbersome than in the SM case and we do not present them here. We have found that the CP asymmetry generated is at most 10^{-9}, even for NHL as heavy as 1 TeV. Therefore, it can not account for the experimental observation.

However, in the case of the neutral leptons there is an additional CP-violating source, because their mass matrix contains space dependent phases which can not be rotated away at two adjacent points by the same set of unitary transformations [16]. This effect vanishes in the limit of zero wall thickness that we have considered previously. Notice also that it is not present for charged leptons because their mass matrix has only an overall dependence on the doublet scalar vacuum expectation value (as occurs in the SM), while the neutral lepton mass matrix has a non-trivial dependence on both singlet and doublet scalar vacuum expectation values. Work is still in progress, but preliminary results indicate that the CP asymmetry obtained due to this mechanism is big enough to generate the observed n_B/s ratio if the NHL masses are $\geq M_Z$ [11].

1. G. Steigman, *Ann. Rev. Astron. Astrophys.* **14** (1976) 339
2. A.D. Sakharov, *JETP Lett.* **6** (1967) 24
3. V.A. Kuzmin, V.A. Rubakov and M.E. Shaposhnikov, *Phys. Lett.* **B155** (1985) 36
4. M.B. Gavela *et al.*, *Mod. Phys. Lett.* **A9** (1994) 795; *Nucl. Phys.* **B430** (1994) 382. P. Huet and E. Sather, *Phys. Rev.* **D51** (1995) 379
5. For a recent review, see A.G. Cohen, D.B. Kaplan and A.E. Nelson, *Ann. Rev. Nucl. Part. Science* **43** (1993)
6. See for instance J. Bernabéu *et al.*, *Phys. Lett.* **B187** (1987) 303 and references therein. L. Randall and T. Wynter, *Phys. Rev.* **D50** (1994) 3457
7. G.C. Branco, M.N. Rebelo and J.W.F. Valle, *Phys. Lett.* **B225** (1989) 385
8. A.G. Cohen, D.B. Kaplan and A.E. Nelson, *Nucl. Phys.* **B373** (1992) 453
9. M. Dine *et al.*, *Phys. Lett.* **B283** (1992) 319; *Phys. Rev.* **D46** (1992) 550
10. G.R. Farrar and M.E. Shaposhnikov, *Phys. Rev. Lett.* **70** (1993) 2833;

11. P. Hernández and N. Rius, in preparation.
12. H.A. Weldon, *Phys. Rev.* **D26** (1982) 2789
13. M.C. González-García and J.W.F. Valle, *Mod. Phys. Lett.* **A7** (1992) 477; E ibidem **A9** (1994) 2569. E. Nardi, E. Roulet and D. Tommasini, *Phys. Lett.* **B344** (1995) 225
14. R. Kobes, G. Kunstatter and K. Mak, *Phys. Rev.* **D45** (1992) 4632
15. M.B. Gavela *et al.*, *Nucl. Phys.* **B430** (1994) 345
16. P. Huet and A.E. Nelson, *Phys. Lett.* **B355** (1995) 229; UW/PT 95-07

RENORMALIZATION GROUP CALCULATION OF THE GALAXY-GALAXY CORRELATION FUNCTION

J. PÉREZ-MERCADER*, T. GOLDMAN†, D. HOCHBERG ‡,*a
AND R. LAFLAMME†

*Laboratorio de Astrofísica Espacial y Física Fundamental
Apartado 50727, 28080 Madrid, Spain

† Theoretical Division, Los Alamos National Laboratory
Los Alamos, New Mexico 87545, USA

‡ Departamento de Física Teórica, Universidad Autónoma de Madrid
Cantoblanco, 28049 Madrid, Spain

The observable Universe is described by a collection of equal mass galaxies linked into a common unit by their mutual gravitational interaction. The partition function of this system is cast in terms of Ising model spin variables and maps exactly onto a three-dimensional stochastic scalar classical field theory. The full machinery of the renormalization group and critical phenomena is brought to bear on this field theory allowing one to calculate the galaxy-to-galaxy correlation function, whose critical exponent is predicted to be between 1.530 to 1.862, compared to the phenomenological value of 1.6 to 1.8

It is an empirical fact that the Universe today is dominated by matter which is in gravitational interaction and is organized into spatial configurations consisting of galaxies, groups of galaxies, superclusters, and even larger structures [1,2]. It is also true that one of the outstanding problems in modern cosmology is to provide a quantitative description of this structure from first principles [3]. One of the standard ways in which structure, that is, the departure from randomness and homogeneity, is described, is by means of correlation functions. These objects are a basic tool for unravelling structure in physical systems ranging from gases and liquids to solids. Experimental input is typically deduced from analysis of the scattering of neutrons or x-rays. By contrast, in cosmology, this input is provided by direct observation. From observation, the galaxy-to-galaxy correlation function ($\xi_{gal}(r)$) is known to scale as $\xi(r) \sim r^{-\gamma}$, where $\gamma \sim O(1.6\,\mathrm{to}\,1.8)$, instead of $\gamma = 1$, which is what one

[a] Based on a talk presented by D. Hochberg at the International Workshop on Elementary Particle Physics, Valencia, Spain, June 5 to 9, 1995.

would naively expect [4] for a homogeneous distribution of matter in a three dimensional space.

Given these observational facts, we shall demonstrate how to *calculate* the exponent γ from first principles [5]. We take our cue by analogy with the well established theory of liquids and interacting gases [6], where the relative arrangements of the atoms and molecules (galaxies, in our case!) are accurately described in terms of the principles and methods of classical statistical mechanics. We will apply the same techniques to a system made up by many galaxies in gravitational interaction, subject to fluctuations emerging from the intrinsic properties of the gravitational interaction [5]. We will obtain the canonical partition function for this system and compute the two-point function and its corrections due to these fluctuations.

Begin by considering the continuous mass density $\rho(\mathbf{r})$ describing the spatial distribution of galaxies. The departure from perfect homogeneity is $\delta\rho(\mathbf{r}) = \rho(\mathbf{r}) - \bar{\rho}$, where $\bar{\rho}$ denotes the average density. The galaxy-galaxy correlation function is defined [3] as $\xi_{Gal}(\mathbf{r}_i - \mathbf{r}_j) =< \delta\rho(\mathbf{r}_i)\delta\rho(\mathbf{r}_j) > /\bar{\rho}^2$, the angular brackets denoting a suitable averaging. This is the joint probability distribution function for finding a galaxy at position $\mathbf{r}_i$ given that there is one at $\mathbf{r}_j$. In the static, nonrelativistic and weak gravitational field limit, (we discuss the effects of expansion later) the gravitational interaction energy for this system is

$$H_{int} = -\frac{G}{2}\int\int d^3\mathbf{r}_i\, d^3\mathbf{r}_j\, \rho(\mathbf{r}_i)\frac{1}{|\mathbf{r}_i - \mathbf{r}_j|}\rho(\mathbf{r}_j), \tag{1}$$

where the integral excludes the point $\mathbf{r}_i = \mathbf{r}_j$, and the factor of $1/2$ appears because each galaxy is counted twice in the integration. It is reasonable to consider the gas of N galaxies as made up of discrete spatially localized points of equal mass m_o at position $\mathbf{r}_j$, so that $\rho(\mathbf{r}) = \sum_{j=1}^{N} m_j \delta(\mathbf{r} - \mathbf{r}_j)$. The interaction energy then becomes

$$H_{int} = \frac{-1}{2}\sum_{i,j}\frac{m_i G m_j}{|\mathbf{r}_i - \mathbf{r}_j|} \equiv \frac{-1}{2}\sum_{i,j} m_i\, L_{ij}^{-1}\, m_j, \tag{2}$$

where m_j is the mass of the galaxy at site $\mathbf{r}_j$, and $i \neq j$. In this "granular" picture, the contrast function $\delta\rho(\mathbf{r}_i)/\bar{\rho}$ is equal to 1 if there is a galaxy at $\mathbf{r}_i$ and is equal to -1 if there is a void. Because we have assumed all N galaxies have equal mass, there is a natural correspondence between the m_i and a two-valued ($\pm$) "spin" variable s_i:

$$m_i = \frac{m_o}{2}(s_i + 1). \tag{3}$$

A galaxy at site $\mathbf{r}_i$ corresponds to "spin-up", while a void or deficit corresponds to "spin-down". Because of (3), we can express the Boltzmann factor and associated partition function of this gravitational Ising system in terms of the spin variables as:

$$Z_{Ising}^{Grav}[\beta] = \sum_{\{s_i\}} exp\left(\frac{1}{4}\sum_{i,j} s_i A_{ij}^{-1} s_j + \frac{1}{4}\sum_j s_j h_j\right), \tag{4}$$

where $A_{ij}^{-1} = \frac{\beta G m_0^2/2}{|\mathbf{r}_i - \mathbf{r}_j|}$, and $h_j = \sum_i \frac{\beta m_0^2 G}{|\mathbf{r}_i - \mathbf{r}_j|}$, and we have dropped an irrelevent constant term. β plays the role of an inverse temperature. The sum in (4) is over all possible spin configurations $s_1 = \pm 1, s_2 = \pm 1, \cdots, s_N = \pm 1$.

To continue, we make use of the Hubbard-Stratonovich (or gaussian) transformation [7]

$$\pi^{N/2} \quad (\det A)^{-1/2} exp\left[\frac{1}{4}\sum_{i,j} s_i A_{ij}^{-1} s_j\right] = \tag{5}$$

$$\int_{-\infty}^{\infty} \Pi_{m=1}^{N} dx_m \, exp\left(-\sum_{m,n} x_m A_{mn} x_n + \sum_m s_m x_m\right),$$

which maps exactly the discrete system described by the spin variables s_i onto a system described by the continuous real variable $-\infty < x_m < \infty$. Applying (5) to the partition function in (4) yields

$$\begin{aligned} Z_{Ising}^{Grav}[\beta] &= C\int [d\phi] \exp[-\frac{\beta}{2}\int (\phi(\mathbf{r}) - h(\mathbf{r}))A(\mathbf{r},\mathbf{r}')(\phi(\mathbf{r}') - h(\mathbf{r}'))d\mathbf{r}d\mathbf{r}' \\ &+ \int d\mathbf{r} \log\cosh(\beta\phi(\mathbf{r}))] \\ &\equiv \int [d\phi]\, e^{-\beta\mathcal{H}[\phi,h]}, \end{aligned} \tag{6}$$

performing the exact configurational sum over the $\{s_i\}$ and after taking the continuum limit which rigorously converts the discrete system into a continuous three dimensional 1-component scalar field theory. The linear term in (4) results in the highly nonlinear log cosh appearing in (6). C is an inessential factor and $h(\mathbf{r}) = \frac{1}{2}\int d\mathbf{r}'\, A^{-1}(\mathbf{r},\mathbf{r}')$. The continuous scalar field ϕ is the order parameter for this system; its n-point Green functions correspond to the statistically averaged product of n times the contrast evaluated at n-sites. The operator A in the continuum limit is given by

$$A(\mathbf{r},\mathbf{r}') = \delta^{(3)}(\mathbf{r}-\mathbf{r}')\frac{-\nabla^2}{2\pi\beta G m_0^2}, \tag{7}$$

since $\nabla^2 \frac{1}{|\mathbf{x}|} = -4\pi\delta^{(3)}(\mathbf{x})$.

As is well known [8], the connected two-point function for the spin system and the field theory are the same. Moreover, the unavoidable fluctuations in the field ϕ result in an *anomolous* dimension η shifting the canonical dimension of the field such that when $|\mathbf{r} - \mathbf{r}'| \to \infty$, the two-point function for the hamiltonian in (6) scales as

$$\lim_{|\mathbf{r}-\mathbf{r}'|\to\infty} < s_i s_j >= \lim_{|\mathbf{r}-\mathbf{r}'|\to\infty} \xi_{Gal}(|\mathbf{r} - \mathbf{r}'|) \sim |\mathbf{r} - \mathbf{r}'|^{-(d-2+\eta)}. \tag{8}$$

Here d is the dimension of space ($d = 3$) and η is the critical exponent for the pair correlation function whose value (0.0198 – 0.064) (see Table 1) differs from zero due to the fluctuations in ϕ.

Thus, our calculation shows that for large separations, the galaxy-galaxy correlation function must scale as

$$\xi_{Gal}(|\mathbf{r}_i - \mathbf{r}_j|) \sim r^{-(d-2+\eta)}, \tag{9}$$

with [9] $1.0198 \leq 1 + \eta \leq 1.064$ for $d = 3$, and without taking the background expansion into account.

Thus far, our calculation has been performed for a static ensemble of galaxies, but the Universe is expanding and the effects of this expansion will modify the values of the critical exponents. From the theory of dynamical critical phenomena, it is phenomenologically known from computer simulations that the way time enters into the correlation function is by a modification of its argument, viz.,

$$\xi(r) \to \xi(r; t) = F(r/l(t)), \tag{10}$$

where $l(t) \sim t^{\varsigma}$ is a function of time and here, r denotes the comoving coordinate, $r_{comoving}$, which is related to the physical coordinate via

$$r_{physical} = a(t)\, r_{comoving}, \tag{11}$$

with $a(t)$ the scale factor for the background spacetime. From computer simulations in condensed matter physics, it is established that $\zeta = 1/3$ for systems with a conserved order parameter, while $\zeta = 1/2$ when the order parameter is not conserved.

For a matter dominated expansion, the case we are interested in, $a(t) = t^{2/3}$ and thus for a given physical separation

$$t \sim (r_{comoving})^{-3/2}. \tag{12}$$

This means that the function $l(t)$ scales as

$$l(t) = t^{\varsigma} = (r_{comoving})^{-3\varsigma/2}. \tag{13}$$

Table 1: Calculated values for the galaxy-galaxy correlation function critical exponent

Method of Calculation	γ_{Static}	$\gamma^{C}_{Expanding}$	$\gamma^{NC}_{Expanding}$
Series estimates	1.056 ± 0.008	1.584 ± 0.012	1.848 ± 0.014
$O(\epsilon)$	0	1.5	1.75
$O(\epsilon^2)$	1.0198	1.530	1.785
$O(\epsilon^3)$	1.037	1.555	1.815
$O(\epsilon^4)$	1.029	1.543	1.801

Putting together these facts yields the scaling behavior of $\xi(r;t)$:

$$\begin{aligned}\xi(r;t) &= \left(r_{comoving}/r_{comoving}^{-3\zeta/2}\right)^{-(d-2+\eta)} \\ &= \left(r_{comoving}\right)^{-(d-2+\eta)(1+3\zeta/2)}. \end{aligned} \tag{14}$$

In other words, the critical exponent resulting from the expansion of the Universe and dynamical critical phenomena effects is

$$\gamma = (d-2+\eta) \times (1+3\zeta/2). \tag{15}$$

The predicted (calculated) value for γ is between 1.530 and 1.596 for $\zeta = 1/3$ (for a conserved order parameter) and between 1.75 to 1.862 for $\zeta = 1/2$ (non-conserved order parameter). Table 1 displays these exponents for both the static and expanding cases (C, NC, denote conserved and nonconserved order parameter, respectively). The static values are taken from Ref[9]. The last four entries refer to expansions in powers of $\epsilon = 4 - d$. These values are to be compared with the values inferred from current galaxy catalogs, which range from 1.5 for the APM survey, to 1.8 for the Lick survey [2,1].

We have thus accomplished what we set out to do, namely, provide a first-principles calculation of the galaxy-galaxy correlation function using standard methods of statistical mechanics and the theory of critical phenomena. Naturally, one might (and should) ask *why* this calculation is successful, in other words, what does a system of galaxies have to do with a spin model? The answer is provided by the universality hypothesis, according to which rather distinct physical systems will exhibit *identical* critical behavior provided the systems in question possess the same space dimensionality and have order parameters with the same dimensionality (or number of independent components). We see that this is the case here: $d = 3$ for both systems, while the scalar field and the magnetization are one-component order parameters for the galaxy system and the 3d Ising model, respectively. In other words, both systems belong to the same universality class. Thus, we must expect them to have the same values for their critical exponents.

Acknowledgements

D.H. would like to thank the organizers for the invitation to participate in this Workshop and for their kind and generous hospitality. Thanks are also due to the participants for their enthusiastic response to this talk, as evidenced by their many thoughtful comments and questions.

References

1. The Lick survey is based on plates taken by C.D. Shane and C.A. Wirtanen, *Publ. Lick Obs.* **22** Part 1 (1967).
2. S.J. Maddox et al., MNRAS **242** (1990) 43P.
3. See, e.g., P.J.E. Peebles, *Principles of Physical Cosmology* (Princeton University Press, Princeton, 1993)
4. As we shall demonstrate below, the galaxy-galaxy correlation function can be expressed as a scalar field propagator. In units of length, a scalar field in d-dimensional space has a canonical dimension of $-(d/2-1)$. The above statement about γ follows at once.
5. J. Pérez-Mercader, T. Goldman, D. Hochberg and R. Laflamme, Los Alamos preprint LA-UR-95-1055.
6. See, e.g., D.L. Goodstein, *States of Matter* (Dover, New York, 1985).
7. See, e.g., J. Binney, N. Dowrick, A. Fisher and M. Newman, *The Theory of Critical Phenomena: An Introduction to the Renormalization Group*, (Oxford University Press, Oxford, 1992). Also, D. Amit, *Field Theory, the Renormalization Group and Critical Phenomena*, revised second edition, (World Scientific, Singapore, 1984).
8. K.G. Wilson, Phys. Rev. **179**, 1499 (1969).
9. M.N. Barber, Phys. Rep. **29C**, 1 (1977).

www.ingramcontent.com/pod-product-compliance
Lightning Source LLC
LaVergne TN
LVHW070825171026
850215LV00042B/42
* 9 7 8 9 8 1 0 2 2 5 5 4 4 *